The Practitioner's Guide to Psychoactive Drugs

THIRD EDITION

The Practitioner's Guide to Psychoactive Drugs

THIRD EDITION

Edited by

Alan J. Gelenberg, M.D.
University of Arizona Health Sciences Center
Tucson, Arizona

Ellen L. Bassuk, M.D.
The Better Homes Foundation
Newton Centre, Massachusetts
and Harvard Medical School
Boston, Massachusetts

and

Stephen C. Schoonover, M.D.
Schoonover Associates
Newton Centre, Massachusetts

PLENUM MEDICAL BOOK COMPANY
New York and London

Library of Congress Cataloging-in-Publication Data

The Practitioner's guide to psychoactive drugs / edited by Alan J.
  Gelenberg, Ellen L. Bassuk, and Stephen C. Schoonover. -- 3rd ed.
      p.   cm.
    Includes bibliographical references and index.
    ISBN 0-306-43461-X
      1. Psychopharmacology. 2. Psychoactive drugs.   I. Gelenberg,
  Alan J. II. Bassuk, Ellen L., 1945-   . III. Schoonover, Stephen
  C., 1947-   .
      [DNLM: 1. Psychotropic Drugs.   QV 77 P8952]
  RC483.P726   1990
  615'.78--dc20
  DNLM/DLC
  for Library of Congress   90-14325
                                                                      CIP

First Printing — January 1991
Second Printing — November 1991

© 1991, 1983, 1977 Plenum Publishing Corporation
233 Spring Street, New York, N.Y. 10013

Plenum Medical Book Company is an imprint of Plenum Publishing Corporation

Printed in the United States of America

To Sherry Gelenberg
—AJG

In memory of Molly Bassuk and
to Sarah and Danny Schoonover
—ELB & SCS

Contributors

Joseph Biederman, M.D.   Associate Professor of Psychiatry, Department of Psychiatry, Harvard Medical School, Boston, Massachusetts 02115; Director, Psychopharmacology Unit, Massachusetts General Hospital, Boston, Massachusetts 02114

Andrew W. Brotman, M.D.   Director, Social and Community Psychiatry, Massachusetts General Hospital, Boston, Massachusetts 02114; Department of Psychiatry, Harvard Medical School, Boston, Massachusetts 02115; Director, Outpatient Services, Erich Lindemann Mental Health Center, Boston, Massachusetts 02114

Lee S. Cohen, M.D.   Attending Psychiatrist, Pregnancy Consultation Service, Clinical Psychopharmacology Unit, Massachusetts General Hospital, Boston, Massachusetts 02114; Assistant Professor of Psychiatry, Harvard Medical School, Boston, Massachusetts 02115

Alan J. Gelenberg, M.D.   Professor and Head, Department of Psychiatry, University of Arizona Health Sciences Center, Tucson, Arizona 85724

Shelly F. Greenfield, M.D.   Resident in Adult Psychiatry, McLean Hospital, Belmont, Massachusetts 02178; Department of Psychiatry, Harvard Medical School, Boston, Massachusetts 02115

Thomas G. Gutheil, M.D.   Associate Professor of Psychiatry, Department of Psychiatry, Harvard Medical School, Boston, Massachusetts 02115; Program in Psychiatry and the Law, Massachusetts Mental Health Center, Boston, Massachusetts 02115

Vicki L. Heller, M.D.   Attending Obstetrician, Department of Obstetrics and Gynecology, Beth Israel Hospital, Boston, Massachusetts 02215; Instructor in Obstetrics and Gynecology, Harvard Medical School, Boston, Massachusetts 02115

David B. Herzog, M.D.   Director, Eating Disorders Clinic, Massachusetts General Hospital, Boston, Massachusetts 02114; Associate Professor of Psychiatry, Department of Psychiatry, Harvard Medical School, Boston, Massachusetts 02115

Steven M. Mirin, M.D.   Acting General Director and Psychiatrist in Chief, McLean Hospital, Belmont, Massachusetts 02178; Associate Professor of Psychiatry, Department of Psychiatry, Harvard Medical School, Boston, Massachusetts 02115

George B. Murray, M.D.   Director, Private Consultation Service, Assistant Professor of Psychiatry, Harvard Medical School, Boston, Massachusetts 02115

Jerrold F. Rosenbaum, M.D.   Chief, Clinical Psychopharmacology Unit, Massachusetts General Hospital, Boston, Massachusetts 02114; Associate Professor of Psychiatry, Harvard Medical School, Boston, Massachusetts 02115

Carl Salzman, M.D.   Associate Professor of Psychiatry, Harvard Medical School, Boston, Massachusetts 02115; Director of Psychopharmacology, Massachusetts Mental Health Center, Boston, Massachusetts 02115

Stephen C. Schoonover, M.D.   President, Schoonover Associates, Newton Centre, Massachusetts 02159

Paul H. Soloff, M.D.   Associate Professor of Psychiatry, University of Pittsburgh, Western Psychiatric Institute and Clinic, Pittsburgh, Pennsylvania 15213

Ronald Steingard, M.D.   Assistant Director, Pediatric Psychopharmacology Unit, Child Psychiatry Service, Massachusetts General Hospital, Boston, Massachusetts 02114; Instructor in Psychiatry, Harvard Medical School, Boston, Massachusetts 02115

Jeffrey B. Weilburg, M.D.   Director, Neuropsychiatry Section, Psychopharmacology Unit, Massachusetts General Hospital, Boston, Massachusetts 02114; Assistant Professor of Psychiatry, Harvard Medical School, Boston, Massachusetts 02115

Roger D. Weiss, M.D.   Clinical Director, Alcohol and Drug Abuse Program, McLean Hospital, Belmont, Massachusetts 02178; Assistant Professor of Psychiatry, Department of Psychiatry, Harvard Medical School, Boston, Massachusetts 02115

Foreword

In the eight years since the publication of the second edition of this *Guide*, psychopharmacotherapy has made many advances not only through the discovery of new medications but by the effective directing of their use to an ever-increasing variety of clinical disorders. These welcome developments are reflected in the concurrent growth and development of the *Guide* itself, which now enters adulthood with renewed vigor.

Under the thoughtful and scholarly leadership of Dr. Alan Gelenberg, the third edition has undergone a significant transformation designed to meet the needs of the modern clinician. The panel of contributors is nearly double that of the former edition with the addition of nine new authors, who have helped in the major revision and rewriting of the text and in a broadening of the topics included. As a consequence, the reader is assured of a thorough and thoroughly up-to-date coverage of current psychopharmacology that is both accurate and aimed at clinical utility. Having reached maturity, the third edition, while maintaining the lineaments of its earlier versions, is a considerably expanded and strengthened guide to treatment. Although now more encyclopedic in content, the new *Practitioner's Guide to Psychoactive Drugs* retains the virtues of a clinical *vade mecum* that informed its predecessors and have earned it a place by the patient's bedside for well over a decade. One may confidently anticipate its long and flourishing career in the years ahead.

John C. Nemiah, M.D.

Boston

Preface

Since its birth four decades ago, the field of psychopharmacology continues to mature. Born with a series of serendipitous, heady, and dramatic breakthroughs in the 1950s, the field has become more sophisticated and has expanded to offer many more options to patients who suffer from medication-responsive psychiatric illnesses.

But recent years have brought disappointments as well as the developments. The earliest of the "newer" antidepressants proved to be less effective and more toxic than originally hoped. Nomifensine and bupropion were withdrawn from clinical use [although bupropion (Wellbutrin®) was recently reintroduced in the United States]. Psychopharmacologists continue to search for the ever-elusive biological tests that will help tailor therapy to specific syndromes; the dexamethasone suppression test and other neuroendocrine assays have failed to live up to their earlier promises.

Despite these disappointments, psychopharmacology greets the 1990s with more therapeutic options than it did 10 years ago. Recently introduced for general use within the United States, clozapine (Clozaril®) holds promise for some schizophrenic patients previously refractory to standard medication. Fluoxetine (Prozac®) has become a widely used antidepressant, largely because of its efficacy and different side effects. Clomipramine and several other new antidepressants bring at least partial relief to those suffering from obsessive compulsive disorder.

Practitioners are increasingly exploiting psychopharmacologic assistance for patients with other mental disorders. Clinicians now recognize the distinction between panic and generalized anxiety disorders, and through clinical research are determining appropriate roles for high-potency benzodiazepines and antidepressants in panic disorders. Buspirone (BuSpar®) appears to be a non-benzodiazepine alternative for generalized anxiety, although its place in our therapeutic guide remains unclear as of this writing.

Although less dramatic than the discovery of new landmark drugs, studies of dose–response relationships among older agents allow physicians to extend their benefits to previously untreated patients. For example, lower doses of antipsychotic agents, for both the acute phases of psychosis and for maintenance treatment, afford comparable efficacy and are less toxic. At the same time, higher doses of tricyclic antidepressants can relieve previously resistant patients. Maintenance lithium for bipolar patients is more effective when higher blood levels are achieved. For newer agents, such as fluoxetine, numerous dose–response relationships are being explored.

The previous two editions of the *Practitioner's Guide* have been widely referred to and read. The third edition's two older siblings adorn the desks and bookshelves of students, residents, other trainees, and a wide range of medical and mental health practitioners. The books tend to be well-thumbed, and thus not only is it now time to update information, but many readers need a clean copy.

We have maintained the format and emphasis that have contributed to the popularity of the earlier editions. The book's organization is essentially the same. We have added sections on the use of medications in patients with borderline personality disorder and eating disorders. As drugs and alcohol ravage our population—disproportionately afflicting the young, the poor, and the mentally ill—psychopharmacologic approaches are being studied and applied in attempts to block the craving for these chemicals. These approaches are reflected in an expanded section on substance abuse.

Since publication of the last edition, Dr. Gelenberg has moved from Boston, Massachusetts to Tucson, Arizona to become Professor and Head of the Department of Psychiatry at the University of Arizona Health Sciences Center. In addition to his involvement with administration, teaching, research, and clinical practice, he continues to write the *Biological Therapies in Psychiatry* newsletter; he is also Editor-in-Chief of the *Journal of Clinical Psychiatry.* Dr. Gelenberg expresses his gratitude and appreciation to the many colleagues in Boston who have contributed to this book and who have been sources of support, stimulation, and friendship throughout his career. He particularly wishes to acknowledge the love and encouragement of his wife, which have provided the bedrock on which his professional activities proceed.

Drs. Bassuk and Schoonover have also changed their career paths. Dr. Bassuk has become president of the Better Homes Foundation, a national nonprofit organization that serves homeless families. Dr. Schoonover has become president of a management consulting firm.

We hope that the third edition of the *Practitioner's Guide* will continue to provide, in an easy, accessible manner, the most up-to-date information on psychoactive drugs and that its application will mitigate the suffering of many patients.

A.J.G.

E.L.B.

S.C.S.

Tucson and Boston

Contents

II. Major Psychiatric Disorders

2. Depression

Alan J. Gelenberg, M.D., and Stephen C. Schoonover, M.D.

3. Bipolar Disorder

Alan J. Gelenberg, M.D., and Stephen C. Schoonover, M.D.

4. Psychoses

Alan J. Gelenberg, M.D.

5. Anxiety

Jerrold F. Rosenbaum, M.D., and Alan J. Gelenberg, M.D.

III. Psychoactive Substance Abuse

7. Psychoactive Substance Use Disorders

Steven M. Mirin, M.D., Roger D. Weiss, M.D., and
Shelly F. Greenfield, M.D.

IV. Special Topics

8. Geriatric Psychopharmacology

Carl Salzman, M.D.

9. Pediatric Psychopharmacology

Joseph Biederman, M.D., and Ronald Steingard, M.D.

10. Psychotropic Drug Use in Pregnancy

Lee S. Cohen, M.D., Jerrold F. Rosenbaum, M.D., and
Vicki L. Heller, M.D.

11. Temporolimbic Epilepsy

Jeffrey B. Weilburg, M.D., and George B. Murray, M.D.

12. Eating Disorders

Andrew W. Brotman, M.D., and David B. Herzog, M.D.

13. Borderline Disorders

Paul H. Soloff, M.D.

14. Medicolegal Psychopharmacology

Thomas G. Gutheil, M.D.

I

State of the Art

1

Introduction: The Practice of Pharmacotherapy

STEPHEN C. SCHOONOVER, M.D., and ALAN J. GELENBERG, M.D.

I. HISTORICAL PERSPECTIVE

Experimentation with drugs that alter mood, thinking, or perception represents a timeless human enterprise. Although largely aimed at the relief of suffering, the utilization of various pharmacological agents also reflects different cultural, religious, and political ideologies. Frequently, implementing effective pharmacological treatments depends more on social issues, economic considerations, and serendipity than on clinical factors. A historical account of the discovery and introduction of psychoactive drugs in the United States highlights these patterns.

The medical community has only recently employed psychoactive drugs to treat mental illness. In the 1840s clinicians prescribed bromides as sedative agents. These chemicals effectively diminished anxiety, but their prolonged use caused significant central nervous system complications. The experience with bromides was portentous; they effectively alleviated psychological symptoms but had hidden long-term dangers. In the latter part of the 19th century, other sedative agents were synthesized, including paraldehyde, urethane, sulfonal, and chloral hydrate. During this same period, Sigmund Freud suggested that cocaine was a useful psychoactive drug, and Emile Kraepelin started the first laboratory for testing drugs in humans.

At the beginning of the 20th century, researchers introduced the barbiturates. Because they more effectively relieved anxiety than any previously used drug, they

STEPHEN C. SCHOONOVER, M.D. • Schoonover Associates, Newton Centre, Massachusetts 02159. ALAN J. GELENBERG, M.D. • Department of Psychiatry, University of Arizona Health Sciences Center, Tucson, Arizona 85724.

largely displaced other agents. Their mood-altering properties, complicated by their addictive potential, promoted widespread use. Since that time, over 2500 barbiturates have been synthesized, but only 50 have been used clinically. No additional clinical advances occurred until the 1950s. However, during the intervening period, researchers unwittingly discovered many of the major drug classes now in use. In 1931, Sen and Bose first reported giving a rauwolfia alkaloid, originally developed for its hypotensive and sedative properties, to psychotic patients.[1] Although the results were unremarkable, they were the first to link the properties of these reserpine-like drugs with their potential effectiveness in psychiatric patients.

Shortly thereafter, Charpentier developed promethazine, a phenothiazine derivative, as an antihistamine and sedative. Clinical trials proved unremarkable, but luck was again close at hand. During this same period, three other important pharmacological groups were synthesized and used clinically—the benzodiazepines, lithium salts, and the tricyclic antidepressant imipramine (Tofranil® and others). By the 1940s, researchers had discovered each of the major chemical groups currently used in psychopharmacology. All were originally synthesized for purposes other than their subsequent clinical use. It took many more years before researchers and clinicians defined guidelines for their administration.

The revolution in psychopharmacology really began in 1949. Promethazine, although disappointing when used alone, effectively potentiated the sedative properties of barbiturates. Based on this finding, researchers developed other phenothiazines to aid in anesthesia. In 1949, Charpentier synthesized chlorpromazine (Thorazine® and others). By 1952, several individuals had noted the beneficial effects of chlorpromazine on mania, paranoia, and other psychoses.[2] They observed that patients improved primarily because of a decrease in disorganized thoughts, feelings, and behaviors rather than because they were sedated. Once researchers identified the "neuroleptic" properties of these drugs, they began to develop similar antipsychotic agents.

Chlorpromazine, which was introduced for general use in the United States in 1954, created a revolution in patient care.[3] Thereafter, other phenothiazines were rapidly developed. In 1958, while attempting to find a more effective analgesic agent, Janssen synthesized haloperidol (Haldol® and others), a butyrophenone with strong neuroleptic properties. Since then, other researchers have discovered additional classes of antipsychotic drugs including the thioxanthenes, the dibenzoxazepines, and the dihydroindolones.

In 1940, lithium chloride was given to cardiac patients as a salt substitute, but many developed severe toxic reactions. This initial experience significantly retarded its acceptance as a psychopharmacological agent in the United States. Despite the apparent toxicity of lithium, Cade continued experimenting. He gave lithium to animals to determine if it increased the solubility of uric acid. Noting its sedative properties, he gave the salt to several manic and agitated patients. In 1949, based on these results, Cade reported lithium's antimanic effects.[4] However, because of continued concerns about toxic reactions, the drug was not approved for use in the United States until 1970.

In 1950, the search for sedative–hypnotics safer than the barbiturates led to the discovery of propanediol carbamates [e.g., meprobamate (Miltown® and others)]. Unfortunately, these compounds proved as troublesome as the drugs they replaced. Physiological tolerance, dependence, and severe withdrawal reactions made them a poor choice for treating anxiety. In 1957, Sternback synthesized the benzodiazepine chlordiazepoxide (Librium® and others).[5] The finding that it calmed animals without producing marked sedation spurred clinical trials in humans. Since then, various benzodiazepines have largely replaced other drugs for the treatment of anxiety and insomnia. In part, this stems from both their clinical effectiveness and broad margin of safety. However, it also reflects an age-old social phenomenon—an epidemic urgency to find relief from anxiety, tension, and worry. Recently, the use of these agents has decreased somewhat, although in the United States the triazolobenzodiazepine alprazolam (Xanax® and others) has gained widespread popularity.

As with most other psychotropic agents, the antidepressant properties of various drugs were identified fortuitously. For example, clinicians noticed that tuberculosis patients treated with isoniazid (INH® and others) and iproniazid experienced elevated mood; this led to further investigation of the monoamine oxidase inhibitors (MAOIs). In 1952, these agents were administered to patients with depression, and by the late 1950s, this class of drugs became part of the common clinical practice. However, when authors reported severe, life-threatening hypertension in patients who failed to eliminate amine precursors from their diet, these agents fell into disfavor. Better knowledge about the "MAOI diet," clearer clinical guidelines, and evidence of their effectiveness in depression and phobic states have rejuvenated their use.

The tricyclic antidepressant imipramine was initially synthesized in the 1940s as a promazine analogue. In 1958, Kuhn successfully treated a group of severely disturbed patients with imipramine.[6] (The drug did little for psychotic individuals but significantly helped those who were depressed.) Thereafter, researchers and clinicians defined guidelines for the use of imipramine to treat "endogenous" depressions. Soon afterwards, several other cyclic antidepressants with similar properties were synthesized. More recently, researchers have developed new antidepressants with novel structures and varying pharmacology, but with nagging problems of toxicity.

The effectiveness of antianxiety drugs and antidepressants stimulated researchers to pursue many other avenues of investigation. Interest in brain chemistry grew rapidly. Researchers developed and began testing hypotheses about symptom etiology. Moreover, an increase in the reliability and validity of diagnostic categories resulted in better-designed studies. These diverse areas of investigation led to the development of clinically useful treatments, various models as frameworks for further research, and a more complex view of the interaction between biological and nonbiological factors. Researchers now realize that new clinical approaches must account for the interaction of biology, experience, and environment on the expression of illness.

II. CURRENT TRENDS IN PSYCHOPHARMACOLOGY

A. Definition of Clinical Syndromes

Although many diverse diagnostic frameworks exist, they often do not help the practitioner decide whether or not to prescribe drugs. In the past, diagnosis often was based on presumed cause (e.g., endogenous vs. reactive depressions). More recently, the third edition of the *Diagnostic and Statistical Manual of Mental Disorders* (DSM-III) was developed to define discrete diagnostic groups according to symptom profiles and inclusion and exclusion criteria. This framework provides reliable diagnoses and consistent criteria for research and encourages accurate communication among professionals. However, it does not specify drug-responsive disorders. Some researchers, studying the outcome of specific drug treatment, have delineated **symptom profiles** that respond to pharmacotherapy. Unfortunately, despite these advances, many patients still do not receive effective medication trials. Some individuals with similar presenting symptoms have disorders requiring different drugs, whereas other patients with disparate symptoms may respond to similar treatments. To remedy this problem, investigators are trying to widen their knowledge of drug-responsive presentations and ideally develop **biological methods of diagnosis.**

B. Development of Brain Chemistry Models and New Pharmacological Agents

A significant advance in psychopharmacology has been the evolution of biological models of disease based on increased knowledge of brain chemistry. Despite their limitations, the **dopamine hypothesis** for schizophrenia and the **catecholamine hypothesis** for depression have provided useful starting points for research. For example, biochemical models of depression have undergone various modifications to account for evidence that norepinephrine (NE), serotonin (5-HT), and acetylcholine may each have a role in depression. The brain chemistry models have stimulated the search for drugs that produce hypothetically effective neurohumoral changes and fewer adverse reactions. On the other hand, new compounds whose effects are not consistent with postulated biochemical alterations [like the antidepressant bupropion (Wellbutrin®), which does not block the reuptake of either 5-HT or NE] promote the revision of existing hypotheses. At present, these advances have only limited pragmatic usefulness for the clinician, but the future promises the development of more definitive approaches.

C. Pharmacokinetics

Ideally, the practitioner should know the pharmacokinetics of each agent including its complete metabolic pathway. However, the characteristics of many psychoactive compounds and the limitations of current laboratory technology often make these goals impossible. The kinetics of many drugs are not known or are not readily available to the clinician because of problems developing affordable, accurate assays. Moreover, many agents have active metabolites that also must be measured, since they, too, are clinically effective. With medications that are protein bound, the percentage of "free drug" may represent only a small fraction of the total; sometimes, current methods do not differentiate the bound from unbound portions.

In addition to studying the kinetics of drugs (see Section IV.B.4), pharmacologists are attempting to define clinical correlates of biochemical tests. For example, they have tried to correlate factors such as clinical effectiveness and unwanted effects and toxicity with the blood level of the drug (and any active metabolites). In addition to actual measurements of the serum concentrations of the drugs themselves, radioactive receptor-binding assays are helping to evaluate the total effects of certain classes of drugs including dopamine-blocking activity (for antipsychotic drugs) and anticholinergic activity (for antiparkinson drugs). Moreover, in patients taking MAO inhibitors, some researchers have noted a relationship between the extent of platelet monoamine oxidase inhibition and clinical response.

D. Design of Drug Studies

The methodology of clinical investigations in pharmacology has improved vastly over the past few decades. The introduction of **double-blind techniques, blind raters, "active placebos," and sophisticated analyses** has brought the light of science to previously insoluble clinical dilemmas. Nevertheless, many methodological problems still beset clinical pharmacology. Drug studies are expensive: it may take as much as $100 million to bring a drug to market. This means that if a drug does not appear to have potential for profit, it may never be adequately studied. Furthermore, the validity of a clinical study depends on the diligence of an investigator in selecting appropriate patients and adhering rigorously to a study protocol. Moreover, the heterogeneity of many psychiatric disorders, such as unipolar depression, can explain divergent results when different investigators study the same antidepressant drugs. Despite these difficulties, however, a surprising consensus has emerged from research studies in recent years, and if sufficient funding is available for future studies, prospects appear bright.

III. PERSPECTIVES ON CLINICAL PRACTICE

In clinical practice, each practitioner must develop various skills such as the ability to identify drug-responsive syndromes and to implement effective medication trials. However, becoming a proficient clinical pharmacologist is often difficult. Our wish to believe that available drugs actually can cure psychiatric disorders sometimes obscures recognition of the limitations of these compounds. Guidelines for prescribing these drugs are still unclear, and many patients continue to experience persistent symptomatology despite their use. Since both acute and chronic psychiatric illnesses vary greatly from patient to patient, discriminating the effects of the drugs from other factors, such as stress or psychotherapy, is sometimes impossible. Despite the confounding features of psychiatric illness, pharmacotherapy cures some patients and alleviates suffering in many others. Therefore, what must the practitioner know and do to provide effective care?

A. Matching Medications with Drug-Responsive Syndromes

Diagnostic distinctions may confuse the clinician. Many syndromes reflect groups of illnesses or clusters of symptoms. More specific delineation of discrete disorders responsive to particular medications requires further understanding of brain chemistry. However, the practitioner currently can rely on careful observation, mental status signs, presenting symptoms, family history, and the course of psychological symptoms to define various clinical syndromes. Once the clinician defines the nature of the disorder, he should determine if it is drug responsive. In addition, he often can identify features that may indicate a decreased response to pharmacotherapy (such as rapid cycling of manic episodes in patients with bipolar affective disorders).

B. Properties of Medication

The clinician should know the various properties of each medication prescribed, including its **clinical action, adverse reactions, kinetics, methods and routes of administration, preparation and dosage, toxicology, and cost.** The characteristics of a psychoactive agent contribute to the practitioner's management of a drug trial. For example, although the mechanisms of action may be largely unknown, clinical experience has helped to define the time required for the onset of clinical effects (e.g., for antidepressants, 2 to 3 weeks in therapeutic doses). Caretakers also must consider the possibility of short- and long-term adverse reactions and toxic responses. The clinician must not only understand these potential risks but know what to do when they occur.

C. Attributes of the Clinician

The practitioner's attitudes and behaviors may have profound effects on the outcome of pharmacotherapy. A positive result may be encouraged or enhanced by optimism and faith in the drug. At a procedural level, the therapist's attitude is reflected by his persistence in completing drug trials and his consistency, emotional availability, and attentiveness to the patient's concerns. Similarly, a clinician who devalues the need for medication, expresses pessimism about the outcome, or develops ambivalent or even hostile feelings toward the patient may negatively affect the treatment course.

D. Attributes of the Patient

The characteristics of each patient always influence compliance and the course of treatment and often affect outcome. In fact, some studies indicate that less than one-half of those taking drugs over a long period reliably follow directions and that lack of compliance accounts for the majority of treatment failures. Therefore, the clinician must consider the psychosocial variables that affect therapy, such as the individual's attitudes toward and understanding of his illness, psychodynamic issues, character style, ability to relate in a therapeutic relationship, coping skills, social supports, attitude toward drug taking, and motivation. For example, if the patient has a negative self-image or if those close to him disapprove of pharmacotherapy, compliance with treatment recommendations is often poor. Each individual also tends to manifest his own pattern of drug taking that usually becomes evident during the early phases of drug treatment. The most typical manifestations include:

1. Overuse or inappropriate use because of
   a. Impulsiveness.
   b. Misunderstanding of the risks of the drug, its mechanism of action, or the regimen prescribed.
   c. Inadvertent polypharmacy (e.g., taking a substance such as alcohol without realizing that it may potentiate the effects of other agents).
2. Underuse (which is most common in clinical practice) because of
   a. Fear of adverse reactions.
   b. Concern about dependence and addiction.
   c. Inconvenience of the medication regimen.
   d. Misunderstanding the prescribed regimen.
3. Inappropriate use because of
   a. Struggles with the clinician for control.
   b. Tendency toward self-regulation of feeling states.

In addition to its effects on drug taking, the patient's attitude toward his therapy may also influence the biological course of the syndrome. Active support from both the clinician and persons close to the patient may have similar influences.

E. Pharmacotherapy and Psychotherapy

Often pharmacotherapy and psychotherapy are considered competing ideologies. The report of The Group for the Advancement of Psychiatry, *Phar-*

macotherapy and Psychotherapy: Paradoxes, Problems and Progress, published in 1975, emphasized the synergism between these two forms of treatment.[7] However, defining the critical psychotherapeutic factors has been difficult. Nonspecific factors (e.g., the therapist's values, the therapist's regard for the patient, the patient's feelings about the therapist) often dominate a specific type or style of treatment. In addition, psychotherapeutic techniques are difficult to delineate and to implement consistently enough to study. Despite significant methodological problems, researchers and clinicians have developed various beliefs about combining pharmacotherapy and psychotherapy.

1. In some severe disorders, particularly acute affective and psychotic episodes, psychotherapy may contribute little to short-term outcome.
2. In many situations, drugs may be necessary but not sufficient for optimal outcome. The clinician should combine medication with a nonbiological approach (e.g., antipsychotic medication and milieu therapy for schizophrenic episodes; MAO inhibitors or cyclic antidepressants and psychotherapy for panic states and phobias; antidepressant medication and milieu therapy or psychotherapy for acute depressive episodes).
3. In depressive episodes, medication and psychotherapy each affect a different part of the disorder. Psychotherapy improves the patient's interpersonal relationships; pharmacotherapy improves mood and reduces symptoms. Combined effects appear to be greater than those produced by either treatment alone.[8]
4. For patients with less severe symptoms (particularly depression and anxiety), psychotherapy alone may be as effective as pharmacotherapy. For example, several recent studies have shown that Beck's cognitive treatment is as effective as drugs for treating depression.[9] Well-designed studies are needed to define the interaction between psychotherapy and pharmacotherapy and the effects of each modality when used alone. For instance, although some patients' hypomanic symptoms diminish with increased therapeutic contact, most individuals (perhaps 90% or more) develop subsequent affective episodes if their lithium is stopped.

When a specific psychotherapy is combined with drug treatment, the interaction may be easier to see. However, to some extent, the success of each drug trial rests on basic psychotherapeutic skills. The clinician must tolerate the patient's symptoms while maintaining his faith in the therapeutic task. Since drugs often have conflicting meanings for both patient and clinician, each may experience various discomforting feelings, including ambivalence, fear, and anger. For the practitioner, many pitfalls exist. In the most severe cases, therapists sometimes blame or ignore patients because of ineffective results or adverse reactions. In other instances, pills may be used as a substitute for talking. This can take two forms—inappropriate trials of medication or appropriate medication without meaningful human contact. Even the most sincere practitioner sometimes administers halfhearted drug trials that are born out of acquiescence to the patient or that result from a sense of medical obligation (without conviction about success).

Focusing on drug giving also may cause the therapist to ignore the patient's viewpoint. Psychiatric illnesses often are stigmatizing. Patients may stop using available supports. Just as often, persons close to the patient may become distant or turn away. These same tendencies may operate in the therapeutic relationship. Sometimes, both transference and countertransference feelings compound the clinician's difficulties in establishing an empathic, caring human bond with the patient. Therefore, we must always remember that pharmacotherapy entails a psychotherapeutic interaction.

In the early years of modern psychopharmacology, while psychoanalysis still dominated American psychiatry, drug treatment was viewed with disdain by many. Insight was the gold standard, and "pill pushing" a poor second best, suppressing symptoms temporarily at the price of self-knowledge. As biological psychiatry has grown, however, as the numbers of drugs and our sophistication in their use have increased, as brain imaging and other techniques of neuroscience have become available, the pendulum has swung. Driven in part by financial constraints on long-term psychotherapy, a trend has developed and accelerated that seeks a drug approach to every psychiatric symptom. Biologically oriented psychiatrists, patients and family members with a similar bent, and managed health care systems focused on the "bottom line" all conspire to push medication at the expense of nonbiological approaches. Too often, the result is counterproductive and frustrating, with polypharmaceutical regimens piled one on the other, commonly resulting in iatrogenic complications and additional diagnostic obfuscation.

To its credit, modern psychiatric research increasingly is focusing on cognitive and behavioral therapies, structured psychosocial treatments, and the interface of nonbiological and biological approaches. The clarifications to be derived from these studies (e.g., interactions between axis I and axis II disorders) will help clinicians and patients out of these unifocal predicaments. In the meantime, clinicians themselves need to retain strong bases in both biological and psychobehavioral constructs and techniques as they approach psychiatric patients.

IV. GENERAL CLINICAL GUIDELINES

A. Patient Evaluation

Each clinician should develop a comprehensive approach to medicating patients. This does not require memorizing all the details of psychopharmacology but rather developing a framework for care that includes guidelines for assessment, thoughtful differential diagnosis, methods for implementing a drug trial and monitoring adverse reactions, and access to essential information.

The first step in providing quality care involves taking a thorough psychiatric history and review of systems and administering appropriate parts of the mental status examination. When the patient is too disorganized or withdrawn, the clinician should interview friends or family members. On the basis of these findings, the clinician must differentiate functional from organic causes. Often caretakers feel overwhelmed by the array of possible medical disorders mimicking or contributing to primary psychiatric illnesses. Organic disorders sometimes present with psychological symptoms (see Tables 1, 2, and 3), whereas medicable psychiatric disorders sometimes present as an organic brain syndrome (i.e., disorganized schizophrenic states, bipolar affective disorders, and "pseudodementia" of depression).

Confusion between organic and functional disorders frequently leads to extensive and unnecessary laboratory testing. Studies show that generally, the clinician can medically screen a patient by taking a careful medical history, reviewing organ systems, completing a physical examination, and obtaining laboratory studies to follow up on positive findings. Several factors may complicate the interpretation of the data. Emotional difficulties caused by a stressful event or significant environmental causes tend to obscure biological factors. In addition, rigid or distorted character defenses sometimes alter or hide the usual presenting symptoms of medicable syndromes.

B. Principles of Drug Use

1. Use Nonbiological Treatments When They Are as Effective as Pharmacotherapy

Many psychological conditions respond to support, crisis intervention, and various therapies (e.g., behavioral treatment for simple phobias or perhaps cognitive or insight-oriented therapies for moderately severe depressions). Each clinician must consider that drug therapies may have significant physical and psychological

**TABLE 1. Common Organic Causes
of Anxiety States**

Endocrine disorders
    Cushing's disease
    Hyperthyroidism
Metabolic disorders
    Hypoglycemia
    Hypocalcemia
Drugs and medications
    Caffeine
    Amphetamines
    Withdrawal states from addictive drugs
    Steroids

TABLE 2. Organic Causes of Psychosis

Space-occupying lesions of the CNS
   Brain abscess (bacterial, fungal, TB, cysticercus)
   Metastatic carcinoma
   Primary cerebral tumors
   Subdural hematoma
Cerebral hypoxia
   Anemia
   Lowered cardiac output
   Pulmonary insufficiency
   Toxic (e.g., carbon monoxide)
Neurological disorders
   Alzheimer's disease
   Distant effects of carcinoma
   Huntington's chorea
   Normal-pressure hydrocephalus
   Temporal lobe epilepsy
   Wilson's disease
Vascular disorders
   Aneurysms
   Collagen vascular diseases
   Hypertensive encephalopathy
   Intracranial hemorrhage
   Lacunar state
Infections
   Brain abscess
   Encephalitis and postencephalitic states
   Malaria
   Meningitis (bacterial, fungal, TB)
   Subacute bacterial endocarditis
   Syphilis
   Toxoplasmosis
   Typhoid
Metabolic and endocrine disorders
   Adrenal disease (Addison's and Cushing's disease)
   Calcium
   Diabetes mellitus
   Electrolyte imbalance
   Hepatic failure
   Homocystinuria
   Hypoglycemia and hyperglycemia
   Pituitary insufficiency
   Porphyria
   Thyroid disease (thyrotoxicosis and myxedema)
   Uremia
Nutritional deficiencies
   B_{12}
   Niacin (pellagra)
   Thiamine (Wernicke–Korsakoff syndrome)

(*continued*)

TABLE 2. (Continued)

Drugs, medications, and toxic substances
  Alcohol (intoxication and withdrawal)
  Amphetamines
  Anticholinergic agents
  Barbiturates and other sedative–hypnotic agents (intoxication and withdrawal)
  Bromides and heavy metals
  Carbon disulfide
  Cocaine
  Corticosteroids
  Cycloserine (Seromycin®)
  Digitalis (Crystodigin®)
  Disulfuram (Antabuse®)
  Hallucinogens
  Isoniazid (INH® and others)
  L-DOPA (Larodopa® and others)
  Marijuana
  Reserpine (Serpasil® and others)

morbidity. In addition, patients often request medications (particularly sedative–hypnotic agents) for inappropriate reasons.

2. Do Not Deny a Patient Appropriate Medication

In many cases, clinicians overlook the signs of a medicable syndrome or withhold medication because symptoms seem to be "reactive." Patients often suffer needlessly and occasionally harm themselves or others.

3. Choose the Drug with the Best Risk/Benefit Ratio

The practitioner should prescribe the medication with the fewest adverse reactions and the greatest clinical effects.[10]

4. Understand the Pharmacokinetics of Psychotropic Agents

A clinician should understand the kinetics of psychotropic agents, since the clinical characteristics of a drug are determined, in part, by biological factors that affect its ability to reach and exit from its site(s) of action. These pharmacokinetic factors—**absorption, distribution, metabolism, and excretion**—have been studied and understood increasingly in recent years and have enhanced our capacity to use drugs rationally.

Knowledge about the kinetics of a drug can help the clinician guide its actions in a patient and also may be useful in selecting a particular compound. The rate of drug **absorption** from the gastrointestinal tract, for example, largely determines the

TABLE 3. Organic Causes of Depression

Neurological disorders	Effects of tumors
Alzheimer's disease	Cancer of the bowel
Cerebral arteriosclerosis	Cancer of the pancreas
CNS degenerative disorders	Carcinomatosis
Huntington's chorea	Cerebral metastases and tumors
Multiple sclerosis	Oat cell carcinoma
Normal-pressure hydrocephalus	Drugs, medications, and poisons
Parkinson's disease	Alcohol
Postconcussion syndrome	Amphetamine withdrawal
Subdural hematoma	Antipsychotics
Metabolic and endocrine disorders	Barbiturates
Addison's disease	Bromides
Cushing's disease	Carbon disulfide
Diabetes	Carbon monoxide
Hepatic disease	Cocaine
Hyperparathyroidism	Digitalis
Hyperthyroidism	Heavy metals
Hypokalemia	Lead poisoning
Hyponatremia	Methyldopa
Hypopituitarism	Opiates
Pellagra	Oral contraceptives
Pernicious anemia	Other sedatives
Porphyria	Propranolol (Inderal® and others)
Uremia	Reserpine (Serpasil® and others)
Wernicke–Korsakoff syndrome	Steroids
Wilson's disease	Miscellaneous disorders
Infectious diseases	Anemia
Brucellosis	Cardiac compromise
Encepalitis (viral)	Chronic pyelonephritis
Hepatitis	Epilepsy
Influenza	Lupus erythematosus
Mononucleosis	Pancreatitis
Postencephalitic states	Peptic ulcer
Subacute bacterial endocarditis	Postpartum state
Syphilis	Rheumatoid arthritis
Tuberculosis	
Viral pneumonia	

speed of onset of action after a single oral dose. Drugs that are absorbed rapidly produce a faster and more intense onset of clinical effects, whereas the reverse is true for more slowly absorbed compounds. Termination of a drug's effects after a single oral dose is largely determined by the rate and extent of the drug's **distribution.** For example, highly lipid-soluble compounds tend to be rapidly and extensively distributed throughout the body's tissues, which indicates a relatively brief duration of clinical effects following a single dose. Distribution, rather than the elimination half-life of the drug, determines its duration of action after a single dose, but after repeated dosing, the elimination half-life becomes clinically important. The elimination half-life determines the rate at which a drug accumulates in body tissues: drugs with longer half-lives accumulate more gradually (steady-state concentrations are achieved after approximately four half-lives). In addition, drugs with long half-lives disappear from the body more gradually following discontinuation, whereas those with shorter half-lives are eliminated more rapidly (which for some drugs can result in the rapid appearance of an intense withdrawal syndrome). The half-life also can guide a clinician in choosing the frequency of dosing intervals: drugs whose half-lives exceed 24 hr usually can be administered once each day.

5. Learn the Differences between Preparations

Parenteral preparations generally have a more rapid onset of action, whereas all oral preparations, whether in capsule, tablet, or liquid form, provide equivalent amounts of medication with equal speed. Therefore, clinicians should avoid prescribing concentrates except for patients who have trouble swallowing medication. These preparations usually cost more than others and may cause mucosal irritation and contact dermatitis.

Occasionally alternative preparations may be useful. For example, depot preparations of antipsychotic medication may benefit some patients over the long term.

6. Prescribe the Simplest Drug Regimen to Increase Compliance

Most patients, particularly those who require chronic medication, do not take it reliably. To increase compliance, clinicians often can give a single bedtime dose. In drugs with a long half-life (like cyclic antidepressants) that are administered in gradually increasing doses, once-a-day medication can be started at the beginning of treatment. Other drugs with a long half-life, such as antipsychotic agents, also may be given in a single dose, usually after several weeks of administration. Although patients often prefer to take benzodiazepines in divided doses, the long-acting agents within this drug class can be given once a day with the same clinical effect. Even lithium (Eskalith® and others) probably can be administered once daily, although there is still controversy about this practice.

7. Avoid Polypharmacy Whenever Possible

Combinations of psychoactive drugs often are not more effective than a single agent. Because of additive properties or drug interactions, polypharmacy may increase the incidence of adverse reactions. Combinations of medication are often promoted for commercial reasons. Moreover, patients already on drugs may have persistent symptoms and demand immediate relief. Clinicians, feeling frustrated, may respond by dispensing additional medications. Despite the evidence against polypharmacy, the average state hospital patient still takes three different medications. However, in the following situations, combinations of drugs may be indicated:

1. When a potential beneficial synergistic effect outweighs the risk of an adverse reaction. For example, sometimes cyclic antidepressants used alone do not alleviate the patient's depression. In some situations, however, their synergistic effect with other drugs such as antipsychotics, MAO inhibitors, methylphenidate (Ritalin® and others), thyroid hormone, or lithium may relieve the patient's symptoms.
2. When the diagnosis calls for a particular combination. For instance, the clinician may administer a combination of cyclic antidepressant and an antipsychotic agent to depressed patients with psychotic symptoms.
3. When adverse reactions caused by a psychotropic drug require treatment. For example, antipsychotic-drug-induced extrapyramidal reactions should be treated with antiparkinson drugs; severe anticholinergic symptoms secondary to cyclics, antipsychotics, or antiparkinson drugs can be treated acutely with physostigmine (Antilirium®).
4. When changing from an acute regimen to maintenance treatment, the use of two preparations may overlap. In schizophrenia, depot fluphenazine (Prolixin® and others) is started while the patient still receives an oral antipsychotic agent. Sometimes, in an acute manic patient who is taking haloperidol, lithium may be started for maintenance therapy before the antipsychotic medication is discontinued.
5. When a patient has two significant symptoms that do not respond to one drug. Sleeping medication may be indicated in depressed patients with persistent insomnia, even after they have been started on an antidepressant regimen.

8. Provide the Most Cost-Effective Treatment

Theoretically, the clinician should help patients save money by prescribing drugs by their generic names. Unfortunately, the bioavailability (i.e., the amount of drug absorbed into the plasma) of different preparations may vary significantly. In practice, the clinician should try to prescribe the least expensive and most appropriate drug first. If this is done, he should discuss possible problems of bioavailability with the patient. In addition, he should encourage the patient (and/or family) to inquire about possible changes in the brand when prescriptions are refilled and to monitor closely the appearance of adverse reactions or the reemergence or worsening of clinical symptoms.

Because of the possible changing effectiveness of the preparations administered, the practitioner may also have to adjust the dosage. By now, most psychotropic medications are available in generic preparations.

9. Exercise Special Care with Medically Ill Patients

Prescribing psychoactive drugs to medical patients may require special measures. These patients often are older—the risk of adverse reactions increases, and tolerance to medication decreases. The elderly often are taking medication that may interact with psychoactive agents. The medical condition itself may represent a relative contraindication to drug treatment or may obligate the clinician to alter the dosage or methods of administration significantly. Because of these limitations, the clinician should:

1. Monitor possible drug interactions, adverse and toxic reactions, and special metabolic problems.
2. Assess the role of the medical condition in the etiology of the psychological symptoms.
3. Administer the lowest possible effective dose.

10. Establish an Ongoing Therapeutic Relationship

After evaluating the attributes of the patient and his illness and choosing the appropriate drug, the clinician should encourage an active collaboration. At the most basic level, this involves talking openly with the patient and providing accurate information. However, the practitioner's job entails many subsequent steps. He should help the patient acknowledge and resolve negative feelings and distorted beliefs about medication. In addition, the clinician must clarify roles and responsibilities for both himself and the patients. In practice, this process often requires time and the development of a trusting relationship.

At a pragmatic level, the clinician must negotiate a collaborative treatment plan with the patient that evolves from an open discussion of possible risks (i.e., common and serious adverse reactions) and potential benefits of pharmacotherapy and possible alternative treatments. Sometimes, if a disorganized patient cannot give informed consent, the caretaker must involve relatives, friends, guardian, or lawyer. The clinician should record in the patient's chart discussions about treatment, overall care, and any significant changes (e.g., dosage changes, severe adverse reactions and their treatment, mental status changes such as suicidal ideation or psychotic symptoms). The structure of treatment varies from case to case. However, at the least, the clinician should see the patient intermittently and be available (often by telephone) during periods of acute distress. Even after acute symptoms subside, a structured treatment program helps to increase medication compliance and may improve the patient's attitude. In addition, the practitioner should periodically re-evaluate the need for medication and develop criteria for discontinuing medication.

11. Complete Each Drug Trial

A full clinical drug trial consists of administering adequate therapeutic doses of medication for an appropriate length of time. The trial culminates with one or a combination of the following:

1. Acceptable clinical result.
2. Intolerable adverse effects.
3. Poor response after appropriate levels are reached or the drug is administered for a time period specific for the illness (e.g., about 3 to 6 weeks for antidepressants or antipsychotics).

By following these guidelines, caretakers can usually provide effective and safe care. Moreover, just as any clinical talent, skill in pharmacotherapy requires a combination of knowledge, practice, positive attitude, humility, and persistence.

V. CONCLUSION

As drug-responsive syndromes become more clearly defined, practitioners increasingly use pharmacotherapy as the primary treatment for many psychiatric disorders. Based on clinical experience and new research findings, medication options have greatly increased. Many new agents with fewer adverse reactions, lower toxicity, and somewhat different effects on brain chemistry have become available. These advances have improved patient care, and the future holds even greater promise. Increased understanding of brain chemistry and pharmacokinetics will significantly change clinical practice. And, as investigators define the interactions between biological and nonbiological interventions, more comprehensive treatment approaches will evolve.

REFERENCES

1. Sen G., Bose K. C.: Rauwolfia serpentina, a new Indian drug for insanity and high blood pressure. *Indian Med World* 2:194–201, 1931.
2. Laborit H., Huguenard P., Alluaume R.: Un nouveau stabilisateur vegetatif, le 4560 RP. *Presse Med* 60:206–208, 1952.
3. Lehmann H. E., Hunrahan G. E.: Chlorpromazine, a new inhibiting agent for psychomotor excitement and manic states. *Arch Neurol Psychiatry* 71:227–237, 1954.
4. Cade J. F. J.: Lithium salts in the treatment of psychotic excitement. *Med J Aust* 2:349–352. 1949.
5. Sternback L. H., Randall L. O., Gustafson, S. R.: 1,4-Benzodiazepines: Chlordiazepoxide and related compounds, in Gordon M. (ed): *Psychopharmacological Agents.* New York, Academic Press, 1964, vol 1, 137–224.
6. Kuhn R.: The treatment of depressive states with G22355 (imipramine hydrochloride). *Am J Psychiatry* 115:459–464, 1958.
7. Group for the Advancement of Psychiatry: *Pharmacotherapy and Psychotherapy: Paradoxes, Problems and Progress,* report 93. New York, Group for the Advancement of Psychiatry, 1975, vol 9.
8. Weissman M. M.: The psychological treatment of depression: Evidence for the efficacy of psychotherapy done in comparison with, and in combination with, pharmacotherapy. *Arch Gen Psychiatry* 36:1261–1269, 1979.
9. Beck A. T., Rush A. J., Shaw B. F., et al: *Cognitive Therapy of Depression.* New York, Guilford Press, 1979.
10. Gelenberg A. J.: New perceptions on the use of tricyclic antidepressants. *J Clin Psychiatry* 50:7(Suppl):3–10, 1989.

II

Major Psychiatric Disorders

2

Depression

ALAN J. GELENBERG, M.D., and STEPHEN C. SCHOONOVER, M.D.

I. INTRODUCTION

A. Psychological Models

Melancholia has accounted for severe human suffering throughout history. Numerous individuals have poignantly described the various symptoms and feelings that accompany severe depressions. However, it was not until the beginning of the 20th century that both psychological and constitutional theories were developed. Abraham postulated intense oral fixation in depressed individuals and differentiated grief from morbid depression.[1] Moreover, he viewed depression as aggression turned toward the self. Freud embraced these ideas but elaborated the concept of object loss.[2] He contended that the real or fantasied loss of an ambivalently loved person precipitates an ego regression with introjection and intense anger toward the lost object. This process results in self-deprecating, guilty feelings. Many authorities have argued that depression is not simply the result of aggression turned inward. Overt expressions of hostility do not relieve depressive symptoms. Weissman et al. emphasize that depression and aggression are separate affects that frequently coexist in depressed individuals.[3]

Other investigators have broadened our theoretical understanding of depression. Bibring viewed depression as a discrete ego state resulting from an individual's inability to achieve his ego ideal.[4] He felt that aggression was not the driving force behind depression; depleted self-esteem was the primary issue and secondarily led to hostility.

In addition to models that emphasize loss and self-esteem, some researchers and clinicians have discussed the role of deprivation during critical developmental

ALAN J. GELENBERG, M.D. • Department of Psychiatry, University of Arizona Health Sciences Center, Tucson, Arizona 85724. STEPHEN C. SCHOONOVER, M.D. • Schoonover Associates, Newton Centre, Massachusetts 02159.

stages. The Harlows, Spitz, Ainsworth, Bowlby, the Robertsons, and Mahler have contributed to our understanding of early attachments and the process of separation–individuation.[5–10] Early deprivation resulting from the loss of or separation from a major attachment (i.e., mothering figure) may have permanent ramifications if it occurs during a critical period and is not replaced by a meaningful relationship. Spitz observed that infants separated from their mothers during the first year of life developed various symptoms that progressed from apprehension and crying to withdrawal, motor slowing, despair, and, finally, detachment and even retarded physical and emotional development. This "anaclitic depression" was also described by Robertson and Bowlby in older children and by Harlow in primates. Clinical studies have since corroborated the importance of developmental deficits in the etiology of depression.

More recently, self-psychology has added another framework for understanding mood disorders.[11] Patients who have a poor sense of self and a lack of soothing self-objects frequently feel overburdened, helpless, worthless, and depressed. In clinical practice, those with severe narcissistic disorders and borderline states often develop adult "anaclitic" conditions marked by feelings of emptiness, loneliness, and lowered mood, even in the presence of others.[12]

Changes in psychoanalytic theory and practice have contributed to a broader view of depression: developmental deficits leave the individual exquisitely sensitive to feelings of deprivation and loss, which are stirred up by dynamically important life events and precipitate low self-esteem, loss of self-love, and self-deprecation.

Along with psychodynamic formulations, other authors have developed various psychological models to explain depressions. From foundations in Ellis' rational emotive therapy and ego psychology, Beck et al. envision depression as primarily a disorder of thinking.[13] Like the psychoanalysts, Beck postulates that distorted views of the self, the world, and the future result from early learning. This negative set consists of disordered cognitive "schemata" and cognitive "errors" that are stimulated by events in the patient's life and cause the patient to feel hopeless and helpless in the face of life's difficulties. Beck, in a way, bridges the psychodynamic and behavioral models.

Learning theory and behavioral psychotherapies brought a new focus to the study of depression. Seligman et al. introduced the concept of "learned helplessness."[14] He hypothesized that people get depressed when they feel a loss of control over positively reinforcing experiences. Seligman et al. think that some individuals exposed to unavoidable aversive stimuli will respond with passivity and helplessness to similar "stresses" in the future. In addition, this reaction can become a generalized personality trait; the individual believes that all adaptive efforts will eventually fail. Lewinsohn et al., like Seligman, view the lack of reinforcing stimuli as central to the development of depression.[15] However, they state that the important deficit is the low number of positive reinforcements "contingent" on the efforts of the individual. Moreover, according to this model, patients exposed to punishing or nonreinforcing responses from their environment may stop behaving in ways that will yield rewarding responses.

Other behavioral causes for depression have also been postulated. Some investigators feel that passive, self-deprecating, or depressed behaviors may become linked to other more fundamental sets of feelings. For example, depressed behaviors may elicit attention or affection, or they may cause others to gratify need-satisfying impulses (secondary gain).

Some authors have focused on social and existential themes. They emphasize not only the importance of relationships but also the broader social context. Problems with role status, individual purpose and meaning, and cultural myths may contribute to a person's sense of alienation and despair.[16]

B. Biological Models

More recently, investigators and clinicians have turned their attention to the biology of depression. Theoretical formulations followed the fortuitous discovery of agents that alleviated severe depression. Researchers found that the tricyclic antidepressants and monoamine oxidase inhibitors (MAOIs) had significant effects on various brain neurotransmitters. In particular, Schildkraut and others hypothesized a relative synaptic **biogenic amine deficit** in depression and an excess in mania.[17]

Since the early reports, investigators have studied many aspects of brain functioning for evidence of the biological nature of depression, including neurotransmitter effects, metabolism and reuptake of monoamines, neuroendocrine feedback mechanisms, cell membrane and cation balance, brain peptides, and receptor phenomena [including the roles of adenylate cyclase and cyclic adenosine monophosphate (AMP)]. Deficits of both catecholamines (norepinephrine) and indoleamines (5-HT) have been suggested in depressed patients. More recently, researchers have proposed that both **serotonin-deficit** and **norepinephrine-deficit** depressions exist.[18] In addition, the interaction between these two neurotransmitter systems (particularly in the switch between mania and depression) has been described.[19] Several authorities postulate a permissive theory in which a serotonin deficit results in a vulnerability to affective illness; a lowered catecholamine level precipitates depression, and a raised level results in mania. **Although the changes in monoamines during a depression may be significant, investigators now think that various relationships between the parts of the central nervous system must be considered in the etiology of severe depression.** This feeling stems not only from research illustrating changes in various brain subsystems but also from the clinical effectiveness of new drugs that might not affect norepinephrine or serotonin levels (e.g., the experimental drugs iprindole and mianserin, both unavailable in the United States). Ultimately, not only might we learn that one or more neurotransmitter systems are involved in the pathophysiology of depression, but it is likely that altered sensitivity and stability of both presynaptic and postsynaptic receptors may also be involved.

C. Integrative Model

Any cogent theory of depression must integrate information from various levels of observation. Akiskal and McKinney envision **depression as a psychobiological "final common pathway of the various interlocking processes at chemical, experiential, and behavioral levels."**[20] They think that many factors including biological vulnerability, developmental deficits, and physiological and psychosocial stressors can contribute in varying degrees to a functional impairment of diencephalic centers that help to maintain mood, motor activity, appetite, sleep, and libido.

Even with the remarkable progress in our understanding of depression, the roles and relationships of various contributing factors to depressive symptoms await clarification. Therefore, we must treat each patient's problem individually by combining biological and nonbiological interventions.

II. DIAGNOSTIC CONSIDERATIONS

A. Clinical Presentations

To define disorders for which pharmacotherapy is strongly indicated, the clinician must differentiate normal from abnormal mood. Individuals frequently experience sadness and depression as a normal response to stress, disappointment, and crisis. Others exhibit depressive symptoms as part of their character style or in response to unconscious themes. Some individuals even appear to have mood changes related to natural rhythms (like the seasons). Many episodes of lowered mood require little or no intervention. However, some reactive, recurrent, or chronic depressions mentioned above require support, psychotherapy, or even pharmacotherapy.

Some individuals with abnormal mood develop more severe, autonomous depressive symptoms (i.e., "endogenous" or "melancholic") that respond to medication. These patients usually have a persistently lowered mood accompanied by feelings of worthlessness, hopelessness, helplessness, guilt, self-deprecation, blame, and pessimism. They also may have self-destructive impulses and impaired functioning. Somatic preoccupations and vegetative signs, including anorexia and weight loss, constipation, anhedonia, decreased libido, psychomotor retardation and/or agitation, and sleep disturbance (including fitful sleep and early awakening) are frequent in more severe depressions. Many individuals also manifest diurnal mood swings—the most serious symptoms occur in the morning, with improvement as the day progresses. In the most severe cases, patients may experience somatic or paranoid delusions.

The third revised edition of the *Diagnostic and Statistical Manual of Mental Disorders* (DSM-IIIR) specifies criteria for major depressive illness (see Table 1).[21] Some individuals, however, who present atypically also respond to medication. They include chronic complaining patients ("crocks"), individuals with "reactive" depressions, elderly with "pseudodementia," and those with hypochondriasis and chronic pain syndromes, phobic and anxiety states, "hysteroid dysphoria," Briquet's disease (hysteria), some severe psychoses (particularly with paranoia), and catatonic states.

Moreover, although severe depressions occur more commonly in patients with significant character problems, drug-responsive conditions are often missed by clinicians. These patients frequently appear lonely, empty, angry, manipulative, or "object hungry" and regressed rather than classically depressed. In addition, patients with severe obsessional character traits may mask a major depression with ruminations, compulsions, and intellectual defenses. In these individuals, the clinician must diagnose an underlying depression by a careful assessment over time or from a successful trial of antidepressant medication. Sometimes nonpsychiatric medications or medical conditions (see Table 3 in Chapter 1) may contribute to or cause major depressive episodes that require medication for their resolution.

The interface between medication-responsive depressions and character

TABLE 1. Major Depressive Illness[21]

Dysphoric mood (depression) and/or loss of interest or pleasure
At least four of the following for at least 2 weeks:
    Poor appetite or weight loss or increased appetite or weight gain
    Insomnia or hypersomnia
    Loss of energy or fatigue
    Psychomotor agitation or retardation
    Feelings of worthlessness or excessive or inappropriate guilt
    Diminished ability to think or concentrate
    Recurrent thoughts of death or suicide
Melancholic subtype includes at least five of the following:
    Loss of interest or pleasure in all, or almost all, activities
    Lack of reactivity to usually pleasurable stimuli
    Diurnal mood swings (i.e., worse in morning)
    Early awakening
    Marked psychomotor agitation or retardation
    Poor appetite or weight loss
    No significant personality disturbance prior to major depressive episode
    One or more previous major depressive episodes with complete or nearly complete recovery
    Previous response to somatic therapy

pathology is fascinating, complex, and only just beginning to yield its secrets to the probes of empirical research. Two areas of recent study have been chronic depression and the borderline personality disorder. Individuals suffering from chronic depression often are considered to have a character disorder rather than a medication-responsive condition. However, when an individual fulfills the criteria for major depression, particularly of the melancholic subtype, chronicity alone does not weigh against medication response. In fact, Kocsis and colleagues have shown that the drug–placebo difference in such patients is as robust as it is in those with acute depression.[22]

In an independent area of clinical and research efforts, patients with borderline personality disorders frequently may qualify for a diagnosis of major depression. Even when they fulfill diagnostic criteria for a melancholic subtype, however, these patients tend to be less responsive to antidepressant medication. In fact, the work of Soloff and others suggests that antidepressant drugs in borderline patients may actually cause problems (see Chapter 13). Nevertheless, clinicians should consider an empirical trial of at least one antidepressant when a borderline patient shows severe depressive symptoms.

B. Classifications

Historically, depressions have been subtyped according to dichotomous categories (e.g., endogenous–reactive and neurotic–psychotic). Recently, this view has been refined and broadened to include primary–secondary and unipolar–bipolar. In general, endogenous depressions correspond to the melancholic subtype of major depressive illness. The abandonment of the endogenous–reactive dichotomy has been important, since it implied that some depressions do not have exclusively environmental causes and require pharmacotherapy, whereas others result from life stresses and should be "worked through" psychotherapeutically. This clinical stance would preclude the use of pharmacotherapy for patients with "reactive depressions" and the use of psychotherapy for "endogenous" depressions. In fact, the decision to medicate with antidepressants relies not so much on presumed causality but on the presence of particular symptom clusters. The neurotic–psychotic classification of depressions is equally misleading, since the definitions of these terms are variable and do not correlate well with drug response.

More recently, diagnosticians have reclassified depressions. The primary–secondary dichotomy avoids some of the problems of the older continua. In primary depressions, there is no previous history of psychiatric illness except for episodes of depression or mania; secondary depressions occur in patients with a previously documented psychiatric disorder (other than an affective episode) or a medical illness. Investigators originally thought that primary depressions might preferentially respond to antidepressants. However, secondary depressions generally have similar symptoms and responses to medication. Clinically, patients with secondary depressions have a somewhat greater incidence of somatic and psychotic symptoms, phobias, anger, and suicidal behaviors. Winokur, in particular, believes that secondary depressions represent an acute process superimposed on a chronic set of problems.

The unipolar–bipolar dichotomy represents a further subclassification of primary mood disorders. Bipolar disorders must have a history of both depression and mania. In contrast, a unipolar depressive disorder must include a history of depressions only, although the number and frequency of depressions require further definition. By studying families, Winokur has further subdivided primary depression into pure depressive disease, depressive spectrum disease, and sporadic depressive illness (nonfamilial). Pure depressive disease occurs most often in older males who have one or more first-degree relatives with depressions. This condition appears equally in men and women. Depressive spectrum disease, in contrast, occurs most frequently in women under 40 years of age. Their female relatives have an increased frequency of depression, and first-degree relatives have a high frequency of alcoholism or antisocial characteristics. Males with depressive spectrum disease usually have alcoholic or antisocial symptoms.

Sporadic depressive disorders comprise the largest group of unipolar depressions (about 40%). Although unipolar illnesses can be divided into relatively homogeneous groups, the clinical ramifications of these findings await clarification. In addition, researchers must further correlate epidemiologic, familial, and clinical data with biochemical findings.

The DSM-IIIR is the currently adopted diagnostic framework for psychiatric disorders. However, because DSM-IIIR must insure a high degree of reliability and validity, some more atypical presentations of depression are excluded. Despite limitations, it defines a group of major depressive disorders (see Table 1) that are generally:

1. Recurrent and severe (70 to 80%).
2. Responsive to antidepressant medications.

C. Biological Diagnosis

Treating depression with drugs will change remarkably when we have biological tests similar to those in infectious disease: these will enable us to subdivide an illness by pathogen and select an appropriately targeted agent. Indeed, as Klerman has quipped, for psychiatry this is our "search for the holy grail." It stands to reason that there may be biological subtypes of depression and that these will prove differentially responsive to pharmacological therapies. When practical markers become available, they will represent a great leap forward, shortening the lag time before remission and making a series of empirical trials of antidepressant drugs a historical curiosity.

To date, unfortunately, we have no such biological testing available to practitioners,[23] and it has not been for lack of effort from investigators, attention in the scientific literature, or research funding. Many hopes have risen and been subsequently dashed. Candidates for biological markers have included urinary metabolites of central monoamine neurotransmitters [e.g., 3-methoxy-4-hydroxyphenylglycol (MHPG)], provocative challenge tests (e.g., amphetamines), and neuroendocrine markers [e.g., the Dexamethasone Suppression Test (DST) and the thyrotropin-releasing hormone (TRH) stimulation test]. Neurophysiological markers, such as the latency before the first rapid eye movement (REM) period in the sleep cycle, also are being actively studied. Even if they prove valid and are shown to have advantages over clinical phenomenology, they are too cumbersome to be of practical utility. If preliminary research suggesting that DST-positive depressed patients (i.e., those who do not show normal suppression of cortisol following a test dose of dexamethasone) have a lower placebo response rate than other similarly depressed patients is substantiated, such testing might be useful in the design of clinical research studies. Like all other proposed markers, however, the DST has not helped clinicians to select patients for antidepressant drug therapy.

For the time being, therefore, clinicians must rely on patients' behavior (as observed and as reported by others) and on self-reports in determining whether drug therapy is indicated and, if so, with which agent. Of course, when history and subsequent physical examination suggest the possibility of an underlying medical disorder (e.g., an endocrinopathy), appropriate lab testing should be conducted. For the future, we eagerly await the arrival of definitive tests to tease apart the neurobiology of depression.

Currently, most psychiatrists prescribe a trial of an antidepressant, typically a tricyclic initially [but increasingly fluoxetine (Prozac®)], for patients with major depression. This is most appropriate when a patient simultaneously has the melancholic subtype and no delusional features. Probably, these patients are most responsive to cyclic and perhaps also to MAOI antidepressants.

There is some evidence that nonmelancholic depressed patients (sometimes termed "atypical") do selectively well with MAOI antidepressants. Although variously defined, this subgroup of patients suffers from a vegetative syndrome opposite to that seen in melancholics: initial insomnia, mood worse later in the day, food craving, and weight gain. At the same time, psychological features distinguish them from melancholics: their mood is more reactive (as opposed to autonomous), and they are more sensitive to interpersonal rejection and more plagued by phobic, anxious, and panic symptoms. Although some research has shown that rigorously defined "atypical" depression responds best to an MAOI antidepressant, second-best to a tricyclic, with placebo in third place, there is some evidence that this win–place–show relationship pertains only to patients with concomitant panic symptoms. The "plain" atypical depressive patient without panic symptoms may respond equally well to all three treatments; i.e., this may be a syndrome as responsive to placebo (or nonbiological treatments) as to any available antidepressant. Knowledge of the truth must await more rigorous research. In the meantime, clinicians should consider trials of antidepressant drugs, perhaps a cyclic agent first (because they are easier to use), with an MAOI as a second option.

The concomitant presence of psychotic features roughly halves the estimated response rate to an antidepressant drug (when used alone), whether a cyclic agent or an MAOI. Certain types of personality features, such as hypochondriasis and histrionic traits, bode poorly for antidepressant response. Prior resistance to biological therapy in multiple previous depressive episodes also may be a poor prognostic sign.[24]

III. GENERAL THERAPEUTIC MEASURES

A broader understanding of the relationships among brain chemistry, behavior, thinking, and feeling has fostered a more open environment for investigation. In fact, as we clarify the multidetermined nature of depressive illness, we also have started to develop a range of therapeutic techniques for depression. Research has demonstrated the effectiveness, if not the necessity, of using somatic treatments in severe mood disorders. However, recent studies also have shown that various psychological techniques may represent an effective alternative for treating some depressed patients.[13,25]

A. Milieu and Crisis Techniques

Many depressed patients manifest intense suffering, suicidal impulses, and poor self-care, requiring immediate and intensive care. Others require hospitalization because of uncertain diagnosis or medical problems that may confound somatic treatment. In these cases, inpatient care offers various therapeutic advantages and possibilities. At the most basic level, the ward milieu contains symptoms and provides protection.

Frequently, depressed individuals exhibit self-destructive thoughts and behaviors that develop, in part, as a result of self-punitive thinking and disengagement from social supports. A therapeutic milieu can interrupt the depressive cycle and provide at least some measure of protection. Patients intent on dying, however, usually succeed unless immediately protected against self-harm. Because severe suicidal impulses often last only a few days, close observation and enforced human contact often help the patient live through the acute crisis. An inpatient milieu offers other therapeutic advantages as well, including relief from a self-demeaning or disorganizing interaction with persons close to the patient, emotional support and self-validation, structured interpersonal interactions, correction of cognitive distortions, maladaptive patterns, and self-punishing rituals, and the practicing of new behaviors. These psychological interventions can also be provided in other settings (e.g., day care) and by other personnel (e.g., home visit and crisis teams). In fact, most intensive care settings employ crisis techniques that have a focused, here-and-now, time-limited, reality-based orientation. The therapeutic approaches include mobilizing the patient's natural supports, adopting an active and supportive (and sometimes directive) role for the therapist, reviewing, labeling, and correcting maladaptive behaviors and cognitive distortions, and learning new coping and problem-solving techniques. In many ways, this approach integrates aspects of cognitive therapy, behavioral techniques, and the short-term psychodynamic therapies.

B. Psychotherapy

Depression is a set of symptoms with complex biopsychosocial determinants. For the most severe depressions, biological factors rather than environmental or psychological factors seem to determine the course of the illness. These syndromes require pharmacotherapy and may be only slightly responsive to specific psychotherapeutic interventions. However, for depressions of moderate severity, various types of short-term psychotherapy can play an important role in treatment.

Weissman et al., in the Boston–New Haven Collaborative Studies, found that both antidepressants and structured, short-term interpersonal psychotherapy worked better than a mock treatment for patients with nonpsychotic major depressive illness. Both drug therapy and combined drug and psychotherapy provided equal symptom relief and prevention of relapse. Psychotherapy alone offered somewhat less symptom relief but more improvement in social adjustment.[26]

As part of the research effort, the investigators developed a method of therapy that has been specified in detail, tested in adequately controlled studies, and applied consistently in concert with pharmacotherapy.[27,28] The technique, including 12 to 16 once-a-week sessions focusing on **current interpersonal relationships,** is comprised of several structured phases. Initially, the clinician uses a semistructured evaluation interview to clarify symptom history and to review typical depressive symptoms and the natural course of depressive illness with the patient. The goal of the assessment is to define interpersonal problems of four distinct types: grief reactions, interpersonal conflicts, role transitions, and interpersonal deficits.

The core treatment sessions focus on present, not past, relationships and behaviors. During the process, therapists actively help patients understand their range of feelings, delineate past model relationships that were positive experiences, and select and refine potentially helpful interpersonal strategies and actions. The clinician uses both directive and interpretive techniques. However, the focus of intervention is on here-and-now interpersonal relationships, not historical, intrapsychic, or ego-structural problems. Perhaps most important, interpersonal psychotherapy has demonstrated an important synergy between biological and psychosocial interventions for moderately severe depressions. In concert with drugs, this psychotherapy seems to decrease symptoms better than drugs alone during the acute illness, increase social functioning during recovery, and produce better treatment outcomes overall.[29]

Cognitive therapy, developed by Beck and Rush, represents another important contribution to the treatment of depressive illness. Instead of an interpersonal relationship focus, their therapy confronts the distorted cognition of depressed patients.[13] The therapy aims to alter punitive, undermining negative thoughts, images, and underlying assumptions about themselves and others. **Cognitions** are those thoughts and images elicited in specific situations. In depressed individuals, cognitions include several kinds of logical errors, such as overpersonalization, ignoring positive parts of situations and overgeneralizing or exaggerating negative aspects of situations.

Beck et al. also emphasize that patients' hopelessness and helplessness stem, at least in part, from longstanding negative views of themselves, the world, and the future (**cognitive triad**) and that depressed individuals habitually perpetuate those views, called **schemata.** These chronic underlying assumptions, derived from early developmental experience, color the way cognitions are used day to day. In the case of depressed patients, undermining schemata and negative cognitions fuel deep, self-denigrating perspectives.

Therapy to correct these misattributions is generally short-term, consisting of one or two sessions per week for 14 weeks. Initially, treatment focuses on helping the patient specify and record cognitions and schemata that are self-deprecating. Subsequently, the therapist outlines possible new or revised cognitions that are more self-empowering and more aligned with objective reality. Later in therapy, the clinician helps the patient identify the schemata, or hidden assumptions, that underlie depressive cognitions.

The final part of therapy focuses on generating new cognitions and schemata and applying them to actual and anticipated life events. Patients may complete this phase of care during the last session of the initial 14-week therapy or may require additional once-or-twice-a-month sessions extending over several months. The most important goals of cognitive therapy include helping the patient recognize self-denigrating cognitive structures and develop practical alternative viewpoints and actions. Increasingly studied and applied, cognitive therapy appears to be effective for moderately severe depression and may have a synergistic effect when used together with antidepressants.

More behaviorally oriented therapists use similar techniques to treat depression.[29] In addition to identifying and confronting cognitive structures, behavioral therapists employ formal cognitive restructuring, skills training (including assertiveness training), and manipulating the environment.

In general, evidence indicates not only that structured psychotherapies, including various forms of short-term dynamic psychotherapy and crisis intervention, help many depressions but also that they might be equal to pharmacotherapy for some milder depressions. However, we must be careful in drawing definitive conclusions pending further investigation. Studies often have employed antidepressants and specific psychotherapeutic methods inadequately. No investigations comparing psychotherapy and pharmacotherapy have addressed the problem of "fit" between a patient and a particular drug or a specific model of therapy. In addition, the accumulating evidence in psychotherapy research indicates that many factors other than the type of therapy significantly promote positive outcome, including a patient's positive attitude toward therapy and the therapist, a therapist's positive feeling about the patient, and the patient's motivation.

Which depressed patients should receive psychotherapy? Every patient should be involved in a therapeutic relationship. Individuals taking drugs should receive emotional support and encouragement and the opportunity to address their feelings about the depressive episode. Moreover, for selected individuals with significant character pathology (e.g., borderline states, narcissistic personality disorders, masochistic or depressive characters) who develop mild to moderately severe depressions in response to chronic disappointment and low self-esteem, long-term insight-oriented psychotherapy might be the treatment of choice. These patients, however, frequently develop more severe and autonomous depressive symptoms accompanied by vegetative signs; such symptoms may improve with medication. In addition, some therapies may be equal to or better than medication for individuals who become depressed in response to severe stress, life crises, real or fantasied losses, or from a severely negative view of the self, world, or others.

C. Electroconvulsive Therapy

1. Introduction

Although its mechanism of action remains unknown, **electroconvulsive therapy (ECT) effectively treats most patients with severe depression.** In fact, 60,000 to 100,000 individuals receive this treatment each year, most without significant morbidity. However, questions of efficacy, concerns about the development of an organic brain syndrome, and occasional abuses have stigmatized this procedure.[30] In addition, antidepressant medication offers an effective and often more acceptable alternative to ECT. Thus, some patients do not receive ECT even when it is the treatment of choice. In the last decade, however, investigators have attempted to define more precisely the various clinical effects and indications and contraindications of ECT.

2. Mechanism of Action

Although no consensus exists about the therapeutic mechanism of ECT, researchers have documented profound central nervous system changes resulting from its use. Some experts believe that it alters neurotransmitter availability similarly to the effects of antidepressant medications. However, electroshock not only increases serotonin levels and increases MAO activity, dopamine, and cyclic AMP but also may lower catecholamine levels. Alternatively, ECT may produce its effects by increasing the sensitivity of postsynaptic neurotransmitter receptors.[31]

3. Indications

Numerous controlled studies demonstrate that ECT works better than placebo and at least as well as cyclic antidepressants in the treatment of severe depression. In addition, ECT may be more effective than cyclics alone for treating psychotic depressions, particularly when accompanied by agitation. Electroconvulsive therapy also alleviates the symptoms of extreme agitation and catatonia in schizophrenics who respond poorly to medication, making these patients more amenable to other treatment.

Electroshock therapy offers significant advantages over medication in certain clinical situations. The strongest indications include:

1. Previous good response to ECT without significant untoward effects.
2. Strong family history of positive response to ECT.
3. Depression characterized by psychotic features, particularly somatic or nihilistic delusions.
4. Relative or absolute contraindication to medication, such as in patients with significant cardiovascular disease or other medical conditions that preclude an adequate trial of drugs, including pregnancy or history of neuroleptic malignant syndrome.
5. Need for immediate containment of symptoms (e.g., severely suicidal patients, severe mania, extremely agitated or assaultive patients, and individuals with severe retarded depression) that impair self-care and feeding.
6. Poor response to trials of medication.
7. Severe, persistent behavior problems or mildly disorganized thinking in schizophrenic patients.

Electroconvulsive therapy should be strongly considered for psychotically depressed patients and those who have not responded to medication. In this situation, the clinician may start or continue an antidepressant drug, since ECT and cyclics may work synergistically. However, the anesthesia procedure occasionally may produce additional risk of cardiotoxicity from the combination of ECT with antidepressants and antipsychotics. Therefore, a clinical decision should be made together with an anesthesiologist. In addition, **patients receiving lithium (Eskalith® and others) have a greater risk of developing significant central nervous system (CNS) impairment during a course of ECT;** therefore, lithium treatment should be discontinued before ECT and restarted after treatment has ended.[32,33] Many patients require a psychoactive drug after ECT to prevent relapse. Although fewer than a quarter of patients on a cyclic antidepressant or lithium relapse after 6 months, more than half of individuals without maintenance pharmacotherapy subsequently have a recurrence. However, the guidelines for starting long-term drug therapy following ECT require further clarification. In occasional cases, maintenance ECT is needed and helpful.

Sometimes ECT does not provide adequate relief from depressive symptoms. Factors that correlate with poor outcome include hypochondriasis, hysterical character structure, absence of vegetative symptoms (i.e., sleeping, eating, bowel function, motor activity), absence of a family history of depression, absence of delusions, presence of reactive factors, and presence of a marked personality disorder. Clinically, however, this list of negative criteria does not provide much guidance, since individuals with these attributes also respond poorly to other treatment modalities.

4. Contraindications and Adverse Reactions

Several medical conditions may limit the use of ECT. Increased risk generally stems from the anesthesia, the induced seizure, or the transient increase in CNS pressure and blood flow. Relative contraindications include **recent myocardial infarction, active pulmonary inflammation, central nervous system tumors, increased intracranial pressure, and recent carbon monoxide poisoning.** (ECT occasionally has resulted in brain damage in these cases.)

The risks that attend preparatory anesthesia include significant respiratory depression, allergy to premedications, and cardiac failure. The main adverse effects of ECT are confusion and short-term memory loss (which may last five times as long in elderly patients). Both risks are significantly reduced with unilateral treatments.[34]

5. Technique

The clinician should inform the patient about what he will experience during and after the procedure: he will be asked to lie down, a sedative will be administered, and he may wake up with some discomfort (and rarely with a sense of being paralyzed). In addition, each individual should know the possible risks (particularly memory loss) and treatment alternatives. The practitioner should obtain informed consent (from a relative or guardian, if necessary) and should carefully document discussions and procedures.

The procedure begins with the administration of an anticholinergic agent (scopolamine or atropine), a short-acting anesthetic agent (usually a briefly acting barbiturate), and a muscle relaxant [succinylcholine (Anectine® and others)]. The barbiturate usually takes about 1 min to work. During anesthesia, because of muscle paralysis, assisted ventilation with oxygen is required. A physician then applies a brief electrical stimulation. The electrical impulse is titrated to produce a generalized CNS seizure of about 40 to 60 sec duration. The strong muscle relaxant protects the patient from physical injury but paralyzes the respiratory and limb muscles. Therefore, to monitor the seizure, the clinician can tighten a blood pressure cuff on one arm, which will prevent the succinylcholine from affecting that limb, and allow the seizure activity to be observed. Alternatively, the clinician may use an electroencephalogram (EEG) tracing to observe the spike-and-wave pattern. After the seizure, recovery usually takes about 10 to 20 min.

The number and frequency of treatments vary with the clinical needs of the patient. Therapy should be administered until the patient experiences complete recovery. **The typical regimen usually consists of every-other-day or twice-weekly treatments for a total of five to ten sessions.** Clinical research and enhanced ECT technology have improved our understanding and technique over recent years. Newer ECT devices make it easier to monitor seizure duration electroencephalographically, and noninvasive fingertip cuffs allow anesthesiologists to monitor arterial blood gases throughout the treatment. For all patients and particularly those receiving unilateral ECT, it is important to verify that a seizure has occurred. Moreover, when using the unilateral method, a suprathreshold stimulus provides better efficacy than a lower-energy stimulus that barely triggers a convulsion. Some practitioners locate the threshold stimulus needed to trigger a seizure on a patient's first treatment, then subsequently administer stimulation at 100% greater energy for bilateral electrode placement and 150%–200% greater energy for unilateral electrode placement.[35] An alternative approach is to monitor the seizure duration, which optimally should be between 40 and 120 seconds. The energy of stimulation is then adjusted upward or downward to keep the seizure duration within that time range. Increasingly, bilateral stimulation is being recognized as more effective and more rapid for most patients (particularly the elderly), with unilateral treatments reserved for those particularly sensitive to the cognitive side effects of ECT or who have already shown deterioration in cognition and memory.

IV. PHARMACOTHERAPY

Twin coincident discoveries in the mid-1950s spawned two distinct classes of pharmaceuticals that, for the first time, could reach to the core of the depressive syndrome, alleviating the helpless, hopeless, and worthless feelings that often drive patients to suicide. Earlier, the only drugs appropriate for treating patients with major depression were those that alleviated symptoms more peripheral to the depressive syndrome. Barbiturates and bromides, for example, could be prescribed to alleviate depressive symptoms such as anxiety, agitation, and insomnia. In a search for new antituberculous agents, iproniazid (withdrawn from clinical use), a structural analogue of isoniazid (INH® and others), turned out to have mood-elevating properties. It also inhibited the enzyme monoamine oxidase (MAO), which catabolizes the monoamine neurotransmitters dopamine, norepinephrine, and serotonin. Thus, a new group of antidepressants was born: the MAO inhibitors. They not only represented a therapeutic breakthrough but also gave rise to speculation about mechanism of action and, reasoning backwards, the pathophysiology of depression itself.

Several years thereafter, research with structural analogues of the newly discovered phenothiazines gave rise to imipramine (Tofranil®). Although its two-dimensional configuration closely resembles that of its phenothiazine parents, minor changes in the central "bridging" ring in this tricyclic compound, which gave the class its name, resulted in alterations in its three-dimensional configuration. This, in turn, resulted in major pharmacological and clinical differences from the phenothiazines. The remainder of the 1950s and the following two decades saw no comparable breakthrough in antidepressant pharmacotherapy. Rather, a parade of "me-too" antidepressants entered the American market, driven largely by the financial needs of their manufacturers. Discovered through the use of preclinical paradigms that virtually ensured similarity to the prototypes, these agents possessed only minor to moderate differences in side-effect profiles.

As the years went by, other countries began to introduce substantially different drugs for treating depression. Then, in 1979, trimipramine (Surmontil®) was marketed in the United States. There was some suggestion that it possessed a unique neurochemical and clinical profile, but the manufacturer has failed to pursue these possibilities with definitive research; the best "hard" evidence suggests that trimipramine is just a seventh "me-too" tricyclic. The years 1980, 1981, and 1982 saw the sequential introduction of three somewhat different antidepressants: amoxapine (Asendin®), maprotiline (Ludiomil® and others), and trazodone (Desyrel® and others), respectively. Although each was introduced with considerable advertising ballyhoo, subsequent research and clinical impressions suggest that these agents offer less than initially met our hopeful eyes.

First, none has proved more effective than the 1950s-vintage antidepressants. For that matter, we know of no antidepressant in clinical or experimental use in the world today that can make that claim. Nor has any new antidepressant provided a substantially different spectrum of activity against the depressive syndromes. Even worse, despite the failure of clinical research to find differences from older antidepressants in degree of efficacy, clinical impressions suggest that some of the newer antidepressants are not as effective as the tricyclics.

Do any of the new antidepressants act faster than the old ones? Desipramine (Norpramin® and others), the demethylated daughter metabolite of imipramine, was introduced with the claim that it had a faster onset of action than other antidepressants. (In desipramine's case, the demethylation step takes a few milliseconds, hardly noteworthy for a depressed patient.) Both amoxapine and maprotiline were introduced with a similar claim, which also seems unfounded.

Adverse effects and toxicity have shown differences—but not particularly in favor of some new antidepressants. Amoxapine, which is a demethylated metabolite of the antipsychotic drug loxapine (Loxitane® and others), has classic neuroleptic activity, which carries with it a full range of extrapyramidal effects—acute dystonic reactions, akathisia, Parkinson's syndrome, and dyskinesias (possibly irreversible) —as well as such neuroendocrine features as hyperprolactinemia, galactorrhea, and amenorrhea. In overdose, amoxapine has been associated with a higher rate of treatment-resistant seizures, acute tubular necrosis, cases of irreversible brain damage, and possibly a higher mortality rate than overdoses with tricyclics.

Because maprotiline is tetracyclic rather than a tricyclic compound, it was hoped that it would have some advantages. However, it is pharmacologically and clinically practically indistinguishable from the tricyclics, with the exceptions of causing a higher rate of seizures and rashes. Trazodone has greater distinctiveness from the tricyclics, but although it has a lower profile of anticholinergic effects, this highly sedating antidepressant has been associated with occasional cases of male priapism and exacerbation of preexisting myocardial irritability. On the positive side, though, trazodone appears to be safer in overdose.

The subsequent introduction into the U.S. market of nomifensine (no longer available) and then bupropion (Wellbutrin®) were fraught with even greater complications. After less than a year, nomifensine was withdrawn worldwide because of its association with a high rate of immunologic abnormalities, most notably hemolytic anemia, which is potentially fatal. Although this experience highlighted potential pitfalls in introducing new drugs, many clinicians have grieved the total withdrawal of this agent, since some patients are uniquely responsive to it. It is possible that neurochemical features distinguishing nomifensine from older compounds accounted for these cases of distinct efficacy: most commonly mentioned is its enhancement of central dopaminergic activity.

Bupropion spent a short time on the American market before it was withdrawn because of its association with a high incidence of epileptic seizures. In 1989 it was reintroduced, but with warnings about its effects on the seizure threshold.

Fluoxetine was introduced into the United States at the beginning of 1988 and, as of this writing, has been enjoying a rather remarkable commercial and clinical success. It has an acceptable level of efficacy, possibly distinct from the tricyclics, and has a different and favorable profile of adverse effects.

There are many other compounds, of varying degrees of similarity and difference, being studied in this area. The chlorinated analogue of imipramine, clomipramine (Anafranil®), was finally marketed in the United States in 1990. Clomipramine's "claim to fame" is its demonstrated efficacy against obsessive compulsive disorder, apparently independent of its antidepressant activity. Clomipramine is encumbered, however, by an apparently higher incidence of seizures and some other adverse effects.

Dothiepin, another tricyclic antidepressant, is the most widely prescribed antidepressant drug in the United Kingdom. It appears comparable to imipramine and the other tricyclics, but with a "gentler" profile of adverse effects. Mianserin is a tetracyclic antidepressant popular in the United Kingdom. Apparently safe in overdose, mianserin may be associated with a higher incidence of bone marrow suppression. Both dothiepin and mianserin are unavailable in the United States.

Zimelidine (withdrawn from clinical use) is a drug with selective properties in enhancing serotonergic neurotransmission. Very popular in Europe, it was withdrawn worldwide some years ago because it was associated with a "flulike" syndrome, progressing at times to Guillain–Barré. All other selective serotonergic agents, particularly those that (like fluoxetine) resemble zimelidine, are being followed for this possible reaction.

Fluvoxamine (not available in the United States) is another serotonergic antidepressant likely to be introduced into the United States. It is unclear in what ways it is different from fluoxetine. The natural compound S-adenosylmethionine (SAM) is being studied for antidepressant properties and, at present, appears to have efficacy with minimal adverse effects. The selective inhibitor of the B type of the enzyme monoamine oxidase, deprenyl, is free of the "cheese effect" by which exogenous tyramine causes hypertensive crises. However, it is unclear whether deprenyl is effective as an antidepressant. The triazolobenzodiazepine alprazolam (Xanax®) is considered by some clinicians to have greater antidepressant effectiveness than older benzodiazepines, such as diazepam (Valium® and others) or chlordiazepoxide (Librium® and others). Whether this is true, however, remains to be defined by future research.

At present, clinicians using medication to treat depressed patients still have a substantial "wish list." It includes drugs with substantially higher levels of efficacy. Depending on the population studied, anywhere from 10% to 40% of patients remain unhelped by our current agents. Drugs with a different antidepressant profile also would be welcome, as would agents with fewer adverse effects and greater safety. There is always the eternal hope for antidepressant drugs that work faster. Still, the drugs that we currently have give us treatment options only dreamed about in the first half of this century. Used thoughtfully, they can provide millions of patients safe passage through a hazardous and potentially fatal illness. It is reasonable to hope that the future will bring drugs with greater efficacy, safety, and specificity.

A. Cyclic Antidepressants

1. Chemistry

The chemical structures of the cyclic antidepressants currently available in the United States are shown in Fig. 1. The chemical structure of the tricyclic antidepressants differs from the phenothiazines only in the substitution of a carbon–carbon double bond for the sulfur atom in the bridge between the two benzene rings. This minor difference, however, changed the three-dimensional conformation of the molecules, which in turn altered their pharmacological and clinical properties.

With two methyl groups on the terminal nitrogen atom of the side chain, imipramine, amitriptyline (Elavil® and others), trimipramine, doxepin (Sinequan® and others), and clomipramine are called tertiary amines. The demethylation of imipramine and amitriptyline result, respectively, in desipramine and nortriptyline (Pamelor® and others), which also are active antidepressants. With a hydrogen substituted for one of the methyl groups on the terminal nitrogen, desipramine and nortriptyline, along with protriptyline (Vivactil®), are known as secondary-amine tricyclics. By and large, the secondary-amine tricyclics have weaker anticholinergic properties and other neurochemical differences that affect their clinical profile and possibly their blood level relationship as well. Much as the addition of a chlorine atom to the phenothiazine promazine (Sparine®), resulting in chlorpromazine (Thor-

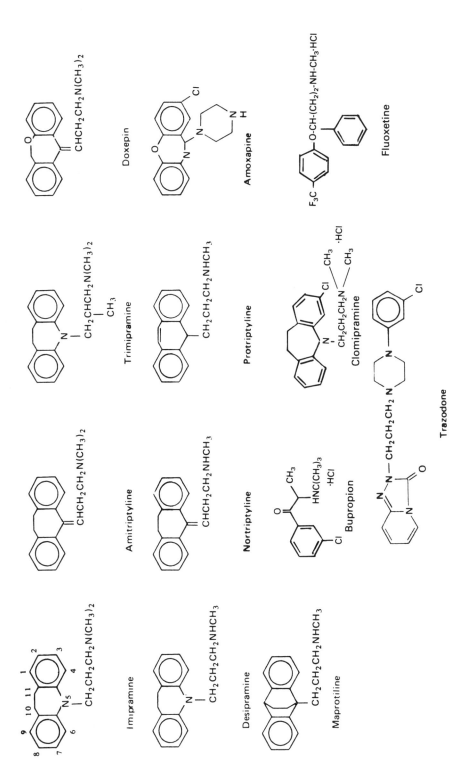

FIGURE 1. Cyclic antidepressants.

azine® and others), produced an agent with distinctive pharmacological and clinical properties, so the addition of a chlorine atom to imipramine, producing clomipramine, has altered its pharmacological and clinical properties. This has resulted in an agent that is more highly serotonergic and possibly has dopamine-blocking properties, and one that appears to be effective in alleviating obsessions and compulsions.

Amoxapine also has a three-ring structure, but it is a demethylated metabolite of the antipsychotic loxapine, which it resembles pharmacologically and clinically; both are termed dibenzoxazepines. Maprotiline has a fourth ring and is, therefore, termed a tetracyclic, but it has few pharmacological or clinical differences from the tricyclics. Trazodone is a phenylpiperazine derivate of triazolopyridine—chemically, pharmacologically, and clinically distinct from the tricyclic antidepressants.

Fluoxetine also is distinct chemically. It is known as a straight-chain phenylpropylamine; this selectively serotonergic new antidepressant appears to have distinct neurochemical and clinical properties.

Different from other antidepressant drugs is bupropion. This drug is structurally related to phenylethylamine, which probably accounts for some of its stimulant properties. Relatively weak in blocking neuronal reuptake of serotonin and norepinephrine, bupropion does inhibit dopamine uptake.

2. Mechanism of Action

Investigators have developed several interesting hypotheses to explain the therapeutic action of cyclic antidepressants. Since no model completely accounts for all of the available facts, many experts think that biologically dissimilar depressions probably exist. Until recently, researchers primarily focused on the way antidepressants affected the availability of amine neurotransmitters.[36] Antidepressants were thought to work by blocking the reuptake of norepinephrine or serotonin. However, recent findings refute or at least complicate this general hypothesis. Some effective antidepressants (such as mianserin or bupropion) do not significantly block the reuptake of norepinephrine or serotonin.[37] Moreover, antidepressants require a few weeks to work, even though their CNS-blocking effects are immediate.

Sulser presents a divergent viewpoint.[38] He proposes that depressions result from increased norepinephrine activity and that antidepressants promote a "downregulation" of activity by decreasing β-adrenergic receptor sensitivity. Other research has described how the cyclics may take a few weeks to increase postsynaptic receptor sensitivity or decrease presynaptic inhibition by desensitization of the α receptor (resulting in a larger release of norepinephrine per nerve impulse).[39] Even the existence of CNS antidepressant substance has been proposed.[40] This has been supported by the recent finding that binding sites fairly specific for imipramine exist.[41] Janowsky et al. also suggest that acetylcholine has an important role in affective disorders. They feel that an imbalance between acetylcholine and norepinephrine or serotonin is important in producing depression and mania (e.g., a predominance of acetylcholine leads to depression). They therefore propose that antidepressants work, at least in part, because of their anticholinergic properties.[42] The current evidence emphasizes that we probably cannot expect to find a common mechanism for all antidepressants. In addition to the existence of depressive subtypes, each distinct syndrome may result from a complex interaction of presynaptic, neurohumoral, and postsynaptic receptor properties of specific neural pathways. As Maas postulates, antidepressants may act in various complicated ways to restabilize an unbalanced system.[43]

3. Adverse Reactions

The clinician should know a drug's adverse reactions and a patient's susceptibilities. Some of these effects cause minor discomfort, but others may interfere with drug trials or produce serious morbidity. Of the common reactions, cardiac, anticholinergic, and CNS effects produce the most clinical morbidity. Untoward effects occur more often when cyclics are coadministered with other drugs or when a patient has a coexisting physical disorder.

Overall, tricyclic antidepressants resemble one another in their adverse-effect profile. In general, secondary-amine tricyclics are less anticholinergic and sedating and produce less weight gain and postural hypotension. Maprotiline's and amoxapine's adverse effects are comparable to those of the tricyclics. However, amoxapine also resembles the neuroleptics in producing both extrapyramidal reactions and neuroendocrine changes.

Trazodone presents a somewhat distinct clinical profile, certainly different from that of the tricyclic antidepressants. Highly sedating, trazodone produces postural hypotension and some gastric irritation. It also is associated with occasional cases of priapism and can exacerbate preexisting myocardial irritability. It is unlikely, however, to produce anticholinergic effects and appears very safe in overdose.

Fluoxetine has a distinctive side-effect profile characteristic of many of the new generation of serotonergic antidepressants. Minimally sedating (and possibly even "activating" for some patients), fluoxetine produces few anticholinergic effects. It can be associated with gastric irritation (nausea, heartburn), unwanted CNS irritability (anxiety, agitation, or insomnia), and headache. Impressively, at usually therapeutic doses fluoxetine appears well tolerated by many patients.

Bupropion, a stimulating (rather than sedating) antidepressant, is associated with a higher rate of seizures than are other antidepressants. These seizures appear to be dose related and may be a greater risk for patients with bulimia. As is the case with other CNS stimulants, bupropion causes excitement, agitation, increased motor activity, insomnia, and tremor in some patients. It resembles fluoxetine and some other new antidepressants in causing more weight loss than weight gain, along with occasional nausea, vomiting, and anorexia. Among the newer antidepressants, it is associated with a relatively low incidence of anticholinergic effects but possibly a higher frequency of headaches and rashes.

a. Gastric

Irritation of the gastric mucosa, resulting in heartburn, nausea, and rarely in vomiting, is a relatively uncommon side effect of tricyclic antidepressants, which, by and large, counteract excessive stomach acidity (via anticholinergic and antihistaminic activity). Some of the newer serotonergic antidepressants are more commonly associated with these side effects. Current examples include trazodone, fluoxetine, and bupropion, as well as fluvoxamine and other drugs that are likely to be marketed in the coming years. Patients troubled by gastric irritation can be advised to take their medicine with food, milk, or an antacid. Alternatively, the daily dose might be divided rather than administered at one time.

b. Hematological

i. Leukocytic Effects and Purpura. Cyclic antidepressants only rarely produce alterations in white cell count. Clinicians have reported leukocytosis, leukopenia, Loeffler's syndrome, eosinophilia, thrombocytopenia, and purpura. These conditions should be medically evaluated, although they seldom cause significant morbidity. The patient can usually continue the medication at the same dosage.

ii. Agranulocytosis and Other Hematolytic Effects. Both the cyclics and the phenothiazines rarely cause agranulocytosis, but the incidence is much lower with antidepressants. Death from infection occurs infrequently if the drug is discontinued immediately and the patient placed immediately in reverse isolation. White cell numbers usually return to normal within several weeks of discontinuing the drug. Most reported cases have occurred with imipramine (although this may be an artifact, since it is the drug most often studied).

Agranulocytosis seems to be an allergic response of sudden onset that usually appears 40 to 70 days after initiation of the medication. Patients have a greater risk with advancing age and concomitant physical illnesses. The clinical syndrome is characterized by a low white count (composed almost completely of lymphocytes), normal red count, and infection (usually involving the oropharynx) accompanied by fatigue and malaise.

Routine blood studies do not help with early detection, since the syndrome develops very rapidly. Instead, the clinician should evaluate and treat any sign of infection, especially of the pharynx. He should discontinue the cyclic and never administer it to that patient again. The practitioner might try an antidepressant with a different structure and follow the patient closely for adverse hematological effects.

It is unclear whether antidepressants ever cause aplastic anemia. Nomifensine was withdrawn from the market worldwide because of its association with hemolytic anemia. The tetracyclic mianserin, widely used in England, has been associated with a higher incidence of bone marrow suppression than tricyclic antidepressants. As newer antidepressants are introduced, we will keep a watchful eye on their hematopoietic properties.

c. Hepatic

Fewer than 0.5 to 1% of patients treated with cyclic antidepressants develop liver toxicity. This adverse reaction seems to be a hypersensitivity response similar to that caused by the phenothiazines.

Most often, the toxicity produces abnormal liver function tests without clinical jaundice. Mild, transient jaundice preceded by abdominal pain, anorexia, fever, and transitory eosinophilia may occur during the first 2 months of treatment. In these cases, the patient has abnormal liver function tests. Most often, individuals have a high conjugated bilirubin and congested bile canaliculi without hepatocellular damage. After the drug is discontinued, recovery from this syndrome usually takes several weeks. When a toxic reaction occurs, the clinician should lower the dosage or switch to another type of antidepressant.

d. Endocrine

Cyclic antidepressants have fewer endocrine effects than the phenothiazines. Unlike the phenothiazines, amoxapine, and possibly clomipramine, the other cyclics have produced no clear-cut cases of amenorrhea or galactorrhea. Rare patients, however, may develop menstrual irregularities.

Antidepressants may lower blood glucose in patients with diabetes mellitus but do not seem to provoke or worsen diabetes. In rare cases, idiosyncratic nephrogenic diabetes insipidus has occurred.

Patients with preexisting thyroid disease may show altered tolerance to cyclics. Euthyroid patients who are given tricyclics and triiodothyronine (T_3, Thyrolar®), thyroxine (T_4, Synthroid® and others) or thyroid-stimulating hormone (TSH) simultaneously sometimes show quicker resolution of the depression while still remaining euthyroid. In women, estrogen supplementation also may have a salutary adjunctive role along with antidepressants.

e. Ocular

Unlike the phenothiazines, the cyclic antidepressants do not produce ocular pigmentation. Tricyclics do, however, commonly cause **blurred vision.** This is caused by a failure of accommodation, which usually leaves distant vision intact. This anticholinergic effect results from ciliary muscle relaxation. Tolerance sometimes develops over the first few weeks of drug treatment.

In patients with **narrow-angle glaucoma,** tricyclics theoretically could cause significant damage. Anticholinergic properties foster pupillary dilation, which can precipitate an acute episode. The problem arises with undiscovered, unreported, or untreated cases. If open-angle glaucoma, however, is being treated, the patient **can** take tricyclics. To detect patients at risk for the narrow-angle variety, the clinician can perform an examination with a penlight and take a good history, including questions about blurred vision, halos around lights, and eye pain. To do a flashlight test, the practitioner should shine a light laterally across the iris. If it illuminates the entire iris, the patient probably does not have glaucoma. If angle closure exists, the iris bunches and will block the laterally shining light. Angle-closure glaucoma usually requires surgery, whereas the common open-angle type is usually well treated with drugs.

Antidepressants with weak anticholinergic properties, like trazodone and fluoxetine, are not burdened with these ocular problems.

f. Cardiovascular

Cardiovascular effects have been among the most troublesome and also among the most misunderstood reactions to tricyclic antidepressants. Fortunately, sophisticated research in recent years has clarified our understanding of these problems. At the same time, some of the newer cyclic antidepressants have expanded treatment options by offering different cardiovascular profiles.

i. Postural Hypotension. Possibly related to their blockade of α-noradrenergic receptor sites, tricyclic antidepressants interfere with the normal hemodynamic changes that maintain cerebral perfusion pressure on rising from a recumbent position. As a result, with postural change, blood pressure falls, and the patient becomes dizzy or, worse still, develops a syncopal episode. Particularly in the elderly, a fall may cause a bone fracture.

Tricyclics vary along a spectrum in their tendency to cause this side effect; however, the position of each agent on this axis has not yet been systematically worked out. We do know, though, that imipramine causes a substantial degree of postural hypotension, whereas nortriptyline causes much less. It appears that amoxapine, maprotiline, and trazodone similarly cause a fair degree of postural blood pressure decline. In contrast, nomifensine, fluoxetine, and bupropion cause minimal postural change.

The elderly are at high risk for the consequences of orthostatic hypotension (namely, falls). It is less clear if they are more likely to develop hypotension than younger patients. People with heart failure, though, **are** at greater risk, along with patients who are volume depleted or who are taking other drugs that might cause hypotension, such as antihypertensives. Patients with impaired cardiac conduction also may be at higher risk, along with those who have dysautonomias such as that caused by diabetic neuropathy.

Many management strategies have been suggested to help patients control antidepressant-induced orthostatic hypotension, but their number indicates that none is universally successful. Approaches such as advising patients to arise slowly (particularly when getting out of a warm bath) and sleeping with the head of the bed tilted upward can sometimes be useful, and patients may be advised to avoid prolonged bed rest. Adequate fluid and salt intake (or even salt supplements) are useful if not medically contraindicated. Support stockings help some patients. Caffeine occasionally is a useful antidote, and some doctors have reported success with yohimbine (Yocon® and others). Switching to another antidepressant less likely to cause hypotension is an option, and even lowering the dose of the original antidepressant might help. Most complicated, but sometimes successful, is the use of the mineralocorticoid fludrocortisone (Florinef® and others).

ii. Tachycardia. As do any strongly anticholinergic agents, tricyclic antidepressants cause a statistically significant increase in heart rate. For most patients, this is clinically insignificant. For a patient with marginally compensated cardiac output, however, the increased pulse rate can trigger or exacerbate heart failure. If tachycardia becomes a problem, switch to drugs with minimal anticholinergic activity, such as MAO inhibitors, trazodone, fluoxetine, or bupropion.

iii. Conduction Delay. Tricyclic antidepressants slow electrophysiological conduction velocities within the heart. This is most evident in the lower segment of the atrioventricular conduction system, between the bundle of His and the ventricles. This is clinically inconsequential for a majority of patients. Patients at risk include those with second-degree heart block or with right or left bundle branch block, who may develop a bradyarrhythmia when treated with a tricyclic antidepressant. If the physician wishes to prescribe a tricyclic for such a patient, the initial dose should be low, and dosage should be increased very gradually. A cardiologist should follow the patient along with the psychiatrist and should conduct Holter monitoring before and after each dose increment. Antidepressant drugs that do not slow cardiac conduction include fluoxetine, trazodone, bupropion, fluvoxamine, and MAO inhibitors.

iv. Antiarrhythmic/Arrhythmogenic Activity. In overdose, with toxic blood levels, tricyclic antidepressants have long been associated with malignant ventricular arrhythmias, often leading to cardiac arrest. Within a broad range of therapeutic concentrations, however, tricyclic antidepressants possess a quinidine-like (type IA) antiarrhythmic activity. Thus, in patients with frequent atrial or ventricular premature contractions, a tricyclic antidepressant will usually result in diminished extra beats. The physician should be wary, of course, when coprescribing a tricyclic antidepressant with another type IA antiarrhythmic, such as quinidine, procainamide (Pronestyl® and others), or disopyramide (Norpace® and others), lest additive toxicity occur.

There is no evidence that nontricyclic antidepressants possess similar antiarrhythmic capacities. In fact, one must be on guard against arrhythmogenicity. Trazodone, for example, may trigger ventricular tachycardia in patients with preexisting myocardial irritability. Since most drugs introduced to the market have been studied only in patients with excellent health status, it may be some time before possible risks of arrhythmogenicity are fully identified in new antidepressant drugs.[44,45]

Guidelines for Clinical Use

1. Take a careful cardiac history, including a review of symptoms.
2. Discuss possible adverse reactions with the patient, particularly postural hypotension, tachycardia, conduction delay, and arrhythmias.
3. In young patients with no apparent cardiac problems, the

choice of drugs should be based primarily on clinical symptoms (i.e., choice of desired secondary pharmacological properties).

4. In older patients or those with manifest cardiac problems or potential orthostatic hypotension:
   a. Select a drug with little hypotensive properties (e.g., nortriptyline, bupropion, fluoxetine).
   b. Advise patients to change from lying or sitting to vertical position slowly and to try sleeping with their head elevated.
   c. Prescribe adequate fluid and salt intake and/or support stockings.
5. In older patients or those with manifest or potential tachycardia, use drugs with minimal anticholinergic activity, such as trazodone, fluoxetine, or MAO inhibitors.
6. In patients with conduction problems who require an antidepressant, initial medication doses should be low and increased very gradually. Medications with minimal cardiac conduction effects should be used (e.g., fluoxetine, trazodone, bupropion, and MAO inhibitors).
7. In a clinical trial with older patients, monitor pulse, orthostatic blood pressure changes, electrocardiogram (EKG) (every few days), and plasma drug levels and divide dosages during the initial phase and in cases requiring high antidepressant dosages.
8. Avoid coadministration of cardiovascular drugs or those synergistic with antidepressants.
9. Consult a psychopharmacologist and/or cardiologist for symptomatic patients.

g. Neurological

i. Sedation/Stimulation. Tricyclic antidepressants tend to be sedating; tertiary amines are more so than their "daughter" secondary-amine metabolites. Doxepin and amitriptyline are particularly sedating, and so is the nontricyclic trazodone. Most patients prefer taking these drugs in single daily doses at bedtime, allowing peak sedative effects to occur during sleep. Some degree of tolerance develops to sedation. The best strategy is to begin with a low dose and increase the dose slowly (usually every few days) as soon as morning "hangover" sedation has diminished to a tolerable degree.

Less sedating antidepressants, and particularly some of the newer cyclic drugs, tend to produce a degree of stimulation that may be experienced as dysphoric. Thus, patients may have difficulty sleeping or may complain of anxiety of agitation. As with sedation, tolerance often develops to this side effect. Some patients may benefit from the coadministration of a sedating drug, such as a benzodiazepine, at least for a few weeks until these symptoms subside on their own. Such stimulation occurs often with fluoxetine, fluvoxamine, and bupropion, occasionally with the secondary amine tricyclics desipramine, nortriptyline, and protriptyline, and less commonly with imipramine itself.

ii. Headaches. Headache frequently accompanies the use of some of the newer cyclic antidepressants such as fluoxetine, bupropion, and fluvoxamine. Headache occurs more with drugs that have a low anticholinergic profile and with some of the newer "serotonergic" antidepressants. Headache is also a common symptom of the withdrawal reaction from tricyclic antidepressants (see below).

iii. Nocturnal Reactions. Myoclonus, an abrupt jerking of a muscle group either when conscious or during sleep, is an occasional side effect of tricyclic antidepressants. Often, myoclonus occurs as an exaggeration of the physiological myoclonus of falling asleep. Occasional patients taking tricyclics report hypnagogic hallucinations, which may be alleviated by dividing the daily dose rather than administering the drug entirely at bedtime.

iv. Seizures. All antidepressants seem to lower the seizure threshold. This is rarely a problem clinically, although in patients with a preexisting epileptic condition, the physician should try to keep the dose low and make increments slowly. The clinician should also be aware of possible interactions with concurrent anticonvulsant drugs. Patients withdrawing from alcohol or sedatives and those with eating disorders may be at increased risk for seizures.

Infrequently, patients will experience seizures *de novo* when taking an antidepressant drug. The risk is idiosyncratic and dose related, but with tricyclic antidepressants it probably occurs in fewer than 1% of patients. Some of the newer antidepressants, however, increase the risk of seizures. With maprotiline, the risk appears to rise fourfold and to be greater at higher doses. Clomipramine, too, presents a greater risk than other tricyclics. Amoxapine also may have an increased risk of seizures, which are most frequent and severe in overdose. For bupropion, seizure incidence has been calculated to be 0.44% for patients taking up to 450 mg per day, rising to a frightening 2.19% for patients at higher dosages.[46] With new agents, the true risk of seizures may not become apparent until the drug has been available for several years longer. Assuming that benzodiazepines are antidepressants, they raise rather than lower the seizure threshold.

v. Anticholinergic Effects. Anticholinergic agents interfere with the encoding of new memories. Thus, although remote memories will remain intact, a patient may have difficulty recalling events earlier in the day or learning new information. The more anticholinergic the specific agent, the higher the dose, and the more anticholinergic agents the patient is taking, the greater the memory defect is apt to become. The elderly and the demented are at greatest risk. Rarely, a patient develops an atropine-type delirium (see Section IV.A.5).

vi. Movement Disorders. With the exception of amoxapine, cyclic antidepressants do not cause the types of extrapyramidal reactions associated with neuroleptic drugs (see Chapter 4). They do, however, cause a fine "action" or postural tremor, most readily observed when a body part is kept at a sustained posture (e.g., hands outstretched). Functionally, this tremor—an exaggeration of a physiological tremor—is apt to become manifest when a patient drinks from cup and saucer. Finer and more rapid than a parkinsonian tremor (which occurs at rest), this action tremor is made worse by anticholinergic agents (do not use antiparkinson drugs), anxiety, and stimulants but might improve with alcohol (not recommended therapeutically) or β blockers. It is akin to the tremor induced by lithium or associated with thyrotoxicosis. It probably has a hereditary component and worsens with aging.

Amoxapine is a neuroleptic drug and, as such, has been associated with the full range of neuroleptic extrapyramidal reactions (see Chapter 4). These include acute dystonias and dyskinesias, akathisia, Parkinson's syndrome, and late-onset and withdrawal dyskinesias.

There have been suggestions that fluoxetine may cause akathisia, although there is some controversy as to whether the motor restlessness associated with nonsedating antidepressants may not be a distinct condition.[47] There are also anecdotal data suggesting that when combined with a neuroleptic, fluoxetine might enhance the probability of extrapyramidal reactions.[47]

vii. Withdrawal Reactions. Tricyclic antidepressants are associated with a distinct withdrawal syndrome. As with most psychoactive compounds, the likelihood of a withdrawal reaction and its severity vary as a function of the dose and the length of time the patient has taken the drug at the time of withdrawal. It may be mitigated in part by very gradual withdrawal.

Symptoms typically begin when the dose has been reduced to a particularly low level or the drug is stopped entirely. Clinical manifestations that have been attributed to anticholinergic rebound include both peripheral and central effects. Autonomic reactions, the opposite of so-called anticholinergic symptoms, include hypersalivation, nausea, vomiting, abdominal cramps, and diarrhea. Central manifestations typically include headaches, vivid dreams, and sleep disturbances. Especially when patients have taken these drugs chronically, withdrawal symptoms may persist for 2 months or longer.

Gradual withdrawal of the antidepressant is the best preventive and management strategy. At the lower end of the dosage scale, consider dividing tablets in half. An alternative (or additional) strategy is to end the drug by administering a dose every other day, then every 3 days, before discontinuing it entirely. If there is no rush to discontinue the agent, tapering of the drug can be protracted over several months, particularly in patients who have previously taken the antidepressant for years.

When symptoms are particularly uncomfortable, the patient can be returned to a higher dose of the antidepressant and then the taper made more gradual. Other antidotes can include a sedating antihistamine with anticholinergic properties [e.g., diphenhydramine (Benadryl® and others)], a less sedating centrally active anticholinergic agent [e.g., benztropine (Cogentin® and others], or, if peripheral autonomic manifestations predominate, a peripherally active anticholinergic drug [e.g., methscopolamine (Pamine® and others)].

Withdrawal reactions with newer agents have not been described but should be watched for.

viii. Psychiatric Side Effects. It seems that virtually any effective antidepressant can precipitate mania in some patients. Many such patients have a prior history of mania and qualify for a bipolar diagnosis. Occasional patients, however, experience mania only on "provocation" with an antidepressant drug, and some authorities have suggested that these patients fall clinically along a bipolar spectrum.

The psychotic symptoms of schizophrenic patients may be made worse by treatment with an antidepressant drug. When a full-blow depression is present in a schizophrenic patient, however, it is acceptable to administer an antidepressant drug along with an antipsychotic agent. Bupropion has been reported to exacerbate psychotic symptoms in patients with schizoaffective disorder; this may be related to its inhibition of neuronal dopamine uptake.[48]

Patients with organic brain syndromes may become increasingly confused when treated with a strongly anticholinergic drug such as a tricyclic antidepressant. Use of cyclic antidepressants with minimal anticholinergic activity, however, may be considered in some other patients.

h. Cutaneous

Similar to most pharmaceutical compounds, tricyclics and other antidepressants can cause skin reactions that are usually allergic. In reintroducing the drug these cases typically result in the abrupt onset of a worse reaction. Therefore, if a patient requires an antidepressant drug, it is best to choose one as dissimilar chemically as possible from the offending agent. Maprotiline and fluoxetine appear to cause a higher incidence of rashes than tricyclic antidepressants.

i. Autonomic

Like the phenothiazines, tricyclic antidepressants produce numerous autonomic effects, predominantly anticholinergic (see Table 2). Patients most commonly report **dry mouth** (usually early in treatment). Generally, physiological tolerance develops to this adverse reaction. However, in some, dry mouth persists and may cause discomfort, mouth infections, and dental caries (sometimes rampant). In severe cases of xerostomia, the clinician should intervene by:

1. Encouraging adequate hydration.
2. Lowering the dosage of medication if possible.
3. Discouraging the use of sugar-laden drinks, chewing gum, or candy.
4. Considering switching to a less anticholinergic antidepressant (such as fluoxetine, trazodone, or bupropion).
5. Providing alternative means of mouth lubrication
   a. "Sugarless" gums (although these actually have sugars that in high doses can contribute to the formation of caries).
   b. Fluoride lozenges.
   c. Direct lubricants (for list of saliva substitutes, see Table 3).

6. Administering bethanecol chloride (Urecholine® and others), a peripherally active cholinergic agent, 25 mg up to three times per day, in patients whose symptoms do not respond to conservative measures, who would not be able to complete a drug trial because of the severity of symptoms, and who are not at increased risk from cholinergic stimulation (e.g., patients with asthma, potential cardiac compromise, or ulcer disease).

The major cardiovascular anticholinergic effect is **tachycardia.** Tolerance to this reaction usually develops quickly. **Blurred vision,** resulting from failure of accommodation, occurs in up to 20% of patients. This symptom appears at the beginning of treatment and rapidly resolves without dosage adjustment. If a patient has severe blurring, and the dosage cannot be lowered, the clinician can give 1% pilocarpine nitrate eyedrops for a short time.

Psychoactive drugs frequently interfere with sexual desire and function, and antidepressant agents are no exception. The mechanism of action remains obscure, although the interplay of various autonomic effects must bear some responsibility.

Delayed, incomplete, or altered orgasm is reported occasionally by both men and women. Decreased libido also is sometimes described by both sexes. In addition, men may experience difficulties achieving or maintaining a penile erection and/or seminal ejaculation. Antidotes that have been tried with varying degrees of success include the serotonin antagonist cyproheptadine (Periactin® and others), the

TABLE 2. Pharmacological Properties of Cyclic Antidepressants

Sedation	Anticholinergic
High	High
Amitriptyline	Amitriptyline
Trimipramine	Imipramine
Doxepin	Trimipramine
Trazodone	Doxepin
Moderate	Moderate
Imipramine	Amoxapine
Amoxapine	Nortriptyline
Nortriptyline	Maprotiline
Maprotiline	Low
Low	Desipramine
Desipramine	Protriptyline
Protriptyline	Very low
Fluoxetine	Trazodone
Bupropion	Fluoxetine
	Bupropion

TABLE 3. Some Saliva Substitutes[a]

Drug		Form
Carboxymethylcellulose or hydroxyethylcellulose solutions		
Moi-Stir® (Kingswood Labs)		Spray, swabs
Orex® (Young Dental)		Spray
Sal-Ese® (North Pacific Dental)		Liquid
Saliment® (Richmond)		Aerosol
Saliv-Aid® (Copley)		Liquid
Salivart® (Westport)	Preservative-free	Aerosol
Saliva Substitute® (Roxane)		Liquid
VA Oralube® (Oral Dis Res Lab)	Sodium-free	Liquid
Xero-Lube® (Scherer)	Sodium-free	Spray
Mucin-containing solution		
Saliva Orthana® (A/S Orthana, Denmark)		Spray, lozenges, gum
Mucopolysaccharide solution		Liquid
MouthKote® (Parnell)		
Glycerate polymer		Gel
Oral Balance® (Laclede)		

[a]Reprinted from *The Medical Letter*[49] with permission of the publisher.

peripherally active parasympathomimetic bethanechol, and the α-2 antagonist yohimbine.

Although the incidence is unclear, there have been case reports of women and, more recently, men who have experienced an egosyntonic increase in libido while taking trazodone. More troublesome with trazodone is priapism, a dysfunctional and often painful erection of the penis unassociated with sexual desire that requires emergency urological intervention and can lead to permanent impotence.

By blocking the muscarinic subtype of acetylcholine receptor, tricyclic and other anticholinergic antidepressants decrease motility in the gastrointestinal and urinary systems. In most patients, this leads to some degree of constipation and possibly urinary hesitancy. The elderly, those taking other anticholinergic agents, and patients with certain medical conditions may be at increased risk. Paralytic ileus is the extreme case of intestinal hypomotility; patients who are recently postoperative are at greater risk of developing this problem. A man with a prostate condition, as another example, is at greater risk of developing acute urinary retention when treated with strongly anticholinergic drugs.

Beyond being aware of these most serious situations, the physician will want to help patients deal with milder but more common versions of these problems. Constipation often can be alleviated by dietary counseling, especially by encouraging an adequate intake of fiber and fluids. In addition, a stool softener may help. Sugarless candies and gum, which contain sorbitol or xylitol, serve a laxating function as well as lubricating the mouth. In some cases, oral bethanechol can alleviate both urinary retention and severe constipation, but it is important to avoid excessive cholinergic rebound.

In addition to atropine-type side effects, cyclic antidepressants cause other autonomic imbalances. Although excessive sweating seems paradoxical in view of the tricyclics' anticholinergic actions, many patients treated with imipramine in particular experience hyperhidrosis. It is usually more annoying than dysfunctional. Some of the newer antidepressants, e.g., fluoxetine, fluvoxamine, and bupropion, have minimal anticholinergic potency but cause autonomic symptoms such as anorexia and nausea.

j. Drug Interactions

Drug interactions fall into two broad categories: pharmacokinetic and pharmacodynamic. The former occur when a drug affects another drug's absorption, distribution, biotransformation, or excretion. Pharmacodynamic interactions, by contrast, occur when two drugs have additive (or antagonistic) effects at a target site, for example, additive anticholinergic actions.

The dynamic interactions possible when a cyclic antidepressant is combined with another antidepressant or with a different type of drug include tachycardia and impaired cardiac conduction, hypotension, anticholinergic toxicity, sedation, and excessive CNS stimulation.

The new antidepressant fluoxetine has been associated with potential hazards when combined with either tricyclic or MAOI antidepressants. The former combination is a pharmacokinetic interaction, in which coadministration of fluoxetine leads to markedly elevated blood levels of the tricyclic, with the possibility of increased side effects or frank toxicity.[50] The interaction with MAOI antidepressants probably reflects both drugs' capacity to elevate brain serotonin, which manifests itself by a potentially lethal syndrome including hyperthermia, muscular rigidity, and seizures.[51] The drug's manufacturer recommends that fluoxetine be discontinued at least 5 weeks before a patient starts taking an MAOI.

k. Weight Changes

Patients taking tricyclic antidepressants often gain weight. This is often a result of constipation or, less frequently, fluid retention. Most commonly, weight gain reflects increased body fat, the mechanism of which is a subject of ongoing debate and investigation. Patients taking tricyclics may have an increased appetite and sometimes a craving for sweets. Some researchers have suggested that tricyclics decrease the resting metabolic rate.

For some of the newer cyclic antidepressants, changes in body weight are less obvious. With fluoxetine and fluvoxamine, however, possibly because of their serotonergic properties, patients tend to **lose** weight. This may also reflect a decreased threshold for satiety. Alternatively, patients taking these drugs may have an increased metabolic rate. Bupropion may also cause weight loss, possibly related to its stimulant properties. Whatever the mechanism, many patients appreciate this feature of these drugs, since some depressed patients struggle with weight control.

4. Precautions for Administration during Pregnancy

No psychiatric drug should be routinely prescribed during the first trimester of pregnancy. Cyclic antidepressants, however, have been given during this period without harmful effects on the fetus. Occasionally, the newborn of a mother who took tricyclics may exhibit irritability, hyperhidrosis, tachycardia, tachypnea, and cyanosis for several days following delivery. These symptoms do not seem to cause long-term morbidity. Therefore, the pregnant patient with a severe depression accompanying a serious suicidal risk or an inability to care for herself might receive a trial of antidepressants. The clinician and patient should carefully weigh the risks and benefits (see Chapter 10).

5. Acute Toxicity

a. Intoxication Syndromes

Individual patients show a variable tolerance to antidepressants. Usual clinical doses may occasionally cause toxic responses resulting from special sensitivity or from interaction with coadministered medications. Most often, however, patients develop CNS depression and/or cardiotoxicity from an acute or chronic overdose of heterocyclics. The clinician should treat each overdose aggressively because of the possibility of refractory cardiac arrhythmias and circulatory or respiratory collapse. Prolonged observation is mandatory, since recovery from acute ingestions can be delayed because of rapid and extensive tissue distribution.

Symptoms develop 1 to 4 hr after the overdose and depend on the age of the patient, individual tolerance, and dose. Sometimes a large number of antidepressant tablets taken at the same time clump together in the stomach and enter the small intestine as an agglomeration. There may be a sudden release of material from this bezoar, which can allow rapid absorption of great quantities of the drug. Clinically, the patient may progress from ambulation and clear consciousness to coma and cardiorespiratory arrest over a period of minutes. With tricyclics, **doses equivalent to 1.2 g imipramine frequently are toxic** (although doses of < 1.0 g occasionally cause severe reactions and death), and **more than 2.5 g is commonly fatal.** Prior to a marked decrease in consciousness, hallucinations, sensitivity to sounds, delirium, or agitation may occur. A "hyperactive" coma may develop and progress to a deeper nonreactive coma. Hypotension or hypertension, dilated, sluggishly reactive pupils, and hypothermia also occur. Patients sometimes die from the severe hyperpyrexia secondary to central anticholinergic effects. Myoclonic seizures and bilateral plantar responses on neurological examination are common. Arrhythmias, which are difficult to control, include ventricular tachycardia, atrial fibrillation, and atrioventricular and intraventricular block. Manifestations of cardiac effects appear on EKG as bundle branch block, varying degrees of heart block, ventricular extrasystoles, and bizarre QRS complexes.

A large acute overdose may have a somewhat different character. Deep coma often appears rapidly. Reflexes decrease, and bilateral plantar responses occur.

Patients frequently exhibit seizures, arrhythmias, and respiratory arrest. The pupils may be of normal size, but they will react sluggishly or not at all. The pulse rate increases, and blood pressure and body temperature decrease.

There is considerable controversy about the correlation of QRS duration, tricyclic antidepressant blood level, and clinical outcome in overdose. One study failed to find a statistically significant association between serum drug levels and the occurrence of seizures or ventricular arrhythmias but did observe a correlation between QRS duration and overdose outcome.[52] Patients with QRS durations less than 0.1 sec had neither seizures nor ventricular arrhythmias, whereas those with QRS $\geq$ 0.1 sec had a 50% incidence of ventricular arrhythmias. By contrast, another study failed to find either blood antidepressant concentrations or the EKG helpful in predicting seizures or ventricular arrhythmias.[53] Still, all patients in this latter study with tricyclic antidepressant serum levels $\geq$ 1000 ng/ml had QRS intervals $\geq$ 0.1 sec, and two of 102 consecutive tricyclic overdose cases that resulted in fatalities met these criteria.

The authors of the latter study recommend that an antidepressant level $\geq$ 1000 ng/ml with a QRS interval $\geq$ 0.1 sec should be taken as a danger sign but that a lower blood level or a normal QRS interval should not be a license to relax. Rather, they should be considered together with the duration of time that has elapsed since the drug ingestion, the level of the patient's consciousness, the presence of seizures, and the patient's vital signs in deciding on the length of observation and the intensity of management.

Management of serious cyclic overdoses poses serious difficulties, since cardiorespiratory collapse, arrhythmias, and coma present both an immediate and persistent life threat (see Table 4). Any patient with a suspected cyclic antidepressant overdose—no matter how small the amount alleged to have been taken or how alert the patient seems—should be rushed to an emergency room. Many fatalities occur before a patient even reaches medical help. In the hospital, an IV line should be established immediately, and the stomach emptied by gastric aspiration and lavage. Next, a physician should administer 50–100 mg of activated charcoal, either as an oral slurry mixed with water or by orogastric tube. It is too thick to administer via a nasogastric tube. The clinician should avoid mixing it with milk, jelly, or other foodstuffs, which can interfere with the charcoal's adsorptive capacity. Repeated administration of charcoal every 4 hr can help remove additional quantities of the antidepressant, which is recirculated via the enterohepatic system.

Catharsis can further speed elimination of the drug. Some preparations of charcoal come mixed with sorbitol, which serves this purpose. An alternative cathartic is magnesium sulfate, 250 mg/kg per dose as a 20% solution.

An EKG should be performed immediately on arrival of the patient at the emergency room. If the patient has altered consciousness, arterial blood gases should be drawn. Blood should also be drawn to establish antidepressant concentrations (which may later rise if absorption from gastrointestinal tract continues), to complete any other necessary tests to rule out the presence of other toxins (most

TABLE 4. Management of Overdose

Initial measures
    Induce emesis and catharsis (magnesium sulfate 250 mg/kg in 20% solution)
    Implement gastric aspiration and lavage with activated charcoal (Charcodote® powder), 50–100 mg,
        every 4 to 6 hr for 24 to 48 hr
    Complete physical examination and obtain vital signs
    Adequately ventilate; monitor by EKG
    Insert an intravenous line with cardiac pacing capacity
    Insert pulmonary artery catheter
    Obtain blood chemistries including antidepressant serum level
Maintenance procedures
    Monitor vital signs, electrolytes, serum heterocyclic levels, and EKG
    Maintain fluid and electrolyte balance
    Ensure adequate ventilation and skin care if the patient is comatose
Hypertension
    Elevate legs
    Adequately hydrate
    Administer phentolamine (Regitine®) 5 mg (if persistent and severe)
Seizures
    Administer diazepam (Valium® and others), 5 to 10 mg IM or IV p.r.n. (avoid barbiturates)
Central and peripheral anticholinergic syndrome
    Slowly inject physostigmine (Antilirium®), 1 to 2 mg IV every 30 to 60 min p.r.n.
Hyperpyrexia
    Use ice mattress, ice packs, or cold sponges
Cardiac interventions
    Avoid type A antiarrhythmics [procainamide (Pronestyl® and others), quinidine, and disopyramide
        (Norpace® and others)]
    Use lidocaine (Xylocaine® and others) for ventricular arrhythmia
    Use lidocaine or phenytoin (Dilantin® and others) for heart block arrhythmia
    Use physostigmine for supraventricular arrhythmia; avoid in cases of cardiovascular or respiratory
        compromise
    Implement volume expansion for decreased left atrial pressure
    Use dopamine (Intropin®) for increased left atrial pressure
    Alkalinize the urine by giving IV $NaHCO_3$ (acid imbalance may predispose to ventricular irritability)
Persistent arrhythmia with heart failure
    Use cardiac pacing or cardioversion

overdoses are multiple!), and to identify any coincident medical conditions or complications.

    The crucial time for observation in antidepressant overdoses is the first 6 hr. If a patient has no alterations in consciousness, EKG abnormalities, respiratory depression, seizures, or hypotension, he can be given a final dose of charcoal and be transferred to a psychiatric unit or given an appointment for psychiatric follow-up. On the other hand, the presence of any clinical sign of toxicity should prompt immediate admission to a medical intensive care unit for at least 48 hr of cardiac monitoring.

Treatment of life-threatening consequences of cyclic antidepressant overdoses is complex and incompletely formulated. Because of their high degree of lipophilicity and extensive tissue and protein binding, cyclic antidepressants are minimally removed by dialysis. The key medical treatments are symptomatic and hinge on treating (or preventing) seizures, respiratory arrest, cardiac arrhythmias, and resultant complications, such as renal failure and brain damage.

Seizures are often treated with intravenous diazepam, 5–10 mg repeated as needed. Seizures may be more common with maprotiline and amoxapine overdose. Amoxapine-related seizures in particular appear difficult to treat. Very high doses of diazepam might be helpful.

If a patient lapses into stupor or coma, respiration should be monitored continuously, respiratory arrest anticipated, and ventilatory assistance readily accessible. Appropriate treatment of hyper- or more commonly hypotension must depend on individual circumstances.

Treating cardiac arrhythmias that result from antidepressant overdoses presents a thorny clinical agenda. Alkalinization, whether by intentional hyperventilation or by infusion of sodium bicarbonate, may improve outcome, particularly in the face of acidosis. Other approaches include electrical pacing, β blockers, β-stimulating agents, and phenytoin (Dilantin® and others). Physostigmine (Antilirium®), which can reverse some of the CNS features of atropine-type toxicity (i.e., delirium), is likely to worsen cardiac rhythm, possibly interfere with respiration, and lower the seizure threshold, so it should **not** be used in circumstances of cardiovascular and respiratory compromise. Type IA antiarrhythmic drugs—i.e., procainamide, quinidine, or disopyramide—should be avoided entirely in cases of cardiac overdoses. Their cardiac effects are apt to be additive with those of tricyclic antidepressants.

When treating previously healthy patients suffering from even devastating antidepressant overdoses, intensive care personnel should not give up hope prematurely. With adequate external cardiovascular perfusion and oxygenation, patients can survive intact after even several hours of resuscitation efforts.

The incidence of fatalities following tricyclic overdoses probably lies between 0.5% and 4%, depending on the amount ingested, the presence of other drugs, and the patient's prior medical status. How do the "second-generation" antidepressants compare? Maprotiline may cause an increased incidence of seizures, but its mortality figures appear comparable. Amoxapine may cause a higher rate of fatalities, certainly produces more seizures, and has been associated with acute tubular necrosis and brain damage following recovery. Trazodone, when taken alone in overdose, appears comparatively benign. Data on fluoxetine overdose are too preliminary as of this writing but are encouraging so far. If alprazolam or other benzodiazepines turn out to have antidepressant properties, they should be comparatively safe when taken in overdose. However, some evidence suggests that triazolobenzodiazepines (like alprazolam) may not be as safe in overdose as older members of this class. To date, there are too few cases with bupropion to assess relative risk of overdose.

b. Anticholinergic Syndromes

i. Description. Many psychoactive drugs have both peripheral and central anticholinergic effects. They occur with tricyclic and some other antidepressants, antipsychotics, some hypnotics, antihistamines, and antiparkinson agents. Combinations of these drugs, such as a psychoactive drug (antidepressant or antipsychotic or both) and an antiparkinson agent, may produce additive anticholinergic effects. Older patients seem particularly sensitive. Acute overdoses of the abovementioned agents also frequently cause anticholinergic crisis.

The anticholinergic syndrome may present with a mixture or a predominance of either peripheral or central symptoms. In its florid state, **the CNS picture consists of confusion, delirium with disorientation, agitation, visual and auditory hallucinations, anxiety, motor restlessness, pseudoseizures (myoclonic jerks and choreoathetoid movements with EEG seizure activity), and a thought disorder (e.g., delusions). The peripheral syndrome may be manifested by decreased bowel sounds and constipation, urinary retention, anhidrosis (decreased sweating), mydriasis (increased pupillary size), dry mouth, cycloplegia (decreased accommodation), increased body temperature, motor incoordination, flushing, and tachycardia.** When these syndromes are caused by cyclic antidepressants, there is also a high risk of life-threatening arrhythmias. Aliphatic and piperidine phenothiazines [especially thioridazine (Mellaril® and others)] and the tricyclics amitriptyline, doxepin, and imipramine are the most anticholinergic. A combination of the above drugs or of these drugs with other anticholinergic agents greatly increases the risk of an anticholinergic syndrome.

ii. Anticholinesterase Therapy. Anticholinesterase therapy has proven very effective for the treatment of anticholinergic syndromes. All anticholinesterases counteract the peripheral manifestations of the syndrome. However, clinicians should use either **physostigmine** or **pyridostigmine** because they cross the blood–brain barrier and therefore counteract the central symptoms. Most clinical studies have focused on physostigmine, but equivalent doses of pyridostigmine also are effective.[54,55] In general, **the clinician should avoid using physostigmine in patients with unstable vital signs.** In these cases, cardiac arrhythmias from cholinergic stimulation are common. Cholinergic effects can also produce seizures (particularly with rapid injection of physostigmine) and respiratory arrest in selected patients. Although practitioners have tried many different regimens, **physostigmine salicylate, 1 to 2 mg IM or IV,** will relieve symptoms dramatically. The clinician should **infuse the drug very slowly (e.g., 1 mg over 2 min), monitor the cardiac status, and have means for respiratory support available. If no improvement occurs within 15 to 20 min, another dose of 1 to 2 mg should be given. Up to 4 mg may be administered over 10 to 15 min.** The body degrades physostigmine almost completely within $1\frac{1}{2}$ to 2 hr. Since the toxic agents may disappear more slowly, additional 1- to 2-mg doses at 30-min intervals may be necessary, even if the initial treatment is successful.

Although physostigmine treatment provides dramatic and sometimes life-saving relief from heterocyclic-induced arrhythmias, it also has risks from cholinergic stimulation. Excessive acetylcholine can result in tearing, salivation, rhinorrhea, sweating, pallor, bronchial constriction, hypotension, muscle weakness and fasciculations, nausea and vomiting, abdominal cramps, urinary frequency, and bradycardia. The antidote is atropine.

6. Pharmacokinetics

Cyclic antidepressants (except protriptyline) are generally rapidly absorbed from the gastrointestinal tract, typically peak in the plasma within 2–8 hr, and remain unbound for about 30 min. Eighty to 90% of the drug is protein bound, although individual differences in binding can produce a fourfold variance in the amount of free drug. Tissue distribution and the first step in metabolism of tertiary cyclics, demethylation, occur rapidly. Cyclics typically have a high volume of distribution because they are quite soluble and actively bind to various tissues. Patients usually achieve a steady state for each dosage after 1 to 4 weeks (depending on the drug used). Liver metabolism, including demethylation, hydroxylation, and glucuronide conjugation, also accounts for large variations in serum levels. Individual differences in patients' microsomal enzyme activity produce steady-state plasma concentrations (bound and unbound) that may vary as much as 40-fold. These characteristics suggest that individual patients may require widely varying doses to produce clinically effective serum levels.

During the initial pass through the liver, 30–70% of the drug is degraded ("first-pass effect"), as only 30–70% of orally administered antidepressant is "biodegradable." Doxepin may produce lower steady-state plasma levels than other tricyclic antidepressants per milligram ingested. This might reflect lower plasma protein binding and greater first-pass degradation. Protriptyline, by contrast, might produce higher plasma levels per milligram, possibly because of comparatively low first-pass metabolism. The first demethylation of tertiary amines results in an active compound, whereas a second demethylation and hydroxylation at the 2-position of the central ring of tricyclics produce degradation products (probably therapeutically inactive but possibly with cardiotoxic effects). Glucuronidation occurs after hydroxylation and makes the derivative water soluble. Initially, about 50% of tricyclics are excreted through the bile, but because of an active enterohepatic circulation, two-thirds of the drugs are eventually eliminated in the urine. Clearance of the cyclics is relatively slow, but they have a wide range of half-lives (see Table 5).

7. Plasma Levels

Knowledge about the correlation between plasma levels of antidepressant drugs and their clinical effects remains limited. Correlations of efficacy are fairly good with nortriptyline, reasonable with imipramine, and limited or controversial with the remaining cyclic compounds.

TABLE 5. Mean Half-Life
of Heterocyclics[a]

Drug	Time (hr)
Trazodone	5
Imipramine	16
Amitriptyline	16
Desipramine	22
Nortriptyline	24
Amoxapine	30
Maprotiline	47
Protriptyline	126
Fluoxetine	48
Clomipramine	24
Bupropion	12

[a]Half-life varies greatly among patients and tends to increase significantly with age.

For nortriptyline, research evidence suggests a curvilinear ("therapeutic window") correlation, with maximum antidepressant efficacy achieved between approximately 50 and 150 ng/ml; above and below that "window" patients are less likely to benefit. With imipramine, there is evidence of a straight linear correlation: maximum benefit comes with levels of imipramine and desipramine above 200 ng/ml; there is no obvious upper limit at which efficacy begins to decline, short of toxicity. In both dose–response studies and blood level studies, with imipramine the rule seems to be that more is better. With other tricyclic and newer antidepressants, the meaning of blood levels is insufficiently understood to recommend routine monitoring (Tables 6 and 7).

Generally, however, there are broad "intuitive" ranges of antidepressant levels that clinicians can use for guidance in selected situations. With most antidepressants, for example, a concentration much under 50 ng/ml probably is subtherapeutic,

TABLE 6. Reasons for Monitoring Cyclic Plasma Concentrations

To document serum levels of cyclics in "nonresponders" after a few weeks of generally adequate doses of medication
To determine if a reticent patient is taking medication
To monitor plasma levels in overdose cases
To monitor plasma levels in medically ill patients
To document serum levels within the "therapeutic window" for nortriptyline (i.e., 50–150 mg/ml)
To determine if a potentially therapeutic dose of imipramine has been reached (ie., >200 mg/ml)
To minimize dosage levels in elderly patients
To determine plasma level in the context of a severe adverse reaction or persistent, bothersome side effects

TABLE 7. Precautions in Obtaining Heterocyclic Plasma Levels

Generally use Venoject® (Kimble-Terumo, Inc.) or glass syringes
Do not use rubber stoppers
Take blood samples 12 hr after the last drug dosage
Promptly centrifuge sample to avoid hemolysis
Use a laboratory with proven reliability

whereas one much above 500 ng/ml could be toxic. Very low levels could signal that a patient is inconsistently compliant with the medication regimen. Or, kinetic factors could be at fault: the patient may have a problem with absorption, for example, or may be a particularly rapid metabolizer. At the other end of the kinetic spectrum, a patient's sensitivity to side effects at relatively modest drug doses may mean he is an unusually slow metabolizer, achieving high blood levels that can be measured despite comparatively low doses.

At times, a physician may want to increase the daily dose of an antidepressant beyond usual ranges. Thus, a patient who has responded minimally to imipramine up to 300 mg/day but whose side effects are minor may benefit from gradual dosage increments, perhaps 50 mg/day every week. In such a case, the doctor may find it reassuring to order a plasma level determination (of imipramine and desipramine) along with an EKG to ensure against unanticipated toxicity. If the blood level is still comparatively low (say, under 200 ng/ml), the physician is justified in this instance in gradually elevating the imipramine dose until either the depression remits or the patient is unable to tolerate side effects. With each successive increment in dose, it would be prudent to order a repeated blood level and EKG. Some patients may require (and tolerate) imipramine doses in excess of 500 mg/day.

In treating toxicity, plasma level determinations can help to establish when the peak occurred and how high the peak was and to follow the egress of the drug. Plasma antidepressant levels in excess of 1000 ng/ml have been correlated with severe and life-threatening toxicity, but even lower levels should not be taken as a guarantee of safety, since absorption may still be progressing in some cases. There is no clear correlation between blood levels and toxicity because of interindividual differences in sensitivity and also the fact that a patient's response to a given blood level when concentrations are rising may not be identical to what happens at the same blood level on the way down.

8. Preparations and Dosage

The wide range of effective clinical doses of cyclic antidepressants results in part from individual differences in absorption, plasma binding, first-pass metabolism, and liver enzyme activity. The tricyclic compounds amitriptyline, imipramine, trimipramine, and desipramine and the tetracyclic maprotiline have usual dose ranges of 150 to 300 mg/day. The tetracyclic maprotiline has comparable potency, but because of its dose-related tendency to provoke seizures, its daily dose should be capped at 225 mg. Nortriptyline is about twice as potent, whereas protriptyline is about four times as potent. Doxepin is somewhat less potent than the other dimethylated tricyclics. Amoxapine and trazodone are about half as potent as imipramine, with clinically effective doses ranging from 150 to 600 mg/day, although there is a hint that trazadone is more effective at lower than higher doses. Fluoxetine response appears optimal at 20 mg/day, with increased side effects (but no greater antidepressant efficacy) at higher doses. Dosage recommendations with bupropion are tailored to reduce seizure risk. Started at 100 mg bid or 75 mg tid, patients should be instructed to take bupropion doses at least 6 hr apart. The daily intake of bupropion may be raised by 75 to 100 mg every 3 days to a daily maximum of 450 mg, taken in no fewer than three divided doses each day. Concentrates and parenteral and slow-release preparations offer no clinical advantages over standard preparations (see Table 8).

B. Monoamine Oxidase Inhibitors

1. Chemistry

Based on their chemical structure, monoamine oxidase inhibitor (MAOI) antidepressants are subdivided into hydrazines and nonhydrazines (Fig. 2). The two hydrazines marketed in the United States as antidepressants are phenelzine (Nardil®) and isocarboxazid (Marplan®). Structurally they resemble the prototype MAOI, iproniazid, which was withdrawn from the market because of hepatocellular toxicity. The one nonhydrazine MAOI antidepressant is tranylcypromine (Parnate®), which resembles amphetamine structurally and clinically.

2. Mechanism of Action

Monoamine oxidase inhibitors inhibit the various subtypes of MAO throughout the body (gut, liver, brain, platelets, and blood vessels). They elevate body levels of epinephrine, norepinephrine, 5-HT, and dopamine by irreversibly binding to the degradation of these substances. Investigators hypothesize that the increased availability of CNS norepinephrine and serotonin may account for the antidepressant activity of MAO inhibitors, although it is conceivable that their antidepressant mechanism may involve MAO and, indeed, that the MAOIs may each act differently.

Recently, two types of monoamine oxidase have been identified (types A and B). Type B (accounting for about 80% of CNS MAO) degrades mostly dopamine and phenylalanine, whereas type A (accounting for 20% of CNS MAO but most of the gastrointestinal MAO) primarily degrades serotonin and norepinephrine. Although the MAOIs that are currently available affect both enzymes, experimental

TABLE 8. Preparations and Dosage of Heterocyclic Antidepressants

Heterocyclic	Brand name	Preparations	Initial daily dosage	Usual therapeutic dosage
Amitriptyline	Elavil® and others	10-, 25-, 50-, 75-, 100-, 150-mg tablets; 25-, 20-mg capsules; vials of 10 ml with 10 mg/ml	25 to 75 mg	150 to 300 mg
Imipramine	Tofranil® and others	10-, 25-, 50-, 100-mg tablets; 75-, 150-mg capsules; ampules of 20 ml with 25 mg/2 ml	25 to 75 mg	150 to 300 mg
Doxepin	Sinequan® and others	10-, 25-, 50-, 75-, 100-, 150-mg capsules; solution of 10 mg/ml	25 to 75 mg	150 to 300 mg
Nortriptyline	Pamelor® and others	10-, 25-mg capsules; solution of 10 mg/5 ml	20 to 40 mg	75 to 150 mg
Desipramine	Norpramin® and others	25-, 50-, 75-, 100-, 150-mg tablets; 25-, 50-mg capsules	25 to 75 mg	75 to 200 mg
Protriptyline	Vivactil®	5-, 10-mg tablets	10 to 20 mg	20 to 60 mg
Trimipramine	Surmontil®	25-, 50-mg capsules	25 to 75 mg	75 to 300 mg
Amoxapine	Asendin®	30-, 100-, 150-mg tablets	50 to 150 mg	150 to 600 mg
Maprotiline	Ludiomil® and others	25-, 50-mg tablets	25 to 75 mg	75 to 225 mg
Trazodone	Desyrel® and others	25-, 50-, 100-mg tablets	50 to 100 mg	150 to 600 mg
Fluoxetine	Prozac®	20-mg capsules	20 mg	20 mg
Bupropion	Wellbutrin®	75-, 100-mg tablets	100 mg b.i.d. or 75 mg t.i.d.	450 mg

agents (such as clorgyline) that selectively inhibit type A MAO seem to be more effective in treating depression than selective inhibitors of type B MAO [such as selegiline (Eldepryl®) or some available drugs used for other purposes, including pargyline (Eutonyl®), procarbazine (Matulane®), or furazolidone (Furoxone®)].

3. Adverse Reactions

The MAOI antidepressants produce a wide range of distinctive adverse effects. The hypertensive crisis that results from their interaction with cheese, other foodstuffs, and a number of drugs has been the most feared. However, it is the less dramatic side effects, such as orthostatic hypotension, weight gain, sexual dysfunc-

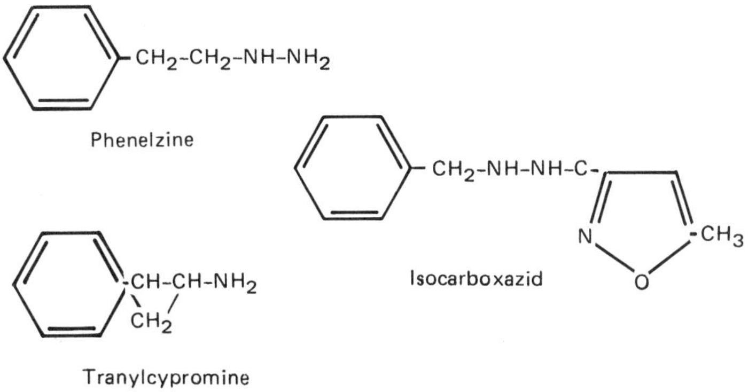

FIGURE 2. Chemical structures of the 3 MAOIs labeled as antidepressants in the United States.

tion, sleep disturbances, and edema, that most often lead to intolerance and drug discontinuation.

a. Hypertensive Crisis

Dramatic and life-threatening hypertensive crises in patients taking MAOI drugs generated a fear of these agents among physicians and their patients. In some countries, they were taken off the market entirely. In the United States, MAOIs were eclipsed by the tricyclic antidepressants.

Better understanding of the factors that can precipitate a hypertensive crisis in an MAOI-treated patient has allowed safer administration of these drugs. Usually, a sharp rise in blood pressure follows the ingestion of an exogenous pressor substance. In the face of MAO inhibition in the gut, blood vessels, and liver, the exogenous pressor is free to gain entry to the systemic circulation. At the same time, peripheral catecholamine stores are fuller than usual and, therefore, available for sudden release, which can lead to hypertension. How high a patient's blood pressure will go, and whether or not a stroke may result, are functions of several variables: the amount of pressor ingested, the distensibility of the patient's vascular tree, and the vulnerability of the patient's cerebral vasculature to a sharp rise in pressure. One patient may be more vulnerable than another, and a given patient can "cheat" on an MAOI diet one time with impunity yet suffer catastrophic consequences on a subsequent occasion.

Table 9 details our current knowledge about foods most likely to contain high levels of pressor substances. The most common of these is tyramine, a degradation product of the amino acid tyrosine. Because a majority of such food–medication interactions have been attributed to the ingestion of cheeses, this is often termed the "cheese effect." Many pressors, however, lurk in soups, sauces, gravies, and dressings. It is prudent to warn patients to avoid foods that they are not preparing themselves and, therefore, are unable to ensure the freshness and contents of. In

TABLE 9. Food and Beverage Products That Can Cause Hypertension with MAOIs[a]

Cheeses
1. High tyramine content: boursault, Camembert, cheddar, Gruyère, Stilton. For all fermented cheeses, tyramine content may be highest near rind and fermentation holes.
2. Moderate tyramine content: Gouda, Parmesan
3. Low tyramine content: American, ricotta, cottage cheese, cream cheese
Other foods
1. High tyramine content: aged and processed meats, sausages, lox, pickled herring (but a recent study found no tyramine in pickled herring excepting brine), bean curd, possibly any overripe produce
2. Moderate tyramine content: salted herring, figs, raisins, broad beans (fava beans), concentrated yeast extracts (but not Brewer's tablets or flakes), pickles, sauerkraut, coffee, chocolate, cocoa, soy sauce, sour cream, snails, avocado, banana peels, licorice
3. Questionable tyramine content: liver (fresh liver probably safe), protein extracts (often in soups), oriental foods. Beware of any gravy, soup, sauce, or dressing of unknown content.
Beverages
1. High tyramine content: some imported beers and ales
2. Moderate tyramine content: most sherry and beer
2. Low tyramine content: champagne, most Italian red wines, Riesling, Sauterne, "hard" liquor (but reactions have occurred with whiskey; cause unknown)
3. Questionable tyramine content: Chianti, liqueurs, nonalcoholic beers
Additives
1. Questionable effects: cyclamates, monosodium glutamate

[a]General rule of thumb: try to eat only fresh foods; avoid overripe or prepared foods; and favor foods that have been prepared simply.

general, the fresher and less processed a food, the less likely it is to contain pressors, which usually are degradation products of protein and increase with the age of the food.

For medications, various substances (outlined in Table 10) can cause hypertension in an MAOI-treated patient. These include prescription and over-the-counter medications as well as drugs of abuse. Some patients have developed hypertensive crises when switched abruptly from one MAOI to another: most notably from phenelzine to tranylcypromine. Therefore, a waiting period of 1 to 2 weeks is recommended.

Although most cases of hypertensive crises in MAOI-treated patients are attributable to the ingestion of exogenous pressors, as detailed above, occasional patients have developed spontaneous hypertensive episodes shortly after a dose of phenelzine or tranylcypromine.[56] The proximity to dosing in reported cases suggests a direct pharmacological effect, perhaps resulting from a sudden release of catecholamines.

A severe, sudden headache typically heralds a hypertensive crisis (although patients taking MAOIs sometimes experience headaches unrelated to blood pressure rises). Accompanying signs and symptoms include flushing, palpitations, retroorbi-

TABLE 10. Drugs That Can Cause Adverse Interactions with MAOIs[a]

Drugs that may cause hypertension with MAOIs
  Stimulants
    Amphetamine (Benzedrine® and others)
    D-amphetamine (Dexedrine® and others)
    Methylamphetamine (Desoxyn®)
    Methylphenidate (Ritalin® and others)
    Cocaine and possibly other illegal drugs
  Sympathomimetics
    Procaine preparations (Novocain® and others, which often contain epinephrine)
    Ephedrine (Tedral® and others)
    Epinephrine (Adrenalin® and others)
    Over-the-counter preparations that contain phenylpropanolamine, including
      Dimetane®
      Coricidin®
      Vicks Formula 44D Decongestant Cough Mixture®
      CoTylenol®
      Alka-Seltzer Plus Cold Medicine® (Alka-Seltzer® acceptable)
      Acutrim®
      Allerest®
      Cheracol Plus® (Cheracol D Cough Formula® acceptable)
      Sine-Off®
      Triaminic®
      Dexatrim®
    Over-the-counter preparations that contain pseudoephedrine, including
      Actifed®
      Sudafed®
      Contac®
      CoTylenol® (Tylenol® acceptable)
      Vicks Formula 44M®
      Vicks Formula 44D®
      Vicks Nyquil®
      Robitussin-PE® (Robitussin® acceptable)
      Sine-Aid®
      Sinutab®
      Tylenol Maximum-Strength Sinus Medication® (Tylenol® acceptable)
    Over-the-counter preparations that contain phenylephrine, including
      Dimetane Decongestant® (Dimetane® acceptable)
      Dristan Advanced Formula® (tablets and coated caplets)
      Neo-Synephrine®
      Nostril®
      Vicks Sinex®
      Robitussin Night Relief®
  Another MAOI
Drugs that can cause other severe adverse reactions with MAOIs
  A number of drugs cause a syndrome that can include such features as hyperpyrexia, hyperreflexia, muscle rigidity, seizures, hypotension, and death. Meperidine (Demerol® and others) is particularly dangerous. Some cyclic antidepressants [most notably clomiprime (Anafranil®)

(*continued*)

TABLE 10. (Continued)

and possibly fluoxetine (Prozac®)] have caused death. It is less clear but possible with the over-the-counter antitussive dextromethorphan (contained in many combination products) and L-tryptophan.

Surgical precautions

1. If possible, discontinue MAOI 2 weeks prior to elective surgery.

2. When a MAOI is administered, warn the anesthesiologist about possible complications regarding the use of pressor agents, although many MAOI-treated patients have received general anesthesia without complications.

*a*From Harrison et al,[65] with permission. Copyright 1989, Physicians Postgraduate Press.

tal pain, nausea and vomiting, and photophobia. If intracerebral bleeding occurs, the patient may collapse suddenly.

Traditional treatments for MAOI-associated hypertensive crises have involved α-blocking agents such as phentolamine (Regitine®) or chlorpromazine. β-blockers also have been used, and there are recent reports of effective treatment with the calcium channel blocker nifedipine (Adalat®, Procardia®) administered sublingually.[57]

b. Hypermetabolic Crisis

A different type of profound medical crisis can occur in MAOI-treated patients who ingest certain other substances that may elevate brain levels of serotonin (Table 10). Clinical manifestations include profound hyperthermia, neuromuscular irritability, altered consciousness (through coma), seizures, and possibly death. Offending agents have included meperidine (Demerol® and others), fluoxetine and some tricyclic antidepressants, most notably clomipramine. Because of its very long half-life, fluoxetine's manufacturer recommends an interval of 5 weeks between the discontinuation of fluoxetine and initiatial treatment with a MAOI. Milder syndromes have been associated with the antitussive dextromethorphan (contained in many over-the-counter "cold" preparations) and tryptophan. The antispasticity agent dantrolene (Dantrium®), useful in malignant hyperthermia of anesthesia and possibly in the neuroleptic malignant syndrome, may be considered an antidote.

c. Postural Hypotension

Symptomatic orthostatic hypotension may be more troublesome to patients taking MAOIs than to those receiving tricyclic antidepressants. Contraactive strategies can include those mentioned earlier for hypotension associated with cyclic antidepressants. Additional strategies include the use of triiodothyronine, methylphenidate (Ritalin® and others), D-amphetamine (Dexedrine® and others), or salt tablets.

d. Cardiac Effects

The MAOIs do not have direct cardiac effects or slow cardiac conduction. Thus, they may be considered alternatives to tricyclic antidepressants for patients with bundle branch block or second-degree AV block.

e. Sexual Dysfunction

Sexual psychology and physiology are probably more frequently impaired by MAOIs than by tricyclics. Problems include decreased libido and inhibited orgasm in both sexes and impaired erection and ejaculation in males. Proposed antidotes have included neostigmine (Prostigmin® and others), bethanechol, and cyproheptadine.

f. Edema and Weight Gain

Weight gain, often associated with increased food intake, is sometimes observed in MAOI-treated patients. This is more common with phenelzine and isocarboxazid than with the amphetaminelike tranylcypromine. If diet management is insufficient, the dose of the antidepressant may have to be lowered or the patient switched to a different agent.

Bipedal edema also is a common side effect of MAOIs. A thiazide or loop diuretic may be tried, but if this side effect is refractory, the patient may need to be changed to a different agent.

g. Neuropsychiatric

i. **Sleep Disturbances.** Frequently patients discontinue MAOI agents because of the combination of restless nocturnal sleep plus irresistible afternoon somnolence. Possibly more common in patients with "bipolar spectrum" depressions, these side effects have been observed with phenelzine and tranylcypromine. There is a suspicion they may be less common with isocarboxazid.[58] This syndrome tends to resist attempts at management, whether with sleeping pills at night or stimulants during the day.

ii. **Psychiatric.** Occasional patients develop manic episodes during MAOI therapy. More common in patients with histories of mania, such pathological euphoria occasionally develops *de novo* on provocation with an antidepressant. Some patients treated with MAOIs also develop confusional states, sometimes while taking the drugs and occasionally as part of a withdrawal syndrome.

iii. **Other Neurological Effects.** Exaggerated nocturnal myoclonus in patients taking MAOIs may respond to a change in medication schedule from bedtime to daytime or to divided doses or occasionally to clonazepam (Klonopin®), 0.5 to 2 mg at bedtime. Paresthesias, often experienced as "pins-and-needles" sensations, are sometimes secondary to pyridoxine (vitamin B_6) deficiency induced by MAOIs and may respond to supplementation with pyridoxine, 50 to 150 mg daily by mouth.

h. Hepatic

Liver dysfunction caused by antipsychotic drugs (and more rarely by tricyclic antidepressants) typically is of the cholestatic type, with a clinical and laboratory profile resembling that of outflow obstruction. In contrast, hepatic injury caused by MAOIs is characteristically hepatocellular, carrying a potential for severe liver necrosis. The prototype MAOI, iproniazid, was withdrawn from clinical use because of hepatotoxicity affecting about 1% of patients, with a mortality among those afflicted that may have exceeded 20%.[59] This reaction is more likely with the two hydrazides, phenelzine and isocarboxazid, than with tranylcypromine. Although the currently available hydrazides are less likely than iproniazid to cause it, the mortality remains comparably high.

After 1 to 6 months of drug treatment, patients developing MAOI-induced hepatotoxicity typically experience anorexia, weakness, malaise, and insidious jaundice, with elevated bilirubin and transaminases eight to 100 times the upper limit of normal. Patients with markedly elevated bilirubin are at greater risk of succumbing to the reaction. Although recovery usually is complete, rarely chronic disease and cirrhosis result. The only safeguard is a clinical index of suspicion: if prodromata appear, the MAOI should immediately be discontinued while the physician orders liver function tests.

i. Other Adverse Reactions

Some MAOI-treated patients report feeling more constipated, whereas others develop anorexia and symptoms of gastrointestinal irritation. Leukopenia, skin eruptions, and photosensitivity are reported rarely.

j. Withdrawal Syndrome

After abrupt MAOI withdrawal, patients have manifested agitation, irritability, pressured speech, insomnia or excessive sleepiness, visual, olfactory, or tactile hallucinations, disorientation, labile mood, paranoid delusions, aggressiveness, slurred speech, myoclonus, hyperreflexia, ataxia, choreoathetosis, and catatonia. Dilsaver has pointed out the similarity between this syndrome and that associated with the abrupt discontinuation of long-term psychostimulant treatment.[60]

4. Contraindications

The clinician should administer MAOIs only after a careful medical and psychiatric history and review of systems. In various medical situations, however, the clinician should use MAOIs with caution. These include **liver disease, advanced renal disease, pheochromocytoma, cardiovascular disease, hypertension, asthma, or chronic bronchitis** (since pressor agents like epinephrine or theophylline may be necessary). In addition, MAOIs should seldom be given to patients who take drugs that can produce dangerous synergisms.

5. Use during Pregnancy

The safety of MAOIs during pregnancy is unsure. The practitioner should consider alternatives when an antidepressant is required (see Chapter 10).

6. Toxicity

Experience with overdoses of MAOI drugs remains limited but is beginning to expand as their use increases. Mortality has been reported following doses of phenelzine, 375 to 1500 mg, and tranylcypromine, 170 to 650 mg. Because clinical signs may not become manifest for 12 hr or more after ingestion, patients should be intensively monitored for an extended period in a medical emergency setting.[61]

Headache, dizziness, and precordial pain may be early symptoms. Signs of hypermetabolism predominate and can include hyperthermia, increased heart and respiratory rates, muscular rigidity, metabolic acidosis, decreased plasma oxygen, and increased carbon dioxide. Irritability of the nervous system may be manifested by restlessness, brisk reflexes, and at times seizures. Hypertension may occur, but hypotension, sometimes occurring late in the course, appears to be a poorer prognostic sign.

The MAOIs may delay gut motility, so gastric lavage should be performed even hours after the ingestion. Acidification can speed and increase the amount of MAOI eliminated in the urine. In contrast, with tricyclic antidepressants, hemodialysis can effectively remove much of the MAOI. Similar to the treatment of cyclic overdoses, activated charcoal and magnesium citrate can help reduce absorption.

In one case, IV dantrolene produced prompt and dramatic improvement in muscular rigidity, trismus, tachycardia, hyperthermia, and metabolic acidosis that resulted from a massive phenelzine overdose.[62] In another report, physicians in Australia recommended pulmonary artery catheterization to monitor hemodynamic response and the use of selected α and β blockade to treat hyperadrenergic signs.[63] These authors also recommend treatment of hyperthermia with active cooling techniques and further point out that prolonged observation and support may be required for the approximately 2 weeks it takes for a patient's supply of the enzyme MAO to regenerate after discontinuing the MAOI.

7. Pharmacokinetics

Monoamine oxidase inhibitors are rapidly and fully absorbed from the gastrointestinal tract. They undergo biotransformation in the liver and are excreted very quickly through the intestinal tract and, to a lesser extent, via the kidneys. Their half-life in the body is very short. However, the MAOIs have long-lasting pharmacological effects, since they permanently inactivate enzymes. Therefore, MAO inhibition continues for up to several weeks after discontinuation of the drug. The body must resynthesize the enzymes before normal metabolism of body amines resumes, a process taking 1 to 2 weeks.

Little is know about the metabolic fate of the three MAOI antidepressants, either in animals or in humans. Some research has focused on acetylation of hydrazides as a possible predictor of clinical response. At least for phenelzine, the primary degradative process appears to be oxidation rather than acetylation. Furthermore, the genetic polymorphism of the liver enzyme system that determines whether individuals are rapid or slow acetylators appears to have no clinical usefulness in predicting treatment response or side effects. Although patients with a degree of platelet MAO inhibition greater than 80% by 2 weeks of therapy are more likely to have a favorable treatment outcome with phenelzine, ensuring adequate dosage is probably equally valuable and certainly less expensive. There is no evident correlation of platelet MAO inhibition and clinical response with either tranylcypromine or isocarboxazid.

8. Preparations and Dosage

Each MAOI is produced in a single preparation (see Table 11). Usual doses vary widely among patients but generally are about 1 mg/kg of body weight per day for phenelzine but less for isocarboxazid and tranylcypromine.

C. Evaluation of the Depressed Patient

The depressed patient can present with many different clinical symptoms and signs. Mood disturbances frequently accompany major losses and other stressful events, particular character styles, and certain psychodynamic constellations. More often, depressed patients come to both medical and psychiatric clinicians with physical complaints. This may take many forms, from a request for a physical examination, complaints of fatigue, weight loss, or insomnia to specific somatic symptoms (often careful questioning reveals that several organ systems are involved). Along with or instead of physical manifestations, the patient frequently has both emotional (e.g., lowered mood, anhedonia, negativism) and psychological symptoms [e.g., irritability, loss of interest, ruminations, poor concentration, suicidal preoccupations, and even pseudodementia (see Chapter 8)]. In addition to defin-

TABLE 11. Preparations and Dosages of Monoamine Oxidase Inhibitors

Generic name	Trade name	Preparation	Usual dosage range (mg/day)
Isocarboxazid	Marplan®	10-mg tablets	10 to 30
Phenelzine	Nardil®	15-mg tablets	45 to 90
Tranylcypromine	Parnate®	10-mg tablets	10 to 30

ing the specific nature and history of the mood disorder, the clinician should also evaluate the other important components of the patient's life that might affect treatment: his environmental supports and important relationships, family history, coping style, and ego strengths.

In Each Depressed Patient, the Practitioner Should

1. Look for life-threatening symptoms (i.e., poor self-care, suicidal and homicidal impulses).
2. Take appropriate measures to address suicidal or homicidal impulses.
3. Assess the patient's diet.
4. Define medication-responsive symptom clusters.
5. Obtain a history of the present illness, focusing on possible situational, constitutional, and biological contributions to the mood disturbance.
6. Inquire about a history of previous affective episodes, their course, treatment, and response(s) (including adverse reactions).
7. Inquire about a family history of affective episodes, their course, treatment, and response(s).
8. Define the patient's strengths, including coping mechanisms and environmental supports.
9. Determine the patient's attitude toward help and medication.
10. Take a complete medical history (including a review of organ systems and documentation of both prescribed and nonprescribed drugs) (see Table 12).
11. Complete or arrange a physical examination (and perhaps other base line studies such as an EKG when clinically indicated).
12. Avoid the nonselective use of extensive medical screening and biochemical tests for major depressive illness.
13. Get appropriate consultation if psychological or medical symptoms are not clarified by the evaluation.
14. Inform the patient (and perhaps persons close to him) about the findings of the evaluation, the recommended treatment(s), benefits, risks, and possible alternatives.
15. Document conversations and recommendations.
16. Establish rapport with the patient and enlist him (and often family members) in the treatment.

After completing a careful evaluation, the clinician should decide how to manage acute symptoms, possible precipitants, and sequelae of the affective episode (e.g., stigmatization, lowered self-esteem, alienation from social supports).

TABLE 12. Drugs Associated with Depression[a]

Antihypertensives
  Guanethidine (Ismelin® and others)
  Methyldopa (Aldomet® and others)
  Reserpine (Serpasil® and others)
  Hydralazine (Apresoline® and others)
  Propranolol (Inderal® and others)
Sedative–hypnotic agents
  Alcohol
  Chloral hydrate (Noctec® and others)
  Benzodiazepines
Steroids
  Oral contraceptives
  Cortisol (Solu-Cortef® and others)
  ACTH (Acthar®)
Antipsychotic medications
Analgesics
  Opiates
Antiinflammatory agents
  Phenacetin (Emprazil® and others)
  Phenylbutazine (Butazolidin® and others)
  Pentazocine (Talwin®)
  Estrogen withdrawal (Premarin® and others)
Antiinfectious agents
  Sulfonamindes
  Clotrimazole (Lotrimin®, Mycelex®)
  Ethionamide (Trecator-SC®)
  Griseofulvin (Fluvicin® and others)

Antiinfectious agents (*cont.*)
  Metronidazole (Flagyl®)
Antineoplastic agents
  Plicamycin (Mithracin®)
  Azathioprine (Imuran®)
  Bleomycin (Blenoxane®)
  L-Asparaginase (Elspar®)
Cardiac drugs
  Digitalis (Crystodigin® and others)
  Procainamide (Pronestyl® and others)
  Propranolol (Inderal® and others)
  Clonidine (Catapres®, Combipres®)
Stimulants
  Amphetamine (Benzedrine® and others)
  Fenfluramine (Pondimin®)
Other drugs
  L-DOPA (Larodopa® and others)
  Amantadine (Symmetrel® and others)
  Methysergide (Sansert®)
  Acetazolamide (Diamox® and others)
  Carbamezapine (Tegretol® and others)
  Choline (Trilisate® and others)
  Disulfiram (Antabuse®)
  Physostigmine (Antilirium®)
  Ethambutol (Myambutol®)
  Indomethacin (Indocin® and others)

[a]Association does not mean causality. Some agents (e.g., reserpine and steroids) have a clear relationship with depression, whereas many other drugs do not.

D. Drug Therapy for the Depressed Patient

1. Initiating Treatment

After reaching a diagnostic decision and assessing the physical status of the depressed patient, how does the practitioner choose a drug for individuals with syndromes generally responsive to cyclic antidepressants or monoamine oxidase inhibitors? If the patient or a close relative has previously responded to a particular medication, use that drug (if it can be given safely). If the caretaker is not primarily concerned about cardiovascular or anticholinergic effects, he can start therapy with virtually any tricyclic. If sedative properties are desired, amitriptyline or doxepin works well. Otherwise, imipramine or a secondary amine can be used.

There is an attraction to using newer drugs, and sometimes they actually bring new benefits to patients. Nevertheless, the several-decades track record accumulated by the older tricyclic antidepressants is impressive and reassuring. There are apt to be few surprises. Therefore, the physician may wish to start a new patient on an old drug. The MAOIs are old, too, but for most depressed patients, tricyclics allow freedom from the dietary and medication constraints that are absolutely essential during MAOI therapy.

Which tricyclic should a clinician choose for a patient without prior experience (or family history) as guidelines? In the United States, secondary-amine tricyclics such as desipramine and nortriptyline have become popular for good reasons. They are less sedating, less anticholinergic, and less apt to provoke postural hypotension than their tertiary-amine "parents."

However, it can be argued that imipramine should initially be administered. This prototype tricyclic appears to have a positive linear correlation between dose (and blood levels) and clinical response: as noted earlier, more is better. Thus, if you prescribe imipramine for a patient, increase the dose progressively until you have reached the limits of a patient's tolerance or until the depression has been vanquished. The advantage of using a drug with such a straightforward dose–response relationship is that you will never have to wonder if you have exceeded a "therapeutic window": short of toxicity, increasing the dose will enhance the likelihood of clinical success.

Both imipramine and amitriptyline can usually be started in doses of 50 to 75 mg/day (see Table 8 for equivalent doses of other cyclics). The clinician should start elderly individuals, children, and small patients on lower doses. **Most people can take the entire dose of medication at bedtime.** If bothersome adverse reactions develop (e.g., oversedation, nightmares, or frightening dreams), or if multiple doses are preferred by the patient for emotional reasons, he can take the drug two or three times a day. Most patients will become partially tolerant to the anticholinergic effects and then, when an optimal therapeutic dose is achieved, can return to a single dose at bedtime. Usually, patients have fewer adverse reactions and less need for coadministered sleeping medication when the drug is given once in the evening. A bedtime regimen also results in less motor and mental slowing and better compliance. Medication is increased gradually (i.e., 25 mg/day or 50 mg every other day) until a therapeutic effect is reached or intolerable adverse reactions develop.

Individuals usually experience an antidepressant effect at 150 to 300 mg/day with imipramine or amitriptyline. However, some patients require much higher doses (even 600 to 700 mg) to benefit clinically. This may entail a more prolonged drug trial and careful monitoring for adverse reactions. In these cases, the practitioner should monitor blood levels and EKG (with doses exceeding 300 mg/day) and carefully document risk, benefits, and information imparted to the patient.

If the clinician has concerns about cardiac or anticholinergic effects (e.g., in the elderly or cardiac patient), antidepressants other than imipramine or amitriptyline should be tried first (perhaps trazodone, desipramine, nortriptylene or fluoxetine) and medication should be administered more slowly with more careful monitoring of adverse reactions. Although the antidepressant effectiveness of the triazolobenzodiazepine alprazolam remains to be scientifically validated, this anxiogenic agent may be considered in elderly or debilitated patients without profound or melancholic depressive symptoms. Likewise, elderly or medically ill patients may sometimes benefit from cautious use of a stimulant such as methylphenidate.

Increasingly, clinicians are turning to the new antidepressant fluoxetine as a first-line choice. As of this writing, fluoxetine is the most widely prescribed antidepressant in the United States. It has a favorable side-effect profile, significantly different from that of the tricyclics. In addition, its dose–response relationship makes it easier to manage them than the tricyclics: most adults respond to 20 mg per day. The elderly or frail patient might be started on 5 or 10 mg a day which, since the drug is only available as 20-mg capsules, must be managed in one of two fashions. Either the patient takes the drug every two to four days, or the capsule can be opened, the drug dissolved in water, and one-quarter to one-half of the solution taken each day. If fluoxetine is started first, at least several weeks should elapse from its discontinuation before an MAOI antidepressant is prescribed.

In all cases of acute outpatient therapy, the clinician must balance practical reality against suicide risk. The seriously self-destructive patient should be hospitalized. With less suicidal individuals, the practitioner should closely monitor treatment by frequent contact with the patient (and family members) and by careful assessment of depressive symptoms and suicidal thoughts and plans. **Initially, to prevent lethal overdose, prescriptions should be entrusted to a family member.**

The clinician should inform individuals receiving antidepressants about their high lethality. For example, borderline patients may sometimes take a few pills in a suicide attempt of low intent but, because they do not understand the dangerousness of these drugs, may develop serious medical problems. Although some authorities have recommended limiting the amount of medication prescribed at any one time (using some arbitrary threshold of lethality, such as 1 g), this may produce little more than a false sense of security for the clinician. A patient intent on suicide can hoard medication to ensure a sufficient dose. Instead, the clinician does best to focus on identifying patients at risk for suicide and maintaining close contact with them and their significant others. Sometimes telephone contact suffices. A depressed patient newly started on antidepressant medication should never be sent away with a prescription and a follow-up appointment a month later. At the very least, telephone contact should be made within a few days of initiating treatment so that side effects can be discussed, reassurance given, and suicidal potential reassessed.

2. Completing a Medication Trial

Although many antidepressants have been introduced with the promise of faster onset of action, these claims have not withstood the test of time. It seems that all antidepressant drugs currently known—tricyclics, MAOIs, and newer agents— take some weeks to become maximally effective. The conventional wisdom is that the lag time is 2 to 3 weeks, but this is oversimplified.

Prospective studies have found differences in symptom responses between drug responders and nonresponders as early as the first week of treatment. In other words, patients who will go on to become responders show the beginnings of improvement in a wide range of depressive symptoms by the seventh day of antidepressant drug therapy. Typically, the patient may be the last to know (or to acknowledge) that he is better; a significant other, or the clinician himself, may be more perceptive. Sleep, by the way, is not a useful symptom for distinguishing responders from nonresponders; the nonspecific sedating effects of these drugs can improve sleep (or cause excessive drowsiness) independent of treatment response.

The upshot of these observations is that a clinician should look for treatment response after as little as 1 week of therapy. The absence of any signs of improvement should suggest a dosage adjustment—usually upward. The advantage of using a drug with a straight linear correlation of dose and response (e.g., imipramine) is that upwards is always the right direction for adjusting dosage.

Some patients show an early and dramatic improvement in depressive symptoms—perhaps a complete remission within the first week. Others show a variable and stuttering course of symptom improvement. Both patterns may be more likely to be placebo responses than a true biological antidepressant reaction. The caveat here is that these patients are more likely to lose their apparent benefit from the antidepressant. Anticipating this, the clinician can be prepared to counteract a tendency to frustration, despair, or anger by suggesting the need to reevaluate the drug and its dose and also to emphasize the role of interpersonal and intrapsychic factors.

While awaiting treatment response, the physician should be in frequent contact with patient and possibly family members—to provide support and reassurance and to assess suicide potential. Family and patients also should be advised from the outset about the lag time before the drugs become effective and should be informed about common side effects. Education about unwanted effects of these drugs can turn an annoyance into "cement" for the therapeutic alliance: the patient can treat dry mouth, for example, as evidence that the drug is in his system and working; it also affirms the knowledgeability of his doctor.

The maximal efficacy of an antidepressant may not plateau for 6 to 8 weeks. Thus, if a patient's improvement has progressed from 80% to 90% between weeks 4 and 5 of treatment, the clinician can stay with the same dose for at least another week, awaiting continued improvement. On the other hand, a plateau in improvement at anything less than 100% or a return to *status quo ante* should prompt a reassessment of the drug dose.

The goal of antidepressant drug therapy should be complete remission of symptoms. Needless to say, character problems will not usually improve, but sometimes symptoms attributed to a patient's personality may reflect untreated depression. Still, there are patients with dysthymic characters who later develop major depression (so-called "double depression"), for whom antidepressant drugs can reverse the major depression but return the patient to a stable depressed character type.

The clinician would do best to continue adjusting the dose of the original antidepressant drug until either the acute depression remits completely or the patient is unable to tolerate side effects. If side effects are the limiting variable, and dosage reduction brings loss of efficacy, then the physician must switch to an alternate antidepressant. In this circumstance, one chooses a drug with fewer of the offending side effects, for example, less anticholinergic, less hypotensive, less likely to promote weight gain.

Failure to treat effectively despite adequate dosing for a sufficient time period is more troublesome. There is no evidence that switching among tricyclics is apt to produce efficacy when one tricyclic has failed. In some patients who have achieved partial but incomplete remission, lithium, often starting with as little as 300 mg a day, may provide a needed "jump start." A newer antidepressant might be tried, such as fluoxetine, or the patient can be switched to an MAOI (with at least a week's washout following a tricyclic and at least 2 weeks after fluoxetine or another MAOI). Electroconvulsive therapy always remains an important backup option.

When a patient has proved refractory to several adequate trials of biological antidepressant treatments, a diagnostic reassessment is in order. Covert medical and neurological factors, substance abuse, character and intrapsychic elements, and interpersonal contributions all must be weighted in the balance. When continuing antidepressant drug treatment appears indicated, one of a number of combinations may be considered (see Section IV.D.10).

3. Continuation Therapy

Having successfully ameliorated the acute symptoms of depression, the practitioner should now embark on what is termed continuation therapy. Typically lasting about 6 months, continuation therapy protects against a relapse of the depression. Presumably, we are "holding the fort" against depressive symptoms while the underlying process runs its course.

For continuation treatment, the clinician should usually continue the same dose that was effective during the acute phase of therapy. If, however, the patient is plagued by annoying but persistent side effects, the physician may gingerly try to back off on the dose, although both patient and doctor should be acutely aware that this enhances the risk of relapse. At times a switch to a different agent might be considered, but again the benefits in comfort must be weighted against the risk of recrudescent symptoms.

After about 6 months of successful antidepressant therapy, the physician evaluates the patient for possible discontinuation of the drug. If depressive symptoms have been **successfully and completely suppressed for at least 4 months,** a gradual taper may be attempted. If, however, there has been a "flickering" of symptoms (whether during a missed dose, a time of stress, or for no apparent reason), it is most prudent to continue the continuation therapy for at least another few months. Tapering the medication after an acute episode of depression should not take place until and unless depression has been virtually absent for at least 4 months.

To avoid antidepressant withdrawal symptoms (see Sections 3.g.vii and 3.j), tapering should be very gradual. Withdrawal symptoms after only 6 months of therapy are not likely to be as intense as those following many years of treatment. Still, if there is no medical urgency about discontinuing the antidepressant, why not prolong the taper over a month or more? Toward the end of the taper, tablets can be broken in half or administered every other day to make the final phases even more gradual.

4. Maintenance Therapy for Chronic or Recurrent Depressions

The course of bipolar illness is, by its very nature, recurrent. Among patients with unipolar depression (i.e., major depression with no history of mania), a recurrent course may affect between 70% and 85%. Probably worse still, 10% to 20% of patients with unipolar depressions develop a chronic course.

Depression is a serious illness. It carries a high morbidity of economic losses and disruption of families as well as other vital interpersonal linkages. Also, depression carries a substantial risk of mortality, usually from suicide, less commonly from neglect. For these reasons, and because of the intense pain associated with depressive episodes, some patterns of mood disorder require maintenance rather than intermittent treatment with medication.

Whether early or late in the course of a depressive illness, if the physician cannot wean the patient from antidepressant medication at the end of continuation therapy because of reemerging symptoms, this suggests that a chronic course is developing. In such cases, if antidepressant medication has successfully suppressed all or most of the depressive symptoms, there is no reason not to continue it indefinitely, possibly with periodic attempts to taper. Obviously, the physician seeks to find the lowest effective dose of the antidepressant, but because most patients can take these agents chronically with impunity, there is no pressing need to compromise efficacy in the interest of lowering the dose. (This situation is different from the concern with neuroleptics that higher doses increase the risk of tardive dyskinesia during the course of maintenance treatment.)

For a patient with multiple recurring episodes of depression, the question of when to consider or initiate maintenance therapy will hinge on the number of episodes as well as their frequency and severity. To be sure, the more numerous, frequent, and severe the depressive episodes have been, the earlier and more vigorously the doctor should pursue a course of maintenance. The need for such an approach should be discussed actively with the patient and possibly with significant others. The benefits in terms of prevention of relapse, with all the attendant morbidity and mortality, are obvious; the risks are few and usually center around uncomfortable rather than hazardous side effects.

There are no known risks that increase with long-term exposure to these drugs. The only effect of the passage of many years is that the patient ages, which may bring a need to reassess the dose of the drug and its interaction with illnesses that may have developed and medications to treat them.

For most patients, the choice of maintenance drug is the drug that was success-ful in acute and continuation treatment. Occasional patients may have required ECT during an acute episode but will benefit from maintenance antidepressant drugs. The dose most likely to be effective over the long haul is the same that treated the acute symptoms. If a patient is uncomfortable at that dose, however, one may try to lower it. As always, efficacy of the drugs has a high priority.

If an antidepressant is ineffective in the prophylaxis of recurrent depressive episodes, the practioner may consider maintenance lithium therapy. This is especial-ly true if the patient has a history of episodes suggesting hypomania or a family bipolar history. Most unipolar patients, however, will not require lithium, which probably is less effective than antidepressants in maintenance therapy. If lithium is to be used, there are no good studies about the optimal plasma range. We prefer the dose levels most effective in bipolar patients, 0.8–1.0 mEq/liter. Of course, for bipolar patients, lithium is the maintenance therapy of choice and should be aug-mented as needed with antidepressant drugs.

5. Depressive Episodes in Bipolar Disorders

Bipolar patients usually are on lithium (Eskalith® and others) maintenance therapy, which is more effective in preventing manic than depressive recurrences. When minor depressive symptoms emerge, time, support, and psychotherapy may suffice. Emergence of a full-blown episode of major depression, however, can be serious and usually requires prompt and vigorous intervention.

For a major depressive episode in a bipolar patient, a tricyclic antidepressant may be added to the maintenance lithium regimen. Despite the lithium, however, there is a finite risk of triggering a switch into mania, although in some patients this would happen whether or not an antidepressant was prescribed. To minimize this risk, discontinue the antidepressant drug soon after the patient has achieved a remission from depression. This also lowers the possibility that prolonged anti-depressant therapy can speed up the rate of a bipolar patient's mood cycles.

Some physicians believe the MAOI tranylcypromine to be particularly effec-tive in treating the bipolar depressed patient, especially when depression is accom-panied by psychomotor retardation, apathy, and withdrawal. There is less experi-ence with some of the newer antidepressants in bipolar patients, although there have been suggestions that bupropion may be beneficial for this population. Electrocon-vulsive therapy is highly effective and should particularly be considered in emergen-cy cases. (Most psychiatrists discontinue lithium during the course of ECT to avoid neurotoxicity.)

6. Some Dichotomies: Mild versus Severe, Acute versus Chronic, Agitated/Anxious versus Retarded

When the symptoms of major depression are present consistently over a period of weeks, most psychiatrists will consider antidepressant drug therapy. If the patient simultaneously has a number of melancholic symptoms, the likelihood of improvement with an antidepressant drug (or ECT) becomes even better. These predictive factors tend to be independent of the degree of severity or longevity of the depression. **Neither mildness nor chronicity should preclude consideration of an adequate antidepressant trial, nor do they suggest use of lower doses. The only way to know if a patient's depressive symptoms will remit on antidepressant drug therapy is to try it!** Obviously, if depressive symptoms yield to psychodynamic interpretations or changes in living or work arrangements, antidepressant drug therapy may become moot. But analogously, if symptoms remit on prescription of an antidepressant drug, major life changes or long-term psychotherapy may become unnecessary. **Until a vigorous and adequate antidepressant trial has been attempted and failed, it is impossible to say whether a patient's symptoms are "characterologic" or "neurotic" rather than drug responsive.**

An appropriate role for antidepressants in treating the patient with dysthymia remains to be elucidated. Indeed, dysthymia most probably reflects a range of disorders, which might include those both medication responsive and unresponsive. Again, the only way to distinguish the two is empirically: most if not all dysthymic patients deserve at least one vigorous antidepressant trial.

In the early days of modern psychopharmacology, there were hopes and expectations that antidepressants that were more or less sedating could be tailored to depressive subtypes such as agitated, anxious, or retarded. This has not proved to be the case. Nevertheless, some patients may appreciate the sedation provided by drugs such as amitriptyline, doxepin, or trazodone—usually prescribed at bedtime—whereas others will prefer the more stimulating effects of fluoxetine or tranylcypromine. Sometimes the use of an adjunctive benzodiazepine for the first few weeks of antidepressant therapy can enhance sleep improvement or help a patient relax during the day. It is unclear whether the triazolobenzodiazepine alprazolam has intrinsic antidepressant properties in addition to its proven anxiolytic activity, but some patients with milder depressions might be tried on it. The clinician must remember, however, that abuse and dependency have been reported as problems, particularly with alprazolam.

Patients with borderline personality disorder often experience episodes of major depression. Unfortunately, this axis I diagnosis often takes precedence over the borderline personality diagnosis on axis II of the DSM-IIIR. Biologically oriented psychiatrists then pursue treatment of the depression vigorously, only to be surprised and perplexed by the patient's apparent refractoriness. Although one or even several antidepressant trials may be warranted in such patients, there is evidence that, by and large, they are less responsive to typical antidepressant therapy (see Chapter 13). For borderline patients, other medications may be in order, along with long-term supportive psychotherapy.

7. "Atypical" Depression

The term "atypical" is used variably in clinical psychopharmacology. By and large, clinicians employ this concept to refer to major depression with a cluster opposite to that of melancholia. Thus, patients with atypical depression experience more depression early in the day, have greater difficulty falling asleep, and may experience food (often carbohydrate) craving with accompanying weight gain. Psychologically, they are often rejection-sensitive, anxious and phobic, and have a more "reactive" (rather than autonomous) mood. There is evidence that atypical patients may do better on MAOI than tricyclic therapy, although the data also can be read to suggest that this difference holds only for those atypically depressed patients who have concomitant panic attacks. Absent panic attacks, the patient with this form of depression may respond equally well to a tricyclic or an MAOI and no better to either than to placebo! Possibly, this "plain vanilla" atypical depression is not a drug-responsive condition.

It may be easier to start such patients on a tricyclic antidepressant because of the greater ease of the regimen. An MAOI should certainly be considered in patients who do not respond or who cannot tolerate a tricyclic. A role for newer antidepressants is unclear, although patients for whom weight gain is a substantial problem may appreciate the weight loss often associated with fluoxetine or fluvoxamine. The clinician must be prepared, however, for the lack of antidepressant responsiveness in some of these patients, although as mentioned above, those with accompanying panic attacks may be benefited, particularly by an MAOI. Electroconvulsive therapy is not usually thought of in this condition.

8. Major Depressive Episode with Psychotic Features

Major depression accompanied by delusions, hallucinations, or other failures in reality testing is a serious and often life-threatening condition. In a majority of cases it is unresponsive to tricyclic or other antidepressants alone. Electroconvulsive therapy produces substantial clinical improvement in over 80% of cases. A combined regimen of a tricyclic and an antipsychotic agent (doses adjusted independently and clinically) tends to have a success rate not much lower than ECT. There are few data on continuation and maintenance therapy for this condition, but it tends to be recurrent.

9. Depression in Schizophrenic Patients

Schizophrenic patients sometimes develop depressive symptoms and syndromes. These require a careful psychiatric differential diagnosis.

What appears (and even feels to the patient) like depression can at times be a reflection of neuroleptic-induced bradykinesia. The motor slowing of this parkinsonian symptom may be interpreted by the patient as low motivation and low spirits. To assess this possibility empirically, the practitioner can prescribe an antiparkinson drug for a few days or a week and perhaps lower the dose of the patient's antipsychotic agent.

From a psychosocial perspective, the clinician also must assess the apparently depressed schizophrenic patient regarding such issues as autistic withdrawal or demoralization. Most commonly, these reactions are apt to be responses to the patient's interpersonal milieu, and sensitive inquiry might reveal emotional issues with which the patient is grappling. Such problems are best handled by counseling, emotional support, and structured psychotherapy. We believe that schizophrenic patients often benefit from a frank acknowledgment of their cognitive and interpersonal limitations, assistance in grieving these realities, and then support in making the most of their assets and strengths.

For many schizophrenic patients, the chronic course of their illness is characterized by interpersonal withdrawal as part of a "defect state" or negative syndrome. Some such negative symptoms have been reported to improve more on treatment with clozapine than with standard neuroleptics, and there are limited data suggesting that the diphenylbutylpiperidine pimozide (Orap®) might help some patients. Unfortunately, there is no scientifically validated treatment for the defect state in schizophrenia.

When a schizophrenic patient develops a full-blown major depressive disorder, antidepressant drugs can sometimes be helpful, used in the same fashion prescribed for any patient with this syndrome. There is a finite risk of worsening the underlying psychosis, but this should be mitigated by the coadministration of a neuroleptic.

10. Resistant Depressions

The key to success in antidepressant pharmacotherapy is adequacy in time and amount: enough drug for a long enough period. The guidelines earlier should yield success in over three-quarters of patients with episodes of major depression. As noted previously, with a positive, linear correlation between dose and efficacy, as in the case of imipramine, the clinician should "push" the dose until meeting the limits of patient tolerance or enjoying the fruits of success. With nortriptyline, where there is reasonable evidence of a curvilinear "therapeutic window" relationship, a treatment-resistant patient can have blood drawn for a serum nortriptyline assay, and the clinician may adjust the dose to bring the blood level within the "therapeutic range" of 50–150 ng/ml. For all other antidepressant drugs, the best that a blood level has to offer is a broad "intuitive" sense of whether or not a patient may be extremely high or low.

Although it is conceivable that some patients might do better with 40 or 60 mg daily of fluoxetine than on the standard adult antidepressant dose of 20 mg per day, controlled studies have failed to find any greater antidepressant efficacy for doses above 20 mg daily. In fact, some patients who fail to improve at a dose of 20 mg daily of fluoxetine may actually benefit from a reduction in the daily dose.

If a patient has failed an adequate trial of a tricyclic, an MAOI, a substantially different antidepressant such as fluoxetine, and certainly ECT, the patient may then truly be categorized as treatment resistant. First of all, a diagnostic reappraisal is in order.

From a psychosocial perspective, a number of factors may complicate or antagonize response to antidepressant therapy. Patients with some forms of character pathology, such as the borderline personality, tend to respond poorly to antidepressant medication (discussed earlier and in Chapter 13). Alternate medications and verbal interactions may be preferable (see Chapter 13). Individuals for whom depressive behaviors are part of their lifestyle and interpersonal coping strategies also may do poorly on antidepressant therapy, although they should still be given at least one trial of adequate pharmacotherapy. A sensitive clinician will always inquire about neurotic and character mechanisms that may be "driving" depressive symptoms. An interview with a significant other can shed new light on the meaning of an individual's symptoms and on the accuracy of the history. Unresolved grief is another phenomenon that sometimes appears to be a treatment-refractory depression.

Many biological factors can similarly render a patient unresponsive to antidepressant medicines. Consider possible adverse interactions with other medications a patient may be taking: prescribed, over-the-counter, or substances of abuse. A wide range of medical illnesses—especially endocrinopathies, infections (including AIDS), brain diseases, or neoplasms—can similarly produce a depression that might fail to respond to antidepressants. To consider all of these and many other biological possibilities, return to the medical history and review of systems, obtain additional data from past records, other treating physicians, or family members, see that an adequate physical examination has recently been performed, and order any laboratory testing indicated by the foregoing.

What if nothing new turns up from either a psychosocial or a biological perspective? At this point, the physician can consider a wide variety of add-ons, unfortunately few of which have been studied systematically. Probably best studied and most frequently used today in clinical practice is adjunctive lithium. As of this writing, interpreting the data on lithium supplementation is unclear. Some patients appear to respond dramatically and quickly to minimal doses of lithium, others more gradually to somewhat higher doses. Perhaps partial responders are most likely to progress to complete response, or maybe some patients are hastened to what would ultimately be a complete recovery anyway. Despite these doubts, the usual augmentation doses are between 300 and 1200 mg daily, maintained for several weeks. One approach is to begin with 300 mg bid. and, if there is no response within a week, to increase to 900 mg daily. Some clinicians prefer to bring patients to a "standard" blood lithium level approaching 1.0 mEq/liter. Most physicians try to stop the dose of lithium after several weeks, whether it is effective (in which case perhaps no longer needed) or ineffective (in which case, something else should be tried).

Adjunctive thyroid hormones have been tried, although our reading of the data suggests that, in patients chemically euthyroid, this approach is less likely to meet with success. The usual strategy is to initiate treatment with 10–25 μg of liothyronine (T_3, Cytomel®), increasing the dose every few days to a maximum of 50–75 μg/day. Some recent studies suggest that thyroxine (T_4) may be an effective supplement in treatment-resistant depressed patients. Other adjunctive strategies with even weaker data bases include the coadministration of reserpine (Serpasil® and others), an antipsychotic drug, or, in women, conjugated estrogens. The amino acid precursors of serotonin and norepinephrine, respectively tryptophan and tyrosine, also have been used adjunctively with varying reports of success. For patients with seasonal affective disorder, spending upwards of 2 hr a day in front of high-intensity full-spectrum lights has been reported to yield dramatic results, although the scientific validation of phototherapy as an antidepressant treatment remains insufficient.

There is no scientific evidence that the combination of a tricyclic and an MAOI antidepressant provides more efficacy than either drug alone. Still, there is a wealth of anecdotal reports suggesting that some patients appear to respond better to the combination than to monotherapy. Many patients have taken this combination with impunity, although a number of caveats are in order. Obviously, as with any regimen involving an MAOI, the usual food, beverage, and medication prohibitions apply. Certain medications are best avoided in this combination, more to prevent the hypermetabolic than the hypertensive crisis. Imipramine, desipramine, and especially clomipramine are best avoided as the tricyclic. Tranylcypromine may be chancier than phenelzine on the MAOI side. It is best to stop previous medications for at least a week and then to begin a low dose of a tricyclic, e.g., 50 mg of amitriptyline, trimipramine, or doxepin. Isocarboxazid or phenelzine is then added in low doses, and, if tolerated, the doses of the tricyclic, alternating with that of the MAOI, are "walked up" every few days gradually and with cautious monitoring of the patient's symptoms and vital signs.

Patients with difficult-to-treat depressions are typically managed with different drugs sequentially or often with combination therapies. This means a need for heightened awareness of potential interactions. Switching between cyclics and MAOIs, or between two different MAOIs, presents potential hazards. At least a week's washout should intervene when changing from a tricyclic to an MAOI, and probably 2 weeks between an MAOI and a tricyclic. The very long half-lives of fluoxetine and its active metabolite suggest the need for an even longer washout between fluoxetine and an MAOI, since cases of potentially lethal hypermetabolic syndrome (presumably a serotonin syndrome) have occurred following this switch of medicines. When patients have been switched from phenelzine to tranylcypromine without sufficient washout, cases of hypertensive crises and strokes have developed. A pharmacokinetic interaction has been described between fluoxetine and tricyclic antidepressants in which fluoxetine coadministration markedly elevates the plasma level of the tricyclic.

For further discussion and details on the treatment of the treatment-resistant depression, the reader is referred to reference 64.

11. Stimulants

Psychostimulants, such as the amphetamines, were used before the tricyclics and MAOIs to treat depressed patients. Often, they did little more than convert psychomotor retardation to anxiety or agitation. Additionally, tolerance developed to their effects, anorexia and insomnia were unwanted reactions, and they could promote dependency and abuse. Furthermore, there was little scientific validation of their antidepressant efficacy, and coupled with increasing government regulation over the prescription of these substances, they fell into disfavor and disuse among the medical community.

More recently, however, there has been heightened interest in the use of stimulants, growing largely out of the experience of consultation–liaison psychiatrists. Because of their very different profile of clinical response and adverse reactions, these drugs have been found safer and more useful than standard antidepressants, particularly in the medically ill and the elderly. Low doses of methylphenidate or D-amphetamine, administered for up to several weeks, can be helpful in turning the corner in depressive symptoms in some such patients. Despite the paucity of scientific validation of these clinical impressions, many psychiatrists practicing in general hospitals are pleased to have them available.

Earlier research suggesting that stimulant response could be used to predict antidepressant response has not withstood the tests of science and time. In some instances, clinicians have used stimulants to get a "leg up" on the lag time while waiting for a standard antidepressant to work or to "jump start" a treatment-resistant patient. Another use of stimulants as adjuncts to antidepressant medications is in patients suffering from drug-induced postural hypotension.

V. CONCLUSION

The 1980s saw the introduction of a number of new antidepressant drugs. In some cases, the only thing new that they provided was toxicity. Other drugs, however, appeared to broaden our range of therapeutic options, both in terms of conditions that are treatable and also different adverse-effect profiles.

Research also has told us more about the appropriate use of old drugs, such as tricyclics and MAOIs. One of the most important findings in recent years is that many depressed patients go untreated or undertreated with antidepressant medications, making vigorous pharmacotherapy a watchword for the last decade of this millennium.

As we look to the future, health care professionals anticipate the advent of biological tests that will allow us to select treatments with greater specificity. We also look forward to new agents greater in efficacy but safer in practice.

Whatever the fruits of neuroscience, there always will be a role for sensitive and skillful interpersonal interactions in the approach to depressed human beings. The response of a patient to an antidepressant pharmaceutical is independent of whether the etiology of that patient's condition is biological, psychosocial, or some combination of the two. To paraphrase Maimonides, seeing in the patient the suffering human being often is the greatest challenge for the clinician, yet the greatest blessing we may bestow.

REFERENCES

1. Abraham K.: *Notes on the Psychoanalytic Investigation and Treatment of Manic–depressive Insanity and Allied Conditions (1911)*. New York, Basic Books, 1960, pp 137–156.
2. Freud S.: *Mourning and Melancholia*, Standard Edition (1916). London, Hogarth Press, 1971, vol XIV.
3. Weissman M. M., Fox K., Klerman C.: Hostility and depression associated with suicide attempts. *Am J Psychiatry* 130;450–455, 1973.
4. Bibring E.: The mechanism of depression, in Greenacre P. (ed): *Affective Disorders*. New York, International Universities Press, 1965, pp 13–48.
5. Harlow H., Harlow M.: Learning to love. *Sci Am* 54:244–272, 1966.
6. Spitz R.: Anaclitic depression. An inquiry into the genesis of psychiatric conditions in early childhood. *Psychoanal Study Child* 2:313–342, 1942.
7. Ainsworth M.: Object relations, dependency and attachment. *Child Dev* 40:969–1025, 1969.
8. Bowlby J.: Childhood mourning and its implications for psychiatry. *Am J Psychiatry* 118:481–498, 1961.
9. Robertson J., Robertson J.: Young children in brief separation: A fresh look. *Psychoanal Study Child* 26:264–315, 1971.
10. Mahler M. S.: On sadness and grief in infancy and childhood: Loss and restoration of the symbiotic love object. *Psychoanal Study Child* 16:332–351, 1961.
11. Kohut H.: *The Analysis of the Self*. New York, International Universities Press, 1971.
12. Kernberg O.: *Borderline Conditions and Pathological Narcissism*. New York, Jason Aronson, 1975.
13. Beck A. T., Rush A. J., Shaw B., et al: *Cognitive Therapy of Depression*. New York, Guilford Press, 1979, pp 1–34.
14. Seligman M., Maier S., Geer J.: The alleviation of learned helplessness and the dog. *J Abnorm Soc Psychol* 73:256–262, 1968.
15. Lewinsohn P., Biglar A., Zeiss A.: Behavioral treatment of depression, in Davidson P (ed): *The Behavioral Management of Anxiety, Depression and Pain*, New York, Brunner-Mazel, 1976, pp 91–146.
16. Decker E.: *The Revolution in Psychiatry*. London, Collier–Macmillan, 1964, pp 108–135.
17. Schildkraut J. J.: Catecholamine hypothesis of affective disorder. *Am J Psychiatry* 122:509–522, 1965.
18. Maas J. W.: Biogenic amines and depression: Biochemical and pharmacological separation of two types of depression. *Arch Gen Psychiatry* 32:1357–1367, 1975.
19. Bunney W. E., Murphy D. L., Goodwin F. K., et al: The "switch process" in manic–depressive illness. I. A systematic study of sequential behavioral changes. *Arch Gen Psychiatry* 27:295–317, 1972.
20. Akiskal H. S., McKinney W. I. Jr: Overview of recent research in depression; integration of ten conceptual models into a comprehensive clinical frame. *Arch Gen Psychiatry* 32:285–305, 1975.
21. American Psychiatric Association: *Diagnostic and Statistical Manual of Mental Disorders*, 3rd ed revised (*DSM-IIIR*). Washington, American Psychiatric Association, 1987.
22. Kocsis J. H., Frances A. J., Voss C., et al: Imipramine treatment for chronic depression. *Arch Gen Psychiatry* 45:253–257, 1988.
23. Brotman A. W., Falk W. E., Gelenberg A. J.: Pharmacologic treatment of acute depressive subtypes, in Meltzer H (ed): *Psychopharmacology: The Third Generation of Progress*. New York, Raven Press, 1987, pp 1031–1040.
24. Brotman A. W., Falk W. E., Gelenberg A. J.: Pharmacologic treatment of acute depressive subtypes, in Meltzer H (ed): *Psychopharmacology: The Third Generation of Progress*. New York, Raven Press, 1987, pp 1031–1940.
25. Weissman M. M.: The psychological treatment of depression. *Arch Gen Psychiatry* 36:1261–1268, 1979.

26. Weissman M. M., Prusoff B. A., DiMascio A., et al: The efficacy of drugs and psychotherapy in the treatment of acute depression episodes. *Am J Psychiatry* 136:555–558, 1979.

27. Herceg-Baron R. L., Prusoff B. A., Weissman M. M., et al: Pharmacotherapy and psychotherapy in acute depressed patients. A study of attrition patterns in a clinical trial. *Compr Psychiatry* 20:315–325, 1979.

28. Weissman M. M., Klerman G. L., Prusoff B. A., et al: Depressed outpatients: Results one year after treatment with drugs and/or interpersonal psychotherapy. *Arch Gen Psychiatry* 38:51–55, 1981.

29. Kovacs M.: The efficacy of cognitive and behavior therapies for depression. *Am J Psychiatry* 137:1495–1501, 1980.

30. Frankel F. H.: Current perspectives on ECT: A discussion. *Am J Psychiatry* 134:1014–1019, 1977.

31. Grahme-Smith D. S., Green A. R., Costain D. W.: Mechanism of the antidepressant action of electroconvulsive therapy. *Lancet* 1:254–256, 1978.

32. Small J. G., Kellams J. J., Milstein V., et al: Complications with electroconvulsive treatment combined with lithium. *Biol Psychiatry* 15:103–112, 1980.

33. Mandel M. R., Madsen J., Miller A. L., et al: Intoxication associated with lithium and ECT. *Am J Psychiatry* 137:1107–1109, 1980.

34. Squire L. R., Slater P. C.: Bilateral and unilateral ECT: Effects on verbal and nonverbal memory. *Am J Psychiatry* 135:1316–1320, 1978.

35. Fink, M.: An adequate treatment? Editorial. *Convulsive Therapy* 5:311–313, 1989.

36. Baldessarini R. J.: The basis for amine hypothesis in affective disorders. A critical evaluation. *Arch Gen Psychiatry* 32:1087–1093, 1975.

37. Wilson E. G., Petrie W. M., Ban T. A.: Possible lack of anticholinergic effects with mianserin: A pilot study. *J Clin Psychiatry* 41:63–65, 1980.

38. Sulser F.: New perspectives on the mode of action of antidepressant drugs. *Trends Pharmacol Sci* 1:92–94, 1979.

39. Crews F. T., Smith C. B.: Presynaptic alpha-receptor subsensitivity after long-term antidepressant treatment. *Science* 202:322–324, 1978.

40. deMontigny C., Aghajanian G. K.: Tricyclic antidepressant: Long-term treatment increases responsivity of rat forebrain neurons to serotonin. *Science* 202:1303–1306, 1978.

41. Briley S., Langer S. Z., Raisman R., et al: Tritiated imipramine binding sites are decreased in platelets of untreated depressed patients. *Science* 202:303–305, 1980.

42. Janowsky D., El-Yousef K., Davis M., et al: A cholinergic adrenergic hypothesis of mania and depression. *Lancet* 2:632–635, 1972.

43. Maas J. W.: Neurotransmitters in depression: Too much, too little or too unstable? *Trends Neurosci* 2:306–308, 1979.

44. Jefferson J. W.: Cardiovascular effects and toxicity of anxiolytics and antidepressants. *J Clin Psychiatry* 50:368–378, 1989.

45. Glassman A. H., Roose S. P.: Cardiovascular effects of tricyclic antidepressants. *Psychiatr Ann* 17:340–347, 1987.

46. Davidson J.: Seizures and bupropion: A review. *J Clin Psychiatry* 50:256–261, 1989.

47. Lipinski J. F., Mallya G., Zimmerman P., Pope H.: Fluoxetine-induced akathisia: Clinical and theoretical implications. *J Clin Psychiatry* 50:339–342, 1989.

48. Preskorn S. H., Othmer S. C.: Evaluation of bupropion hydrochloride: The first of a new class of atypical antidepressants. *Pharmacotherapy* 4:20–34, 1984.

49. Treatment of xerostomia. *Med Lett* 30:74–76, 1988.

50. Vaughan D. A.: Interaction of fluoxetine with tricyclic antidepressants. Letter to editor. *Am J Psychiatry* 145:1478, 1988.

51. Sternbach H.: Danger of MAOI therapy after fluoxetine withdrawal. Letter to editor. *Lancet* 2:850–851, 1988.

52. Boehnert M. T., Lovejoy F. H. Jr.: Value of the QRS duration vs the serum drug level in predicting seizures and ventricular arrhythmias after an acute overdose of tricyclic antidepressants. *N Engl J Med* 313:474–479, 1985.

53. Foulke G. E., Albertson T. E.: QRS interval in tricyclic antidepressant overdose: Inaccuracy as a toxicity indicator in emergency setting. *Ann Emerg Med* 16:160–163, 1987.

54. Crome P., Dawling S., Braithwaite R. A., et al.: Effect of activated charcoal on absorption of nortriptyline. *Lancet* 2:1203–1205, 1977.

55. Granacher R. D., Baldessarini R. J.: Physostigmine: Its use in acute anticholinergic syndrome with antidepressant and antiparkinson drugs. *Arch Gen Psychiatry* 32:375–379, 1975.

56. Gelenberg A. J.: MAOIs: Assorted concerns. *Biol Ther Psychiatry* 11:33–36, 1988.

57. Clary C., Schweizer E.: Treatment of MAOI hypertensive crisis with sublingual nifedipine. *J Clin Psychiatry* 48:249–250, 1987.

58. Teicher M. H., Cohen B. M., Baldessarini R. J., et al: Severe daytime somnolence in patients treated with an MAOI. *Am J Psychiatry* 145:1552–1556, 1988.

59. Zimmerman H. J., Ishak K. G.: The hepatic injury of monoamine oxidase inhibitors. *J Clin Psychopharmacol* 7:211–213, 1987.

60. Dilsaver S. C.: Monoamine oxidase inhibitor withdrawal phenomena: Symptoms and pathophysiology. *Acta Psychiatr Scand* 78:1–7, 1988.

61. Tollefson G. D.: Monoamine oxidase inhibitors: A review. *J Clin Psychiatry* 44:280–288, 1983.

62. Kaplan R.F., Feinglass N. G., Webster W., et al: Phenelzine overdose treated with dantrolene sodium. *JAMA* 255:642–644, 1986.

63. Breheny F. X., Dobb G. J., Clarke G. M.: Phenelzine poisoning, case report. *Anesthesia* 41:53–56, 1986.

64. Rosenbaum J. F., Gelenberg A. J.: Drug treatment of resistant depression, in Karasu T. B. (ed): *Treatments of Psychiatric Disorders. A Task Force Report of the APA*. Washington, American Psychiatry Association, 1989.

65. Harrison W. M., McGrath P. J., Stewart J. W., Quitkin F.: MAOIs and hypertensive crises: The role of OTC drugs. *J Clin Psychiatry* 50:64–65, 1989.

3

Bipolar Disorder

ALAN J. GELENBERG, M.D., and
STEPHEN C. SCHOONOVER, M.D.

I. INTRODUCTION

Lithium (Eskalith® and others), a naturally occurring salt, has been used for various medical purposes since the 1800s. Initially, it was given to patients who suffered from urinary calculi and gout. Later, it was combined with bromides and used as a sedative. In 1949, Cade observed that lithium calmed agitated psychotic patients: ten manic patients responded, six schizophrenic and chronically depressed psychotic patients did not, and one patient's symptoms reappeared after lithium was stopped.[1] During this same period, lithium was introduced in the United States as a salt substitute for cardiac patients, but it caused numerous toxic reactions and several deaths. Thus, even with the emergence of convincing evidence that lithium was safe and effective in manic depressive illness, it was not accepted in the United States until 1970.

Lithium currently is employed to treat **acute hypomanic** or **manic episodes** and **recurrent mood disorders.** Despite its effectiveness, however, its mechanism of action remains unclear. Some authorities postulate that lithium corrects an ion-exchange abnormality.[2] Other researchers have shown that it increases neuronal release of norepinephrine, serotonin, and dopamine, increases the reuptake and metabolism of norepinephrine, alters serotonin receptor sensitivity, and may decrease the availability of acetylcholine. Forn and Valdecasas hypothesize an alteration of the adenylate cyclase system that affects membrane permeability by changing the functioning of postsynaptic receptor sites.[3]

The major mood disorders affect 10% to 20% of people in a lifetime. Bipolar disorder and recurrent depressions are among the most common and treatable disorders of this type. In most patients, the episode recurs, with episodes becoming more

ALAN J. GELENBERG, M.D. • Department of Psychiatry, University of Arizona Health Sciences Center, Tucson, Arizona 85724. **STEPHEN C. SCHOONOVER, M.D.** • Schoonover Associates, Newton Centre, Massachusetts 02159.

severe and frequent with age. Clinicians have successfully treated these disorders with pharmacological agents, particularly lithium salts. In fact, **50% to 80% of patients with "typical" bipolar illness respond to lithium.**

Other periodic disorders with an affective component that may be related to manic–depressive illness (e.g., schizoaffective disorder) can sometimes be effectively treated with lithium too. Less clearly defined periodic disturbances are also occasionally responsive to lithium and include periodic catatonia,[4] periodic alcoholism (particularly accompanied by depressive affect),[5] emotionally unstable character disorders,[6] cyclothymic personality disorders,[7] and obsessive compulsive states.[8] A possible role for lithium in the treatment of schizophrenic patients remains to be elucidated.

In practice, how quickly and how well does lithium work? In acute hypomanic or manic episodes, lithium is frequently effective within 1 to 2 weeks but may require several more weeks or even a few months to contain the affective episode fully. Sometimes an antipsychotic agent [such as haloperidol (Haldol® and others)] should be used acutely to manage behavioral excitement and acute psychotic symptoms. After the acute symptoms are diminished, lithium maintenance therapy decreases the number, severity, and frequency of affective episodes. Even with drug treatment, some patients experience various symptoms, intermittent periods of distress, or unwanted effects. To achieve optimal results, the clinician may need to experiment with various dosage levels and continue maintenance therapy for a year or longer. In addition, patients with acute depressive symptoms tend not to respond to lithium alone.

Initially, the clinician should fully evaluate the patient to make an accurate diagnosis and then develop a comprehensive approach to care. Therefore, to establish a framework for safe and effective clinical care, we discuss the following practical guidelines:

1. Defining symptom profiles and diagnosing patients responsive to medication.
2. Delineating clinical effects, pharmacokinetics, and adverse reactions.
3. Specifying acute and maintenance medication regimens for bipolar affective episodes and recurrent unipolar depressions.
4. Developing a medication philosophy that emphasizes risks and benefits and effective collaboration with the patient within a caring therapeutic relationship.

II. BIPOLAR DISORDER

The bipolar type of mood disorder is characterized by episodic mood swings. Different patterns include:

1. Episodes of mania alternating with depression. These mood "switches" may occur frequently (every few days) or may be widely spaced.

2. Episodes of mania followed by an intervening period of normal mood and then by the onset of depression. In this subgroup, intermittent depressive episodes followed by a period of normal mood may occur without the development of hypomania or mania.
3. Rare cases of unipolar mania with episodes varying in length (up to several decades in some untreated cases).[9]

The clinician can use the severity and pattern of a patient's illness as a base line for monitoring the effectiveness of pharmacotherapy. Because of the range of clinical presentations and the wide variation in the course of bipolar illness, the clinician must carefully follow each individual.

A. Clinical Presentation

Bipolar disorder occurs in about 0.4–1.2% of the general population. Patients usually develop their first mood episode (usually mania or hypomania) between the ages of 20 and 40 (average 32), although teen-age presentations are not uncommon. Most individuals with bipolar illness have their first episode before age 30. (Late onset of mania should prompt a neurologic and medical differential diagnosis, since mania may be secondary to many such conditions.) In mania, elevated mood is common but not universal. Sometimes depressive symptoms break through a facade of emotional lability. Pressure of speech, flight of ideas, increased motor activity, and decreased sleeping time often accompany the sense of elation and well-being. These patients are very self-involved but often can sense the vulnerabilities of others and become inappropriate and intrusive. If mania progresses, the patient's sense of humor may border on anger and irritability. Pressured speech may deteriorate into fragmented sentences and clang associations, and a sense of well-being may escalate into grandiosity. In the most extreme form, the patient becomes psychotic. Auditory hallucinations, delusions, and paranoia are frequent, and incoherence, sensory distortions, agitation, and combativeness may accompany the most disorganized states.

Clinicians have described some cases of prolonged mania or intermittent mania without depression. Most commonly, however, the illness is bipolar. The first hospitalization usually is for mania. The depression typically is indistinguishable from other forms of major depression, except that they are typically more frequent and shorter in duration. The third revised edition of the *Diagnostic and Statistical Manual of Mental Disorders* (DSM-IIIR) defines the symptom criteria, which include persistently lowered mood, psychomotor retardation, feelings of helplessness, hopelessness, and worthlessness, vegetative symptoms, and various somatic concerns.

Occasionally, a mood episode in a bipolar patient is not clearly mania or depression, but rather a mixture of both. Recent evidence suggests that this mixed picture may be less treatment responsive and have a poorer prognosis than simple mania or depression. Similarly, "cycling episodes," in which patients move from one mood pole to the other without an intervening period of euthymia, also have a bad outcome.[10]

B. Course of the Illness

Both manic and depressive episodes are usually time limited (typically 3 to 9 months). As patients grow older, however, the episodes may occur more frequently and last longer. Most often, the manic and depressive episodes of bipolar illness are shorter and more frequent than in pure mania or major depressive illness. Patients with rapid cycling (four or more episodes per year) between mania and depression have a more guarded prognosis. Most bipolar individuals become relatively symptom-free between episodes. However, one-quarter to one-third of patients exhibit significant symptoms or decreased functioning between episodes. Although general trends exist, the course of bipolar disorder varies considerably in different individuals. Many patients have episodes every few years; others have clusters of episodes; some show worsening in both frequency and intensity over a period of years.

In general, recurrent episodes are the rule rather than the exception: at least 80% of patients with an initial episode of mania will have one or more future episodes.[11] The effects can be devastating. An estimated one-quarter of bipolar patients attempt suicide (a figure exceeded only in schizophrenia and schizoaffective disorder),[12] and probably even more of them engage in a variety of hazardous behaviors. In addition, work, family, and social contacts deteriorate, initially around episodes but later even between episodes. A study sponsored by the U. S. Public Health Service reports that without adequate treatment, the average women whose bipolar disorder begins at age 25 can expect to lose approximately nine years of life, 14 years of such major activities as work, school, and children, and 12 years of normal health.[13]

Although the trends gleaned from longitudinal studies can provide a general set of expectations about the course of bipolar disorders, the clinician must assess the nature and course of each patient's illness individually. An appropriate treatment plan can be developed only after evaluation of the impact and meaning of the illness in the patient's life along with objective clinical symptoms.

III. GENERAL MEASURES FOR TREATING BIPOLAR DISORDER

Many approaches, including electroconvulsive therapy (ECT), pharmacotherapy (antipsychotic drugs, lithium, and antidepressants), milieu therapy, and various psychotherapies are used to treat bipolar illness.

A. Electroconvulsive Therapy

Electroconvulsive therapy is an effective treatment for either mania or major depression (see Chapter 2). For manic patients, bilateral application of electrodes may be more effective. Although there are contradictory data about lithium administration during ECT, to avoid neurotoxicity it is most prudent to discontinue lithium during ECT.

Circumstances That May Warrant the Use of ECT

1. Previous good response to ECT.
2. Medical contraindication to drug therapy.
3. Immediate threat of suicide or homicide or serious withdrawal or regression that interferes with self-care.
4. Failure of successive trials of pharmacotherapy.
5. Acute disorganization or combativeness that requires immediate behavioral containment.

B. Milieu Therapy

To ensure a manic patient's safety and to contain symptoms, inpatient care may be necessary. These patients, however, frequently disrupt the ward milieu. The patient's infectiously elevated mood may stimulate reinforcement from the staff and other patients, and irritability and hostility can be highly provocative. Structuring daily activities, correcting cognitive distortions, and limiting external stimulation by staff and patients may reduce the patient's excitement. Caretakers should remember that milieu therapy and pharmacotherapy are synergistic. A safe and stable environment promotes medication compliance, and the salutary effect of drugs increases insight, emotional availability, understanding, and formation of social attachments.

C. Group Therapy

Fieve has described "lithium groups" designed to discuss issues related to maintenance therapy.[14] This approach is not only time efficient for clinicians but also advantageous for patients. Sharing similar concerns about medication usage can be both supportive and growth producing while it increases compliance.

Families and "significant others," too, can benefit from learning about mood disorders and sharing common approaches to dealing with their loved ones. Support and practical counseling are valuable, and people close to the patients can learn to recognize prodromal signs of impending relapse, noninflammatory ways of coping with symptoms, to predict side effects, the importance of avoiding sleepless periods, and more.

Although group therapy can be helpful during the maintenance phase of treatment for patients with mood disorders, it can be counterproductive during episodes of acute illness, particularly for hospitalized patients. A depressed patient might become irritable with forced socialization. A manic patient, in contrast, can become more excited with interpersonal stimulation and, in turn, may be provocative to others.

D. Psychotherapy and the Therapeutic Relationship

An appropriate role for psychotherapy in the treatment of patients with bipolar illness is a subject of ongoing research. Even assuming that disorders of mood, particularly bipolar illness, reflect genetic diatheses, nonbiological treatments can augment pharmacotherapy. Perhaps psychotherapy acts synergistically to diminish the likelihood of recurrences. Lithium might diminish the amplitude of reaction to life events and stresses, while psychotherapy may allow a patient to avoid interpersonal crises and, when they occur, to deal with them more adaptively.

Data about whether or not psychoanalytically oriented therapies are appropriate for bipolar patients are inconclusive. Anecdotal case evidence suggests that bipolar patients stabilized on medication who are otherwise appropriate for insight therapy can be so treated. For most patients, however, here-and-now approaches combining cognitive, interpersonal, behavioral, and psychoeducational components with appropriate balancing of individual, group, and family treatments can be useful.

Beyond that, a sustained relationship with a psychotherapeutically oriented clinician can improve the outcome of drug treatment. Patients will be more likely to tolerate adverse effects and to work with the doctor to find optimal drug levels and contraactive options. Family members will be better educated to spot early signs of toxicity and to report prodromata of relapse or recurrence. Patients in regular contact with a therapist are less likely to drop out of treatment and more likely to be picked up early in the course of a new mood episode.

IV. PHARMACOTHERAPY

A. General Clinical Considerations

Pharmacological agents have markedly decreased the morbidity associated with bipolar illness, recurrent depressions, and several other cyclic psychiatric disturbances. The antipsychotic drugs provide an effective means of controlling behavioral symptoms and thought disorder that accompany mania and hypomania. Antidepressant medications usually contain depressive episodes. Lithium can diminish the symptoms of acute mania and, over the long term, often suppresses both manic and depressive episodes. Carbamazepine (Tegretol® and others) and various other agents are being tried as adjuncts or second-line options for patients with mood disorders. However, researchers have not yet established the safety and efficacy of these drugs for this indication.

Antipsychotics and lithium are both useful for treating mania.[15,16] Antipsychotics, particularly haloperidol, can rapidly diminish increased motor activity and also suppress other symptoms of this disorder. These agents have definitive advantages over lithium for treating acute behavioral excitement. However, the antipsychotic drugs only initially suppress the symptoms. Questioning reveals that these patients continue to feel internally disorganized. Lithium, in contrast, has more specific antimanic effects than the antipsychotics and usually diminishes manic symptoms within 5 to 14 days. However, the time required to control the manic symptoms fully with lithium may vary from days to several months.

If mild, a depressive episode in a bipolar patient may be "ridden out" with continuing lithium maintenance therapy and psychological support. More severe depressions, however, require biological treatment. As noted in the previous chapter, tricyclic agents often suffice, although there is the ever-present hazard of precipitating a switch into mania. Some psychiatrists prefer a monoamine oxidase inhibitor (MAOI) antidepressant, particularly tranylcypromine (Parnate®). Bupropion (Wellbutrin®) may be an alternative. Electroconvulsive therapy remains a highly effective alternative. Sometimes, antipsychotic drugs may be required to contain the symptoms of a psychotic depression. Whether lithium alone helps in acute depressive episodes is unclear.

Eliminating significant mood swings is the optimal clinical response to lithium maintenance therapy. However, mood fluctuations, and even the recurrence of a major episode, do not indicate that lithium treatment has failed. Some clinicians feel that manic or depressive symptoms that occur within a few months of initiating treatment represent a relapse of the initial episode and may require medication adjustment. Patients on a maintenance lithium regimen also may develop mild manic and depressive symptoms that subside spontaneously or respond to interpersonal support or a transient increase in lithium.

Long-term studies have consistently demonstrated that lithium maintenance suppresses mood swings in most bipolar patients.[17] However, individuals must take appropriate amounts of medication consistently. In fact, failures of lithium therapy are quite frequent. Most occur during the first year of treatment and result from inadequate resolution of the initial episode, inadequate regulation of dosage, or lack of patient compliance. About 20% to 30% of patients discontinue lithium therapy on their own.

The Most Common Reasons for Stopping Medication

1. Denial of the illness.
2. Unawareness of the consequences of stopping medication.
3. Feeling no effects from the drug.
4. Intolerable adverse reactions.
5. Longing to be "high."
6. Fear of diminished competence, productivity, or creativity.[18]

By discussion of these and other concerns with the patient, the number of treatment failures attributed to poor patient compliance (three-fourths of the total) should be diminished.

B. Pharmacokinetics

Lithium is administered clinically as the citrate or carbonate. In the body, lithium circulates as a small ion with a single positive electrical charge. It is rapidly and usually fully absorbed after oral administration, but slow-release preparations also are available. Although the exact sites of absorption are unclear, blood levels after single oral doses peak in 2–4 hr (6–8 hr after delayed-release dosage), and complete absorption takes about 8 hr. In practice, the clinician should obtain serum lithium levels 12 hr after the last dose: standard recommendations are based on this reference time.

Unbound to plasma proteins, the lithium ion initially is distributed in the extracellular fluid and then enters different tissues at different rates. A dynamic equilibrium is established between plasma and cells, and at steady state—usually achieved between 3 and 8 days—the plasma lithium will accurately reflect total body lithium and can be used to monitor treatment. The apparent volume of distribution of lithium is about the same as total body water (approximately 70% of body weight), but at equilibrium different organs concentrate lithium to differing degrees. In the liver, the lithium concentration is lower than that in extracellular fluid, whereas in muscle, bone, and thyroid, it is two to four times higher, and in brain about the same as in extracellular fluid—roughly 40% of the concentration in plasma.[19]

Lithium is not metabolized. About 95% is excreted by the kidneys; small amounts appear in sweat and feces. Lithium is fully filtered and about 80% reabsorbed in the proximal tubules. Small amounts are reabsorbed in the loop of Henle, but most of the remaining is excreted in the urine. Under normal body conditions, the amount of lithium filtered and reabsorbed is constant. Lithium excretion is, therefore, proportionate to plasma concentrations, and lithium clearance is relatively constant for each patient. With a single dose of lithium, 50% is excreted within 5 to 8 hr. In patients on maintenance lithium, the half-life is about 24 hr. However, depending on the age and kidney status of patients, the half-life may vary widely (approximately 18 hr in young, healthy individuals to 36 hr in older patients).

Effective regulation of lithium depends on the sodium and fluid balance of the body. When lithium is initially administered, a cation balance among lithium, sodium, and potassium occurs; sodium and potassium excretion increases. Clinically, the balance between lithium and sodium is important for the following reasons:

1. Sodium depletion can result in marked lithium retention and possible toxicity.
2. High levels of lithium can lead to increased sodium excretion.

In cases of toxicity, the loss of body sodium obligates more lithium retention by the kidney, prolonging the toxicity. Decreased fluid intake also can diminish lithium excretion and could lead to intoxication.

C. Adverse Reactions

1. Endocrine

The long-term use of lithium may cause various endocrine abnormalities.

a. Thyroid

Lithium therapy can produce significant changes in thyroid functioning. It inhibits several steps in the process of hormone synthesis and degradation, including iodine uptake by the thyroid gland, iodination of tyrosine, the release of T_3 and T_4, the peripheral degradation of thyroid hormones, and the stimulating effects of TSH (thyroid-stimulating hormone). Patients can usually compensate for the initial decrease in thyroid hormones. However, approximately 5% of patients treated with lithium develop signs of hypothyroidism; another 3% develop a diffuse nontender goiter.[20] This latter effect usually occurs in women between 5 months and 2 years after commencing treatment. In addition, about 30% of patients develop consistently increased levels of TSH indicated by elevated hormone or abnormal clinical signs and symptoms.[21] Each of these thyroid abnormalities typically remits when lithium is stopped.

Clinically, hypothyroidism can be confused with a depressive disorder. Therefore, the practitioner should always search for the clinical manifestation and laboratory findings indicative of thyroid disease.[22]

Since lithium affects thyroid function so profoundly, what measures should the clinician take during the various stages of maintenance therapy?

Before Starting Treatment

1. Screen for signs and symptoms of preexisting thyroid disease.
2. Obtain laboratory tests, including TSH (thyroid-stimulating hormone), thyroxine [T_4 (Synthroid® and others)], T_4I (free thyroxine index), and T_3RU (resin uptake).

During Treatment

1. Obtain a TSH measurement every 6 months during the first year and yearly thereafter.
2. Follow up any elevated TSH with a complete battery of thyroid tests.
3. Maintain a clinical index of suspicion continuously.

4. Initiate thyroid replacement therapy if hypothyroidism occurs (discontinuing lithium is not necessary).

Although incompletely studied, there is clinical evidence that even in the absence of diminished peripheral thyroid hormones, treating elevated TSH (diminished thyroid reserve) with thyroid supplementation may avoid the development of hypothyroidism and also might allow greater stability of mood.

In rare instances, lithium appears to trigger cases of hyperthyroidism. A high index of suspicion is the best guideline, with appropriate testing of thyroid function to confirm the clinician's impressions.

b. Diabetes Mellitus

Lithium has many and incompletely understood effects on carbohydrate metabolism, including altered glucose tolerance. Rarely, patients may develop diabetes mellitus.[23] In these cases, the clinician must decide whether or not the risk from diabetes outweighs the morbidity of the individual's affective disorder.

Generally, diabetic patients tolerate lithium well. Sometimes, however, this agent causes altered sensitivity to insulin, requiring a dosage readjustment.

2. Renal

As noted earlier, the body almost entirely depends on the kidneys for the elimination of lithium. In the patient who has virtually no glomerular filtration, such as one totally dependent on dialysis, a dose of lithium will remain within the body from one dialysis until the next. By extension, patients with renal filtration below normal will have a proportionately longer than normal elimination half-life of lithium.

Aside from the kidney's being the primary organ of lithium excretion, it is also the site of some of the most common side effects, polydipsia and polyuria, which affect roughly 60% of patients taking lithium. Although these effects usually are reversible, in occasional patients they may persist after lithium has been discontinued, suggesting some degree of structural impairment. In addition, there has been concern since the mid-1970s that lithium causes renal structural damage. There have also been rare reports of nephrotic syndrome.

a. Decreased Concentration

Presumably by counteracting the effects of antidiuretic hormone (ADH or vasopressin) on the renal tubule, lithium decreases the kidney's maximal concentrating capacity, analogous to nephrogenic diabetes insipidus. The obligatory water loss goes up, often to 3 liters or more per day. Such patients must not restrict their water intake, for they will be subject to dehydration with attendant lithium intoxication and potential renal shutdown. Instead, they must drink as much fluid as it takes to maintain hydration—a situation that at times can lead to a life of commuting between water cooler and water closet.

For some patients a lower dose of lithium may decrease the extent of polyuria, but lower levels can bring reduced protection against relapse. There also is evidence that a once-a-day dosing schedule of lithium can reduce polyuria.[24] The use of diuretics (paradoxical though it sounds, with the mechanism poorly understood) and a diet low in protein and solutes are alternative treatment options.[25]

b. Structural Changes and Glomerular Filtration Alterations

Beginning in 1977, reports began to appear of structural damage in patients treated with lithium. The incidence of this finding varied depending on patients studied, and so did the nature of the lesions and relation to therapy. By and large, the changes described were of a tuberointerstitial nature; although similar changes have been described in patients with bipolar illness before treatment with lithium, they probably occur at a greater frequency in patients who have been so treated. Mellerup et al. estimate a 10% incidence of structural changes in lithium-treated patients.[26] Unquestionably, lithium intoxication can lead to acute functional and structural alterations.

Fortunately, the most vital kidney function, glomerular filtration, is minimally impaired by long-term lithium treatment. Even patients treated for several decades appear to have negligible decreases in glomerular filtration rates beyond the decline expected with age. Additionally, to date no case of renal failure has been unequivocally linked to lithium. Thus, the "kidney scare" of the late 1970s and early 1980s has turned out to be a tempest in a teapot.

c. Renal Function

Earlier recommendations for extensive testing of kidney function—including 24-hr urine collections and concentrating capacity tests—have proven unnecessary.[22] Instead, pretreatment assessment should include a history and necessary laboratory testing to rule out renal pathology or, if present, to assess its nature and degree. First, the clinicians should complete personal and family medical histories and review of systems. Laboratory tests should include a serum creatinine and blood urea nitrogen (BUN) along with routine urinalysis.

Once lithium therapy has been initiated, a serum creatinine determination every 6 months should suffice. Beware, however, of progressive rises in serum creatinine, even within the normal range, as these may herald a decline in glomerular filtration rate. Similarly, a rise in serum lithium without an increase in dose may reflect decreasing glomerular filtration. Patients on maintenance therapy should be questioned periodically about changes in urine volume, and their families should be counseled about the importance of adequate hydration—especially under conditions of heat, increased physical activity, gastroenteritis, or other situations that may alter fluid and electrolyte balance.

d. Other Kidney Effects

Rare cases of nephrotic syndrome have been described in patients treated with lithium. Although this condition is usually detected by asymptomatic proteinuria on routine urinalysis, occasional patients develop edema. Almost all cases have been reversible after discontinuing lithium, although at least one patient required continued steroid therapy.[27]

Because lithium and sodium are handled similarly in renal reabsorption mechanisms, sodium depletion can lead to increased lithium reabsorption and, consequently, elevated blood levels. Thus, using salt-lowering diuretics or instituting a restricted-sodium diet can raise blood lithium levels in a lithium-treated patient. Limited research on strenuous exercise suggests that, contrary to initial expectations, blood levels of lithium are more likely to decrease than to increase, presumably as a result of a relatively greater loss of lithium than sodium in sweat.

3. Hematological

Mild to moderate elevations of the white blood cell count (typically 12,000 to 15,000 cells/mm³) occur commonly with both acute and chronic administration of lithium.[28] Neutrophils are most affected. This effect is clinically benign and reversible and can develop at various doses and at any stage of treatment.

4. Cardiovascular

Lithium has significant effects on the heart. However, adverse reactions in clinical practice are rare. Perhaps because of its effects on potassium balance, lithium causes electrocardiogram (EKG) T-wave flattening or inversion in 20% or more of individuals.[29] Sometimes with normal doses, U-waves also appear. Since these are common changes, the clinician should obtain a baseline EKG prior to starting therapy.

Adverse cardiac reactions are rare even in patients with known heart disease. **Several types of conduction problems have appeared during lithium therapy,** including first-degree atrioventricular block, irregular or slowed sinus node rhythms (particularly in elderly patients), and increased numbers of ventricular premature contractions.[30,31] Some individuals have also developed severe congestive heart failure, cardiomyopathy, and ventricular tachycardia. However, the degree to which lithium therapy contributed to the development of these disorders was unclear.

During toxic states, some patients have developed persistent ventricular tachycardia, atrial fibrillation, advanced A–V block, and vascular collapse.[32]

Because Cardiac Reactions May Produce Serious Morbidity in an Occasional Patient

1. Take a careful cardiac history.
2. Obtain a baseline EKG.
3. Follow the patient's cardiac status at least by monitoring clinical signs and symptoms.
4. Fully evaluate any conduction abnormalities that occur during

therapy.
5. Consider lowering dosage in patients with conduction changes if continued lithium therapy is a high priority. Similarly, in individuals with irregular or slowed sinus rhythm, consider a pacemaker if other measures fail to provide cardiac stability.
6. Discontinue lithium in those who develop heart failure, ventricular tachycardia, or cardiomyopathy.

5. Cutaneous

Lithium occasionally causes bothersome skin problems. A pruritic, maculopapular rash may appear during the first month of therapy. This allergic response has a variable course that sometimes can progress to a serious dermatitis. Other observed dermatological effects include acneiform lesions, hyperkeratotic papules, xerosis cutis, cutaneous ulcers, thinning and drying of scalp hair, exacerbation or appearance of psoriasis, chronic folliculitis, and anesthesia of the skin. Only the more serious responses require the discontinuation of lithium. Many of these reactions respond to conservative treatment, remit over time, or may not reappear if lithium is stopped and then restarted.

6. Gastrointestinal

Gastrointestinal (GI) reactions occur frequently during initial lithium therapy or as a result of dosage increase and include **gastric irritation, anorexia, abdominal cramps, nausea, vomiting, and diarrhea.** Gastrointestinal symptoms sometimes herald impending toxicity. Individuals, however, vary greatly in their sensitivity to changes in dosage or preparation and in the persistence of their symptoms.

To Cope With Gastrointestinal Reactions

1. Check the serum lithium level of any patient who develops gastrointestinal symptoms after the start of or suddenly during treatment.
2. Treat gastric irritation by administering lithium after meals, lowering the dose, further dividing dosages, or using a sustained-release preparation.
3. Treat other GI symptoms by changing or further dividing the dose or switching to another preparation (tablets, capsules, elixir, or sustained release).
4. If a patient develops a viral gastroenteritis, monitor lithium blood levels closely and be prepared to discontinue lithium temporarily.

7. Central Nervous System and Neuromuscular

Lithium commonly produces central nervous system (CNS) and neuromuscular effects at therapeutic doses. The most frequent reactions occur at the beginning of treatment and include **mental dullness, decreased memory and concentration, headache, fatigue and lethargy, muscle weakness, and tremor.** These symptoms most often remit quickly and usually do not require changing the dosage.

A fine rapid hand tremor accompanies lithium therapy in 15–70% of patients, depending on dose. It is an action or postural tremor (made worse by sustained posture or purposeful movements) and may persist from the beginning of treatment, appear at any point, or recur.[33] However, it usually appears when therapy starts and decreases over time. Central nervous system stimulants (including caffeine), anxiety, muscle tension, and, occasionally, antiparkinson drugs worsen the tremor, whereas sedative–hypnotic agents may improve it. Patients with other forms of action tremor (i.e., familial, idiopathic, senile) develop lithium-induced tremors more frequently. Rarely, tremulousness may involve the upper extremities, face, or eyelids. Lowering the dose of lithium can provide relief. However, if the tremor still persist, **propranolol (Inderal® and others) 20 to 160 mg per day,** may help.[33] An alternative might be potassium supplementation.

Lithium also causes various other troublesome symptoms, including dysarthria, vertigo, ataxia, tinnitus, nystagmus, autonomic slowing of bladder and bowel function, visual distortion, muscle irritability (e.g., twitching, fasciculations, facial spasm, increased tendon reflexes, clonus, and choreoathetosis), and even a full organic brain syndrome. When any such signs appear abruptly in the course of lithium therapy, they may herald toxicity and should prompt a blood-level assay.

Extrapyramidal reactions, although sometimes occurring without any apparent predisposition, are more likely in the elderly, those taking antipsychotic medication, and those with toxic blood levels of lithium. Cogwheeling and more generalized muscle rigidity are more common symptoms.[34] Sometimes these reactions appear during the first months of therapy at therapeutic levels of lithium. However, they more often accompany serum lithium levels greater than 1.5 mEq/liter. With levels above 3.0 mEq/liter, patients show profound neurotoxic effects that can progress to seizures and incontinence, stupor, coma, brain damage, and death.

In addition to the adverse reactions already mentioned, rare cases of irreversible brain damage with usual doses of lithium may occur (see Chapter 4). This occurs most often in patients who are elderly, schizophrenic, or impaired neurologically.

To Minimize CNS Effects

1. Take an appropriate history for schizophrenia, possible central nervous system disease, tremors, and antipsychotic medication usage.
2. Note and carefully follow any mild, initial neurological symp-

toms (i.e., mental dullness, poor concentration, weakness, lethargy, or tremors).
3. Discontinue lithium and obtain a serum level in patients who develop any significant neurological symptoms (particularly an organic brain syndrome).
4. Treat hand tremors by reassurance, lowering of dose, or administration of 20 to 160 mg of propranolol per day.

8. Ocular

Lithium therapy rarely causes adverse reactions involving the eyes. Sometimes tearing, itching, burning, or blurring may occur during the first few weeks of treatment. With decongestant eyedrops and time, most of these reactions abate. Occasionally, more significant reactions develop, including exophthalmos, worsening of cataracts, and two reported cases of bilateral papilledema. However, the role of lithium in these disorders remains unclear.

9. Weight Gain

Reports vary widely about weight gain during lithium therapy. Twenty to 60% of patients gain more than 20 lb.[35] Although the etiology of this effect is unclear, many causes have been hypothesized, including altered carbohydrate and lipid metabolism, increased fluid intake and retention, improved appetite with the resolution of affective episodes, diminished thyroid functioning, and increased intake of high-caloric fluids. Since this adverse reaction is frequent, patients should be warned and instructed about possible dietary restrictions. For those who gain weight, limiting calories usually results in appropriate weight loss.

10. Other Adverse Reactions

These include metallic taste and pretibial edema.

11. Pregnancy

Various congenital abnormalities have been reported in babies exposed to lithium in utero, particularly anomalies of the heart and great vessels (Ebstein's anomaly). Therefore, lithium therapy should be avoided during pregnancy, particularly the first trimester. In addition, infants born to mothers taking lithium occasionally develop temporary adverse reactions antepartum. Also, lithium's concentration in breast milk may significantly affect nursing infants (see Chapter 10).

D. Drug Interactions

Lithium and other drugs, particularly the diuretics, have various synergistic effects (Table 1). Thiazide diuretics commonly cause increased levels of lithium by decreasing clearance. This reaction can occur quickly, resulting in significant increases in blood levels and, finally, in toxicity. The potassium-retaining diuretics, spironolactone (Aldactone® and others), ethacrynic acid (Edecrin®), and triamterine (Dyrenium® and others), may also cause moderate increases in the lithium level over time, as may amiloride (Midamor® and others). Some other diuretics, notably osmotic drugs and carbonic anhydrase inhibitors, result in decreasing lithium levels as a result of increased excretion. The clinician should be especially cautious if he administers lithium and methyldopa (Aldomet® and others) together; they may cause both hypertension and toxic symptoms at normal blood levels.

Antipsychotic agents are commonly prescribed with lithium for acutely disorganized patients. This practice is usually safe. However, some individuals have developed acute neurotoxicity and/or permanent brain damage (described most

TABLE 1. Drug Interactions

Increased levels of lithium
   Amiloride (Midamor®)
   Thiazide diuretics
   Ethacrynic acid (Edecrin®)
   Triamterene (Dyazide®, Dyrenium®)
   Spironolactone (Aldactazide®, Aldactone®, and others)
   Phenylbutazone (Butazolidin® and others)
   Indomethacin (Indocin®)
   Ibuprofen (Motrin®)
   Mefenamic acid (Ponstel®)
   Naproxen (Naprosen®)
   Sulindac (Clinoril®)
   Zomepirac (Zomax®)
   Antipsychotic agents?
   ACE inhibitors
Decreased levels of lithium
   Osmotic diuretics
   Carbonic anhydrase inhibitors
   Caffeine
   Theophylline (Tedral® and others)
   Theobromine diuretic (Athemol®)
Increased adverse reactions
   Cardiovascular toxicity: hydroxyzine (Atarax®, Vistaril®, and others)
   Somnambulism: antipsychotic agents
   Toxic symptoms with normal blood levels: methyldopa (Aldomet® and others)
   Hypertension: methyldopa (Aldomet® and others)

often with haloperidol). These reactions occur rarely and may reflect lithium toxicity only or a neuroleptic malignant syndrome (see Chapter 4). Whether or not a combined toxic effect occurs is still unclear.

Other adverse drug reactions include increased cardiovascular toxicity with hydroxyzine (Atarax®, Vistaril®, and others), somnambulism with antipsychotic agents, increased levels of lithium with prostaglandin synthetase inhibitors [phenylbutazone (Butazolidin® and others), indomethacin (Indocin® and others), ibuprofen (Motrin® and others), mefenamic acid (Ponstel®), naproxen (Naprosyn®), sulindac (Clinoril®), and zomepirac (no longer available)], and increased excretion of lithium (resulting in decreased blood levels) with xanthines [e.g., caffeine, theophylline (Tedral® and others), and theobromine (Athenol® and others)].

When bipolar patients are treated with ECT, it is best to discontinue lithium therapy for the duration of the treatment series. Although controversy remains, there is some evidence that coadministering lithium and ECT leads to an increased incidence of confusional states. In patients receiving neuromuscular blockers for ECT or surgical procedures, lithium might sometimes prolong the recovery period for muscle function.

To Minimize Possible Drug Synergisms

1. Avoid polypharmacy whenever possible.
2. Lower the dose of lithium in patients starting on drugs known to elevate blood lithium levels.
3. Monitor serum lithium levels more often in patients on osmotic diuretics or a carbonic anhydrase inhibitor.
4. Administer furosemide (Lasix® and others) as a diuretic if it fits the patient's medical needs (this drug may not affect blood levels).
5. Discontinue lithium and obtain a blood level in any patient who develops symptoms of an organic brain syndrome while taking antipsychotic medication.
6. Avoid abruptly discontinuing caffeinated beverages (e.g., coffee, tea, colas) and monitor blood levels more closely in patients who markedly change their consumption of caffeine.
7. Avoid the use of methyldopa and hydroxyzine with lithium.

E. Toxicity

As noted previously, lithium causes a wide range of unwanted effects. Many appear at the onset of therapy, but patients soon become tolerant to them, and they cease to be a problem. As in all clinical pharmacology, patients' sensitivity to any side effect at a given serum lithium concentration will vary. For some patients, a concentration necessary for an effective result will produce unacceptable adverse reactions. For most patients, side effects are tolerable at therapeutic levels, but the appearance or reappearance of certain problems—such as tremor, gastrointestinal disturbances, or increased urination—could herald an increase in blood level to toxic concentrations.

Toxic blood levels of lithium can occur acutely by accidental or intentional overdose or chronically through the use of excessive doses, decreased renal function, drug interactions, diminished salt intake, or other fluid and electrolyte abnormalities. Some patients will develop toxic symptoms and signs at doses usually considered therapeutic, whereas other patients will tolerate fairly high plasma concentrations without ill effect. Most probably, these individual variations reflect differences in brain levels achieved per given serum concentration, but they also may reflect differences in end-organ sensitivities.

Patients should not have serum levels above 1.5 mEq/liter, and certainly alarm bells should sound at 2.0 and above. Twelve-hour levels above 3.0 should be considered an emergency and dealt with promptly.

Gastrointestinal symptoms may initially appear, followed or accompanied by central nervous system depression. This may include somnolence, sluggishness, the various hallmarks of an organic brain syndrome, dysarthria, seizures, choreoathetoid movements, increased muscle tone, and increased deep tendon reflexes. Cardiovascular collapse marked by lowered blood pressure, irregular cardiac rhythm, decreased urine output, and conduction abnormalities (and EKG changes) may be life threatening.

Acute intoxication causes significant CNS depression. Patients develop pyramidal tract signs and impaired consciousness or coma. Individuals who take too much medication chronically gradually develop CNS impairment. Sluggishness and drowsiness may progress over a period of days. Often, gastrointestinal symptoms, slurred speech, ataxia, and coarse tremor accompany these changes. If the initial signs of chronic intoxication are overlooked, a more florid CNS syndrome may develop, most often manifested by hyperpyrexia and stupor or coma. It may also include neurological asymmetries, nystagmus, stiff neck, and hyperextension of the extremities.

The possibility of lasting neurological impairment as a result of lithium intoxication most probably is a product of the maximal blood concentration and the duration of toxic levels. Various permanent neurological sequelae are possible, most commonly involving cerebellar and cognitive dysfunctions.[36]

Lithium toxicity is generally managed by supportive measures. If the toxicity occurs as part of an acute medication regimen or minor overdose, blood levels never exceed 3.0 mEq/liter, and kidney function is intact, careful observation usually suffices. Since the lithium is excreted rapidly, the syndrome most often abates within a few days. However, in a large or chronic overdose, very large total body lithium stores may accumulate. In these cases, the patient often suffers persistent (several days or longer) life-threatening CNS depression and cardiovascular impairment.

Reasons for Obtaining a Lithium Level

1. Symptoms and signs of toxicity, including nausea, vomiting, diarrhea, increasing fatigue, mental dullness, tremor, or ataxia.
2. Signs of mania, including euphoria, irritability, hyperactivity, inappropriate actions, inability to complete tasks, persistent insomnia.
3. Symptoms and signs of depression, including lowered mood, self-deprecating behavior, guilt, decreased activity, insomnia, weight loss, poor appetite, suicidal thoughts.
4. Changes in dose initiated by the clinician or patient. (Obtain level about 5 days after dosage is changed to allow steady state.)
5. The development of medical disease, particularly those that cause fever or diarrhea.
6. Significant increases in sweating (e.g., move or visit to a warmer climate or marked increase in exercise).
7. The institution of steroids, diuretics, antipsychotic drugs, sodium bicarbonate, ACE inhibitors, or nonsteroidal antiinflammatory drugs.
8. Question of pregnancy.
9. Change in salt intake or diet.
10. Signs of thyroid deficiency.

Management of Serious Toxic States

1. Rapidly assess (including clinical signs and symptoms, serum lithium levels, electrolytes, and EKG), monitor vital signs, and make an accurate diagnosis.
2. Discontinue lithium.
3. Support vital functions and monitor cardiac status.
4. Limit absorption
   a. If alert, provide an emetic.
   b. If obtunded, intubate and suction nasogastrically (prolonged suction may be helpful, since lithium levels in gastric fluid may remain high for days).
5. Prevent infection in comatose patients by body rotation and pulmonary toilet.
6. In all cases, vigorously hydrate (ideally 5 to 6 liters per day); monitor and balance the electrolytes.
7. In moderately severe cases
   a. Implement osmotic diuresis with urea, 20 g IV two to five

times per day, or mannitol, 50 to 100 g IV per day.
  b. Increase lithium clearance with aminophylline, 0.5 g up to
     every 6 hr, and alkalinize the urine with IV sodium lactate.
  c. Ensure adequate intake of NaCl to promote excretion of
     lithium.
8. Implement hemodialysis in the most severe cases. These are
   characterized by
  a. Serum lithium levels between 2.0 and 4.0 mEq/liter with se-
     vere clinical signs and symptoms (particularly decreasing
     urinary output and deepening CNS depression).
  b. Serum lithium levels greater than 4.0 mEq/liter. Most pa-
     tients completely recover from lithium toxicity; several may
     die; some develop permanent neurological damage.

F. Preparations and Dosage

In the United States, lithium comes mostly as a carbonate salt in oral preparations of 300 mg (8.12 mEq). This dosage and composition usually meet most
clinical needs (see Table 2). However, a few patients exhibit sensitivities to various
preparations.

In Individuals Who Develop Adverse Reactions

1. Provide once-a-day regimen (or even alternate-day regimens in
   elderly patients).
2. Further divide doses by administering portions of scored tablets.
3. Switch between tablets and capsules (or vice versa).
4. Administer lithium citrate syrup (to try a different salt) or give
   smaller doses.
5. Give slow-release tablets to minimize serum level fluctuations.

TABLE 2. Lithium Preparations and Dosages

Lithium carbonate 300-mg capsules
    [Eskalith® (S, K & F), Lithonate® (Reid-Rowell), PFI-LITH® (Pfizer),
        Lithium carbonate (Roxane)]
Lithium carbonate 300-mg tablets
    [Lithane® (Miler), Lithium Carbonate (Roxane), Lithotabs® (Reid-Rowell)]
Lithium carbonate 450-mg slow-release tablets
    Eskalith CR® (S, K & F)
Lithium carbonate 300-mg slow-release tablets
    [Lithobid® (CIBA)]
Lithium citrate concentrate of 8 mEq/5 ml
    [Cibalith-S® (CIBA), Lithium Citrate Syrup (Roxane)]

V. CLINICAL APPLICATIONS

A. Diagnostic Evaluation

When employing lithium for an acute episode, the clinician should have at the least a working diagnosis compatible with this therapy. As noted earlier, the diagnoses of bipolar, manic, and schizoaffective disorder clearly justify lithium treatment during the acute phase, whether lithium is used alone or in conjunction with a neuroleptic. At times, the physician may prescribe lithium on a trial-and-error basis for another psychiatric indication, usually based on the patient's insufficient response to another treatment. At other times, the overlapping symptoms of several possible psychiatric diagnoses—for example, acute mania versus schizophrenia—will lead the physician to try lithium on an empirical basis.

In addition to making a primary diagnosis, the practitioner should familiarize himself with the factors that help predict outcome. He also should assess current stressors and precipitating events; psychodynamic, characterologic, and interpersonal issues; and ego strengths and environmental resources. The development of an initial rapport that contributes to a growing therapeutic relationship occurs simultaneously with this information-gathering process.

If the diagnosis of bipolar disorder is unclear, the clinician might justifiably hesitate before recommending long-term maintenance lithium therapy. An acute episode with manic features could also be a brief reactive psychosis, a drug-precipitated reaction, the beginning of schizophrenic illness, or a manifestation of a medical or neurologic disorder. Especially when this is a first episode in a relatively young patient without a family history of mood disorders, the practitioner may elect to follow the patient carefully over time to chart out the course of future episodes, if any. Contacts should be regular; psychosocial, medical, and neurologic data gathering open and ongoing; and the physician available between visits, should intercurrent symptoms arise.

When a bipolar diagnosis is clear cut, however, maintenance lithium therapy might be considered from the very first episode, especially if it is a manic one. Although the course of episodes in bipolar disorder varies from patient to patient, episodes tend to become more frequent and severe over time, recurrences are the rule, and it is a lifetime disorder. Furthermore, each episode carries a risk of mortality and a substantial likelihood of psychosocial and economic morbidity. Worse, there is some suggestion that the occurrence of each episode might actually enhance the risk of future episodes, analogous to the "kindling" model of epileptic seizures. All of these facts and hypotheses add up to a strong argument for early maintenance treatment in the course of a bipolar disorder, once the diagnosis is apparent.

Factors Predicting Lithium Responsiveness

1. Positive response
   a. Cyclothymic personality features

b. Family history of mania
c. Response of family member to lithium
d. Prior manic episode
e. Onset of illness with mania
f. Euphoria and grandiose delusions during acute manic epi-
    sode
g. Diagnosis of primary mood disorder
2. Negative response
a. Rapid cycling (more than four episodes/year)

B. Medical Evaluation

The clinician should complete a medical history, including a review of organ systems and physical examination, before starting lithium therapy. The patient should be screened for pregnancy, thyroid disease, epilepsy, renal disease, cardiovascular disease, and evidence of brain damage. If the patient has any of these conditions, lithium should be administered more cautiously.

It bears emphasis that a first onset of mania after age 40 should prompt a thorough medical and neurologic differential diagnosis. Primary mania seldom appears *de novo* in midlife or later, but a wide range of brain insults can precipitate secondary mania, often indistinguishable by behavior alone from a primary mood disorder. At a minimum, the physician should ensure a thorough physical and neurologic examination, routine laboratory and neuropsychological testing, and an electroencephalogram (EEG) and head computerized tomography (CT). When a specific and treatable condition is found, it should be addressed specifically, and the manic syndrome should be treated on a symptomatic basis: often antipsychotic drugs will be satisfactory. If, however, there is no specific treatment for the underlying organic condition, a psychiatrist may treat the manic symptoms much as he would primary mania, i.e., with lithium, neuroleptics, and, if necessary, other alternatives that will be discussed later, particularly anticonvulsants.

Appropriate Medical Screening

1. Base line complete blood count
2. Base line EKG
3. Urinalysis, BUN, electrolytes
4. Serum creatinine
5. Serum TSH, T_4, T_3RU (or other measure of binding globulin), and free T_4 (measured or calculated)

C. Initiating Treatment

The therapeutic relationship provides a foundation for all drug therapies. In acute situations, a caretaker may use his position to establish quick rapport and, when necessary, to confront or set limits. His initial approach should include various crisis intervention techniques, such as mobilization of environmental resources. Involvement of persons close to the patient generally increases medication compliance. However, maintenance therapy requires the formation of a trusting collaboration between clinician and patient. With many patients, open discussion about the illness sets the tone for the ongoing dialogue. Most individuals have some trouble accepting their illness and its possible ramifications. Therefore, the clinician should try to understand how mood episodes have impact on the patient's life. He should explore how each episode has interfered with the patient's relationships and major roles.

During the early phase of treatment, patients with bipolar illness most often want to know:

1. Where did I get it?
2. Is it transmitted to my children?
3. Is my problem permanent?
4. What is the treatment?
5. What are the risks of drug therapy?
6. How effective is therapy?

These queries frequently encompass more fundamental, but less obvious, concerns about control, dependency, personal defectiveness, and "lovableness" in personal relationships. Therefore, the clinician should try to clarify both the explicit and implicit concerns and requests before providing answers that might close off continued discussion. No matter what course is taken, however, discussion of the potentially inherited nature of bipolar illness and its possible course should be coupled closely with explanations about the effectiveness of lithium maintenance and the importance of the therapeutic relationship. Establishing an open relationship may take many months. The clinician should encourage the patient to become an active partner in maintenance therapy. For example, the patient and the family members should be sensitized to emerging hypomanic and depressive symptoms. The clinician, patient, and family together should delineate a list of symptoms that are early manifestations of the patient's affective illness and require immediate clinical contact (such as sleeplessness, spending extra money, or anorexia). Also, the patient and family should be enlisted to detect medical conditions that might affect the lithium level and signs and symptoms of toxicity.

The discussion of these aspects of treatment often stimulates other questions, including:

1. How long will treatment last?
2. How will we know when to stop lithium?
3. What are the risks of treatment?
4. How will I feel on the medicine?
5. Does lithium interact with other drugs?

Again, these questions should be answered only after determining what they mean to the patient and how he will use the information. In addition, the clinician has a

TABLE 3. Adverse Reactions to Lithium

Initial	Potentially indicative of toxicity
Nausea and vomiting	(*cont.*)
Diarrhea	Confusion
Lethargy	Vertigo
Drowsiness	Coarse tremor and twitching
Muscle weakness	Muscle weakness
Fine tremor	Choreoathetosis
Increased thirst (polydipsia)	Persistent but benign
Increased urination (polyuria)	Increased thirst
Potentially indicative of toxicity	Increased urination
Nausea, vomiting, diarrhea	Fine tremor
Drowsiness and mental dullness	Weight gain
Slurred speech	Edema

human and legal responsibility to discuss adverse reactions and potential toxicity (see Table 3), alternative treatments, limitations of maintenance therapy, and long-term prognosis. Although medication-free periods may be tried, there is a real risk of recurrences, and no one seems to "lose" a mood disorder over time; if anything, they worsen.

D. Bipolar Disorder

1. Acute Manic Episode

Hypomanic symptoms can often be treated in an outpatient setting, sometimes by psychosocial support, appropriate family intervention, and environmental manipulations, with lithium as the sole medication. As the patient's excitement, poor judgment, and embarrassing or dangerous behaviors mount, and as frank mania with psychotic features develops, hospitalization will usually be required to protect the patient and initiate, restart, or adjust the medication regimen. In such cases, it is probable that the patient will require an antipsychotic drug in addition to lithium, even though, as we mentioned earlier, this might increase the risk of neurotoxicity in occasional patients. **Haloperidol** works well for acute behavioral excitement and the disorganized thinking of manic psychosis, usually takes effect promptly, and can be parenterally administered. However, all antipsychotic drugs can probably contain acute psychotic states.

As in the treatment of schizophrenic psychoses (see Chapter 4), the usual effective daily dose for haloperidol is between 5 and 15 mg. Coadministered lithium probably will act synergistically to normalize a patient's mood, but not for a week or more. What should the clinician do if a patient is highly excited and possibly dangerous? Once in vogue clinically, the concept of "rapid neuroleptization"—that is, the use of high, frequent, and parenteral doses of a high-potency antipsychotic drug—is no longer in favor. These ultrahigh neuroleptic dose regimens were shown to be no more effective, but they produced a higher rate of adverse effects, especially extrapyramidal reactions and possibly neuroleptic malignant syndrome.

For the highly excited manic patient, then, clinicians turn more now to adjunctive benzodiazepine medications. Some have claimed clonazepam (Klonopin®) to be an effective antimanic drug in its own right, but scientific data are limited and inconclusive, and our own clinical impression has been that clonazepam functions more as a highly sedating agent and, therefore, a good adjunct to a neuroleptic. Lorazepam (Ativan® and others) has the advantage of availability in parenteral form, which is distinctive in its rapid and predictable absorption when administered intramuscularly. Whichever benzodiazepine is chosen, it will allow enhanced behavioral control of the acutely excited manic patients without the risks of high antipsychotic doses. At the same time, benzodiazepines raise rather than lower the seizure threshold, do not add cardiovascular complications, treat rather than exacerbate extrapyramidal reactions, and will not engender the neuroleptic malignant syndrome.

For the acutely excited manic patient, physical restraints and ECT, preferably bilateral, are available. Inpatient staff should be encouraged to minimize stimulation of these patients, even to the point of isolating them from other patients.

Lithium may more specifically resolve manic symptoms than antipsychotic drugs. Unfortunately, lithium usually takes 5 to 14 days to work, maybe a month or more for a full response. When a crisis approach and/or a therapeutic relationship helps to contain the patient, lithium alone may suffice. However, individuals who need behavioral control may need an antipsychotic agent coadministered with lithium. Each clinician should remember that antipsychotic drugs suppress mania but leave many patients internally disorganized. Therefore, the clinician may find it necessary to prescribe combined or overlapping drug regimens for a substantial period of time. Usually, the clinician determines if the lithium is having an adequate antimanic effect by gradually decreasing the antipsychotic after achieving both behavioral control and therapeutic serum levels of lithium (i.e., from 0.8 to 1.3 mEq/liter).

In general, *if the patient is stable,* start withdrawing the neuroleptic after about 5 days of lithium therapy. Tapering the drug usually requires 1 to 2 weeks. The clinician can assess whether or not manic symptoms are reappearing by carefully evaluating the patient's feelings and thoughts. Many manic patients, in fact, will report racing thoughts even when they appear calm.

Initially, practitioners should prescribe lithium in divided doses, follow the serum levels, and then increase the amount as necessary. Although various formulas have been proposed by which steady-state lithium plasma levels can be predicted based on the level after a single lithium dose, in practice this is not necessary. Instead, the physician may begin therapy with a modest dose, such as 300 mg of lithium carbonate b.i.d., and then measure the plasma level after a week's treatment. Of course, in an elderly, frail, medically ill, or renally impaired patient, the dose can be started with as little as 300 mg daily (or even every other day). If the plasma level after a week of therapy at the initial dose is below 0.8 mEq/liter, the clinician can increase the dose by one capsule daily and check the level again after another week. If symptoms or signs suggesting toxicity occur before a week since the most recent dosage increment, a 12-hr plasma level should be checked to assure that toxic blood levels have not been reached.

The above gradual approach to initiating lithium therapy should mitigate adverse effects and make the patient's introduction to lithium a gentler experience. Also, although this approach takes more time, the slow onset of lithium's therapeutic effects usually means that it is not the critical variable in "breaking the back" of a manic episode. Having said this, there are times when a clinician may feel desperate to move speedily in controlling a florid and dangerous manic episode. If such is the case in a young, healthy, and robust patient, one may start with 900 or 1200 mg of lithium daily in divided doses and check the blood level after only 4 days. Such a strategy may lead to "overshoot" in plasma lithium level, and if any 12-hr blood level comes back above 1.5 mEq/liter, the doctor should promptly lower the lithium dose. With a more rapidly escalating regimen, medical and nursing staffs must be doubly sensitive to early signs of toxicity, in which case the next dose of lithium should be held while awaiting an immediate blood-level determination.

Many clinicians have opined that acutely manic patients require a much higher dose of lithium to achieve a given blood level than the same patients do when they have returned to euthymia. The validity of this observation remains untested scientifically, and if it is true, it is unclear whether or not it may simply be an epiphenomenon of increased motor activity. Clinically, the prudent response is to use sufficient lithium to achieve the desired blood level, but to be observant during the postmanic phase for signs of toxicity that might accompany a rise of blood lithium level into a toxic range.

After the first week, lithium levels should be determined weekly for a few weeks and then about once a month. Once steady state has been achieved, the frequency of dosing may be reduced to just once or at most twice daily, even with standard lithium carbonate preparations. When the manic episode resolves, the clinician must make a decision about maintenance therapy.

2. Hypomania

Clinicians often can begin lithium treatment in hypomanic patients without hospitalization. Lithium is relatively easy to start once the patient has been medically cleared. Give 300 mg of lithium carbonate twice or three times a day for 1 week and then draw a serum level. Inform the patient about the blood-drawing procedure and instruct him to allow 12 hr between the last dose and the serum measurement. For each dosage change, a steady-state level is achieved after about 1 week. After drawing the first serum level, the clinician can usually estimate a maintenance dose. In maintenance therapy, 0.8 to 1.0 mEq/liter is the usual effective range. When side effects are intolerable, they may diminish at reduced plasma concentrations of lithium. However, with reduced levels comes diminished protection against relapse and recurrence, particularly mania.[37] Few if any patients will experience prophylactic benefit from lithium at blood levels below 0.4 mEq/liter. During the first week of therapy, the initial administration of lithium has few risks because of rapid clearance and low total dosage. Therefore, telephone contact or short office visits usually suffice unless the patient requires additional support and reassurance. Gastrointestinal symptoms are common during initial treatment and during dosage increases. If these symptoms are severe, however, smaller doses or a different preparation may help. If the dose is changed, another serum level should be obtained a week later. Once the desired level is reached, monthly checks are adequate. The clinician should remember that hypomanic and manic persons might retain more lithium than when they are euthymic. Therefore, before the number of serum levels is diminished, the patient's mood should be stable. For fully stabilized patients, the clinician can obtain lithium levels as infrequently as every 3 months. In any patient taking lithium, and particularly in those stopping medication, the caretaker should educate the patient and relatives about the signs of incipient mania. Some patients may want to carry a wallet card indicating their condition and medication.

Lithium can provide dramatic relief for both manic and hypomanic symptoms. However, a maintenance regimen for the prevention of future episodes may require the administration of lithium for 12 to 18 months before the clinician can fully determine its effectiveness.

3. Maintenance Therapy

The practitioner must shift his stance when introducing the possibility of a maintenance regimen to a patient, who can and must have a responsible role in treatment. In fact, only the patient can control many of the vital aspects of therapy. Therefore, the clinician should openly discuss with the patient the nature of the mood illness and the clinical effects and pharmacokinetics of lithium. He should also inform the patient about possible adverse reactions, normal precautions, and the importance of monitoring the serum lithium (see Table 3). For example, the patient should be advised to maintain a stable salt intake and contact the physician when there are significant changes in salt balance [e.g., after marked sweating, diarrhea from an illness, institution of a low-sodium diet, when another medication is coadministered (particularly diuretics), or when pregnancy is planned or expected]. The patient also should be told about common adverse effects: hand tremor, gastrointestinal upset, polydipsia and polyuria, and weight gain. Patients should know that changes in caffeine intake may alter lithium levels. In addition, patients should be instructed to stop lithium and contact the physician if any of these symptoms persist or get worse or if dizziness, drowsiness, slurred speech, or ataxia appears. Sometimes we give patients a "log" to highlight important information, reinforce their role, and provide a means of following their treatment. The development of a collaborative relationship usually instills feelings of confidence and competence and improves compliance.

At least in some individuals, symptoms are precipitated by stressful events. Therefore, many patients benefit from treatment that helps to alter a negative self-image or a destructive relationship to their environment. Patients and their families should know that sleep loss can precipitate mania and they should be prepared to take appropriate steps as needed. In all instances, a trusting, constant, collaborative helping relationship is prerequisite. Moreover, formal psychotherapy may sometimes offer the patient a more self-directed means of preventing, suppressing, understanding, or resolving some of the affective symptoms.

Clinicians should intermittently evaluate thyroid and renal functioning. A consensus about appropriate tests and their timing does not exist. However, we recommend a serum creatinine determination every 6 months and a serum TSH at 6 months and then annually thereafter. Beware of gradual creatinine increases, even within the normal range, as these may indicate a gradual deterioration in glomerular filtration. Similarly, a gradual increment in lithium level despite a constant dose may reflect a similar renal deterioration. If TSH is elevated, a full thyroid panel is indicated, and together with an internist, the psychiatrist may consider thyroid supplementation. We hasten to emphasize that, despite periodic laboratory screening, the clinical index of suspicion must remain foremost as a means to detect treatment-emergent medical conditions. Routinely questioning the patient about physical symptoms, along with psychiatric and other mood changes, may prompt targeted laboratory tests outside of the schedule we recommended. Any patient—and particularly those taking maintenance medications—deserves appropriate annual physical examinations, which, depending on a patient's age, may be supplemented with an EKG or additional laboratory testing.

E. Acute Depressive Episode

If depression is relatively minor in a bipolar patient, sometimes it can be "ridden out" with continuing lithium treatment and psychosocial support. When biological intervention is necessary—based on severity of symptoms, dysfunction, and possibly threat to life—maintenance lithium still should be continued, but now an antidepressant must be considered. Tricyclics can be employed, but as always there is the danger of triggering a switch into mania, and there is some concern that tricyclics may speed up the frequency of mood cycling. Some clinicians find MAOI antidepressants, and tranylcypromine in particular, especially effective for the depressed bipolar patient. There are limited data suggesting that bupropion (Wellbutrin®) may be uniquely helpful in the depressed phase of bipolar illness and may stabilize mood rather than trigger manic episodes. The role of other new antidepressants in the treatment of bipolar depression has yet to be established. Electroconvulsive therapy is a highly effective biological antidepressant for patients insufficiently responding to drugs, incapable of tolerating them, or for whom depression is of life-threatening severity.

F. Schizoaffective Disorder

The boundaries between schizophrenia and schizoaffective disorder and between the latter and bipolar disorder are blurry at best and nonexistent at worst. Most probably, the schizoaffective label covers patients who fit in either the schizophrenic or mood-disorder category, as well as others who might have a distinct condition. Since there are no valid predictors—clinical or biological—of pharmacological response, the best strategy is empirical. With lithium as a less hazardous option than neuroleptics for long-term therapy (no tardive dyskinesia), patients with a suggestion of mood disorder should be tried on it first. In some cases, the clinician may conclude that combining lithium with an antipsychotic drug is the maintenance treatment of choice, but because of the additive risks and adverse effects, the clinical record should clearly justify this polypharmacy.

G. Other Possible Indications for Lithium

The U.S. Food and Drug Administration accepts acute mania and the prophylaxis of recurrent episodes of mania as indications for prescribing lithium. The scientific literature also supports a role for lithium in attenuating recurrent episodes of depression among bipolar patients and perhaps among some unipolar patients as well. There are also data suggesting the efficacy of lithium as an adjunct to antidepressants in treatment-resistant patients during acute episodes.

Beyond the treatment of primary disorders of mood, data supporting a role for lithium in other conditions become thin indeed, yet the number of applications proposed in the literature and used in practice is extremely broad. Some patients, for example, who meet criteria for schizophrenia appear to benefit from lithium therapy, most often as an adjunct to neuroleptics.[38] Unfortunately, clinical criteria are not particularly helpful in predicting who among a schizophrenic population is likely to respond to lithium, so lithium trials are most commonly conducted among treatment-resistant schizophrenic patients as well as in those with suggestions of a mood disorder (or schizoaffective disorder) in the patient or family. Likewise, aggressive or cyclic behavior among patients with frank brain damage or mental retardation sometimes improves on treatment with lithium, and again, empirical trials are the only sound way to know if this approach will be effective. Lithium also is among the many treatments that have been proposed for women suffering from premenstrual syndrome (late luteal phase dysphoric disorder), but the scientific difficulties in studying this condition are legendary, and to date no obviously effective therapy has emerged.

H. Rapid Cyclers

Patients who have four or more mood episodes per year have been termed "rapid cyclers" and may respond less favorably to maintenance lithium therapy. Nevertheless, such patients should be tried on lithium treatment, and at least a year's trial should be conducted at blood levels of at least 1.0 mEq/liter before concluding that lithium does not have even partial effectiveness in preventing future attacks.

When rapid cycling continues despite an adequate trial of lithium, there is evidence that it is best to discontinue or avoid antidepressants, particularly tricyclics. [Alternatives when these patients become depressed include ECT, bupropion, and possibly MAO inhibitors.] Other treatments for rapid cyclers include thyroid hormones, carbamazepine, and antipsychotic drugs, preferably in combination with lithium. Rapid cyclers should be carefully worked up for concomitant neurological and medical conditions, particularly endocrinopathies.

I. Alternatives to Lithium for Acute Mania and for Maintenance Therapy

An acute episode of mania usually will remit on dual treatment with lithium and an antipsychotic drug. As mentioned earlier, the concomitant administration of a benzodiazepine can provide necessary sedation for particularly excited patients. When mania continues unchecked, and/or the behaviors become urgent, bilateral ECT may become a treatment of choice.

It is a rare manic patient who will not improve with the above suggestions, and when faced with such a patient, the clinician would do well to seriously reconsider the entire psychobiological spectrum of diagnoses. In particular, rule out concomitant neurological and medical conditions, substance abuse, and character pathology masquerading as mania.

Alternative treatments for acute mania are not well studied or particularly satisfactory. Options mentioned include the anticonvulsants carbamazepine and valproic acid (Depakene® and others), probably better as adjuncts than as sole treatments. Valproic acid, in particular, is more likely to be effective in patients with some evidence of organic brain disease. Insufficiently studied for this indication are the antihypertensive clonidine (Catapres®), benzodiazepines alone, calcium-channel blockers such as verapamil (Calan® and others), and the amino acid L-tryptophan.

When a bipolar patient responds insufficiently to lithium maintenance therapy, treatment rapidly becomes empirical. For partial but incomplete response, adjunctive therapy with an antipsychotic agent or antidepressant may prove useful. Obviously, the former is more appropriate where manic symptoms require suppression, the latter to alleviate depression. Needless to say, antipsychotic agents carry the risk of tardive dyskinesia, which may be more problematic among patients with mood disorders, and antidepressants might be associated with an increased frequency of cycling.

All of the alternatives mentioned above for the treatment of acute mania may be considered for maintenance treatment, too. The data are sparse, however, and once again, if any of these agents is effective, it is more likely to be as an adjunct than as a sole treatment. To be sure, this is even more troublesome for patients who cannot tolerate lithium than for those who are incompletely responsive. Carbamazepine is widely believed to be as effective as lithium, but the number of patients studied in well-controlled trials are few, and clinical experience with carbamazepine has not found it to be as satisfactory as lithium.[39]

More often than not in clinical practice, bipolar patients are maintained on lithium combined with one or more other psychotropic drugs. Whether this common practice derives from thoughtful attention to patients' medication responses or reflects a lack of rigor in pharmacotherapy is unclear. In all cases clinicians should assess the need for multiple medications, justifying it in their own minds and then reporting these observations and thoughts in the patients' records. The clinician must remember that long-term treatment with antipsychotic drugs carries with it the risk of tardive movement disorders, which may be even greater in patients with mood disorders. There is also the concern that antidepressant drug therapy could speed up the cycling of bipolar patients.

VI. CONCLUSION

Although less common than recurrent major depression, bipolar illness is impressive in its manifestations, more frequently requires hospitalization, and carries considerable hazards in morbidity and mortality. Its response to biological treatment, however, is equally impressive, and for most patients with this psychiatric condition, lithium is little short of a godsend. As noted earlier, though, for patients who cannot tolerate or do not benefit fully from lithium, treatment alternatives are empirical and often less than fully satisfactory.

All patients with bipolar disorder require a stable relationship with the treating physician. This is no different from any other chronic medical illness. Many patients also require some form of psychosocial intervention to enjoy maximum mood stability over the lifetime course of this condition. Approached sensitively and knowledgeably, treating patients with bipolar disorder can be both successful and gratifying.

REFERENCES

1. Cade J. F. J.: Lithium salts in the treatment of psychotic excitement. *Med J Aust* 11:349–352, 1949.
2. Davis J. M.: Overview: Maintenance therapy in psychiatry: II. Affective disorders. *Am J Psychiatry* 133:1–13, 1976.

3. Forn J., Valdecasas F. G.: Effects of lithium on brain adenyl cyclate activity. *Biochem Pharmacol* 20:2773–2779, 1971.

4. Gjessing L. R.: Lithium citrate loading of a patient with periodic catatonia. *Acta Psychiatr Scand* 43:372–375, 1967.

5. Kline N. S., Wren J. C., Cooper T. B., et al: Evaluation of lithium therapy in chronic and periodic alcoholism. *Am J Med Sci* 268:15–22, 1974.

6. Rifkin A., Quitkin F., Carrillo R., et al: Lithium carbonate in emotionally unstable character disorder. *Arch Gen Psychiatry* 27:519–522, 1972.

7. Gottfries C. G.: The effect of lithium salts on various kinds of psychiatric disorders. *Acta Psychiatr Scand [Suppl]* 203:157–167, 1968.

8. American Psychiatric Association: *Diagnostic and Statistical Manual of Mental Disorders*, 3rd ed revised (*DSM-IIIR*). Washington, American Psychiatric Association, 1987.

9. Nurnberger J R., Roose S. P., Dunner D. L., et al: Unipolar mania: A distinct clinical entity? *Am J Psychiatry* 136:1420–1423, 1979.

10. Keller M. B., Lavori P. W., Coryell W., et al.: Differential outcome of pure manic, mixed/cycling, and pure depressive episodes in patients with bipolar illness. *JAMA* 255:3138–3142, 1986.

11. Winokur G., Clayton P. J., Reich T.: *Manic Depressive Disease.* St. Louis, C. V. Mosby, 1969.

12. Weissman M. M., Leaf P. J., Livingston B. M., et al.: The epidemiology of dysthymia in the community: Rates, risks, comorbidity and treatment. *140th Annual Meeting of the American Psychiatric Association*, Chicago, American Psychiatric Association, 1987.

13. Medical Practice Project: A State-of-the-Science Report for the Office of the Assistant Secretary for the U. S. Department of Health, Education and Welfare. Baltimore, Policy Research, 1979.

14. Fieve R. R.: The lithium clinic: A new model for the delivery of psychiatric services. *Am. J Psychiatry* 132:1018–1022, 1975.

15. Garfinkel P. E., Stancer H. C., Persad E.: A comparison of haloperidol, lithium carbonate and their combination in the treatment of mania. *J Affect Disord* 2:279–288, 1980.

16. Prien R. F., Caffey E. M. Jr, Klett C. J.: Comparison of lithium carbonate and chlorpromazine in the treatment of mania. *Arch Gen Psychiatry* 26:146–152, 1972.

17. Fieve R. R.: Overview of therapeutic and prophylactic trials with lithium in psychiatric patients, in Gershon S., Shopsin B. (eds): *Lithium. Its Role in Psychiatric Research and Treatment.* New York, Plenum Press, 1973, p 336.

18. VanPutten T.: Why do patients with manic–depressive illness stop their lithium? *Compr Psychiatry* 16:179–183, 1975.

19. Cooper T. B.: Pharmacokinetics of lithium, in Meltzer H. Y. (ed): *Psychopharmacology: The Third Generation of Progress.* New York, Raven Press, 1987, pp 1365–1376.

20. Schou M., Anderson A., Eskajaer-Jensen S., et al.: Occurrence of goiter during lithium treatment. *Br J Med* 3:710–713, 1968.

21. Emerson C. H., Dyson W. L., Utiger R. D.: Serum thyrotropin and thyroxin concentration in patients receiving lithium carbonate. *J Clin Endocrinol Metab* 36:338–346, 1973.

22. Jefferson J. W.: Lithium carbonate-induced hypothyroidism—its many faces. *JAMA* 242:271–272, 1979.

23. VanderVelde C. D., Gordon M. W.: Manic–depressive illness, diabetes mellitus and lithium carbonate treatment. *Arch Gen Psychiatry* 21:478–485, 1969.

24. Mellerup E. T., Plenge P., Rafaelson O. J.: Renal and other controversial adverse effects of lithium, in Meltzer H. Y. (ed): *Psychopharmacology: The Third Generation of Progress.* New York, Raven Press, 1987, pp 1443–1448.

25. Gelenberg A. J.: Lithium-induced polyuria: Approaches to management. *Biol Ther Psychiatry Newslett* 6:17–19, 1983.

26. Gelenberg A. J., Wojcik J. D., Falk W. E., et al: Effects of lithium on the kidney. *Acta Psychiatr Scand* 75:29–34, 1987.

27. Gelenberg A. J.: Lithium and the nephrotic syndrome. *Biol Ther Psychiatry* 12:24, 1989.

28. Shopsin B., Gershon S.: Pharmacology–toxicology of the lithium ion, in Gershon S., Shopsin B.

(eds): *Lithium: Its Role in Psychiatric Research and Treatment.* New York, Plenum Press, 1973, pp 107–146.

29. Schou M.: Electrocardiographic changes during treatment with lithium and with drugs of the imipramine type. *Acta Psychiatr Scand [Suppl]* 169:258–259, 1963.

30. Roose S. P., Nurnberger J. L., Dunner D. L., et al: Cardiac sinus node dysfunction during lithium treatment. *Am J Psychiatry* 136:804–806, 1979.

31. Jaffe C. M.: First-degree atrioventricular block during lithium carbonate treatment. *Am J Psychiatry* 134:88–89, 1977.

32. Worthley L. J. C.: Lithium toxicity and refractory cardiac arrhythmia treated with intravenous magnesium. *Anesth Intens Care* 2:357–360, 1974.

33. Jefferson J. W., Greist J. H.: Adverse reactions—neurological tremor, in Jefferson, J. W., Greist, J. H. (eds): *Primer of Lithium Therapy.* Baltimore, Williams & Wilkins, 1977, pp 139–150.

34. Ghandirian A. M., Lehman H. E.: Neurological side effects of lithium: Organic brain syndrome, seizures, extrapyramidal side effects, and EEG changes. *Compr Psychiatry* 21:327–335, 1980.

35. Vestergaard P., Amdison A., Schou M.: Clinically significant side effects of lithium treatment: A survey of 237 patients in long-term treatment. *Acta Psychiatr Scand* 62:193–200, 1980.

36. Schou M.: Long-lasting neurological sequelae after lithium intoxication. *Acta Psychiatr Scand* 70:594–602, 1984.

37. Gelenberg A. J., Kane J. M., Keller M. B., Lavori P., Rosenbaum J. F., Cole K., Lavelle J.: Comparison of standard and low serum levels of lithium for maintenance treatment of bipolar disorder. *N Engl J Med* 321:1489–1493, 1989.

38. Donaldson S. R., Gelenberg A. J., Baldessarini R. J.: The pharmacologic treatment of schizophrenia: A progress report. *Schizophrenia Bull* 9:504–527, 1983.

39. Prien R. F., Gelenberg A. J.: Alternatives to lithium for preventive treatment of bipolar disorder. *Am J Psychiatry* 146:840–848, 1989.

4

Psychoses

ALAN J. GELENBERG, M.D.

I. INTRODUCTION

The focus of this chapter is on antipsychotic agents—drugs also known as neuroleptics, antischizophrenic agents, and major tranquilizers (a misnomer). This pharmacological family has become a mainstay in the treatment of schizophrenia and other psychotic disorders as well as many nonpsychotic conditions. We deal primarily with the phenothiazines and related antipsychotic drugs but briefly discuss various experimental agents. The chapter also describes other approaches to treating psychoses and concludes by discussing future directions.

The most firmly grounded indication for using antipsychotic drugs is in the treatment of schizophrenia. These agents can suppress the more florid and acute symptoms of schizophrenic psychosis such as hallucinations, delusions, other aspects of thought disturbance, and excited, aggressive behavior. In addition, they may alleviate similar symptoms in patients with other syndromes, such as paranoid disorders, schizophreniform disorder, brief reactive psychosis, schizoaffective disorder, atypical psychosis, and psychosis associated with mood disorders such as melancholia and mania. Furthermore, antipsychotic drugs may alleviate psychotic symptoms or excited and assaultive behavior in patients with organic mental disorders, retardation, and childhood psychoses. At times, these ubiquitous chemicals have helped in the treatment of patients with severe pain syndromes and difficult personality disturbances.

The almost never-ending list of disorders treated with antipsychotic agents, which practically spans the third revised edition of the *Diagnostic and Statistical Manual of Mental Disorders (DSM-IIIR)*, far exceeds the experimental data base that should ideally bolster clinical practice. The most convincing data support the use of these drugs to treat schizophrenic patients. Some experimental evidence also exists suggesting that they are effective for other disorders such as affective psycho-

———

ALAN J. GELENBERG, M.D. • Department of Psychiatry, University of Arizona Health Sciences Center, Tucson, Arizona 85724.

ses, but for various illnesses the literature is merely anecdotal. However, even when scientific research has not kept pace with clinical practice, patients continue to require medication; therefore, this chapter discusses the most up-to-date information on the use of these compounds, blending data from scientific research with evidence from clinical experience.

II. OTHER TREATMENTS

A. Nonbiological

A mixture of interpersonal approaches has been developed to treat psychotic individuals. These therapies have been based on varied theoretical frameworks, practical considerations, and common sense. Attempts to dissect experimentally what works from what does not work and under what conditions have been hampered by the heterogeneity of disorders (e.g., schizophrenia) and by difficulties defining and standardizing therapeutic techniques.

In general, clinicians should treat acutely psychotic patients with respect, concern, empathy (tempered with appropriate distance), a low-stimulation environment, and clearly delineated rules and limits. Recommended approaches range from supportive to insight-oriented therapy as well as individual, family, and group treatment. The type of psychotherapy recommended by a physician depends on assessing the patient's needs and resources. When a patient is evaluated, especially for a more insight-oriented approach, particular attention must be given to favorable prognostic factors, prediction of transference regression, and existence of environmental supports. Supportive treatment aims at minimizing dysfunction and maximizing individual strengths by mobilizing ego and environmental resources.

Controlled studies of acute schizophrenic inpatients show that psychotherapy alone does not significantly reduce psychopathology and that drug therapy results in greater short-term improvement. Moreover, shortening the period of acute psychosis by administering drugs can mean a better long-term prognosis.[1] Psychotherapy is far more effective in discharged patients who are followed in a comprehensive aftercare program. The benefits of psychotherapy become more apparent after 1 to 2 years of ongoing treatment and if they are evaluated in terms of psychosocial functioning rather than symptom suppression.[2] In the long term, various therapeutic modalities (e.g., psychotherapy, rehabilitation measures, social casework, public health nursing) should be combined with drug therapy to diminish the likelihood of psychotic relapse and to help the patient maintain routine living patterns.

Depending on a patient's personality, internal and external resources, and the nature of the illness, using a combination of therapeutic modalities may be beneficial. One review concluded that psychotherapy with schizophrenic patients was most effective when it was combined with somatic treatment and focused on everyday issues rather than nonspecific psychological concerns.[3] Recent research suggests that, when used together with maintenance antipsychotic drug therapy, family therapies and social skills training can improve the quality of life for schizophrenic patients.[4] Family approaches geared to reducing the quantity of intense, negative emotional interactions between patients and close relatives ("expressed emotionality") appear to diminish the rate of relapse. In fact, a major reduction in the degree of such emotion-laden contacts might actually diminish the need for antipsychotic drugs. In a complementary way, it is likely that such targeted strategies as social skills training can reduce symptoms and improve interpersonal adjustment for schizophrenic patients. Many patients with mood disorders find the insights gained during psychotherapy helpful in improving their sense of well-being, their marital and family relationships, and their social functioning in general, and possibly in diminishing the impact of their mood swings. Behavioral therapy is often valuable in various psychotic conditions, usually as an adjunct to antipsychotic drugs.

B. Nondrug, Biological

1. Electroconvulsive Therapy

Electroconvulsive therapy (ECT) can play a dramatic, even life-saving, role in the treatment of psychotic depression. It also suppresses acute mania. A wealth of clinical anecdotes testifies to similarly impressive results in many cases of catatonic schizophrenia (however, not all catatonia is schizophrenia, which may cloud the issue[5]).

At the 1985 Consensus Development Conference on ECT sponsored by the National Institutes of Health and Mental Health, Dr. Joyce Small reported that ECT can play a role in treating schizophrenic disorders that are relatively acute and marked by intense affective symptoms.[6] By contrast, patients who have been ill for 5 years or more are relatively unlikely to improve. Antipsychotic drugs appear superior in efficacy to ECT for the treatment of acute schizophrenia, but the combination of drugs and ECT might be better still. Electroconvulsive therapy is clearly superior to placebo in patients who have been ill for less than 2 years. The treatment of acute schizophrenia with ECT seems to require a greater number of sessions than does the treatment of acute depression.

2. Psychosurgery

The frontal lobotomy, used to treat many schizophrenic patients earlier in the century, has largely been abandoned because of an unfavorable risk/benefit ratio. More restrictive psychosurgical procedures, such as cingulotomy, may effectively treat patients with chronic, severe depressions and intractable pain syndromes, but no substantial evidence supports the use of psychosurgery in the treatment of psychosis.[7]

3. Megavitamin Therapy

Although the original idea of using a vitamin (niacin) to treat schizophrenia was based on a rational hypothesis (the transmethylation theory), orthomolecular theories and megavitamin therapy have subsequently become more a cult than a scientifically based clinical approach. The dogma is muddled, the therapeutic approaches chaotic (using a plethora of supposedly "organic" substances), and the scientific grounding virtually nil. Although occasional cases of behavioral disturbances might reflect dietary inadequacies or toxicity (which, in an affluent society, is most likely to reflect food faddism), there is no evidence to suggest that localized cerebral vitamin deficiences or allergies produce schizophrenia or any other form of psychiatric illness.[8] Similarly, although it is conceivable that some neuropsychiatric disturbances may be alleviated by selective dietary chemicals, the approaches used by orthomolecularists and megavitamin therapists are unlikely to be helpful and may, in fact, be harmful (e.g., through the use of contaminated substances sold in "health food" establishments).[9]

III. ANTIPSYCHOTIC DRUGS

A. Introduction and Terms

The introduction of the first antipsychotic drugs—reserpine (Serpasil® and others) and the phenothiazines—in the early to mid-1950s revolutionized psychiatry. For the first time, chemotherapy offered more than sedation to acutely disturbed and psychotic patients. Ultimately, antipsychotic drugs helped to diminish the previously sharp increase in the number of hospital beds in the United States occupied by schizophrenic patients. Many of these individuals could now function in their communities with periodic admission to psychiatric units, often in general hospitals. Psychiatric wards became quieter and less violent. The subspecialty of psychopharmacology was born, and many other important classes of psychopharmaceuticals have emerged. Unfortunately, the advent of antipsychotic drugs has not eliminated the problem of chronic schizophrenia. With the passage of the Community Mental Health Centers Act in 1963, President John F. Kennedy mandated a "bold new approach" to treating the mentally ill. However, because of political and fiscal pressures, many severely ill patients were "deinstitutionalized" to nonexistent community programs. The sad result has been the re-creation of back wards in the community, with many chronic schizophrenic patients poorly medicated, hardly managed, and homeless.[10]

The rauwolfia alkaloid reserpine caused a number of problems, but the effects of the phenothiazine chlorpromazine (Thorazine® and others) were impressive. From this prototypical phenothiazine emerged a host of sister compounds with similar actions and effectiveness but with different potencies (i.e., milligram dosages) and unwanted effects. During the past three decades, various classes of antipsychotic compounds have been synthesized. They sometimes differ structurally from chlorpromazine but are pharmacologically and clinically similar.

Clinicians and researchers have labeled antipsychotic drugs with various synonyms, but the terms often have resulted in both semantic and conceptual confusion. When they were first used, these pharmaceuticals often were called "major tranquilizers." This term developed from the observation that chlorpromazine and reserpine produce somnolence and relaxation and the consequent misbelief that their major action was sedative. However, relatively nonsedating antipsychotic compounds just as effectively combat psychotic symptoms. Moreover, patients often become tolerant to the sedating effects of antipsychotic drugs but not to the antipsychotic effects themselves. Finally, because of the implication that "major tranquilizers" are on a spectrum with, but more powerful than, sedative–hypnotic and antianxiety agents (sometimes called "minor tranquilizers"), this term is best avoided.

Because antipsychotic drugs frequently produce signs of neurological dysfunction, most notably Parkinson's syndrome and other extrapyramidal reactions, the term "neuroleptic" was coined. In fact, researchers originally believed that (1) any drug effective in combating psychosis must produce extrapyramidal effects and (2) in any given patient, the induction of extrapyramidal signs indicated an optimal therapeutic dose. However, based on the following evidence, these assumptions are incorrect:

1. The piperidyl phenothiazine thioridazine (Mellaril® and others) produces a relatively low incidence of short-term extrapyramidal effects, yet is as effective as any other antipsychotic agent.
2. Clozapine (Clozaril®), a dibenzodiazepine compound, is an effective antipsychotic agent that produces few, if any, extrapyramidal effects.
3. Many patients show marked clinical improvement in psychotic symptoms without experiencing parkinsonian signs or related reactions.

Therefore, although all antipsychotic agents on the United States market are, in fact, neuroleptics, an antipsychotic drug need not be neuroleptic—a fact presenting a challenge for future research. Furthermore, **when treating individual patients, clinicians should try to avoid the emergence of neurological effects.** Recent research suggests the intriguing possibility that the first appearance of slight hypokinesia and rigidity might signal the minimally effective neuroleptic dose for the treatment of an acutely psychotic patient[11,12]; this remains to be more fully explored.

B. Effects on Behavior and the Nervous System

In animals, antipsychotic drugs inhibit conditioned avoidance behavior, suppress electrical intracranial self-stimulation, block vomiting and aggression produced by the dopamine agonist apomorphine, and produce cataleptic immobility resembling human catatonia. In early testing, researchers observed that antipsychotic drugs potentiated anesthesia and produced a state called "artificial hibernation." Chlorpromazine, the prototype phenothiazine, did not by itself induce anesthesia but rather promoted sleep and diminished interest in the environment; animals required increased stimulation or motivation to perform tasks. Antipsychotic drugs have relatively little tendency to suppress vital centers in the brainstem: coma, respiratory depression, and cardiovascular collapse are rare, even at very high doses.

In the electroencephalogram (EEG), phenothiazines and other antipsychotic chemicals produce slowing and synchronization and a decrease in arousal-induced changes—effects that are reversed by dopamine agonists. The low-potency agents (e.g., chlorpromazine) also tend to lower the seizure threshold. Clinically, this effect is particularly important in patients predisposed to seizures, such as those with epilepsy, and in individuals undergoing withdrawal from sedative–hypnotic drugs (including alcohol) (see Chapter 11).

Most antipsychotic agents (with thioridazine as an interesting exception) have antiemetic effects. They can protect against the nausea and vomiting that usually follow administration of apomorphine, presumably by blocking the latter's dopamine agonistic effects in the chemoreceptor trigger zone of the medulla.

C. Mechanism of Action

The antipsychotic drugs block dopamine receptors in various pathways within the brain, which probably accounts for their therapeutic effectiveness as well as for some of their more prominent unwanted effects. According to widely held theories, antipsychotic activity depends on the blockage of postsynaptic receptors in dopamine-mediated pathways that run from the midbrain to the limbic system (septal nucleus, the olfactory tubercle, and the amygdala) (see Fig. 1) and to the temporal and frontal lobes of the cerebral cortex. In fact, the effectiveness of antipsychotic drugs in blocking the D-2 subtype of dopamine receptors correlates with their clinical potency (i.e., usual daily doses). Presumably, tolerance to the dopamine-blocking action of antipsychotic drugs does not develop in these mesolimbic and mesocortical pathways, explaining the impression that tolerance does not develop to their antipsychotic efficacy. [Although some authors have postulated the development of tolerance to the antipsychotic effectiveness of these drugs ("tardive psychosis"),[13] at best this is rare, and most clinicians and scholars remain to be convinced that it is a valid phenomenon.]

Antipsychotic drugs also block dopamine receptors in the pathways from the substantia nigra in the midbrain to the head of the caudate nucleus in the basal ganglia (see Fig. 1). Interruption of communication in this nigrostriatal pathway is thought to account for Parkinson's syndrome—bradykinesia, rigidity, and tremor. In this neuronal network, tolerance to the dopamine-blocking action of the drugs does seem to develop. Chemically blocked dopamine receptors are initially underac-

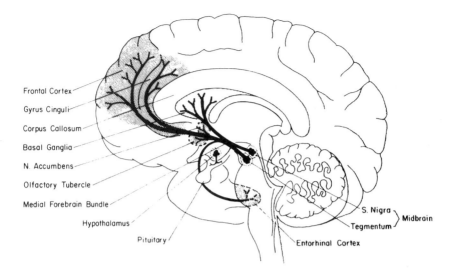

Frontal Cortex
Gyrus Cinguli
Corpus Callosum
Basal Ganglia
N. Accumbens
Olfactory Tubercle
Medial Forebrain Bundle
Hypothalamus
Pituitary
S. Nigra
Tegmentum
Midbrain
Entorhinal Cortex

FIGURE 1. Some dopamine tracts in human brain (longitudinal representation). From Lader,[63] with permission .

tive and then become normally active to overactive, developing what is analogous to denervation supersensitivity. Underactivity of the striatal dopamine receptors presumably results in parkinsonian signs; overactivity has been postulated to cause tardive dyskinesia, a syndrome of abnormal involuntary movements, although this is being called into question, as we discuss later. Acetylcholine and γ-aminobutyric acid (GABA) mediate transmission within adjacent connecting neuronal systems. Treatment of both Parkinson's syndrome and tardive dyskinesia may involve drugs purported to act on any or all of these three neurotransmitters (see Section III.G.1.a).

Researchers believe that a third important dopamine pathway, the tuberoinfundibular system, is affected by antipsychotic drugs. It projects from the arcuate nucleus of the hypothalamus to the median eminence, where it acts to inhibit (directly or indirectly) the release of prolactin from the anterior pituitary (see Fig. 1). By blocking dopamine neurotransmission in this system, antipsychotic drugs cause increased prolactin secretion and hyperprolactinemia, producing unwanted effects and possible long-term toxicity (see Section III.G.6). Although researchers have believed that tolerance did not develop to the prolactin-elevating effect of antipsychotic drugs, it now appears that at least partial tolerance may develop over a period of many months and even years. Some have suggested that the elevation of prolactin in plasma could be used as an index of the clinical effectiveness of antipsychotic drugs; however, maximum prolactin elevation occurs at much lower doses than those usually required to treat a psychotic disorder.

A possible clue to different behavior of the various major dopamine pathways emerges from recent understanding about dopamine autoreceptors. Autoreceptors are present on the body or axon of a nerve cell and respond to its own neurotransmitter by decreasing its synthesis and release. Thus, they serve as part of a single-cell feedback inhibition loop. Interestingly, dopamine autoreceptors do not occur in the mesocortical pathways but are present in the nigrostriatal and mesolimbic pathways.[14] The former does not exhibit tolerance to antipsychotic drugs, but the latter two do. Similarly, we think tolerance does not develop to the antipsychotic actions of these drugs but does to their extrapyramidal and prolactin-raising actions. Thus, it may be the presence or absence of the dopamine autoreceptors within a system that determines whether tolerance to a clinical effect will or will not occur.[14]

Must parkinsonian and hormonal effects accompany antipsychotic activity? The discovery of clozapine, which produced few (or perhaps no) extrapyramidal effects and had only a weak prolactin-elevating action, suggested that the answer is no. Moreover, neurophysiological differences among the dopamine systems can mean that pharmacological agents may act differentially at different dopamine receptors. The goal, then, is to discover new drugs that are more selective dopamine blockers.

The potent effects of antipsychotic agents on the autonomic nervous system explain many of their adverse reactions. They block α-noradrenergic receptors, which probably accounts for their hypotensive action, particularly on a postural basis. At common clinical doses, low-potency antipsychotic drugs tend to be more potent at α-adrenergic receptors and to produce a greater drop in orthostatic blood pressure. The same low-potency agents also cause more sedation, which may reflect α-adrenergic antagonism at brain receptors. Central effects on noradrenergic systems might be involved in antipsychotic activity as well.[15]

Antipsychotic agents block muscarinic acetylcholine receptors, producing other autonomic effects (see Section III.G.2). As a group, antipsychotic drugs tend to be much less potent as anticholinergic agents than the cyclic antidepressants (see Chapter 2). The most potent antimuscarinic drug among antipsychotic agents available in the United States is the phenothiazine thioridazine, which approaches the cyclics in anticholinergic activity. High-potency antipsychotic agents act relatively weakly at cholinergic receptors.

Antipsychotic drugs produce other effects on neurotransmission, although their significance is unclear. Similarly to cyclic antidepressants, they block the reuptake of norepinephrine, but this effect is probably outweighed by their tendency to antagonize norepinephrine receptors. Antipsychotic drugs also block serotonin and histamine receptors; the latter might account for some of their sedative and appetite-increasing tendencies. Other effects involve β-adrenergic receptors, GABA, serotonin, and peptide neurotransmitters.

The hypothalamus mediates various antipsychotic drug effects. In addition to effects on prolactin, antipsychotic drugs inhibit the release of growth hormone (which might have implications for their use in children). Some of their effects on autonomic activity may also result from actions within the hypothalamus. Furthermore, they impair the temperature-regulating mechanisms by making the normally homeothermic mammalian system poikilothermic (i.e., the body temperature drifts toward that of the environment), which has resulted in cases of hypo- and hyperthermia. In the extreme, antipsychotic drugs have created what has been termed the neuroleptic malignant syndrome (see Section III.G.1.d), also thought to involve the hypothalamus. Increase in appetite, yet another hypothalamic effect, often results in weight gain.

D. Classes and Chemistry

Table 1 presents a list of antipsychotic drugs grouped by chemical classes that are currently available by prescription in the United States. The first group, the phenothiazines (chlorpromazine is the prototype), are three-ring (tricyclic) molecules made up of two benzene rings linked by a sulfur and a nitrogen atom. The nitrogen atom, which is attached to a carbon side chain, determines the phenothiazine subtype. A straight chain of carbon atoms attached to the nitrogen indicates an aliphatic phenothiazine (e.g., chlorpromazine). When the amino nitrogen at the end of the chain is incorporated into a cyclic structure, the molecule is a piperidine (e.g., thioridazine). A somewhat different cyclic structure results in the piperazine phenothiazines [e.g., trifluoperazine (Stelazine® and others)]. When a piperazine phenothiazine has a terminal hydroxyl (OH) group, esterifying it with a fatty acid results in a highly fat-soluble hybrid that diffuses into the body's adipose tissue, releasing the parent phenothiazine over a period of weeks. Examples are the enanthate and decanoate esters of the piperazine phenothiazine fluphenazine (Prolixin® and others).

Replacing the nitrogen atom in the central ring with a carbon atom produces a second group of effective antipsychotic substances, the thioxanthenes. They have either aliphatic [e.g., chlorprothixene (Taractan®)] or piperazine [e.g., thiothixene (Navane® and others)] structures and are chemically and pharmacologically similar to the phenothiazines.

The butyrophenones, developed by Janssen, appear structurally quite different from the phenothiazines but are pharmacologically very similar to the piperazines. The only butyrophenone labeled for antipsychotic use in the United States is haloperidol (Haldol® and others), although the anesthetic agent droperidol (Inapsine® and others) also appears to have antipsychotic properties. Closely related to the butyrophenones are the diphenylbutylpiperidines, currently undergoing experimental investigation. This class includes pimozide (Orap®), an extremely potent blocker of the neurotransmitter dopamine, and penfluridol (not available in the United States), which has a prolonged duration of action enabling it to be administered once weekly by mouth.

The dibenzoxazepine drugs are a fourth antipsychotic group with a three-ring structure. The only member currently labeled as an antipsychotic agent in the United States is loxapine (Loxitane® and others). [The demethylation of loxapine has resulted in amoxapine (Asendin®), a compound with antidepressant properties.]

TABLE 1. Currently Available Antipsychotic Drugs[a]

Nonproprietary name	Trade name	Approximate potency[b]	Available as injectable	Chemical structure/representative agent
Phenothiazines				
Aliphatic				
Chlorpromazine	Thorazine® and others	1:1	Yes	Chlorpromazine
Triflupromazine	Vesprin®	4:1	Yes	
Piperidine				
Thioridazine	Mellaril® and others	1:1	No	Thioridazine
Mesoridazine	Serentil® and others	2:1	Yes	
Piperazine				
Trifluoperazine	Stelazine® and others	30:1	Yes	Trifluoperazine
Fluphenazine	Prolixin® and others	50–100:1[c]	Yes[d]	
Perphenazine	Trilafon® and others	10:1	Yes	Thiothixene
Prochlorperazine	Compazine® and others	7:1	Yes	
Thioxanthenes				
Chlorprothixene	Taractan® and others	1:1	Yes	Haloperidol
Thiothixene	Navane® and others	25:1	Yes	
Butyrophenone				
Haloperidol	Haldol® and others	50:1	Yes	
Diphenylbutylpiperidine				Pimozide
Pimozide	Orap®	Not established	No	
Dibenzodiazepine				
Clozapine	Clozaril®	2:1	No	Clozapine
Dibenzoxazepine				
Loxapine	Daxolin®, Loxitane®, and others	7:1	Yes	Loxapine
Dihydroindolone				
Molindone	Moban®	10:1	No	Molindone

[a]Adapted, with permission, from Gelenberg.[16]
[b]Milligram equivalence to chlorpromazine.
[c]Oral only. Potency equivalences not established for long-acting injectable forms. Usual dosage for fluphenazine decanoate is 0.5 to 2.5 ml every 3 to 5 weeks.
[d]Available both as short-acting fluphenazine HCl injection and as long-acting enanthate and decanoate esters.

A closely related compound called a dibenzodiazepine—clozapine—appears to be a nonneuroleptic antipsychotic drug (see Section III.K).

The fifth antipsychotic group, the dihydroindolines, are solely represented by molindone (Moban®). Rauwolfia alkaloids, such as reserpine and tetrabenazine (not available in the United States), are infrequently used as antipsychotic agents and are primarily of historical interest.

For a tricyclic antipsychotic to be effective, three carbon atoms must lie between the amino nitrogen and the nitrogen of the center ring. The addition of an electronegative substance to the benzene ring (e.g., Cl, SCH_3, CF_3) enhances its efficacy, whereas the piperazine group on the side chain increases its potency. As a rule, **molecules with greater milligram potency produce less sedation and hypotension but more acute extrapyramidal reactions** (see Table 2).

E. General Principles of Use

How do clinicians choose among the various antipsychotic agents? Most importantly, they should review the patient's medication history. If the patient has previously responded favorably to a given agent, then try that agent again. In fact, if a patient has been taking a drug for maintenance therapy, and an acute exacerbation occurs during a stressful period, then raising the dose of the drug may diminish the symptoms. However, if a patient has responded unfavorably to a given drug because of either lack of efficacy or unacceptable adverse effects, avoid that agent. On the other hand, if a patient has had no prior experience with antipsychotic drugs, then use the experiences of a family member as a guide. Failing all of these suggestions, the clinician is free to choose among the agents on the basis of his own experience and the spectrum of adverse effects. In general, the high-potency antipsychotic drugs are less sedating, produce less hypotension, and have less effect on the seizure threshold, fewer anticholinergic effects, less cardiovascular toxicity, less weight

TABLE 2. Spectrum of Adverse Effects Caused by Antipsychotic Drugs

Low potency	High potency
Fewer extrapyramidal reactions (especially thioridazine)	More frequent extrapyramidal reactions
More sedation, postural hypotension	Less sedation, postural hypotension
Greater effect on the seizure threshold, electrocardiogram (especially thioridazine)	Less effect on the seizure threshold, less cardiovascular toxicity
More likely skin pigmentation and photosensitivity	Fewer anticholinergic effects
Occasional cases of cholestatic jaundice	Occasional cases of neuroleptic malignant syndrome
Rare cases of agranulocytosis	

gain, and very little effect on the bone marrow and liver. On the negative side, high-potency antipsychotic drugs have a greater incidence of acute extrapyramidal effects. The converse is true of low-potency agents (see Table 2).

F. Pharmacokinetics

The rate and completeness of absorption of drugs determine the rapidity and intensity of onset of their clinical effects. Orally administered antipsychotic drugs are variably absorbed. When given intramuscularly, the drugs produce higher (2 to 10 times) and more reliable blood levels at a more rapid rate. Antipsychotic agents are highly lipophilic [with the apparent exceptions of thioridazine and its metabolite mesoridazine (Serentil®)] and bind tightly to protein and body membranes; they are poorly dialyzed. As with other drugs that cross the blood–brain barrier, they also cross the placental membrane to enter the fetal circulation and are transported into mammalian milk.

The half-life of a drug determines the time required to achieve a steady-state concentration in the body (i.e., the point at which tissue concentrations become stable). Steady state is reached after approximately four half-lives. Drugs with longer half-lives may be administered less frequently, and they disappear from the body more gradually when discontinued following chronic treatment. The half-life of an antipsychotic drug in plasma following a single dose is usually 10 to 20 hr (see Table 3). However, the drug's effects typically persist much longer, presumably because their brain half-lives are longer. When these drugs are administered chronically, the brain and body adipose tissues become saturated; these supplies are then released and excreted very slowly. In addition, when an antipsychotic drug has many metabolites (e.g., chlorpromazine), they are detected in the urine for many months after discontinuation of the drug following chronic administration.

Attempts to correlate plasma concentrations of antipsychotic drugs with their clinical efficacy have yielded contradictory results. This unsatisfying situation reflects problems in research methodology, but also in biochemical assays and questions about active drug metabolites. Although it was once thought that a radioreceptor assay measuring the level of dopamine-blocking activity in serum would replace direct chemical measurements of drug concentrations, this is less clear-cut at present.[18,19] Available evidence hints at the existence of a curvilinear ("therapeutic window") relationship for the butyrophenone haloperidol, a drug with no (or possibly one) active metabolite(s), although the precise upper and lower boundaries on this "window" remain to be delimited.

Regardless of any given antipsychotic drug's specific half-life, virtually all of these agents can be administered in a once-daily dose, usually at bedtime. The clinical effectiveness of this regimen presumably reflects the drugs' prolonged brain effects and gradual elimination from the body. After chronic dosing, biological effects of these agents persist for many weeks. After prolonged use of decanoate preparations, clinical and biological effects can extend beyond 6 months!

TABLE 3. Pharmacokinetics of Antipsychotic Agents[a]

Antipsychotic agents	Time to peak after oral dose (hr)	Elimination half-life (hr)[b]	% Protein binding	Active metabolites	Tentative therapeutic serum concentration[c]
Phenothiazines					
Chlorpromazine	2–4	16–37 (but active metabs probably much longer)	98–99	Many, unclear which of >160 metabs active	30–350 ng/ml?
Fluphenazine (Oral)	—	12–24		None?	0.5–3.0 ng/ml?
Mesoridazine		~6	~25	None	0.8–2.4 ng/ml?
Perphenazine	—	8–21		Mesoridazine Sulphoridazine	1–1.5 ng/ml?
Thioridazine	2 (Concentrate) 4 (Tabs)	7–42	96–99		
Trifluoperazine	3–6	17–18	Unknown	Uncertain (possible demethylated and hydroxylated metabolites)	Not established (1–2.3 ng/ml?)
Butyrophenones					
Haloperidol	2–6	12–40	~90	Reduced Haloperidol?	5–20 ng/ml?
Dibenzoxepines					
Loxapine	1 IM/2 PO	3.4 (Lox) (9 8–OH Lox 30 8–OH Amox)	~90	8–OH Loxapine 8–OH Amoxapine	30–100 ng/ml (Parent Lox plus 8–OH Lox)
Dibenzodiazepines					
Clozapine	1–4	6–33	>90	Uncertain (demethylated metabolites may be active)	Not established (100–600 ng/ml?)
Indolones					
Molindone	0.5	1.5	None	?	
Thioxanthenes					
Thiothixene	1–3	$\alpha = 3.5$ $\beta = 34$	Not studied	None identified	2–15 ng/ml?

[a] From Taylor and Caviness.[17]
[b] Brain elimination half-life may be more germane clinically but none established; brain half-lives probably much longer, accounting for prolonged action after discontinuation.
[c] None well established.

As with most psychotropic drugs, lipophilic parent compounds are oxidized to inactive hydrophilic metabolites largely within microsomal enzymes in the liver. These metabolites are excreted primarily in the urine and to a lesser extent in bile. Metabolism and excretion of antipsychotic drugs are greatest in healthy young people and less at either end of the age spectrum.

The blood levels achieved with antipsychotic drugs following a standard dose vary widely among individuals, which may account for some of the differences in clinical effects.

G. Adverse Effects and Toxicity

1. Neurological

a. Extrapyramidal Syndromes

Antipsychotic drugs may cause four types of extrapyramidal syndromes: acute dystonic reactions, akathisia, Parkinson's syndrome, and, after longer-term use, tardive dyskinesia. This section describes each syndrome and its treatment.

i. Acute Dystonic Reactions. Acute dystonic reactions, including acute dyskinesias (i.e., abnormal involuntary movements of various types) and oculogyric crises, typically occur during the early hours or days following the initiation of antipsychotic drug therapy or after marked dosage increment. Involuntary muscle contractions are common, particularly about the mouth, jaw, face, and neck. The symptoms are episodic and recurrent, lasting from minutes to hours. There may be trismus ("lockjaw"), dystonia or dyskinesias of the tongue, opisthotonus (spasms of the neck that arch the head backward), or eye closure. In oculogyric crises, there is a dystonic reaction of the extraocular muscles, and gaze is fixed in one position.

Acute dystonic reactions are distressing, particularly to the patient, family member, or clinician who is unfamiliar with them. They may be uncomfortable. They are rarely dangerous. However, in rare cases there can be respiratory compromise with the potential for a fatality.

The diagnosis of acute dystonic reaction is usually not difficult if it is clear that a patient has recently begun taking an antipsychotic drug or has had a switch in the type or dosage of medication. At times, however, eliciting this information may be difficult, especially with patients who are taking prochlorperazine suppositories (and say they are not taking any tranquilizer pills), who do not wish to acknowledge use of antipsychotic drugs, or who have sought antipsychotic drugs for illicit use (described recently). Among the many neuropsychiatric syndromes that must be considered in the differential diagnosis of acute dystonic reactions are tetanus, seizures, and conversion reactions.

The highest-potency antipsychotic drugs have the greatest likelihood of producing acute dystonias, the low-potency agents much less, thioridazine and clozapine the least. Young people are at greater risk of developing this syndrome than the elderly, and males do more frequently than females. When the acute dystonic reactions are frequent or severe in a patient, it may be worthwhile to assay the serum concentration of calcium, since rare cases of hypocalcemia have been detected in this way.

The mechanism underlying acute dystonic reactions is unclear. One hypothesis is that the syndrome reflects an acute increase in dopamine neurotransmission in the basal ganglia, which transiently supervenes the blockade of dopamine receptors brought about by the same drugs. Although dystonic reactions do not occur in naturally occurring Parkinson's disease (paralysis agitans), they are observed in postencephalitic Parkinson's syndrome, which shares other features with antipsychotic drug-induced extrapyramidal reactions.

Although the mechanism may be unclear, the treatment of an acute dystonic reaction is straightforward, readily available, and usually dramatically successful. **Parenteral treatment is preferred for initiating drug therapy, with intravenous being more rapid than intramuscular.** The intravenous injection of contraactive medication provides both immediate relief and confirmation of the diagnosis. For this purpose, a number of classes of agents have been employed with considerable success. An injectable anticholinergic antiparkinson drug such as **benztropine (Cogentin® and others),** 1 mg, or **biperiden (Akineton® and others),** 2 mg, may be used. Other clinicians prefer to administer an antihistamine drug such as **diphenhydramine (Benadryl® and others),** 50 mg. Some doctors prefer to use a benzodiazepine intravenously, such as diazepam (Valium® and others), up to 10 mg. This drug is safe and comfortable and does not add additional anticholinergic effects. As with the injection of any other central depressant compound, equipment for support of the airway should be immediately available should an emergency occur. Injection should be slow (5 mg of diazepam per minute), and care must be taken to avoid accidental intraarterial injection. Even injectable caffeine has been used with success.

Following immediate relief of acute dystonic signs, the clinician may wish to begin oral administration of one of these drugs, using the lowest effective dose (for range, see Table 4). If this successfully prevents additional reactions, the contraactive medication can usually be tapered and discontinued within several weeks. Would it be wise to coadminister an antiparkinson drug from the beginning of antipsychotic drug therapy in the hope of avoiding an acute dystonic reaction? Recent data suggest that one may offer partial protection to some patients through this "prophylactic" approach.[19] Therefore, the physician may want to consider this approach for patients at highest risk (i.e., young males receiving high-potency antipsychotic agents) and for those in whom a reaction is likely to be clinically disruptive. For other patients, it makes sense to avoid a drug that may be unnecessary, reserving treatment until signs of dystonia appear.

Physicians, nurses, family members, and anyone else who may observe a patient receiving an antipsychotic medicine should be aware of the possible occurrence of a dystonic reaction early in therapy. Treatment should be readily available. For most patients, the risk of dystonic reactions appears to wane with continued antipsychotic drug therapy.

TABLE 4. Antiparkinson Agents
Used in the Treatment of Neuroleptic-Induced EPS[a]

Generic name	Trade name	Type of drug	Usual dose range (mg per day)	Injectable
Amantadine	Symmetrel®, etc.	Dopamine agonist	100 to 300	No
Benztropine	Cogentin®, etc.	Antihistamine and anticholinergic	1 to 6	Yes
Biperiden	Akineton®, etc.	Anticholinergic	2 to 6	Yes
Diphenhydramine	Benadryl®, etc.	Antihistamine and anticholinergic	25 to 200	Yes
Ethopropazine	Parsidol®	Antihistamine and anticholinergic	50 to 600	No
Orphenadrine	Norflex®, etc.	Antihistamine	50 to 300	Yes
Procyclidine	Kemadrin®, etc.	Anticholinergic	6 to 20	No
Trihexyphenidyl	Artane®, etc.	Anticholinergic	1 to 10	No

[a]Adapted from Wojcik.[23]

ii. Akathisia. Akathisia is another extrapyramidal reaction associated with both antipsychotic drugs and postencephalitic Parkinson's syndrome (rarely with Parkinson's disease). Akathisia is a symptom (i.e., subjective) defined as a **compulsion to be in motion.** Patients describe an inner restlessness, an intense desire to move about simply for the sake of moving. Persons who are virtually paralyzed by the akinesia of postencephalitic Parkinson's disorder have been known to ask other people to move their limbs, just to relieve this intense compulsion. Patients suffering from akathisia are often observed to pace aimlessly, fidget, and be markedly restless. At times, akathisia may cause a worsening of psychosis.[20]

Akathisia can occur early in the course of drug treatment, or it may not appear for several months. It, too, appears to be more prevalent with high-potency drugs. The natural course of akathisia is less clear than that of acute dystonic reaction: at times it appears to wane, yet some patients are troubled by it for a long time. Its mechanism is obscure but may involve a blockade of mesocortical dopamine receptors.

Treatment responses are variable. Nevertheless, akathisia is an important syndrome to recognize, as it may severely complicate a patient's response to antipsychotic drug therapy; perhaps most important, it makes patients extremely unhappy. Approaches to treatment include attempts to lower the dose of the antipsychotic drug, switch to a lower-potency agent, or add a contraactive drug. The same drugs discussed for the treatment of acute dystonic reactions—anticholinergic antiparkinson agents, antihistamines, and benzodiazepines—also may be tried orally in cases of akathisia, although results are less universally successful (see Section III.G.1.a). Whether the dopamine agonist antiparkinson drug amantadine (Symmetrel® and others) may play a role in the treatment of this disorder is unclear.

Recent research has shown that the β-adrenergic blocking drug propranolol

(Inderal® and others) can effectively treat both subjective and objective manifestations of akathisia while producing an acceptable profile of side effects; other β blockers may be safe and effective as well.[19] Another agent that diminishes central noradrenergic activity, the antihypertensive clonidine (Catapres®), also may be effective but appears to cause an unacceptable incidence of sedation and hypotension.[20]

iii. Parkinson's Syndrome.

Parkinson's is characterized by a triad of signs: **tremor, rigidity, and akinesia (or bradykinesia).** The tremor (by definition, a regular and rhythmic oscillation of a body part around a point) is approximately 4 to 8 cycles per second and is greater at rest than during activity. A parkinsonian tremor is often observed in the hands; the thumb rubbing against the pad of the index finger may produce a characteristic "pill-rolling" appearance. The tremor also can involve the wrists, elbows, head, palate, or virtually any body part. In neuroleptic-induced Parkinson's syndrome, the tremor is typically bilateral; unilateral tremors should raise questions about the etiology. Although tremor is very common and may be one of the earlier signs in naturally occurring Parkinson's disease, it is less common than rigidity and akinesia and may not appear until relatively late in the drug-related syndrome.

Rigidity is an increase in the normal resting tone of a body part. It is detectable only by palpation on physical examination. (In other words, a patient does not **look** rigid; he must **feel** rigid.) In testing for rigidity, the physician asks the patient to relax completely and allow body parts to be manipulated without moving them. The examiner then rotates the head on the neck, moves the major joints in the upper and lower extremities, and raises each extremity to note the rapidity with which it falls by gravity. An increased resistance to passive motion and a slow return from a raised position denote the presence of rigidity. When tremor coexists with rigidity, the rigidity may take on the feel of a "cogwheel." In extreme forms of the condition, rigidity may mimic (or actually become) the waxy flexibility with sustained postures characteristic of catatonia.[21] In antipsychotic drug-induced Parkinson's syndrome, rigidity tends to be more common than tremor but less common than the third sign of the triad—akinesia.

Akinesia is a disinclination to move in the absence of paralysis. Literally, akinesia means the absence of motion, whereas bradykinesia, truer to the clinical reality in most patients, implies a slowness of motion. The bradykinetic patient frequently shows a masklike facies, with diminished expressiveness and less frequent eye blinking.

The bradykinetic patient turns his body "en bloc." Instead of turning first with the eyes, then with the neck, followed in turn by shoulders, hips, and lower extremities, the Parkinsonian patient turns as if he were one solid block of wood, without joints. A typical stance includes a flexing of elbows and wrists together with a stooped posture. The patient's gait is typically inclined forward, and he may walk with small, rapid steps (marche à petit pas). In the extreme forms, bradykinesia, as well as rigidity, may shade into catatonic immobility.[21]

In a less severe manifestation, the slowed movements of the parkinsonian patient may appear primarily as apathy, boredom, and a "zombielike" appearance. If other signs of Parkinson's syndrome are not prominent, this social akinesia may be misdiagnosed as depression.[22] Together with the major triad of Parkinson's syndrome (tremor, rigidity, and bradykinesia), associated signs often include seborrhea and drooling.

Parkinson's syndrome usually occurs within weeks to months after the beginning of antipsychotic drug therapy. Although tolerance develops in many patients, the disorder may persist and require ongoing treatment in others. Women and the elderly are affected more commonly. Again, the high-potency agents appear more likely to promote this disturbance. Probably a late-occurring variant of Parkinson's syndrome is the so-called rabbit syndrome, in which a rapid tremor of the mouth and jaw are reminiscent of the facial expression of a rabbit. The rabbit syndrome appears to respond pharmacologically similarly to other forms of Parkinson's.

Parkinson's syndrome reflects diminished dopaminergic input from the substantia nigra in the midbrain to the head of the caudate nucleus in the basal ganglia (see Fig. 1). In this extrapyramidal network, neuronal systems employing dopamine as a neurotransmitter are in balance with other systems mediated by acetylcholine and GABA. In naturally occurring forms of Parkinson's syndrome, the dopaminergic defect is a result of the destruction of dopamine-containing neurons in the substantia nigra. This destruction may occur as a result of viral infection (e.g., encephalitis), poisoning (e.g., carbon monoxide, manganese), or for unknown reasons (i.e., Parkinson's disease). In the drug-induced variety, a blockade of the dopamine receptors within the caudate nucleus appears to be responsible for the disturbance.

The treatment of Parkinson's syndrome reflects our understanding of the pathophysiology. One approach is to decrease the dose of the antipsychotic drug, presumably thereby decreasing the degree of dopamine blockade at the synaptic receptor. Alternatively, a less potent antipsychotic agent may be employed.

Among the original neuroleptics, thioridazine has the lowest incidence of parkinsonian reactions. The reason for this differential effect of antipsychotic drugs in causing Parkinson's syndrome may be related to variations in activity within the several dopamine systems. An alternative explanation is that the agents that produce the fewest extrapyramidal reactions, such as thioridazine, have the most potent effect in blocking acetylcholine, thus acting as if they had inherent antiparkinson activity. Clozapine has by far the lowest incidence of parkinsonism.

Pharmacological contraactive therapy for parkinsonism consists of attempts either to counterbalance the decreased dopamine neurotransmission by blocking acetylcholine transmission [anticholinergic antiparkinson drugs include benztropine and trihexyphenidyl (Artane® and others)] or by increasing dopamine neurotransmission (e.g., by the use of amantadine and others). Other dopamine agonist drugs such as L-dopa and bromocriptine (Parlodel®) may be useful in other forms of Parkinson's syndrome but are seldom employed for the drug-induced variety.

Table 4 lists available drugs that are useful for the treatment of drug-induced Parkinson's syndrome. With constant awareness of the pleomorphic and often subtle manifestations of extrapyramidal reactions, the physician prescribes one of these contraactive drugs as needed for a patient's comfort and optimal functioning, always seeking the lowest effective dose. Doctors should periodically attempt to taper and discontinue a concomitant antiparkinson drug, although recent evidence has shown that many patients on maintenance neuroleptic therapy require parallel maintenance with antiparkinson drugs. For most, this can be carried out with impunity.[20]

With the exception of amantadine, all of the antiparkinson drugs listed in Table 4 have anticholinergic and/or antihistaminic properties. Among these drugs, trihexyphenidyl has a relatively short half-life, whereas benztropine has a relatively long half-life. Amantadine is a dopamine agonist that has few anticholinergic effects. This feature may make it preferable in patients particularly sensitive to anticholinergic reactions.[23] For many patients, the strongly anticholinergic antiparkinson drugs can impair memory, time perception, and cognition: these central nervous system anticholinergic side effects also might prompt the use of amantadine.[24] Amantadine, 100 mg b.i.d. to 100 mg t.i.d., may be effective in some cases of drug-induced Parkinson's syndrome, particularly the more severe variety, when anticholinergic antiparkinson drugs have been ineffectual.[25] Amantadine has a relatively long half-life (approximately 24 hr). It is excreted in the urine unchanged and may lead to toxicity (including psychiatric symptoms) in patients with impaired renal function.

Antiparkinson drugs, as well as other chemicals with anticholinergic activity (e.g., belladonna-containing compounds), have been used for "recreational" purposes.[26] It appears that the anticholinergic activity provides a mood-elevating effect for many people, creating a feeling of euphoria or a "high." Some afficionados of the drug culture actually use the drugs to create a toxic delirium, which they perceive as pleasurable. Many schizophrenic patients treated with antiparkinson drugs become more attached to these drugs than to their antipsychotic agents, possibly because of the relief of extrapyramidal symptoms or perhaps because of directly pleasurable effects. Trihexyphenidyl has been reported to be more commonly abused than other antiparkinson agents. If this is true, it could be related to the pharmacology of the drug or simply to its popularity and, therefore, availability. Anticholinergic toxicity and its treatment are described in Section III.G.2.

iv. Tardive Dyskinesia and Tardive Dystonia. Tardive dyskinesia is characterized by abnormal, involuntary, choreoathetotic movements involving the tongue, lips, jaw, face, extremities, and occasionally the trunk. Not many years after the introduction of antipsychotic agents, patients were described who displayed these movements after a period of drug treatment. However, many questioned the association between these signs and the use of antipsychotic drugs. Instead, they attributed the movements to senile choreas, other brain disorders, schizophrenic stereotypies and mannerisms, or adventitious mouth movements associated with poor dentition and dry mouth. For the most part, clinicians generally ignored the existence of tardive dyskinesia.

By the early 1970s, however, tardive dyskinesia was recognized as more widespread than hitherto appreciated, and a 1981 review even suggested that its prevalence has progressively increased over the years since the introduction of antipsychotic medication.[27] Moreover, epidemiologic evidence and clinical opinion now clearly support a drug-related etiology for tardive dyskinesia, which has become a serious concern among psychiatrists, other physicians, allied mental health professionals—and lawyers.

Tardive dyskinesia typically involves orobuccolingual masticatory movements. These may include lip smacking, chewing, puckering of the lips, protrusion of the tongue, and puffing of the cheeks. A common early sign is wormlike movements of the tongue. Other movements of the face can be observed, including grimacing, blinking, and frowning. Another late-onset condition, presumably caused by antipsychotic drug exposure, is tardive dystonia, characterized by slow, sustained, involuntary twisting movements of limbs, trunk, neck, or face, including involuntary eye closure—blapharospasm.[28]

In the young, movements sometimes begin in the distal extremities and may consist of rapid, purposeless, quick, jerky movements distally—chorea ("dance")—or more sinuous, writhing movements proximally—athetosis. Occasionally, abnormal movements involve the trunk and pelvis. The signs of tardive dyskinesia can range in intensity from minimal to severe. Patients' awareness of the movements similarly varies. Institutionalized chronic schizophrenic patients may deny even very severe movements, whereas a highly functioning patient with a mood disorder could be extremely troubled by the most minimal symptom. The movements may embarrass the patient and interfere with important activities such as eating, talking, and dressing. In rare instances, tardive dyskinesia can impair breathing and swallowing.

Although movements identical to tardive dyskinesia occur with exposure to neuroleptics, a metaanalysis of 21 studies found that neuroleptics multiply the risk of developing dyskinesia by 2.9 times.[29] It is generally believed that for tardive dyskinesia to occur, a patient must have been exposed, more or less continuously, to an antipsychotic drug for at least 3 to 6 months, although it is probable that a rare, highly sensitive patient may develop this syndrome after an even briefer period. Presumably, increasing exposure to antipsychotic drugs increases the risk of developing tardive dyskinesia, although if a patient has not developed it after some length of time, he is probably less vulnerable. The only risk factor that has appeared consistently in studies of tardive dyskinesia is advanced age, although wide interindividual sensitivity to the development of this syndrome probably exists. There is no convincing evidence that patients taking one type of antipsychotic drug, such as high or low potency, are more or less likely to develop tardive dyskinesia. Patients with mood disorders may be at greater risk than those with schizophrenia. It appears that "drug holidays," i.e., regular, abrupt discontinuation of antipsychotic drug therapy, do not reduce the risk of tardive dyskinesia. Although it has been suggested that drug holidays may even increase the risk,[27] this is controversial. Prospective studies suggest a link between cumulative dosage and the development of tardive dyskinesia, underscoring the wisdom of maintaining patients on the lowest effective dosage over prolonged periods of time.

Estimates of the prevalence of tardive dyskinesia have varied widely—from 1% to over 50% of patients currently taking these drugs.[30] However, when the most minimal cases are removed from consideration, an attempt is made at differential diagnosis, and a careful and reliable screening procedure is used, the prevalence of tardive dyskinesia in a group of patients maintained on antipsychotic drugs for a variable period of time (i.e., a typical outpatient schizophrenic population) is likely to be 20% to 30%.[31] (A comparable group of patients not treated with antipsychotic drugs may have a prevalence rate of 5%.) A more important question concerns the incidence of tardive dyskinesia; preliminary data suggest 3–4% per year of patient exposure.[31] The prevalence of tardive dystonia may be 1–2%.[32]

Although the prevalence of tardive dyskinesia is considerable (and actually may be growing, as the population at risk takes neuroleptics for longer and longer periods), for most patients the neurological manifestations are comparatively mild. Many patients deny even an awareness of the dyskinetic movements, and dysfunction is rare. The most severely afflicted patients, and those likely to suffer the most, are patients with tardive dystonia.

The movements of tardive dyskinesia may appear during treatment when a patient is taking a constant dose of an antipsychotic, or they may initially appear when the dosage of the drug is lowered or stopped entirely. Conversely, if movements are present, increasing the dose of an antipsychotic drug can make the movements cease. This latter phenomenon has been referred to as "masking" of the movements and gives rise to the concept of "covert dyskinesia," in other words, dyskinesia observed only when an antipsychotic drug has been discontinued.[33] On the other hand, when an antipsychotic drug is discontinued abruptly, a patient may show transient dyskinetic movements that can disappear in a matter of days or weeks. This phenomenon has been known as withdrawal dyskinesia, and it might conceivably indicate that a patient is vulnerable to a persistent dyskinesia if the medication is reinstituted and continued. Tardive dyskinesia itself is believed to be persistent, although evidence suggests that if the movements are detected early and medication is discontinued, many patients will show gradual improvement over time.[34]

The linkage between early development of Parkinson's syndrome and later development of tardive dyskinesia—as well as possible contributing factors to the development of tardive dyskinesia from earlier use of antiparkinson drugs—is unresolved. What is clear is that anticholinergic drugs frequently exacerbate the movements of tardive dyskinesia once they are present and that administration of anticholinergic agents may even "unmask" latent movements. Conversely, discontinuing an anticholinergic drug in a patient with tardive dyskinesia may improve the movements. Tardive dystonia, by contrast, might actually improve in some cases on administration of an anticholinergic drug.

The most widely held hypothesis about the mechanism of tardive dyskinesia has involved the nigrostriatal dopamine pathway. Chronic blockade of dopamine receptors within the basal ganglia leads first to receptor underactivity, then to overactivity. In other words, the initial receptor blockade from antipsychotic drugs leads to hypoactivity at the receptor neurons. This state is believed to underlie the genesis of Parkinson's syndrome. Prolonged blockade, however, results in a situation analogous to denervation supersensitivity. In fact, chronic neuroleptic administration to animals has been shown to cause an actual multiplication in the number of postsynaptic receptor sites. Thus, long-term administration of antipsychotic drugs leads ultimately to an increase in activity at postsynaptic dopamine receptors. This state of excessive dopaminergic activity is believed to result in the abnormal movements of disorders such as tardive dyskinesia and Huntington's chorea.[35]

The "dopamine excess" hypothesis of tardive dyskinesia is not completely satisfying, however. For example, prolonged exposure to neuroleptic drugs results in dopamine receptor supersensitivity in virtually all experimental animals, but only some patients develop tardive dyskinesia. Also, such supersensitivity is *less* likely to develop in older animals, but tardive dyskinesia is *more* likely to develop in older patients. These and other objections have led many scholars to conjecture that the dopamine supersensitivity hypothesis of tardive dyskinesia is too simplistic to be universally applicable. Perhaps such a mechanism is valid in some patients, but alternative neurotransmitter systems that may go awry and contribute to the development of tardive dyskinesia include those involving GABA, acetylcholine, and norepinephrine.

Tardive dyskinesia (and Huntington's chorea) are in many ways the opposite of Parkinson's syndrome: the former is a disorder of extraneous movements, the latter a paucity of movements; the former is thought to reflect dopaminergic overactivity (and relative cholinergic underactivity), whereas the latter is caused by too little dopamine neurotransmission (and relatively too much cholinergic) (but see above reservations). Pharmacological strategies that tend to make one better frequently make the other worse.

Treatment approaches to tardive dyskinesia are based on our understanding (and beliefs) about the nature of antipsychotic drugs and tardive dyskinesia. First, of course, comes primary prevention. This means (1) avoiding unnecessary exposure of patients to antipsychotic drugs. In particular, the use of these agents to treat relatively benign conditions or those that could respond equally well or better to other agents (e.g., anxiety, nonpsychotic depression) should be avoided whenever possible. (2) When it is necessary to employ antipsychotic drug therapy, use the lowest dose for the shortest period of time. Of course, most schizophrenic patients will require prolonged therapy with these agents, but the clinician should find the lowest effective maintenance dose. Similarly, in the treatment of patients with other chronic disorders such as mental retardation and organic brain syndrome, the use of drugs should be minimized and constantly reevaluated.

Secondary prevention of a disorder means early detection. For tardive dyskinesia, this entails routine screening and monitoring of patients for the presence of abnormal movements. A standard neurological examination can be employed for this purpose, or a clinician may want to use a specific examination procedure and rating scale such as the Abnormal Involuntary Movement Scale designed by the National Institute of Mental Health (see Fig. 2). It is best to perform an examination

INSTRUCTIONS: Complete Examination Procedure (next page) before making ratings. MOVEMENT RATINGS: Rate highest severity observed. Rate movements that occur upon activation one *less* than those observed spontaneously.	CODE: 0 = None 1 = Minimal, may be extreme normal 2 = Mild 3 = Moderate 4 = Severe

(Circle one)

FACIAL AND ORAL MOVEMENTS	1. Muscles of facial expression, e.g., movements of forehead, eyebrows, periorbital area, cheeks; include frowning, blinking, smiling, grimacing.	0	1	2	3	4
	2. Lips and perioral area, e.g., puckering, pouting, smacking.	0	1	2	3	4
	3. Jaw, e.g., biting, clenching, chewing, mouth opening, lateral movements.	0	1	2	3	4
	4. Tongue. Rate only increase in movement both in and out of mouth, NOT inability to sustain movement.	0	1	2	3	4
EXTREMITY MOVEMENTS	5. Upper (arms, wrists, hands, fingers). Include choreic movements (i.e., rapid, objectively purposeless, irregular, spontaneous), athetoid movements (i.e., slow, irregular, complex, serpentine). Do NOT include tremor (i.e., repetitive, regular, rhythmic).	0	1	2	3	4
	6. Lower (legs, knees, ankles, toes), e.g., lateral knee movement, foot tapping, heel dropping, foot squirming, inversion and eversion of foot.	0	1	2	3	4
TRUNK MOVEMENTS	7. Neck, shoulders, hips, e.g., rocking, twisting, squirming, pelvic gyrations.	0	1	2	3	4

(continued)

FIGURE 2. Abnormal involuntary movement scale (AIMS). Asterisk (*) denotes activated movements. From Department of Health, Education, and Welfare, Public Health Service, Alcohol, Drug Abuse, and Mental Health Administration, National Institute of Mental Health.

GLOBAL JUDGMENTS	8. Severity of abnormal movements.	None, normal	0
		Minimal	1
		Mild	2
		Moderate	3
		Severe	4
	9. Incapacitation due to abnormal movements.	None, normal	0
		Minimal	1
		Mild	2
		Moderate	3
		Severe	4
	10. Patient's awareness of abnormal movements. Rate only patient's report.	No awareness	0
		Awareness, no distress	1
		Aware, mild distress	2
		Aware, moderate distress	3
		Aware, severe distress	4
DENTAL STATUS	11. Current problems with teeth and/or dentures.	No	0
		Yes	1
	12. Does patient usually wear dentures?	No	0
		Yes	1

EXAMINATION PROCEDURE

Either before or after completing the Examination Procedure, observe the patient unobtrusively, at rest (e.g., in waiting room).

The chair to be used in this examination should be a hard, firm one without arms.

1. Ask patient whether there is anything in his/her mouth (i.e., gum, candy, etc.) and if there is, to remove it.
2. Ask patient about the *current* condition of his/her teeth. Ask patient if he/she wears dentures. Do teeth or dentures bother patient *now*?
3. Ask patient whether he/she notices any movements in mouth, face, hands, or feet. If yes, ask to describe and to what extent they *currently* bother patient or interfere with his/her activities.
4. Have patient sit in chair with hands on knees, legs slightly apart, and feet flat on floor. (Look at entire body for movements while in this position.)
5. Ask patient to sit with hands hanging unsupported. If male, between legs; if female and wearing a dress, hanging over knees. (Observe hands and other body areas.)
6. Ask patient to open mouth. (Observe tongue at rest within mouth.) Do this twice.
7. Ask patient to protrude tongue. (Observe abnormalities of tongue movement.) Do this twice.
*8. Ask patient to tap thumb, with each finger, as rapidly as possible for 10 to 15 seconds, separately with right hand, then with left hand. (Observe facial and leg movements.)
9. Flex and extend patient's left and right arms (one at a time). (Note any rigidity.)
10. Ask patient to stand up. (Observe in profile. Observe all body areas again, hips included.)
*11. Ask patient to extend both arms outstretched in front with palms down. (Observe trunk, legs, and mouth.)
*12. Have patient walk a few paces, turn, and walk back to chair. (Observe hands and gait.) Do this twice.

FIGURE 2. (*Continued*)

and note the results prior to the initiation of antipsychotic drug therapy (or before too long into the course) and then to repeat the examination every 6 to 12 months while the patient remains on drug therapy. If early movements are detected, the clinician will want to consider lowering the dose of the antipsychotic drug or discontinuing it altogether. In addition, the patient (or next of kin) should be fully apprised of the clinical dilemma, the options, and the clinician's recommendation for further treatment.

What does one do when the presence of abnormal involuntary movements is unequivocal? First, consider the diagnosis (see Table 5). Is any other diagnosis likely? Are the movements characteristic of tardive dyskinesia, or might they be tremors or other movements? Are they chronic mannerisms and stereotypies charac-teristic of the underlying psychiatric disorder, or have these movements only ap-peared following chronic drug treatment? Family history should be explored to rule out the presence of hereditary disorders of the central nervous system such as Huntington's chorea or dystonia. The patient's history also should be reexplored to establish a reasonable linkage between drug therapy and the appearance of abnormal movements. Movements that inexorably progress are more likely caused by a de-generative neurological disorder than by tardive dyskinesia. A neurological exam-ination should reveal only abnormal involuntary movements and specifically those characteristic of tardive dyskinesia; other neurological systems, such as sensory or pyramidal, should not be involved. A physical examination should search for the presence of ancillary signs of other extrapyramidal disorders such as the Kayser–Fleischer rings of Wilson's disease. Finally, laboratory testing can rule out other disorders—e.g., copper and ceruloplasmin levels in Wilson's disease, radiological evidence of caudate degeneration in Huntington's disease—if these conditions are suspected. At times, the psychiatrist may wish to seek consultation from a neu-rological colleague.

If the diagnosis of tardive dyskinesia appears firm, what is the course of treatment? Raising the dose of the antipsychotic drug might succeed in partially (or even completely) suppressing the movements, but the clinician should avoid this strategy if possible. Available evidence suggests that continuing neuroleptic therapy, although unlikely to exacerbate the degree of dyskinesia, may enhance the probability that the dyskinesia will become irreversible.[36,37] Of course, if continu-ing an antipsychotic drug is necessary to control the psychosis or prevent episodic relapses (as is often the case), then the clinician must recommend this course, and it is up to the patient (or guardian) to decide whether to consent.

If abnormal movements are detected early, and it is clinically practical to taper and discontinue the antipsychotic medication, it is more likely that the dyskinesia will ultimately improve (although it may be temporarily exacerbated initially). Although this is ideal from the standpoint of the movement disorder, for most schizophrenic patients the trade-off in exacerbation of psychosis will not be worth the price. Schizophrenic psychosis usually is worse than tardive dyskinesia, and most often the clinician, patient, and family will decide to continue drug therapy, always at the lowest effective dose.

TABLE 5. Tardive Dyskinesia: Differential Diagnosis

Disorder	Associated mental disturbances	Family history	Laboratory tests
Dyskinesias on withdrawal from neuroleptic medication	Yes	No	None
Schizophrenic stereotypies and mannerisms	Yes	No	None
Spontaneous oral dyskinesias associated with aging (including Meige syndrome)[a]	Yes	No	None
Oral dyskinesias related to dental conditions or prostheses[b]	No	No	None
Torsion dystonia	No	Yes	None
Idiopathic focal dystonia [oral mandibular dystonia, blepharospasm, spasmodic "habit spasms" (tics)]	No	No	None
Huntington's disease[c]	Yes	Yes	CAT scan
Wilson's disease	Yes	Yes	Serum copper and ceruloplasmin
Magnesium and other heavy metal	Yes	No	Specific for metals
Fahr's syndrome or other disorders with calcification of the basal ganglia	No	Yes	Skull X-rays
Extrapyramidal syndromes following anoxia or encephalitis	Yes	No	None
Rheumatic (Sydenham's) chorea ("Saint Vitus dance")	No	No	None
Drug intoxications—L-dopa, amphetamines, anticholinergics, antidepressants, lithium, phenytoin	Yes	No	For specific agents
CNS complications of systematic metabolic disorders (e.g., hepatic or renal failure, hyperthyroidism, hypoparathyroidism, hypoglycemia, vasculitides)	Yes	Possible	For specific disorders
Brain neoplasms (thalamic, basal ganglia)	Yes	No	CAT scan, other brain scans, EEG, etc.

[a]Meige syndrome is a disorder of middle age characterized by progressive oral, lingual, and buccal dystonia together with blepharospasm. The movements are indistinguishable from those of tardive dyskinesia, but patients with Meige syndrome need not have had antedating antipsychotic drug exposure, and Meige syndrome is a progressive disorder.
[b]Be sure to ask patients about the state of their mouth and teeth and whether they have any gum in their mouth when you are examining them for tardive dyskinesia.
[c]The abnormal movements of Huntington's disease are primarily chorea with little dystonia or athetosis. The movements are generalized, producing a "fidgety" appearance in the early stages. Although movements of tardive dyskinesia are more stereotyped and abnormal, the movements of Huntington's disease appear to be normal movements at an increased frequency. Patients with tardive dyskinesia have dyskinesias other than chorea. In general, patients with Huntington's disease have more trouble keeping their tongues out of their mouths, whereas tardive dyskinesia patients may have trouble keeping their tongues in their mouths.

Tips for the Diagnosis of Tardive Dyskinesia

Helpful
1. History
   a. Exposure to antipsychotic drug
   b. Nonprogressive disorder
   c. Rule out family history of movement disorders

2. Physical examination
   a. Movements characteristic of tardive dyskinesia
   b. Rule out neurological signs in other systems (e.g., cerebellar, sensory, pyramidal)

Not Helpful
1. What makes movements better (e.g., sleep, relaxation) or worse (e.g., tension, psychological "gain")
2. Whether the patient can voluntarily inhibit movements
3. Whether movements are worsened or "unmasked" by a decrease in the antipsychotic drug dose
4. Laboratory tests—although laboratory tests are not helpful to rule in tardive dyskinesia, they can be helpful in ruling out other disorders.

If the movements continue and are troublesome (and irrespective of whether the patient remains on antipsychotic drug therapy), the clinician will want to alleviate as much of the patient's discomfort as possible. Unfortunately, there is no standard and accepted treatment for tardive dyskinesia. Therapy with a benzodiazepine drug such as diazepam or clonazepam (Klonopin®) may help some patients, either by its sedative or muscle-relaxing properties or possibly via its role in increasing GABAergic tone.

Just as Parkinson's syndrome often responds to strategies that increase dopamine neurotransmission, tardive dyskinesia may be alleviated by treatments that diminish dopaminergic tone. Thus, a dopamine-depleting agent such as reserpine or tetrabenezine may be useful. (As mentioned earlier, the use of dopamine blockers such as antipsychotics themselves is generally avoided.)

Another treatment approach focuses on the cholinergic system. Much as Parkinson's is treated with anticholinergic drugs, so tardive dyskinesia has been approached through the use of cholinomimetics. Deanol was earlier employed for this purpose, but its status as a cholinergic agonist is in question, and its efficacy in tardive dyskinesia is doubtful. Choline chloride has been employed in this fashion, but initial positive results have not been sustained, and its many unwanted effects

**TABLE 6. Comparison of Tardive Dystonia
and Tardive Dyskinesia**

Tardive dystonia	Tardive dyskinesia
Strikes younger	Strikes older
Strikes sooner in the course of neu- roleptic treatment	Strikes later in the course of neuro- leptic treatment
Poor prognosis	Variable prognosis
More males	More females (?)
Patients with mood disorders may be more susceptible	Patients with mood disorders may be more susceptible
Anticholinergics may improve con- dition	Anticholinergics usually worsen condition

make it clinically troublesome. Phosphatidylcholine, contained in the naturally occurring lipid lecithin, also has been used for this purpose. Our own group has found lecithin to produce statistically significant but clinically minimal benefits.[38] If a patient is already taking an *anti*cholinergic drug, the abnormal movements of tardive dyskinesia may be diminished if the anticholinergic is withdrawn; the "downside," however, is that parkinsonian signs may emerge or become worse.

Another agent that has been used in tardive dyskinesia with varying degrees of success is baclofen (Lioresal® and others), which may operate via a GABAergic mechanism. Many other drugs (e.g., propranolol) also have been administered experimentally in patients with tardive dyskinesia, but unfortunately, no consensus has emerged.

As with tardive dyskinesia, treatments for tardive dystonia (Table 6) are un-satisfying. Some patients may improve with anticholinergic drug treatment, and others experience symptomatic relief from reserpine or tetrabenezine (not available in the United States). Clozapine, which appears not to engender new cases of tardive dyskinesia or dystonia, actually might improve the abnormal movements, although it is unclear whether this might not be some form of "masking." In extreme cases of blepharospasm in patients with tardive dystonia, injections of botulin toxin may be therapeutic.

Steps in Diagnosis and Management of Tardive Dyskinesia

Preliminary procedure
1. Complete a baseline neurological examination.
2. Inform the patient and family about the risk of tardive dys-kinesia.

Primary prevention
1. Avoid unnecessary exposure to antipsychotics.
2. Use lowest dose of antipsychotic for shortest time.

Secondary prevention
1. Routinely screen for dyskinetic movements [routine neurological examination or rating scale such as Abnormal Involuntary Movement Scale (AIMS)].
   a. Administer at the beginning of treatment.
   b. Administer every 6 to 12 months.
2. Consider lowering dosage or discontinuing antipsychotic drug if early signs develop.

Management
1. Confirm diagnosis by reviewing symptoms, drug history, and neurological examination.
2. Consider relatively benign drugs first if movements are uncomfortable enough to require treatment.
   a. Benzodiazepines (e.g., diazepam, clonazepam).
   b. ? Clozapine.
   c. ? Baclofen.
   d. ? Propranolol.
3. Refer patient to specialized center.

b. Sedation

It was initially believed that sedation was important for the effectiveness of antipsychotic drugs, but this no longer appears true. Drugs can have a major impact on the primary symptoms of psychosis such as hallucinations and thought disorder without producing somnolence. As mentioned previously, the low-potency antipsychotic drugs such as chlorpromazine, thioridazine, and chlorprothixene tend to be more sedating. Clozapine is especially sedating. If around-the-clock sedation is desired, these drugs can be administered several times daily during initial therapy. If nighttime sedation is indicated, the drugs may be administered once daily at bedtime. Tolerance to sedation tends to develop over a matter of days or several weeks.

c. Seizures

Antipsychotic drugs, particularly the low-potency agents, lower the seizure threshold. For most patients, this is seldom a problem. However, an occasional patient without a history of epilepsy will experience a seizure during treatment with very high doses (or a particularly rapid increase in dose) of a low-potency agent. The problem is more likely to surface in the case of a patient with marginally controlled seizures or in a state of heightened vulnerability such as withdrawal from sedative–hypnotic drugs or alcohol. Each case must be handled individually, but consideration should be given to lowering the dose of the antipsychotic drug, changing to a high-potency agent, and/or adding (or increasing) an anticonvulsant (see Chapter 11). Clozapine lowers the seizure threshold in a dose-dependent fashion.

d. Neuroleptic Malignant Syndrome

The neuroleptic malignant syndrome (NMS) is a serious disorder consisting of fever, muscular rigidity, and stupor, which develops in association with antipsychotic drug therapy.[39] Other features include autonomic dysfunction (e.g., increased pulse, respiration, sweating, and blood pressure instability) and occasionally respiratory distress. Laboratory findings commonly indicate leukocytosis and elevated serum CPK. This syndrome typically develops explosively over a 24- to 72-hr period beginning anywhere from hours to months after initial drug exposure. A patient may have received prior treatment with antipsychotic drugs without showing this pattern.

The neuroleptic malignant syndrome has been associated with various antipsychotic drugs but is more prevalent with high-potency agents, particularly when the dose is being raised quickly. Haloperidol has been associated with a disproportionate share of case reports. Both sexes may be affected at any age, but among reported cases, young adult males predominate. The incidence of this disorder is probably around 0.1% of patients exposed to neuroleptics.[40] Mortality from NMS may be 10–20%, but when the patient survives sequelae are rare.[41]

The mechanism of NMS has yet to be elucidated. Most theorizing has focused on the basal ganglia and hypothalamus, where blockade of dopamine neurotransmission is held responsible (withdrawal of dopamine agonists, as well as institution of dopamine antagonists, has been associated with precipitation of NMS). An alternative site for the pathology of NMS could be peripheral muscles, analogous to the malignant hyperthermia of anesthesia.

Levenson and Simpson have underscored the heterogeneity of cases of neuroleptic-induced extrapyramidal signs plus fever.[42] They emphasize that many such patients have diverse and independently treatable medical or neurological conditions, which require a careful differential diagnosis.

Since a patient may develop NMS on one exposure to a neuroleptic, yet not on another exposure, it is likely that additional factors may play a role in the pathogenesis. Dehydration may be one such factor, and clinicians would do well to attend to hydration status during antipsychotic therapy, particularly in warmer weather when patients are exerting themselves, or if they have inadequate oral intake.

Appropriate management requires immediate discontinuation of antipsychotic drugs and institution of supportive measures (e.g., lowering body temperature, hydration) along with ruling out infection and other possible metabolic abnormalities. Although no contraactive treatment has been clearly proved safe and effective, most attention in recent years has focused on dantrolene (Dantrium®) and bromocriptine.

How to treat subsequent psychotic episodes in a patient with a history of NMS is capturing increasing attention among clinicians.[43] Low-potency neuroleptics may be safer than those of higher potency, although even clozapine has been associated with NMS. When a patient has a history of this syndrome, the clinician should first verify the past diagnosis. If it is validated, the clinician should consider nonneuroleptic treatments such as ECT or, in the case of a mood disorder, lithium (Eskalith® and others). When a neuroleptic is thought to be required, one of lower potency should be considered, and the dose raised gradually. It is of particular importance that all previous signs of NMS should have ended completely at least 2 weeks before the reintroduction of an antipsychotic.

2. Anticholinergic

a. Peripheral

Antipsychotic drugs block a subtype of cholinergic receptor known as muscarinic. This type of receptor, which responds to interneuronal release of acetylcholine, is located on postganglionic neurons of the parasympathetic branch of the autonomic nervous system (as well as autonomic ganglion cells and certain cortical and subcortical neurons). Thus, the atropinelike action of antipsychotic drugs antagonizes the effects of the parasympathetic nervous system. The most strongly anticholinergic among the traditional antipsychotic drugs is thioridazine, which is almost as potent as the cyclic antidepressants. High-potency antipsychotic drugs (e.g., fluphenazine, haloperidol) are comparatively weak in their atropinelike action.

In the skin, anticholinergic drugs can produce warmth, flushing, and dryness; in the eye, dilated pupils, difficulty with visual accommodation, and increased intraocular pressure; in the mouth, dryness; in the lungs, drying of secretions; in the heart, increased rate; in the stomach, decreased acid secretion; in the bowel, diminished motility with potential constipation; in the urinary tract, smooth muscle slowing, which can lead to delayed urination; in the penis, delayed (or retrograde) ejaculation.

In general, the elderly are more sensitive to atropinelike effects than are younger patients. Obviously, certain illnesses are exacerbated by anticholinergic actions of drugs. An abrupt attack of narrow-angle glaucoma is a rare possibility that can occur in a predisposed individual. Fortunately, the more common open-angle glaucoma, particularly when controlled by drugs, is less likely to be made worse by drug administration. Individuals prone to dental problems may suffer from diminished salivation. Patients with respiratory problems could be adversely affected by diminished pulmonary secretions. Similarly, those with cardiac disorders may be compromised by tachycardia as well as by more direct cardiotoxic effects, which are discussed below. Individuals with gastrointestinal disturbances, including those who have recently had abdominal surgery, may be adversely affected by the diminished bowel motility. Similarly, a man with prostatic enlargement could find urination difficult during treatment with an anticholinergic drug.

Some degree of tolerance develops to these effects over weeks and months (and conversely, rebound can occur on drug withdrawal). When peripheral anticholinergic activity of antipsychotic drugs is a problem, the clinician may wish to lower the dose of the drug or switch to a drug with less atropinic activity (i.e., a higher-potency agent). At times, symptomatic treatment may be offered, such as sugarless gum and lozenges for dry mouth, a stool softener or mild laxative for constipation. In occasional patients, the use of a peripherally active parasympathomimetic agent such as bethanechol (Urecholine® and others) may be indicated.

In its anticholinergic activity, clozapine again is something of an anomoly. Its binding potency at the muscarinic receptor is higher than that of older antipsychotic drugs, even thioridazine. However, one of the common side effects of clozapine is hypersalivation: patients typically awaken with a pillow soaked with saliva.

b. Central

Presumably through their action in blocking central muscarinic cholinergic receptors, antipsychotic drugs can at times produce memory difficulties, confusion, and, in the extreme, delirium.

c. Serious Toxicity

An antipsychotic agent alone, particularly a higher-potency compound, is unlikely to produce serious atropine-type toxicity. However, the more potent antimuscarinic antipsychotic drugs (especially thioridazine), or the combination of an antipsychotic drug with an anticholinergic antiparkinson agent, cyclic antidepressant, or other antimuscarinic compound, or a drug overdose can lead to serious toxicity, especially in the elderly and others vulnerable to these conditions. The patient suffering from an atropinelike delirium is typically confused, disoriented, and agitated. Pupils are large (although responsive), mucous membranes are dry, and skin is hot and flushed. Tachycardia and markedly diminished bowel sounds are common.

Intravenously administered physostigmine (Antilirium®), 1 to 2 mg in a single, slow injection, can be both diagnostic and therapeutic, as it will reverse both the peripheral and central signs of atropine-type toxicity. (If successful, physostigmine can be repeated IM every few hours until delirium clears.) Physostigmine is a centrally active inhibitor of acetylcholinesterase, and as such it increases the amount of acetylcholine available to cholinergic receptors, thus circumventing the drug's blockade. The infusion of physostigmine should be slow, with careful cardiac monitoring and readily available means to support respiration, should this prove necessary. A physician who wishes to administer physostigmine should be familiar with the pharmacology of this drug and the hazards of cholinergic excess (which can be reversed by the administration of atropine). As a final note, physostigmine should be administered with the greatest caution to a patient who is showing cardiovascular instability as a result of a drug overdose or other medical problems[44] (see Chapter 2).

3. Cardiovascular and Respiratory

a. Hypotension

Orthostatic hypotension during antipsychotic drug administration has been attributed to a combination of hypothalamic actions and peripheral α-adrenergic blockade. Some degree of tolerance may develop. This reaction is more common with the low-potency drugs (e.g., chlorpromazine) and may be more of a problem with the elderly or with patients with preexisting postural hypotension and vascular instability such as those undergoing a sedative–hypnotic withdrawal reaction. It is likely that parenteral administration of an antipsychotic drug may provoke more severe hypotension than oral ingestion of the same drug. The new drug clozapine produces substantial postural hypotension.

If postural hypotension does develop, it usually can be managed by keeping the patient horizontal. The next step, if necessary, is the administration of intravenous fluids to expand the vascular volume. If these two maneuvers are not effective, then a pure α-adrenergic pressor agent should be administered. Metaraminol (Aramine® and others) will usually suffice, although norepinephrine (Levophed® and others) also may be considered. The use of a mixed α- and β-stimulating drug such as epinephrine (Adrenalin and others) could lead to a paradoxical drop in blood pressure, since the α receptors are blocked, and the unopposed β stimulation can promote further hypotension. Similarly, the use of a primarily β-stimulating drug such as isoproteronol (Isuprel® and others) should be avoided.

b. Cardiac

Antipsychotic drugs have both antiarrhythmic and arrhythmogenic effects on the heart. Antagonism of sympathetic activity in the hypothalamus and a local anesthetic property that stabilizes the cardiac cell membrane [similar to the effects of lidocaine (Xylocaine® and others)] may cause the antiarrhythmic effect. In addition, antipsychotic drugs have direct quinidinelike effects on the myocardium and cardiac conduction system and affect electrolyte balance. Explanations of the arrhythmogenic action of phenothiazines and other antipsychotic drugs similarly include both central activity and a direct effect on the myocardium. Pathologists have detected microscopic and ultramicroscopic lesions in the cardiac tissue of patients who died suddenly during the course of phenothiazine therapy, suggesting a direct toxic effect.

Whether antipsychotic drugs are protective or toxic to the heart probably depends on the agent, the dose, and the underlying state of a patient's cardiac physiology. Low-potency antipsychotic agents, especially thioridazine and clozapine, tend to be more cardiotoxic. On the electrocardiogram (EKG), increased heart rate, prolongation of the QT and PR intervals and T wave, and depression of the ST segment are occasionally observed, particularly with thioridazine, clozapine, and pimozide (Orap®). For most patients, these effects are not clinically troublesome. Occasionally, however, a life-threatening arrhythmia, such as torsades de pointes (polymorphic ventricular tachycardia), can occur. Patients with preexisting cardiac disease should be monitored carefully (e.g., clinical examinations, vital signs, EKGs), and low-potency drugs (especially in excessive doses and particularly thioridazine, clozapine, and pimozide) should be avoided.

Overdoses of antipsychotic drugs (especially thioridazine) can cause cardiac arrhythmias. Because antipsychotic agents have quinidinelike actions, arrhythmias caused by overdoses should not be treated with quinidine or related type 1 antiarrhythmic drugs [procainamide (Pronestyl® and others) and disopyramide (Norpace® and others)]. In the treatment of cardiac arrhythmias that result from overdoses with phenothiazines and related drugs, antiarrhythmics such as lidocaine, phenytoin (Dilantin® and others), and propranolol may have a role. The early use of transvenous pacing also has been recommended.

Combinations of antipsychotic drugs and antidepressants can produce additive cardiotoxic effects. In particular, combinations of thioridazine, clozapine, pimozide, and cyclic antidepressants, especially amitriptyline (Elavil® and others), are best avoided.

Occasional cases of sudden death reported in patients receiving antipsychotic drugs are difficult to interpret and probably result from various causes. However, episodes of potentially fatal ventricular arrhythmias are a possible factor.

4. Ocular

As already noted, the anticholinergic effects of antipsychotic drugs can affect the eye. Increased mydriasis can make a patient more light sensitive. Interference with visual accommodation may result in complaints of blurred vision. Because of the rare possibility of precipitating an attack of angle-closure glaucoma, prior to treatment with any anticholinergic agent patients should be asked about history of visual symptoms—such as eye pain, blurring, and halos—that could suggest previous narrow-angle episodes. Open-angle glaucoma is less likely to be a problem, but the patient should be managed conjointly with an ophthalmologist.

Prolonged treatment with high doses of low-potency antipsychotic drugs has been associated with the deposition of pigment in the lens, cornea, conjunctiva, and retina, often together with skin pigmentation. Except in extreme cases, these are unlikely to interfere with vision. However, if it is necessary to expose a patient to high doses of a low-potency agent for long periods of time, periodic ophthalmologic examinations would be worthwhile, perhaps annually.

Of greater clinical significance is the pigmentary retinopathy that may accompany thioridazine treatment. This disorder, which can result in visual impairment, is unlikely if doses of thioridazine do not exceed 800 mg per day, but even this is relative. Thus, thioridazine is the one antipsychotic agent that should be considered to have an absolute "ceiling dose"—800 mg daily—which must not be exceeded for even brief periods of time. (If a patient does require high doses of an antipsychotic drug, it is best to change to a different one.)

5. Cutaneous

Virtually any drug is capable of producing an allergic rash in occasional patients, typically between 2 and 10 weeks following initial exposure. This is usually maculopapular, erythematous, and itchy, affecting the face, neck, trunk, and extremities (often the palms and soles of the feet). Allergic reactions vary in distribution and severity, the most extreme being exfoliative dermatitis, which can be life-threatening. In most cases, discontinuation of the precipitating agent is followed by prompt remission of symptoms and signs. The itching and rash can be treated symptomatically (with topical steroids). Subsequent antipsychotic therapy should be with a drug from a different chemical group. Occasionally, contact dermatitis occurs in patients hypersensitive to antipsychotic drugs, particularly to liquid preparations.

Patients receiving low-potency antipsychotic medication sometimes become very sensitive to sunlight. The resulting reaction resembles severe sunburn. Management of the acute reaction is the same as for sunburn. Subsequent treatment should include switching to a higher-potency antipsychotic drug or warning the patient to protect himself from sun, whether physically or through the use of a sunscreening preparation [usually one containing *para*-aminobenzoic acid (PABA)].

Occasional patients treated with high doses of low-potency phenothiazines for prolonged periods develop a blue–gray discoloration of the skin, specifically in skin exposed to sunlight. This usually occurs in conjunction with pigmentary changes in the eye. Aside from cosmetic concerns, it is unclear if this reaction has any clinical importance.

As mentioned earlier, seborrheic dermatitis often occurs in conjunction with Parkinson's syndrome.

6. Hormonal, Sexual, and Hypothalamic Reactions

Previously, we discussed how antipsychotic drugs increase prolactin release from the anterior pituitary and consequently cause hyperprolactinemia, probably as a result of dopamine-blocking activity. In females, this occasionally leads to galactorrhea (also, rarely, in males), decreased frequency or flow of menstruation, and in both sexes a diminished libido. Although so far there is no convincing evidence that chronic antipsychotic drug treatment increases the risk of breast cancer, it is not inconceivable that the growth of prolactin-sensitive tumors may be enhanced by the presence of elevated circulating prolactin concentrations. Therefore, women should be asked about personal and family history of breast cancer, and women receiving long-term antipsychotic therapy should undergo periodic breast examinations. Although there is no evidence that chronic antipsychotic drug treatment can result in increased incidence of pituitary adenomas, it remains at least a theoretical possibility that lactotrophes in the pituitary will be stimulated. Therefore, in the face of chronic hyperprolactinemia, patients should be examined occasionally for possible evidence of pituitary enlargement. Persistent amenorrhea and galactorrhea suggest that the clinician should lower the antipsychotic drug dose. At times, the dopamine agonist bromocriptine may be considered, but this should be discussed with an endocrinologist.

Antipsychotic agents appear to lower circulating levels of testosterone through either a direct or indirect mechanism. Together with the effects of elevated prolactin, decreased testosterone may contribute to the diminished libido observed in some males treated with these chemicals.

Effects on blood sugar, growth hormone, and thyroid hormone also have been reported, although it is doubtful that these are of much clinical import. Chlorpromazine impairs glucose tolerance and insulin release, which may be clinically noteworthy in some prediabetic patients.

Another sexual symptom, already mentioned under anticholinergic effects, is interference with ejaculation. This is most frequently observed in young males treated with thioridazine and is manifested by delayed ejaculation or retrograde ejaculation. In the latter case, the patient reports orgasm without emission, followed by urination that has a "foamy" appearance. Some males also report difficulty maintaining and sustaining erection. In these cases, change to a high-potency drug.

Both females and males taking antipsychotic drugs sometimes complain of delayed, altered, or inadequate orgasms. The mechanism is unclear but may include psychological, hormonal, or autonomic factors. Possible therapeutic strategies can include lowering the dose of the antipsychotic agent, switching to a higher-potency compound, or using a peripherally active cholinomimetic, such as bethanechol. The antiserotonin antihistaminic drug cyproheptadine (Periactin® and others) has been reported helpful in reversing orgasmic dysfunction attributed to antidepressant drugs; whether it might have similar value in patients on neuroleptic therapy is untested.

In addition to effects on prolactin and the existence of the neuroleptic malignant syndrome, antipsychotic drugs seem to have many other actions at the level of the hypothalamus. These probably include cardiovascular effects as well as various hormonal and autonomic changes. Temperature regulation is also impaired by phenothiazines and their relatives, particularly the low-potency ones, making patients more vulnerable to hypo- or hyperthermia (depending on the ambient temperature).

Another unwanted effect, probably related to hypothalamic changes, is increased appetite with resultant weight gain. In some patients, this can be most marked and unpleasant. Probably, appetite increase (which may be related to antihistaminic effects) is more common with low-potency drugs. Management may include switching to a higher-potency agent and dietary counseling.

7. Hepatic

The low-potency antipsychotic drugs occasionally produce a syndrome of cholestatic jaundice, probably a combination of a direct toxic effect and an allergic reaction. Typically, within the first month of treatment, the patient develops fever, chills, nausea, malaise, pruritis, and right upper quadrant abdominal pain, followed within days by jaundice. Liver function tests reveal an obstructive pattern with increased alkaline phosphatase and conjugated (direct) bilirubin. Transaminase enzymes also may be elevated, but they do not reach the levels observed in hepatitis.

The recommended treatment for cholestatic jaundice is discontinuing the antipsychotic drug, although it is possible that patients will recover despite continued therapy. In almost all cases, recovery occurs over a matter of weeks and is complete and without sequelae. In predisposed patients, cholestatic jaundice may lead to chronic biliary cirrhosis. Subsequent treatment probably should be with a different antipsychotic drug, preferably a high-potency agent.

The presence of preexisting liver disease does not contraindicate the use of antipsychotic drugs. However, when the liver is impaired, metabolism of antipsychotic drugs may be slowed, and other drugs that use similar enzyme pathways could be affected by the addition of another liver-metabolized agent. In addition, the central action of antipsychotic drugs may worsen a case of hepatic encephalopathy.

8. Hematological

Low-potency antipsychotic agents probably are weakly toxic to some elements of the bone marrow, particularly stem cells of the granulocyte series. Almost all patients can compensate for this and show no more than a transient leukopenia. However, a rare patient goes on to develop agranulocytosis.

Perhaps 1 in 3000 or 4000 patients treated with chlorpromazine (and possibly other low-potency drugs) will develop agranulocytosis. (This occurs rarely, if at all, with high-potency drugs. For clozapine, the incidence of agranulocytosis may be much higher, perhaps as high as 2%.)

The onset is typically within the first 2 to 3 months of drug therapy. (It must be remembered, however, that switching to a different agent "restarts the clock.") Many patients treated with antipsychotic drugs show a transient leukopenia, but resistant stem cells probably allow compensation. However, in a patient vulnerable to developing agranulocytosis, the granulocyte count drops precipitously over the course of several days. The white blood cell count may fall below 1000, and practically all those cells will be lymphocytes.

Routine blood counts will be unlikely to detect the abrupt onset of agranulocytosis unless they are performed two to three times each week for the first several months of treatment. For this reason, they are generally not recommended (clozapine is an exception). Instead, the best approach is to maintain a high clinical index of suspicion. Thus, sore throat, fever, malaise, or other symptoms or signs of infection should prompt **immediate** white blood count with a differential. If the count is low, antipsychotic drugs should be discontinued immediately, and the patient put into reverse isolation to prevent infection. If infection does not supervene, a normal blood count returns within several weeks. If infection does occur, there is substantial mortality. The mortality rate is considerably elevated if the antipsychotic drug is not discontinued.

Agranulocytosis associated with the new and atypical antipsychotic drug clozapine is particularly noteworthy. First, as mentioned above, it is much more common (afflicting perhaps 1–2% of patients who take it). Second, it does not appear to show cross-tolerance with other drugs, even with the chemically similar loxapine.[45] Third, unlike the probable toxic mechanism underlying agranulocytosis induced by chlorpromazine and other phenothiazines, agranulocytosis engendered by clozapine is more likely caused by an immune mechanism.[45] This (with some empirical data) suggests that reexposure of a patient to clozapine could result in an abrupt drop in white blood cell count. Finally, there is evidence of genetic vulnerability to clozapine-induced agranulocytosis, with patients of Ashkenazi Jewish background possibly at highest risk.[45] Since at least one patient has developed agranulocytosis after more than 1 year of clozapine therapy, current requirements include indefinite, weekly white blood cell counts in patients taking this medication.

Antipsychotic drugs probably can affect platelet function, although whether this may have therapeutic or toxic significance is unclear. A variety of immunologic and coagulation changes have been described in patients taking chlorpromazine, sometimes in association with hepatomegaly.

About half of a group of patients who had been treated chronically with chlorpromazine had a positive antinuclear antibody test, and more than 75% showed increased serum concentrations of IgM associated with prolongation of partial thromboplastin time. In addition, a number of autoantibodies have been found in patients treated with chlorpromazine, often in association with splenomegaly.[46]

In a recent study contrasting the immunologic effects of chlorpromazine versus those of haloperidol, 6 of 29 patients taking chlorpromazine but none of 14 taking haloperidol had progressive elevations of serum IgM at the end of five years.[47] By the end of five years, 87% of chlorpromazine-treated and 50% of haloperidol-treated patients had antinuclear antibodies, but none had developed a lupus-like syndrome. Several of the chlorpromazine-treated patients in this series, however, did develop immune syndromes (Waldenström's macroglobulinemia and immune thrombocytopenia), leading these authors to recommend that patients who develop increased serum IgM while taking chlorpromazine be switched to other antipsychotic drugs.

Occasional reports associate antipsychotic drugs with other hematological syndromes, but no relationship has been clearly established.

9. Pregnancy and Lactation

All antipsychotic drugs can cross the placenta and enter the fetus. Considering all epidemiologic data at present, no consistent evidence has emerged suggesting that any antipsychotic drug is a teratogen. However, the possibility of a low-level association cannot be ruled out. Moreover, antipsychotic drugs may affect the developing nervous system, with possible long-lasting neurochemical and behavioral effects.[48] In addition, newborn babies have shown syndromes of neuroleptic withdrawal from the drugs along with extrapyramidal and hypothalamic reactions. For these reasons, if at all possible, antipsychotic drugs should be avoided during pregnancy.

Similarly, antipsychotic drugs can be detected in human milk at levels roughly similar to those found in blood. Although this means that the nursling would receive relatively small amounts each day, concern about the effects of psychotropic drugs on the developing nervous system, referred to above, must be taken into consideration in weighing treatment decisions (see Chapter 10).

10. Withdrawal Reactions

Habituation and addiction are not believed to occur with antipsychotic drugs. Rarely, these agents have been used recreationally.

As noted earlier, tolerance does develop to some of the unwanted effects of these agents. Partial to complete tolerance may develop to sedation, and some degree of tolerance can occur for hypotension and anticholinergic actions. Conversely, when the drug is stopped, rebound reactions may occur. These can include insomnia, nightmares, and other disturbances of sleep, as well as cholinergic rebound such as increased salivation, abdominal cramps, and diarrhea. To make patients more comfortable, discontinue these drugs gradually (e.g., 5–10% of dose per day) rather than abruptly.

Obviously, discontinuation of antipsychotic drugs may be followed by relapse of psychosis. This is generally believed to be a recurrence of the underlying psychiatric disorder. Although it has been suggested that in some patients, psychosis actually is created or exacerbated as a result of supersensitivity created by the drugs,[49] this remains controversial and inadequately proved.

As discussed in the section on tardive dyskinesia, discontinuation of antipsychotic drugs is occasionally followed by transient withdrawal dyskinesias. In other cases, discontinuing the drug unmasks persistent dyskinesia.

11. Overdose

Fortunately, antipsychotic drugs have a high therapeutic index—the ratio of toxic to therapeutic dose. Ingestion of an antipsychotic agent alone in an overdose attempt seldom results in death. (However, it should always be remembered that another agent may have been ingested along with the antipsychotic drug). Possibly more dangerous than other antipsychotics in overdose are thioridazine and loxapine.

When an antipsychotic drug is ingested in an overdose, the anticholinergic effects will probably delay gastric emptying; passage of a nasogastric tube with the application of suction should always be attempted, even many hours after the overdose. Instillation of activated charcoal may further diminish absorption. Attempts at inducing emesis may be unsuccessful because of the antiemetic action of these drugs, particularly the low-potency agents. Because antipsychotic agents have a high lipophilicity coupled with a strong adherence to tissue and proteins, attempts at dialysis are relatively unsuccessful.

Antipsychotic drugs alone seldom produce lethal suppression of vital brainstem centers. However, the low-potency agents are capable of inducing serious blood pressure problems, which, as noted earlier, should be treated with pure α-stimulating drugs. Low-potency drugs also can occasionally produce serious arrhythmias, which, with thioridazine, are most likely to be dangerous. As noted earlier, in the treatment of these arryhthmias, quinidine and other type 1 antiarrhythmic agents (namely, procainamide and disopyramide) should be avoided because of additive effects (see Section III.G.3).

We previously discussed how physostigmine may have a role in reversing some of the central nervous system effects of atropinic toxicity caused by antipsychotic agents. However, physostigmine can worsen some arrhythmias, lower the seizure threshold, and produce respiratory arrest. For this reason, it should be used extremely cautiously, if at all, in conditions of cardiovascular and respiratory instability.[44]

H. Drug Interactions and Combinations

Drugs can interact with each other at two levels—pharmacokinetic and pharmacodynamic. In a pharmacokinetic interaction, one drug interferes with the absorption, distribution, metabolism, or excretion of another drug, effectively raising or lowering blood and tissue levels of the other. In a pharmacodynamic interaction, one drug combines with another to increase or decrease effects at a target organ. There are several reasons why a patient may receive more than one drug at the same time: to enhance a specific therapy, because two or more illnesses must be treated concurrently, or when a second drug must be added to counteract unwanted effects of the first drug.

In general, to minimize potential unwanted interactions, simplify the therapeutic regimen, enhance compliance, and more easily assess specific contributions to a clinical outcome, a physician should prescribe the fewest possible drugs for a patient. However, in a number of clinical situations, a physician will want to coadminister another drug with an antipsychotic agent. We have already mentioned the treatment of extrapyramidal effects with antiparkinson drugs. In such circumstances, several pharmacodynamic interactions are possible. Most antiparkinson drugs (with the partial exception of amantadine) have anticholinergic actions that will add to those of the antipsychotic agent. Many antiparkinson drugs also have antihistaminic effects, which can enhance the sedation caused by antipsychotic drugs. In addition, pharmacokinetic interactions also have been reported; it is possible that antiparkinson drugs lower the blood levels of some antipsychotic agents; however, data about this point are inconsistent.[50,51]

Sometimes it may be justified to coadminister a benzodiazepine together with an antipsychotic agent. As mentioned earlier, a drug like diazepam may effectively counteract acute dystonic reactions or akathisia. There appears to be little potential for a pharmacokinetic interaction between these two classes of drugs. However, an important pharmacodynamic interaction is additive sedation. Similarly, any sedative–hypnotic drug (including alcohol) has this potential interaction with antipsychotic agents. In the extreme, depression of respiration or blood pressure may occur.

At times (e.g., in the treatment of psychotic depression), a physician may elect to treat a patient with both an antipsychotic drug and a cyclic antidepressant. A pharmacokinetic interaction, probably at the level of hepatic drug-metabolizing enzymes, will raise blood levels of both drugs higher than if either had been administered alone. In addition, additive sedative, anticholinergic, and cardiovascular effects are likely.

Lithium and an antipsychotic drug are combined in the treatment of many patients with mania. A syndrome of severe neurotoxicity—including neuromuscular signs, cognitive changes, and hyperthermia—was described in four patients receiving treatment with both lithium carbonate and haloperidol.[52] Similarly, EEG abnormalities occurred in several patients who developed encephalopathy during therapy with a combination of lithium and thioridazine.[53] On the other hand, broad clinical experience and a number of published series have testified to the safety of lithium/antipsychotic drug combinations for a vast majority of patients. It is possible, though, that an occasional patient will experience enhanced neurotoxicity from the combination, and any suggestion that this may be occurring should prompt a rapid lowering of dosages or drug discontinuation.[54]

A pharmacokinetic interaction has also been described between lithium and chlorpromazine: mean plasma chlorpromazine levels were significantly lower during combined therapy.[55] Lithium may slow gastric emptying, which would enhance the gut metabolism of chlorpromazine and diminish its absorption, or it may interfere with the absorption of chlorpromazine by diminishing its transport across gut membranes.

Antipsychotic agents also can interact with drugs used for the treatment of nonpsychiatric conditions. Phenothiazines (like cyclic antidepressants) can interfere with the antihypertensive actions of guanethidine (Ismelin® and others). Low-potency antipsychotic drugs may potentiate the postural hypotension observed with various antihypertensive drugs. Haloperidol has been reported to cause an organic brain syndrome when combined with methyldopa (Aldomet® and others). By decreasing the metabolism of phenytoin (and perhaps of other anticonvulsant drugs), antipsychotic drugs may elevate anticonvulsant blood levels, potentially producing toxicity. Through their tendency to produce extrapyramidal reactions, antipsychotic drugs can counteract the antiparkinson effects of drugs used to treat naturally occurring Parkinson's syndrome.

Caffeine, contained in many beverages and over-the-counter medicines, also may interact with antipsychotic drugs. At the dynamic level, it can counteract sedative effects, possibly necessitating an increased dose of the antipsychotic drug. At a kinetic level, some caffeine-containing beverages might diminish the absorption of antipsychotic drugs from the stomach, but this is controversial.[56]

The rule of thumb is that when any drug is added to any other drug, the possibility of interactions must be considered. For this reason, as stated at the beginning of the section, the clinician should minimize the use of multiple drugs in the same patient at the same time.

I. Laboratory Tests and Monitoring

In contrast to lithium, no routine laboratory tests are essential for all patients receiving antipsychotic agents. Similarly, blood assays of antipsychotic levels have not yet reached the stage of general applicability. However, pretreatment laboratory tests (supplementing a history and physical examination) can be useful to assess a patient's general health status prior to initiating antipsychotic drug therapy and for periodic reassessments during treatment. These may include a complete blood count with a differential, liver function tests, and in men over 30 and women over 40, an electrocardiogram. (In the initial assessment of a psychotic patient, a differential diagnosis may support assessment of other organ systems or metabolic and endocrine functions. In addition, it would be wise to perform routine screening for syphilis, chest disease, and other illness.[57]) During treatment, the clinician should occasionally recheck a patient's health status, although it makes little sense to repeat laboratory tests more frequently than once per year in asymptomatic patients. To recapitulate earlier points: liver disease does not automatically contraindicate the use of antipsychotic drugs; maintaining a high clinical index of suspicion for agranulocytosis and jaundice within the early months of therapy is superior to routine laboratory testing (except for clozapine, which requires weekly blood counts). Routine EKGs may be prudent for patients maintained on pimozide.

As mentioned in the section on tardive dyskinesia, monitoring of patients for early signs of abnormal involuntary movements is wise before and during chronic antipsychotic drug therapy. Prior to treatment and at intervals of 6 to 12 months thereafter, the performance of a basic examination (such as the AIMS) would be worthwhile (see Fig. 2). Because of concerns about breast tumors in patients taking these drugs (see Section III.G.6), women should be periodically examined (and taught to examine themselves, if feasible) for breast lumps.

When patients are taking antipsychotic drugs, various laboratory tests may be artifactually altered. As examples, invalid increases may be reported in serum alkaline phosphatase, serum and urine bilirubin, cerebrospinal fluid protein, serum cholesterol, urine diacetate, serum 17-hydroxysteroids, urine porphyrins, serum transaminases, and urine urobilinogen. In addition, false-positive results may be reported in urinary phenylketonuria and pregnancy testing. Protein-bound iodine (PBI), radioactive iodine, and serum and urine uric acid testing may be inaccurate as well.

J. Clinical Uses of the Antipsychotic Drugs

Based on a medical and psychological evaluation of the patient, the clinician should determine the need for medication and its place within the treatment plan (see Sections I and II in this chapter). The choice of a specific agent depends on the patient's medication history and an understanding of pharmacokinetics, adverse effects, toxicity, and drug interactions. The clinician should discuss his decision to medicate with the patient and/or his family, guardian, etc. (For further discussion of informed consent, competency, etc., see Chapter 14). Pretreatment laboratory tests might be useful to assess a patient's general health prior to initiating drug therapy (see Section III.I).

1. Acute Treatment

In the early placebo-controlled trials of phenothiazines in the treatment of acute schizophrenic psychosis, about one-fourth of patients who received placebo responded reasonably well. Unfortunately, we are unable to predict in advance those psychotic patients who do not require medication. For each patient, a different amount of medication may be required to contain each exacerbation. Also, a patient may take several weeks or more to show optimal antipsychotic effect to a given dose of medication. Nevertheless, it is occasionally possible to allow an acutely psychotic person a short drug-free period to improve prior to instituting the therapy. In an acutely psychotic individual without a long history of psychosis, transferring the patient from a stressful situation to a benign, supportive, and nonconfrontative environment, such as the hospital, may itself be helpful. If the person's behavior is not overtly disruptive, he can be given an opportunity to reconstitute spontaneously over several days. If spontaneous improvement does not occur, if the severity of the psychosis or the nature of the patient's behavior precludes a drug-free interval, or if the patient has a long history of psychotic exacerbations, then antipsychotic drug therapy should be started.

Some physicians used to employ extremely large doses of high-potency antipsychotic drugs early in treatment. This approach, known as rapid neuroleptization, attempted to achieve rapid remission of psychosis and rapid discharge from the hospital. Unfortunately, double-blind studies have been unable to confirm any enhanced efficacy of this form of treatment contrasted with standard dosage regimens. In other words, treating all psychotic patients with "megadoses" does not control psychosis or allow discharge any more rapidly, and it is associated with an increase in untoward effects. Thus, the clinician should treat most patients at the lower end of the therapeutic range for acute psychosis, e.g., 5 to 10 mg/day of haloperidol or fluphenazine. Additional sedation can be achieved through the adjunctive administration of a benzodiazepine [e.g., lorazepam (Ativan® and others)]. Only patients who in the present or past have been unresponsive to such standard doses administered over a period of weeks should be considered for higher doses.

For the highly excited or dangerous patient, the combination of a neuroleptic drug and a benzodiazepine may provide rapid pharmacological control in the context of a calm but structured environment. A high-potency drug, such as haloperidol, can be administered parenterally and is less likely to produce hypotension or seizures than lower-potency antipsychotic agents.[58] A dose of 2.5 to 10 mg can be administered intramuscularly every 1–2 hr, alternating with intramuscular injections of lorazepam (the only benzodiazepine to achieve reliable blood levels when administered through this route), 1 to 2 mg, until the patient is quiet. Vital signs should be monitored at least hourly.

An alternative to haloperidol in a combined regimen would be mesoridazine, 25–50 mg, which is more sedating than haloperidol and less likely to produce extrapyramidal reactions, but a drug that causes more hypotension.

Yet a third option for parenteral neuroleptic therapy is droperidol. Available in the United States labeled only for use in anesthesia, droperidol is a butyrophenone like haloperidol, but more sedating and possibly faster acting. It has been described as very effective in emergency management of acutely excited and violent patients, but because it has no oral preparation currently available, the patient must later be switched to another product.

Early in the treatment course, administer antipsychotic drugs in divided doses, perhaps three or four times per day (more frequently for patients who are severely excited). Once satisfactory levels have been achieved, and the patient has become tolerant to unwanted effects (probably within several weeks), a once-daily dosage usually will suffice. If a patient can be maintained out of the hospital, a once-daily dosage, even from the beginning, may enhance compliance. Particularly when a more sedating agent is employed, a bedtime dose may be optimal, as it can improve sleep, and the peak anticholinergic and hypotensive effects will occur while the patient is asleep.

What if a patient does not appear to respond to an antipsychotic drug even after 6 to 8 weeks? First, the clinician must wonder whether the patient is actually receiving the agent. An outpatient may not have filled a prescription or may be taking the drug irregularly if at all. An inpatient might be "cheeking" tablets or capsules. Questions to be asked at this juncture include: Is anyone observing the patient taking the medication, including an inspection of the patient's mouth? Does the patient have any noticeable "side effects" from the drug, such as extrapyramidal reactions? If compliance is a particular problem, consider changing the dosage regimen, administering a liquid preparation, or using a parenteral agent. Always ask patients how they feel about a drug. They might reveal covert effects such as sexual dysfunction or akathisia, which may be improved by switching to a different agent; or, one may learn of a patient's wish to remain "crazy."

If compliance is not a problem, yet the patient has not improved, he may require a higher dose of medication. Increase the dosage until improvement occurs or limiting adverse effects develop. Although U.S. Food and Drug Administration (FDA)-approved labeling (contained in the *Physicians' Desk Reference* and in the package insert) provides useful guidelines about maximal dosage, the actual upper limit may vary from patient to patient. The one exception to this rule is thioridazine, which must not be administered in dosages exceeding 800 mg per day—even for a brief period of time—because of the danger of pigmentary retinopathy with potential loss of vision.

Some patients metabolize antipsychotic drugs very rapidly in the gut and in the first pass through the liver. A physician might suspect that this is the case in a patient who is taking relatively large doses orally, yet experiences neither clinical benefit nor adverse effects. These patients may respond to higher oral doses or to a parenteral preparation.

Although all antipsychotic drugs are equally effective, some patients appear to do better with one agent than another. Thus, failure to improve with one (after the above factors have been considered) is a reason to change to a different agent.

Clinicians should be wary, however, of a tendency to be "too quick on the prescription trigger," switching agents every few days in a vain attempt to "do something." Only after a drug has been given an **adequate trial (at least 300–600 mg of chlorpromazine—equivalent dose per day for 6 to 8 weeks)** should the clinician switch to a different agent. When it is time to change, it makes most sense to choose a drug in a different chemical class.

One other consideration in the apparently refractory patient is psychosocial factors, which may be "fueling" the psychosis. Antipsychotic drugs raise the threshold at which an individual predisposed to psychosis actually becomes psychotic in response to interpersonal and intrapsychic stresses. A chronic schizophrenic, for example, might remain quite stable and relatively intact (albeit still awkward, isolated, and apathetic) when in a low-stress environment, even with little or no medication. However, given a precipitant (an argument with a parent or being placed in a "double bind"), he may decompensate into a state of disorganized thinking and be unable to distinguish reality from fantasy.

It takes more medication to "cap" a psychosis in the face of ongoing stresses and threats than in more placid circumstances. In fact, if the pressure is sufficient, psychochemistry alone will not bring the psychosis under control. Hence, the clinician must consider psychological and environmental factors when treating an acutely decompensated patient. If a patient remains at home and is unimproved despite adequate chemotherapy, change the environment; e.g., arrange for hospital admission. If he is in the hospital, inquire about ongoing contacts both inside and outside the hospital. Occasionally, a schizophrenic patient's mother will sit on his lap during visiting hours, another patient may be making sexual advances, or a naive staff member might be getting provocatively close. In such circumstances, decreasing the intensity of the patient's conflict may allow a hitherto ineffective drug to become strikingly efficacious.

And how effective is effective for an antipsychotic drug? The drugs are best at suppressing the most florid symptoms of psychosis such as hallucinations, delusions, excited, assaultive, and belligerent behavior, and disorganized thinking. If the patient has a primary disorder of mood, reconstitution should involve what Bleuler called restitutio ad integrum—in other words, a return to an integrated and normal state of personality and functioning. For a schizophrenic patient, however, recovery from psychosis is likely to involve return to a somewhat strange, awkward, and isolated existence.

2. Transition and Continuation Therapy

When psychotic symptoms have remitted successfully, we pass from the acute phase of active symptom suppression to continuation therapy. Continuation therapy entails the continued administration of medication in the hope of avoiding a relapse. This is analogous to continuing antimicrobial therapy past the period of symptom relief and fever defervescence. After the alleviation of an acute psychotic episode, continue therapy for about 6 months.

It is best not to change the type and dosage of antipsychotic medication at the time of an important transition such as discharge from the hospital. If a patient has responded favorably to an antipsychotic drug, continue the same dose of that drug through the return to home, job, and community. Antipsychotic agents presumably provide a buffer for the stresses of the transitional period.

If a person has resumed a normal life without evidence of recurrent psychosis, the physician should gradually lower the dose of the antipsychotic drug. If dosage tapering has proceeded smoothly, within about 6 months consider discontinuing the agent entirely. However, when a psychotic illness has shown itself to be chronic, with frequent recurrences (a common pattern among schizophrenic patients), maintenance therapy may be in order.

3. Maintenance Therapy

a. Definition

Maintenance therapy means long-term treatment. For some patients, usually those with chronic schizophrenia, discontinuing maintenance antipsychotic drug therapy (or even reducing the dose) leads to a rapid return of psychotic symptoms. For these patients, the drugs play an ongoing suppressive role.

For other patients on maintenance therapy, the drugs appear to play a "prophylactic" role. In this group, drug discontinuation raises the probability of a symptomatic relapse at an indeterminate time in the future. For them, most probably the drug raises the threshold to psychosocial stressors that can precipitate a reexacerbation of the illness. In the absence of this protective effect, a milder stress can lead to symptom return.

b. Chronic Schizophrenia

If a group of schizophrenic patients in remission is switched from active antipsychotic medication to placebo, roughly 65% to 80% will relapse within about 1 year. How long a patient will go before relapsing is variable and unpredictable. In any given month, roughly 10% of the remaining unrelapsed medication-free patients may be expected to relapse. At the end of 1 to 2 years, a small cohort of schizophrenic patients will not have relapsed despite the lack of antipsychotic drug therapy; they have less need for maintenance treatment. Unfortunately, we are unable to identify these patients in advance. Patients clinically stable on relatively low doses of antipsychotic drugs may be more likely to maintain remission following complete cessation of the agent. Also, patients in relatively low-stress environments are more likely to remain stable without drugs, although the onset of stress can heighten their vulnerability.

As in the treatment of acute psychosis, growing evidence suggests that maintenance therapy too may be approached with lower neuroleptic doses. For high-potency agents such as haloperidol or fluphenazine, 0.5 to 1 mg daily may suffice to prevent symptom exacerbation and relapse for many chronic patients. In maintenance therapy less may be more: in other words, lower doses can often control symptoms while at the same time diminishing the risks of tardive dyskinesia as well as of less troubling but more annoying side effects that might undermine compliance. Research has shown that lower antipsychotic drug doses do, on average, lead to higher rates of relapse, especially with longer periods of follow-up.[59] Nevertheless, relapses in patients on low-dose antipsychotic drug therapy tend to be less severe than those occurring in patients free of medication: they remit more easily when the drug dose is raised and are less likely to result in social disruption and the need for hospitalization.

Several groups of investigators have been examining an intermittent or "targeted dose" strategy for chronic schizophrenic patients. This involves drug treatment only during an acute episode, with drug discontinuation afterwards. During maintenance therapy patients are carefully monitored, and family members know how to reach a clinician promptly to report prodromal symptoms. This approach relies heavily on therapeutic rapport, and it also assumes that the clinician can identify prodromal symptoms, which may include such nonspecific evidence of distress as anxiety and insomnia. The validity, practicality, and utility of this approach are still being studied.

Perhaps the most practical solution lies in combining the low-dose and intermittent treatment strategy. Patients can be maintained for prolonged periods on very low doses of antipsychotic drugs, but any evidence of a "roughening" of their clinical course should prompt a rapid raising of the dose until symptoms subside. Thus, a patient receiving, say 0.25 cc of fluphenazine decanoate every 3–4 weeks might receive a supplement of 1–5 mg of fluphenazine hydrochloride daily by mouth or an additional 0.25 cc of decanoate until acute symptoms have subsided.

The use of "drug holidays"—i.e., periodic, planned withdrawal of neuroleptic therapy—has fallen out of favor. Initially designed as a means to lower a patient's exposure to these agents, drug holidays have not appeared to lower the risk of tardive dyskinesia (and may, in fact, actually increase it). By contrast, when patients are taking these medicines chronically, a 5-out-of-7-day schedule (i.e., once-daily dosing Monday through Friday, with the weekend "off") does not represent a neuropharmacological "holiday," since the brain will hardly notice the falloff in levels (because of saturation of fat and tissue stores). This approach may make sense for patients in day treatment programs, who can take the medication once each weekday in front of a nurse.

As of this writing, there are three long-acting antipsychotic drugs available by prescription in the United States: the enanthate and decanoate esters of the phenothiazine fluphenazine and the decanoate ester of the butyrophenone haloperidol. Fluphenazine decanoate probably has a longer duration of action than its enanthate sister, which is infrequently used.

Fluphenazine decanoate is usually administered intramuscularly (it may also be given subcutaneously) in doses of 12.5 mg (0.5 cc) through 75 mg (3 cc) every 3 to 4 weeks. Although some patients have received injections greater than 125 mg (5 cc) at a time (requiring the use of several injections), there is a growing recognition that patients can be maintained on the lower end of the dosage scale, with some receiving as little as 1.25 mg (0.5 cc) at each dose. Additionally, the dosing interval can be progressively increased so that patients need only receive injections every 1 to 2 months, based on their clinical responses. Symptomatic exacerbation is best treated through the use of oral or IM fluphenazine hydrochloride (short-acting), although the maintenance dose also may be increased if need be. Since once it is injected a decanoate cannot be removed, it is best to err on the lower side and supplement with shorter-acting preparations. Unfortunately, there is no established conversion factor to equate a dose of oral fluphenazine hydrochloride to a dose of the decanoate, but each can be titrated within the usual therapeutic ranges.

Haloperidol decanoate is recommended for once-monthly administration, but there are probably patients who can take it, too, at less frequent intervals. The concentration of haloperidol in the decanoate preparation is 50 mg/cc. For patients already stabilized on oral haloperidol, the recommended conversion formula is to multiply the daily dose by 10 to 15 and then to administer that dose (in milligrams) of haloperidol decanoate in a monthly injection. Again, the best approach is to "start low and go slow," particularly when using a long-acting preparation.

Encouraging evidence accumulated in the late 1980s suggests that clozapine, besides having a lower profile of extrapyramidal effects, can also alleviate chronic symptoms of psychosis in patients previously resistant to standard neuroleptics. A large collaborative study showed substantial improvement in 30% of patients with treatment-resistant schizophrenia.[60] Additional longitudinal data suggest that, after a year's treatment, more than two-thirds of patients in this category may become responders.[61] Thus, although encumbered by agranulocytosis and other potentially serious side effects, clozapine may bring relief to a substantial subset of previously treatment-resistant schizophrenics.

Because of clozapine's unique side-effect profile, specific approaches to clinical management are required for patients maintained on the drug.[62] As mentioned earlier, patients who have recovered from clozapine-induced agranulocytosis should never again be exposed to this drug. Those with seizures can, however, be reexposed, possibly in conjunction with an anticonvulsant, and certainly at a lower initial dose. Sedation can be managed with a lower dose or a bedtime biasing of the dosage schedule and, as with other psychotropic medications, symptomatic hypotension may be treated with support stockings, increased dietary sodium, or the mineralocorticoid fludrocortisone (Florinef® and others). Hyperthermia is common with clozapine, but usually benign and transient; infection and NMS must be ruled out. Occasional patients who develop nausea and, less commonly, vomiting after weeks or months of clozapine therapy have been treated successfully with metoclopramide (Reglan® and others).

The Use of Fluphenazine

1. Initially administer 0.25 cc of fluphenazine decanoate.
2. Supplement with oral or IM fluphenazine hydrochloride as needed.

3. Give second dose of 0.25 cc of fluphenazine decanoate after 4 weeks.
4. Begin to taper oral fluphenazine.
5. If you can't stop oral fluphenazine without reemergence of psychotic symptoms, increase decanoate dose to 0.5 cc at next 4-week administration.
6. Once patient is stable on fluphenazine decanoate alone, see if a 5- or 6-week interval suffices. (Some patients may need decanoate only every 2 months or less often!)
7. If patient begins to become symptomatic toward the end of a dosing interval, shorten the interval. (Rare patients may need fluphenazine decanoate as often as every 2 weeks.)
8. If a patient begins to become symptomatic around a time of stress, supplement decanoate with oral or IM fluphenazine hydrochloride as needed or increase the decanoate dose for one to two dosing intervals until the stress and symptoms pass.

c. Other Diagnostic Groups

In no other psychiatric disorder is the indication for maintenance antipsychotic drug therapy as clearly established as it is in the treatment of the schizophrenic syndrome. Nevertheless, clinical experience suggests that patients with other diagnoses occasionally benefit from long-term treatment with antipsychotic drugs. For example, an occasional patient with bipolar affective disorder may benefit from maintenance therapy with an antipsychotic agent, either in addition to or as a substitute for lithium. Similarly, a patient with recurrent episodes of psychotic depression might be assisted. We must remember, however, that the risk of developing tardive dyskinesia or dystonia following neuroleptic exposure may be increased in patients with mood disorders. Some patients with uncontrollable behavior caused by mental retardation or organic brain syndrome will be more manageable on maintenance antipsychotic drug therapy; however, it should be demonstrated that nonpharmacological approaches, such as environmental alteration and behavioral therapy, do not alleviate the problem. Patients with borderline personalities may benefit from antipsychotic drugs, usually in relatively low doses, for long-term assistance with uncontrollable waves of emotions and, at times, destructive impulses (see Chapter 13). The butyrophenone haloperidol and the diphenylbutylpiperidine pimozide are often used for chronic symptom suppression in patients with Gilles de la Tourette syndrome. Pimozide might also be useful in treating monosymptomatic hypochondriacal psychosis such as delusions of parasite infestation or dysmorphophobia. In patients taking pimozide, the EKG should be carefully monitored for QT prolongation.

If antipsychotic drugs are to be used for disorders other than schizophrenia, particularly when they are used over a period of many months and even years, the clinical record should reflect in detail the rationale behind the therapy and the care taken in its administration. In light of the long-term risks of antipsychotic drug treatment, a patient's chart should indicate why alternate treatments were not employed. The record should also detail the continued need for antipsychotic agents, attempts made to lower the dose or discontinue their use, and steps taken to monitor for unwanted effects and toxicity, such as tardive dyskinesia. Finally, seek consultation, even a quick "curbside consult" with a colleague, and note it in the record.

K. New Drugs

Since introducing chlorpromazine about 30 years ago, we have seen the development of two other subclasses of phenothiazines and four other distinct groups of antipsychotic drugs. However, with the exception of a spectrum of unwanted effects, there has been little new until clozapine. Despite advertising ballyhoo, most of the drugs introduced since chlorpromazine have fallen more or less into the me-too category: no better, no worse, no different. This is not to say that the availability of a spectrum of antipsychotic drugs does not facilitate clinical practice; it does. The ability to choose among agents allows physicians to minimize unwanted effects for each patient. Also, a patient may respond better to one drug than to another.

What can we expect in the future? In the past, the drug companies, unwilling to gamble on "long shots," tested the same compounds in animal models for potential antipsychotic action, resulting in rediscovery of the same type of agent despite chemical differences in the molecules themselves. Thus, the pharmacological profile of new drugs displayed a monotonous similarity to the old drugs.

An exception to this me-too lockstep is the newly introduced dibenzoxazepine drug clozapine. Clozapine is an effective, low-potency, sedating antipsychotic drug that is distinguished by its lack of extrapyramidal effects. Moreover, clozapine produces little elevation in plasma prolactin. Although other explanations have been given, it is possible that clozapine is a "smart" dopamine blocker, effectively counteracting dopamine receptors in brain centers that run amok in psychosis (mesolimbic and mesocortical pathways?) while having relatively little effect on neurological (nigrostriatal) and hormonal (tuberoinfundibular) tracts. However, clozapine has an unfortunate tendency to produce a high incidence of agranulocytosis as well as hypotension and hypothermia.

Unlike many other stories in pharmacology, in which the discovery of one compound with unique properties was soon followed by a parade of similar moieties, no one has discovered anything else like clozapine. Ideally, we would like another nonneuroleptic antipsychotic drug, but with less toxicity. Since such an ideal agent has not been forthcoming, despite several withdrawals and reintroductions in clinical use in Europe and research in the United States, clozapine has now been introduced for prescription use, although ringed with caveats, warnings, and precautions. Adding to our interest in this new–old drug, however, are recent data suggesting that some patients previously refractory to more standard neuroleptics may respond favorably to clozapine.

The β-adrenergic blocking agent propranolol has been tried, alone and combined with neuroleptics, for the treatment of psychosis for about two decades. Unfortunately, the lack of sufficient systematic research leaves more questions than answers. By itself, propranolol looks less than impressive in alleviating psychosis. It may, however, have a role to play in diminishing assaultive and explosive outbursts, particularly in patients with underlying organic brain disease. As an adjunct to antipsychotic drugs, propranolol might be useful in some treatment-resistant patients, although it is possible that its role is no more complicated than raising the blood level of the antipsychotic agent.

Physicians contemplating the use of propranolol should thoroughly familiarize themselves with this agent and with β blockers in general. For example, patients with asthma and other bronchospastic conditions should usually avoid it altogether. Diabetics and those with heart disease may have an increased risk of toxicity. All patients should have the drug tapered gradually, as abrupt discontinuation may be life threatening. Although many patients take heroic doses with impunity, bradycardia and hypotension can present severe risks.

Carbamazepine (Tegretol® and others) and other anticonvulsants are occasionally tried in psychotic patients who have failed to improve with standard neuroleptics or have evidence that suggests an underlying seizure disorder (whether by history, clinical signs and symptoms, or on the electroencephalogram). Lithium is another agent sometimes considered, alone or as an antipsychotic drug adjunct, particularly for patients with affective symptoms in addition to psychosis. Antidepressant drugs may have a role to play in schizophrenic patients who also have symptoms of major depression, but they are unlikely to help the apathy and anhedonia that make up the negative symptoms of this condition. Other drugs that have been tested recently without conclusive results include the norepinephrine-diminishing antihypertensive agent clonidine and the calcium channel blocker verapamil (Calan® and others).

A number of other substances also are in various phases of experimentation. The outcome will depend on the quirks of nature, the vicissitudes of public funding for psychiatric research, and the degree to which private industry perceives a potential for profit.

IV. CONCLUSION

Psychiatrists continue to hope that better agents will be developed for the treatment of psychosis. These would include more effective drugs, drugs that could reach that minority of schizophrenic patients who remain refractory to currently available agents, substances with fewer unwanted effects and a lower incidence of toxicity, and a generally broader range of clinical options. In particular, we wish for drugs that will not cause tardive dyskinesia, yet are without the other toxicities of clozapine.

Yet, despite their shortcomings, the current crop of antipsychotic drugs have been strikingly impressive in the relief they have provided for many patients and families. Although imperfect, they have largely been able to quell the onslaught of the more virulent aspects of psychosis. And although they have a broad range of untoward reactions and toxicity associated with their use, they are nonaddicting, have a high therapeutic index, and, considering the extent of their usage, are relatively nontoxic.

REFERENCES

1. May P. R. A., Tuma A. H., Yale C., et al: Schizophrenia—a follow-up study of results of treatment. II. Hospital stay over two to five years. *Arch Gen Psychiatry* 33:431–506, 1976.
2. Hogarty G. E., Goldberg S. C., Schooler N. R., et al: Drugs and social therapy in the aftercare of schizophrenic patients. II. Two-year relapse rates. *Arch Gen Psychiatry* 31:603–608, 1974.
3. Epstein N. B., Vlok L. A.: Research on the results of psychotherapy: A summary of evidence. *Am J Psychiatry* 138:1027–1035, 1981.
4. Schooler N. R.: The efficacy of antipsychotic drugs and family therapies in the maintenance treatment of schizophrenia. *J Clin Psychopharmacol* 6:11S–19S, 1986.
5. Gelenberg A. J.: The catatonic syndrome. *Lancet* 1:1339–1341, 1976.
6. Gelenberg A. J.: ECT: Controversies and consensus. *Mass Gen Hosp Newslett Biol Ther Psychiatry* 8:29–32, 1985.
7. Bartlett J., Bridges P., Kelly D.: Contemporary indications for psychosurgery. *Br J Psychiatry* 138:507–511, 1981.
8. Lipton M. A., Mailman R. B., Nemeroff C. B.: Vitamins, megavitamin therapy, and the nervous system, in Wurtman R. J., Wurtman J. J. (eds): *Nutrition and the Brain*, vol 3: *Disorders of Eating and Nutrients in Treatment of Brain Diseases*. New York, Raven Press, 1979, pp 183–264.
9. Gelenberg A. J.: Psychiatric disorders. In: Paige, D. M. (ed): *Clinical Nutrition*, ed 2. St. Louis: C V Mosby, 1988, pp 419–420.
10. Bassuk E. L.: The homelessness problem. *Sci Am* 251:40–45, 1984.
11. McEvoy J. P.: The neuroleptic threshold as a marker of minimum effective neuroleptic dose. *Compr Psychiatry* 27:327–335, 1986.
12. McEvoy J. P., Stiller R. L., Farr R.: Plasma haloperidol levels drawn at neuroleptic threshold doses: A pilot study. *J Clin Psychopharmacol* 6:133–138, 1986.
13. Chouinard G., Jones B. D.: Neuroleptic-induced supersensitivity psychosis: Clinical and pharmacologic characteristics. *Am J Psychiatry* 137:16–31, 1980.
14. Ross C. A.: Buspirone in the treatment of tardive dyskinesia. *Med Hypoth* 22:321–328, 1987.
15. Cohen B. M., Lipinski J. F.: In vivo potencies of antipsychotic drugs in blocking alpha-1 noradrenergic and dopamine D-2 receptors: Implications for drug mechanisms of action. *Life Sci* 39:2571–2580, 1987.
16. Gelenberg A. J.: Treating the outpatient schizophrenic. *Postgrad Med* 64:48–56, 1978.
17. Taylor W. J., Caviness M. H. D.: *A Textbook for the Clinical Application of Therapeutic Drug Monitoring*. Irving, Texas, Abbott Laboratories, 1986, p. 467.
18. Ko G. N., Korpi E. R., Linnoila M.: On the clinical relevance and methods of quantification of plasma concentrations of neuroleptics. *J Clin Psychopharmacol* 5:253–262, 1985.
19. Keepers G. A., Clappison V. J., Casey D. E.: Initial anticholinergic prophylaxis for neuroleptic-induced extrapyramidal syndromes. *Arch Gen Psychiatry* 40:1113–1117, 1983.
20. VanPutten T., Mutalipassi L. R., Malkin M. O.: Phenothiazine-induced decompensation. *Arch Gen Psychiatry* 30:102–106, 1974.
21. Gelenberg A. J., Mandel M. R.: Catatonic reactions to high potency neuroleptic drugs. *Arch Gen Psychiatry* 34:947–950, 1977.

22. Rifkin A., Quitkin F., Klein D. F.: Akinesia: A poorly recognized drug-induced extrapyramidal behavior disorder. *Arch Gen Psychiatry* 32:642–674, 1975.

23. Wojcick J. D.: Antiparkinson drug use. *Mass Gen Hosp Biol Ther Psychiatry Newslett* 2:5–7, 1979.

24. Gelenberg A. J., VanPutten T., Lavori P., et al: Anticholinergic effects on memory: Benztropine versus amantadine. *J Clin Psychopharmacol* 9:180–185, 1989.

25. Gelenberg A. J.: Amantadine in the treatment of benztropine-refractory neuroleptic-induced movement disorders. *Curr Ther Res* 23:375–380, 1978.

26. Smith J. M.: Abuse of the antiparkinson drugs: A review of the literature. *J Clin Psychiatry* 41:351–354, 1980.

27. Jeste D. V., Wyatt R. J.: Changing epidemiology of tardive dyskinesia: An overview. *Am J Psychiatry* 138:297–309, 1981.

28. Burke R. E., Fahn S., Jankovic J., et al: Tardive dystonia: Late-onset and persistent dystonia caused by antipsychotic drugs. *Neurology* 32:1335–1346, 1982.

29. Morgenstern H., Glazer W. H., Niedzwiecki D., et al: The impact of neuroleptic medication on tardive dyskinesia: A meta-analysis of published studies. *Am J Public Health* 77:717–724, 1987.

30. Wojcik J. D., Gelenberg A. J., Labrie R. A., et al: Prevalence of tardive dyskinesia in an outpatient population. *Comp Psychiatry* 21:370–380, 1980.

31. Kane J. M., Woerner M., Borenstein M., et al: Integrating incidence and prevalence of tardive dyskinesia. Read before the IVth World Congress of Biological Psychiatry, Philadelphia, Sept 8–13, 1985.

32. Yassa R., Nair V., Dimitry R.: Prevalence of tardive dyskinesia. *Acta Psychiatr Scand* 73:629–633, 1986.

33. Gardos G., Cole J. O., Tarsy D.: Withdrawal syndromes associated with antipsychotic drugs. *Am J Psychiatry* 135:1321–1324, 1978.

34. Quitkin F., Rifkin A., Gochfeld L., et al: Tardive dyskinesia: Are first signs reversible? *Am J Psychiatry* 134:84–87, 1977.

35. Baldessarini R. J., Tarsy D.: Tardive dyskinesia, in Lipton M. A., DiMascio A., Killam K. F. (eds): *Psychopharmacology: A Generation of Progress.* New York, Raven Press, 1978, pp 993–1004.

36. Casey D. E., Povisen U. J., Meidahl B., et al: Neuroleptic-induced tardive dyskinesia and parkinsonism: Changes during several years of continuing treatment. *Psychopharmacol Bull* 8(22):250–253, 1986.

37. Yagi G., Itoh H.: A ten-year follow-up study of tardive dyskinesia—with special reference to the influence of neuroleptic administration on the long-term prognosis. *Keio J Med* 34:211–219, 1985.

38. Gelenberg A. J., Dorer D., Wojcik J., et al: A crossover study of lecithin for tardive dyskinesia. *J Clin Psychiatry* 51:149–153, 1990.

39. Caroff S. N.: The neuroleptic malignant syndrome. *J Clin Psychiatry* 41:79–83, 1980.

40. Gelenberg A. J., Bellinghausen B., Wojcik J. D., et al: A prospective survey of neuroleptic malignant syndrome in a short-term psychiatric hospital. *Am J Psychiatry* 145:517–518, 1988.

41. Shalev A., Munitz H.: The neuroleptic malignant syndrome: Agent and host interaction. *Acta Psychiatrica Scand* 73:337–347, 1986.

42. Levinson D. F., Simpson G. M.: Neuroleptic-induced extrapyramidal symptoms with fever: Heterogeneity of the "neuroleptic malignant syndrome." *Arch Gen Psychiatry* 43:839–848, 1986.

43. Gelenberg A. J., Bellinghausen B., Wojcik J. D., et al: Patients with NMS histories: What happens when they are rehospitalized? *J Clin Psychiatry* 50:178–180, 1989.

44. Baldessarini R. J., Gelenberg A. J.: Using physostigmine safely. *Am J Psychiatry* 136:1608–1609, 1979.

45. Lieberman J. A., Johns C. A., Kane J. M., Rai K., Pisciotta A. V., Saltz B. L., Howard A.: Clozapine-induced agranulocytosis: Non-cross-reactivity with other psychotropic drugs. *J Clin Psychiatry* 49:271–277, 1988.

46. Zarrabi M. H., Zucker S., Miller F., et al: Immunologic and coagulation disorders in chlorpromazine-treated patients. *Ann Intern Med* 91:194–199, 1979.

47. Zucker S., Zarrabi H. M., Schuback W. H., *et al.*: Chlorpromazine-induced immunopathy: Progressive increase in serum IgM. *Medicine* 69:92–100, 1990.

48. Gelenberg A. J.: Psychotropic drugs during pregnancy and the perinatal period. *Mass Gen Hosp Biol Ther Psychiatry Newslett* 2:41–42, 1979.

49. Chouinard G., Jones B. D.: Neuroleptic-induced supersensitivity psychosis: Clinical and pharmacologic characteristics. *Am J Psychiatry* 137:16–21, 1980.

50. Rivera-Calimlim L., Castaneda L., Lasagna L.: Effect of management on plasma chlorpromazine in psychiatric patients. *Clin Pharmacol Ther* 14:978–986, 1973.

51. Dysken M. W., Javaid J. I., Chang S. S., et al: Fluphenazine pharmacokinetics and therapeutic response. *Psychopharmacology* 73:205–210, 1981.

52. Cohen W. J., Cohen N. H.: Lithium carbonate, haloperidol, and irreversible brain damage. *JAMA* 230:1283–1287, 1974.

53. Spring G. K.: Neurotoxicity with combined use of lithium and thioridazine. *J Clin Psychiatry* 40:135–138, 1979.

54. Gelenberg A. J.: Is it safe to co-prescribe lithium with a neuroleptic? *Mass Gen Hosp Biol Ther Psychiatry Newslett* 2:13, 1979.

55. Rivera-Calimlim L., Kerzner B., Karch F. E.: Effect of lithium on plasma chlorpromazine level. *Clin Pharmacol Ther* 23:451–455, 1978.

56. Gelenberg A. J.: Coffee, tea, and antipsychotic drugs revisited. *Mass Gen Hosp Biol Ther Psychiatry Newslett* 4:42–43, 1981.

57. Gelenberg A. J.: Psychiatric emergencies: The psychotic patient. *Drug Ther* May 1981 pp 25–36.

58. Granacher R. P. Jr: Management of the acutely psychotic patient. Read before the Symposium on Antipsychotic Drug Therapy: Current Concepts and Future Trends, Key Biscayne, Florida, March 28, 1986.

59. Marner S. R., Putten T., Mintz J., et al: Low- and conventional-dose maintenance therapy with fluphenazine decanoate. *Arch Gen Psychiatry* 44:518–521, 1987.

60. Kane J., Honigfeld G., Singer J., Meltzer H.: The Clozaril Collaborative Study Group: Clozapine for the treatment-resistant schizophrenic: A double-blind comparison vs chlorpromazine/benztropine. *Arch Gen Psychiatry* 45:789–796, 1988.

61. Meltzer H. Y.: Duration of a clozapine trial in neuroleptic-resistant schizophrenia. Letter to editor. *Arch Gen Psychiatry* 46:672, 1989.

62. Lieberman J. A., Kane J. M., Johns C. A.: Clozapine: Guidelines for clinical management, *J Clin Psychiatry* 50:329–338, 1989.

63. Lader M.: *Introduction to Psychopharmacology.* Kalamazoo, Upjohn, 1980, p 30.

5

Anxiety

JERROLD F. ROSENBAUM, M.D., and ALAN J. GELENBERG, M.D.

I. INTRODUCTION

A. Anxiety: A Definition

Anxiety is a universal human response to routine stress and emotional conflict that is experienced both psychologically and physiologically. It is important to distinguish "normal" anxiety from "pathological" anxiety or **anxiety disorder.** Pathological anxiety may be distinguished from normal by its autonomy, intensity, duration, or associated behavior. When "autonomous," anxiety appears to have a "life of its own," with minimal basis in identifiable environmental stimuli. The "intensity" of symptomatic distress for pathological anxiety frequently exceeds the patient's capacity to bear the discomfort; the experience, therefore, is unlikely to engender a healthy, adaptive response. When symptoms recur or persist over time, the duration of anxious suffering will typically indicate pathology. Finally, pathological anxiety may trigger such stereotyped behavioral responses as avoidance or lifestyle constriction.

Similar to fear, anxiety symptoms are generally of two types: the first is a sustained state of increased arousal, concern, worry, and vigilance resembling the uncertainty, caution, and alerted stance of an individual passing through a potentially dangerous setting, like a dark, deserted alleyway; the second type resembles the state of alarm, terror, or panic that would be triggered by an actual sudden life threat, as if an assailant had suddenly emerged from that same alleyway. The former, analogous to generalized or anticipatory anxiety, does not appear to be just a milder variant of the latter, which resembles panic attacks. Panic attacks themselves

JERROLD F. ROSENBAUM, M.D. • Clinical Psychopharmacology Unit, Massachusetts General Hospital, Boston, Massachusetts 02114; Department of Psychiatry, Harvard Medical School, Boston, Massachusetts 02115. **ALAN J. GELENBERG, M.D.** • Department of Psychiatry, University of Arizona Health Sciences Center, Tucson, Arizona 85724.

vary in intensity and symptoms and include mild and limited symptom attacks as well as major attacks. Pharmacological and familial studies suggest a clinically relevant distinction between these two anxiety states.

B. Symptoms

Anxiety is experienced cognitively, affectively, physically, and behaviorally.

1. Cognitive

The cognitive experience of anxiety typically involves apprehensive, vigilant, and pessimistic appraisal of circumstances. For a panic attack, the cognitive component is one of catastrophic thinking, with appraisal of the attack as a life-threatening ailment or one likely to lead to insanity or great embarrassment. With generalized or anticipatory anxiety, the cognitive response is more one of rumination, worry, and overconcern.

2. Affective

Anxiety is a dysphoric state of emotional arousal that frequently drives the sufferer to seek relief. Many who have suffered panic attacks, for example, describe the event as one of the most unpleasant experiences of their life; they would go to any length to avoid a subsequent similar experience.

3. Physical

The intensity of physical symptoms of anxiety varies, as does the primary somatic locus of anxious suffering among individuals. Typical symptoms are cardiac (palpitations, tachycardia, chest tightness), respiratory (hyperventilation, dyspnea, shortness of breath), neurological (dizziness, paresthesias, tremulousness), or gastrointestinal (nausea, intestinal cramping, diarrhea). Other autonomic symptoms frequently occur (hot or cold flashes, diaphoresis, dry mouth).

4. Behavioral

Possible behavioral concomitants of anxiety include avoidance, flight, a freeze response, help-seeking behavior, or other maladaptive attempts to diminish the experience, such as excessively dependent behavior or rituals.

II. MODELS

As innate responses for protecting the human organism and enhancing survival, panic and vigilance are normal in the face of threatening stimuli. As "anxiety" or psychopathological symptoms, other factors besides actual physical threat must be implicated as "triggers" or cause. Of several explanatory models, the *biological* places emphasis on the wiring and juices of the central nervous system, the *psychodynamic* on meanings and memories, and the *behavioral* on learning and behavior.

A. Biological

Recent animal and neuronal receptor studies emphasize two central systems as being principally involved in fear and pathological anxiety.[1,2] One critical component involves central noradrenergic mechanisms, with particular importance placed on the locus ceruleus (LC), a small retropontine structure that is the primary central nervous system (CNS) source of the neuromodulator norepinephrine. When this nucleus is stimulated in primates, for example, an acute fear response can usually be elicited with distress vocalizations, fear behaviors, and flight. Alternatively, destruction of the LC may lead to abnormal complacency in the face of threat. Biochemical perturbations that increase LC firing elicit anxious responses in animal and man that are blocked by agents that decrease LC firing, some of which are in clinical use as antipanic agents.

A second system involves "benzodiazepine receptors" with particularly relevant concentrations in limbic system structures, especially the septohippocampal areas. One role of the limbic system is to "scan" environmental inputs for life-supporting and threatening cues and to monitor internal or bodily sensations. Vigilance (or its psychopathological equivalent, generalized anxiety) may reflect limbic system arousal. Benzodiazepine receptors in high concentration in relevant limbic system structures may play a role in modulating limbic alert, arousal, and behavioral inhibition by increased binding of the inhibitory neurotransmitter γ-aminobutyric acid (GABA). As one might expect, there are neuronal connections between the LC and the limbic system that coordinate among surveillance, alert, and alarm. Although it is unlikely that any one system or substance is the source or cause of anxiety, the discovery of neuronal receptors for which benzodiazepines have particularly high affinity suggests the existence of ligands that may regulate the experience of anxiety as either endogenous anxiolytics or anxiogens. The fact that neuronal connections and neurotransmitters mediate emotional experience, however, does not of itself minimize the importance of nonpharmacological therapies for anxiety, which also may ultimately act by mobilizing these endogenous anxiety-regulating factors.

Other important clues to a biological basis of anxiety[3] include the observed familial and probably genetic transmission of disorder, the apparently unique sensitivity of patients with panic disorder to certain physiological provocations (e.g., lactate, CO_2), asymmetric blood flow in limbic structures on positron emission tomography (PET) scans of panic patients, and temperamental differences in very young children of patients with anxiety disorder.

If anxiety proneness has a biological basis, one would assume evolution had a reason to select for this characteristic. The anxious individual may have been important to the survival of the larger group by serving as a sentinel, being more alert to danger and more concerned with situations posing threat such as places of restricted escape. Further, many normative "anxious" responses are necessary for survival, such as separation anxiety. The reaction to acute separation in infant primates, for example, takes a form analogous to that of a panic attack, with intense arousal, distress, and care-seeking behavior. The sense of terror experienced by the infant on separation from the mother may be the prototype panic attack, which is reactivated in later life as the consequences of earlier failures at separation and individuation or of early trauma or in those genetically predisposed.

The simplest model for panic and anxiety may be that fear and anxiety are protective and that any good security system must include a surveillance, alarm, and help-seeking mechanism. Further, as anyone with a home security system knows, otherwise well-designed systems frequently have "bugs" and tend to "false alarm" or are triggered by ordinary events because of an excessively sensitive threshold. The problem remains, however, to find either the "bug" or the mechanism to regulate downward the sensitivity to inappropriate activation.

B. Behavioral

Behavioral understanding of anxiety disorder antecedents focuses on learning. Anxiety symptoms or avoidant behavior become associated with benign settings or objects as a result of generalization from earlier traumatic experiences. For example, a child ridiculed by peers because of a mistake during "show and tell" pairs embarrassment and shame with speaking before groups and thereafter manifests anticipatory anxiety and avoidance of public speaking. Self-defeating cognitive "habits" sustain the syndrome with unrealistic catastrophic thinking "expecting the worst." For panic disorder, misinterpretation of bodily stimuli with exaggerated cognitive and behavioral responses also serves to maintain the response. In this model, anxiety occurs when an individual encounters a signal that "predicts" a painful or feared event. Thus, learning theory presumes some real or perceived prior traumatic experience to condition the anxious response.

C. Psychodynamic

Psychoanalysts view anxiety as a manifestation of intrapsychic conflict and describe a continuum of responses. Secondary anxiety signals the ego to mobilize defenses to deal with a stress or conflict. "The anxiety acts as a warning that danger is impending; but the danger is one that the individual is capable of coping with; it is not yet overwhelming or inevitable."[4] Psychoanalysts also have described how anxiety is central to symptom formation and the development of the neuroses (i.e., obsessive compulsive, phobic, etc.). The symptoms

are the end product of the mixture of elements from both an unconscious impulse and the ego defenses directed against it. The symptoms are a compromise formation, the result . . . of vectors of forces showing the summated effects of all the elements.[4]

For example, a ladylike and very proper middle-aged woman little given to displays of emotion—and certainly not anger—becomes manifestly anxious when she is alone with her husband. The clinician learns that the symptoms started after the woman learned of her husband's recent infidelity. She is brimming with a sense of outrage toward her husband, but these feelings are intolerable to her. When they threaten to surface, she becomes overtly anxious.

More recent psychodynamic observations, emphasizing object relations, point to the role of internalized objects and their critical function in maintaining affective stability under stress. An adequate developmental experience typically generates internalized objects (introjects) that provide a soothing function, regulating the individual's internal milieu, in the face of distress; the absence of these internal regulators allows the organism to fragment and to experience anxiety in various situations.

Phobic disorders whose sufferers may experience panic, anticipatory, or no anxiety symptoms at all (depending on the success of avoidance behavior) serve to illustrate the differing models. The biological view would recognize the stereotyped nature of phobias, that most objects and situations in everyday life that truly threaten us are rarely selected as phobic stimuli. Children, for example, proceed normally through various developmental phobias (stranger, separation, darkness, etc.), but rarely become phobic of objects and situations that are associated with danger (e.g., electric outlets, roads). Further, most phobic stimuli (e.g., snakes, spiders, close spaces) have meaning in the context of biological preparedness[5] and were presumably selected for by evolution. Psychodynamic observations underscore the use of defense mechanisms such as displacement and the contribution of meanings and symbolic representations of phobic objects and situations. Learning theory, on the other hand, would predict some real or perceived prior traumatic experience with the object or situation. As with most human experience, anxiety rarely can be reduced to one explanatory framework. Clinical observation and research indicate contributing roles for all three models as determinants of behavior and as guides to therapeutic strategies. It is likely that an inherited "anxiety proneness" renders the development of an anxiety disorder more likely, but the onset, course, and shape of the disorder are greatly influenced by development, life events, and relationships.

III. DIFFERENTIAL DIAGNOSIS

Most patients with anxiety symptoms are initially evaluated by a primary care physician, internist, cardiologist, or neurologist for various somatic complaints. Given the predominance of physical symptoms in the experience of anxiety, the differential diagnosis of anxiety symptoms is complicated. Anxiety may be a **consequence** of a medical illness. An anxiety disorder (e.g., panic) may have been **triggered** by a medical illness (e.g., hyperthyroidism) but continue on its own after the inciting disorder is treated. Anxiety may be **associated** or linked with certain medical conditions (e.g., panic disorder with mitral valve prolapse, peptic ulcer disease, irritable bowel, hypertension). Anxiety may **mimic** a variety of medical ailments or, in turn, be mimicked by them. Thus, standard medical procedures, including history, physical, and laboratory examination, are indicated in evaluating a patient with anxietylike symptoms. Given the frequency of coexisting medical problems in psychiatric patients, physical symptoms should not be presumed to reflect anxiety. Because anxietylike symptoms accompany most psychopathological states and many organic problems, the differential diagnosis of anxiety is extensive.

A. Medical

A list of all possible organic causes of anxiety could constitute the index of a medical textbook. Endocrine disorders, such as hyperthyroidism, may reproduce severe anxiety symptoms. Similarly, hypoglycemia, whether caused by an insulinoma or by exogenously administered insulin, can result in anxietylike symptoms, as can secreting tumors such as carcinoid and pheochromocytoma. Patients with various neurological abnormalities—including encephalopathies of diverse etiologies, postconcussion syndrome, and seizure disorders—may report anxiety. Others with severe pulmonary disease and hypoxia also report anxietylike symptoms. Excesses of stimulant drugs (including caffeine) and withdrawal from sedative drugs (including alcohol) can be culprits as well.

It would be impossible to memorize all the possible physical causes of anxiety, and it would be a useless exercise to sit before an anxious patient with a checklist. Rather, the clinician should maintain a high index of suspicion and complete a thorough assessment. When taking a history, inquire about past and present medical illnesses and medications, other treatments, and doctors consulted. Inquire also about the use of over-the-counter drugs as well as "recreational" substances. Hallucinogens, for example, may precipitate a "bad trip" or panic attack. Always ask about alcohol and estimate the amount of caffeine consumption.

Complete a careful review of systems. Although the presence of certain symptoms may not have emerged initially when taking a history from an anxious patient, the systems review may reveal a pattern—such as skin and hair changes, heat intolerance, and weight loss (possible hyperthyroidism)—that can suggest further diagnostic evaluation.

The clinician should complete a routine physical examination with specific attention to areas highlighted by the history. Laboratory tests can be ordered on the basis of findings from the history and physical examination and anticipated medication therapy.

B. Psychiatric

Anxiety accompanies most psychiatric disorders. Patients with schizophrenia, other psychoses, affective disorders, borderline states, and personality disturbances may present with severe anxiety. Before prescribing an antianxiety drug, the clinician should complete a thorough evaluation. Psychoses, affective disorders, personality disturbances, and organic brain syndromes should be considered in the differential diagnosis of anxiety. Once the clinician determines the diagnosis, the general rule is to treat the primary disorder specifically—antipsychotic drugs for schizophrenia, antidepressants for depression, antianxiety treatments for generalized anxiety disorders, etc. On the other hand, it is important to recall that psychiatric disorders may coexist; an anxiety disorder may be comorbid with other axis I psychiatric syndromes listed in the third revised edition of the *Diagnostic and Statistical Manual of Mental Disorders* (DSM-IIIR). Among the most difficult psychiatric differential diagnoses is the differentiation of a DSM-IIIR axis I disorder such as an anxiety disorder from an axis II diagnosis in a patient with substantial underlying character pathology such as an individual with avoidant or borderline personality. In the latter case, the individual may describe himself as anxious or depressed when the internal state is more accurately one of emptiness, aloneness, boredom, or fragmentation as a consequence of interpersonal upheavals. Although the presence of an axis II disorder does not exclude treatment of anxiety symptoms, it is important to recognize when character pathology contributes to ongoing symptomatic distress.

IV. ANXIETY DISORDERS

Current nomenclature identifies nine specific subtypes of anxiety disorders: panic disorder with agoraphobia, panic disorder without agoraphobia, agoraphobia without history of panic disorder, social phobia, simple phobia, obsessive–compulsive disorder, posttraumatic stress disorder, generalized anxiety disorder, and anxiety disorder not otherwise specified. Anxiety disorders are the most common form of psychiatric illness in this country, with 15–20% of the population having experienced one of these conditions at some time in their lives.

A. Panic Disorder

Patients with this diagnosis have experienced at least one unexpected panic attack and have either experienced four panic attacks in a 4-week period in their life or one panic attack followed by at least a month of sustained fear of having a subsequent attack. The features of a panic attack include the experience of at least four of the symptoms listed in Table 1 developing suddenly or within 10 min of the onset of the first symptom noticed.

A panic attack is an intensely distressing experience, usually lasting a few to several minutes, with fairly stereotyped physical, cognitive, and behavioral components. Patients with panic disorder may experience these attacks intermittently or in clusters and may develop a number of complications, including persistent anxiousness, phobic avoidance, depression, and alcoholism or other drug overuse.

The physical symptoms of a panic attack include cardiac, respiratory, neu-

TABLE 1. Symptoms of Panic Attacks

1. Shortness of breath (dyspnea) or smothering sensations
2. Dizziness, unsteady feelings, or faintness
3. Palpitations or accelerated heart rate (tachycardia)
4. Trembling or shaking
5. Sweating
6. Choking
7. Nausea or abdominal distress
8. Depersonalization
9. Numbness or tingling sensations (paresthesias)
10. Flushes (hot flashes) or chills
11. Chest pain or discomfort
12. Fear of dying
13. Fear of going crazy or of doing something uncontrolled

rological, gastrointestinal, and other reactions. Cognitively, the patient feels a sense of terror or fear of losing control, dying, or going crazy. Behaviorally, the patient often flees from the setting in which the attack is experienced to a safe, secure, or familiar place or person. Help-seeking behavior is characteristic of a person having panic disorder.

The patient may vividly remember the initial or "herald" attack, which appears in some cases to "turn on" the disorder. Although this first spell is often unexpected ("out of the blue") and intense, later attacks may be either spontaneous and unexpected or preceded by a buildup of anticipatory anxiety. The latter type, called a situational attack, occurs in settings where the patients might sense being at risk for panic or the site of prior attacks, such as a crowded room or stores. Major attacks involve four or more of the symptoms in Table 1. Episodes involving fewer symptoms are considered "limited symptom" attacks.

The typical onset of panic disorder is in early adult life. The illness afflicts women two to three times more often than men. Adults with panic disorder (in particular, those who develop agoraphobia) frequently have a childhood history of separation anxiety symptoms,[6] particularly "school phobia." The disorder is clearly familial and may have a genetic basis.[7]

The onset of the disorder typically follows a major life event (such as a loss, threat of a loss, or other upheavals in work or home situations) or a physiological event such as medical illness (e.g., hyperthyroidism or vertigo) or drug use (e.g., marijuana or cocaine). However, a patient whose herald attack seems to be triggered by the use of cocaine, for example, may continue to have persistent or recurrent symptoms without further drug use. In light of the impressive paroxysmal physical symptoms of a panic attack, most sufferers ascribe the experience to a physical cause and seek evaluation and treatment initially from a nonpsychiatric physician. If the syndrome progresses without proper diagnosis, hypochondriacal behavior may result, followed by demoralization if relief is not obtained.

A panic attack, like an endogenous "false alarm," appears to turn on a state of vigilance or "postpanic" anxiety that resembles generalized anxiety disorder. Between attacks, patients may remain symptomatic with a constant, low-level, anticipatory anxiety that may crescendo into full-blown panic in certain situations, sometimes unexpected. In this state of vigilance, the patient may develop mild or extensive phobic avoidance, usually of travel or places of restricted escape. This behavior may begin immediately following the onset of attacks, after a number of attacks, or never at all. In some cases, phobic avoidance evolves as a progressive constriction, and the patient increasingly avoids a variety of settings, in particular those where attacks have occurred. Thus, agoraphobia frequently occurs as a complication of panic attacks.

The demoralization that accompanies the sustained distress and progressive disability of panic disorder may extend to a major depressive episode with characteristic signs and symptoms. The relationship between panic and depression, however, is a complicated one. Although some patients manifest no depressive symptoms, for others, it is unclear which disorder is primary when the onset of symptoms is concurrent.

Alcohol use can temporarily tame the distress of panic disorder, but the patient soon experiences rebound symptoms, setting the stage for alcohol overuse. Thus, it is not surprising that more than 10% of adult alcoholics may meet criteria for a diagnosis of panic disorder.

The mortality rate of patients with panic disorder may be higher than that of others. Data suggest that there is increased premature mortality from suicide and, for men, from cardiovascular illness.[8] Mitral valve prolapse is also diagnosed much more frequently in panic patients than in either the normal population or patients with generalized anxiety disorder.

Although there are a number of treatment strategies, including pharmacological and behavioral interventions, that can minimize or prevent the symptoms and complications of this disorder, many patients suffer residual symptoms or go many years with their disorder before coming to treatment. Epidemiologic studies suggest that panic disorder and agoraphobia occur in as much as 3% to 6% of the population.

B. Phobic Disorders

There are three general categories of phobic disorders: agoraphobia, social phobia, and simple phobia. **Agoraphobia** is the fear of settings or places of restricted escape, where help might not be available in the event of an emergency, as with a panic attack. Patients suffer travel restrictions and may endure a variety of agoraphobic situations (such as crowds, restaurants, stores, bridges, tunnels, airplanes, or restricted-access highways) with considerable distress or may remain homebound or unable to travel without a companion. Some patients suffer agoraphobia without a clear history of typical panic attacks.

Social phobics experience anxious distress in situations where the patient is the focus of scrutiny by others and fears humiliation or embarrassment. A common circumscribed social phobia is stage fright and fear of public speaking. Other examples include the inability to sign one's name in public or to use public lavatories. Social phobic anxiety may resemble the lower-grade, "generalized" variety or be experienced with the intensity of a panic attack. Social phobia is distinguished from normal anxiety, in part, by the extent of avoidance or social and occupational limitation. Social phobia may also be global with generalized anxiety or panic symptoms in any social setting, whenever the sufferer is in the presence of others, with relative relief available only when alone or with very familiar people.

Social phobias have a 6-month prevalence rate of about 1.5%. They typically have their onset in late adolescence and afflict men and women in roughly equal numbers. This disorder is generally considered underdiagnosed and undertreated. There may be considerable overlap in the presentation of social phobia and panic disorder. Although patients with either disorder may avoid or need to flee from meetings or other social gatherings, social phobics are generally more comfortable when they are alone, in contrast to patients with panic disorder. Similar to panic disorder and agoraphobia, social phobia is frequently complicated by alcohol overuse.

Simple phobias are the irrational fear of and need to avoid specific objects or settings, such as insects or animals, or specific situations (e.g., airplanes, heights). If the object or situation is encountered, the individual suffers intense anxiety that may reach the level of panic.

C. Generalized Anxiety Disorder

In the past, patients who suffered anxiety, whether of the generalized or panic form, had been termed "anxiety neurotics," but observations of patient's responses to drug treatments initiated a "pharmacological dissection" of anxiety disorders. Patients who suffered panic attacks, agoraphobia, or "phobic anxiety states" showed preferential responses to medications that were not "anxiolytic" *per se* (e.g., certain tricyclic and monoamine oxidase inhibitor antidepressants). Standard anxiolytic medications such as benzodiazepines were relatively ineffective in preventing panic attacks but were useful in controlling other forms of generalized anxiety.

Generalized anxiety disorder is diagnosed if the patient has suffered from persistent anxiety for 6 months or more. These episodes are manifested by unrealistic or excessive worry (not related to having a panic attack) and by various symptoms such as motor tension, autonomic hyperactivity, as well as vigilance and scanning behavior that are unrelated to any specific medical or organic cause.

Although generalized anxiety may represent the most prevalent form of anxiety symptoms, its actual prevalence as opposed to reactive anxiety symptoms resulting from transient stressful life situations is unknown. As described (see Section III), anxiousness may be featured in a number of medical and psychiatric conditions; for example, persisting and severe anxiousness may be the presenting symptom of a major depressive disorder. Selection of treatments for persistent anxiety without panic attacks is also controversial. Stress management techniques, other cognitive and behavioral interventions, and various psychotherapeutic interventions may be useful in dealing with relationships and life situations that exacerbate persisting anxiety. On the other hand, substantial numbers of patients with persistent anxiety may obtain long-lasting relief and improvement in function from pharmacotherapy with traditional anxiolytics, antidepressants, and newer antianxiety agents. In lieu of maintenance treatment, short-term pharmacotherapy may help deal with the peaks of stressful anxiety.

D. Other DSM-III Adult Anxiety Disorders

1. Obsessive Compulsive Disorder

Patients with obsessive compulsive disorder (OCD) suffer (1) recurrent unwanted and senseless thoughts that are distressing and intrusive but are unable to be suppressed (obsessions) or (2) compulsions, the need to repeat certain behaviors or rituals in a stereotyped fashion to avoid or diminish intense anxiety.

Although many OCD patients suffer from persistent anxiety symptoms or panic attacks, particularly when confronted with triggers for obsessional rumination or when restrained from carrying out compulsive rituals, the inclusion of OCD among anxiety disorders may not be a good fit for many OCD sufferers. Indeed, OCD may have several forms, including compulsive washing or checking, fears of contamination, primary obsessional slowness, and religious or bizarre ritual behaviors. Obsessive compulsive disorder is frequently an exceptionally disabling syndrome, but, for diagnosis, other causes for obsessive thoughts such as bulimia (with food obsessions) or psychosis must be considered. Depression may also exist with OCD symptoms. Because of the bizarre nature of some obsessions, many patients with the disorder are frequently misdiagnosed as having schizophrenia. The 6-month prevalence rate of OCD is between 1% and 2%. The onset of OCD is generally before the age of 25. As with other disorders, proposed etiologies include psychodynamic, neurochemical (e.g., serotonin dysfunction), neurological (e.g., basal ganglia dysfunction), and behavioral or learning models.[9] There is only a partial overlap between OCD and compulsive personality disorder; OCD also aggregates in families of sufferers of such other disorders as tics and Tourette's syndrome.

2. Posttraumatic Stress Disorder

Patients with this condition may manifest symptoms resembling other anxiety disorders such as panic attacks, but the disorder begins at the time of experiencing or witnessing a life-threatening, violent, or other catastrophic event. Subsequently, the patient reexperiences the traumatic event in vivid memories, recurrent dreams, or other experiences (such as flashbacks). Anxiety symptoms, the tendency to startle easily, sleep dysfunction, and the need to avoid stimuli reminiscent of the original event, all characterize this syndrome.

The actual prevalence of posttraumatic stress disorder (PTSD) is uncertain, but it is estimated that 1% of the general population suffers from it. Of course, this disorder is more prevalent in populations exposed to trauma such as civilians after attack and particularly in wounded Vietnam veterans. PTSD patients have been described as suffering enduring vigilance for and sensitivity to threat with excessive alertness and overreactivity to stimuli. Like panic patients, they manifest increased autonomic arousal, and those who go on to suffer PTSD after trauma may have had more frequent childhood history of behavioral problems as well.

3. Adjustment Disorder with Anxious Mood

This category, although not a DSM-IIIR diagnosis, serves to describe the great number of patients who experience transient periods of pathological anxiety associated with interpersonal, occupational, or other upheavals in their lives. Symptoms are generally time-limited but may achieve sufficient intensity to require treatment.

V. ANTIANXIETY AGENTS

The various anxiety syndromes are mainly treated with two general classes of medication: **antidepressants** (chiefly tricyclic antidepressants and monoamine oxidase inhibitors) and various **antianxiety agents,** most often the **benzodiazepines.** The antidepressant medications are discussed in detail in Chapter 2. The following section focuses primarily on the benzodiazepines. Other agents less commonly used to medicate anxiety are discussed at the end of this section. (See Table 4 and Section VI for a clinical discussion of how these drugs are used to medicate the subtypes of anxiety disorders.)

A. Benzodiazepines

The popularity of this class of antianxiety drugs has generated such articles as "The Benzodiazepine Bonanza" and "Valiumania." From 1964 through 1973, the number of antianxiety drug prescriptions filled in United States drug stores increased dramatically. Most were for the benzodiazepines chlordiazepoxide (Librium® and others) and diazepam (Valium® and others). This trend did reverse, possibly because of adverse publicity among physicians and the lay public. Even with declining general use, over 60 million prescriptions for benzodiazepines were filled in the United States in 1979.[10] Fifteen percent of respondents to one U.S. survey reported using antianxiety agents in the 1970s—a figure comparable to that reported for other developed countries.[11]

Chlordiazepoxide, the first benzodiazepine antianxiety agent synthesized, was marketed in 1960. Three years afterwards, diazepam was introduced and rapidly became the most widely prescribed drug in the world. The popularity of these agents has spawned a number of chemical siblings.

In the following sections, we discuss the eight benzodiazepine drugs currently labeled for antianxiety use in the United States: chlordiazepoxide, clorazepate (Tranxene® and others), diazepam, lorazepam (Ativan® and others), oxazepam (Serax® and others), prazepam (Centrax® and others), halazepam (Paxipam®), and alprazolam (Xanax®). Two other benzodiazepines, flurazepam (Dalmane® and others) and temazepam (Restoril® and others), are marketed for insomnia. Clonazepam (Klonopin®), widely used for treating panic disorder, is labeled for the treatment of epilepsy. The labeled indication for benzodiazepine use (e.g., anxiety versus insomnia) probably has more to do with marketing decisions than with unique pharmacological profiles of the agents.

Benzodiazepines, as noted earlier, are the most widely prescribed antianxiety agents in the world. In fact, they are the most widely prescribed drugs of any type. Their popularity is generally deserved. Clinically, they are highly effective for alleviating acute anxiety, and, in most cases, they probably retain relative efficacy over time. Toxicologically, they are remarkably safe in situations of overdose, and their overall record of safety is unparalleled. Although they are by no means perfect (indeed, much of this chapter is devoted to caveats about their use), benzodiazepine drugs represent a definite step forward in the pharmacological treatment of anxiety.

1. Chemistry

Molecular structures of the eight benzodiazepines currently labeled for anxiolytic use are shown in Fig. 1. There are several subtypes of benzodiazepines with some differences in their pharmacology (see Table 2). Diazepam is a prototype of the 2-ketobenzodiazepines, whose other members include desmethyldiazepam (the active agent during administration of clorazepate, prazepam, and halazepam). These drugs are all biotransformed by hepatic oxidative reactions and have long elimination half-lives.

Oxazepam, lorazepam, and the hypnotic temazepam are 3-OH benzodiazepines. They are metabolized by conjugation and have short to intermediate half-lives. Alprazolam is a triazolobenzodiazepine, which also undergoes oxidation in the liver, but whose rate of elimination is considerably more rapid than that of the 2-ketobenzodiazepines. Chlordiazepoxide is chemically distinct but pharmacologically similar to the 2-keto drugs. Clonazepam is a 7-nitrobenzodiazepine, which has a distinct biotransformation.

2. Pharmacological Effects and Mechanism of Action

Benzodiazepines reduce the effects of anxietylike behavior in animals. For example, in a commonly employed behavioral paradigm, the presence of a light in a rat's cage is followed shortly by a painful electric shock. The animal soon learns the "meaning" of the light and, in its presence, manifests anxiety: shaking, tachycardia, crouching, urinating, and defecating. If he had previously been working on a task (e.g., pressing a bar to obtain food) in the presence of the light (i.e., the anxiety condition), he stops working and remains immobilized. However, administration of

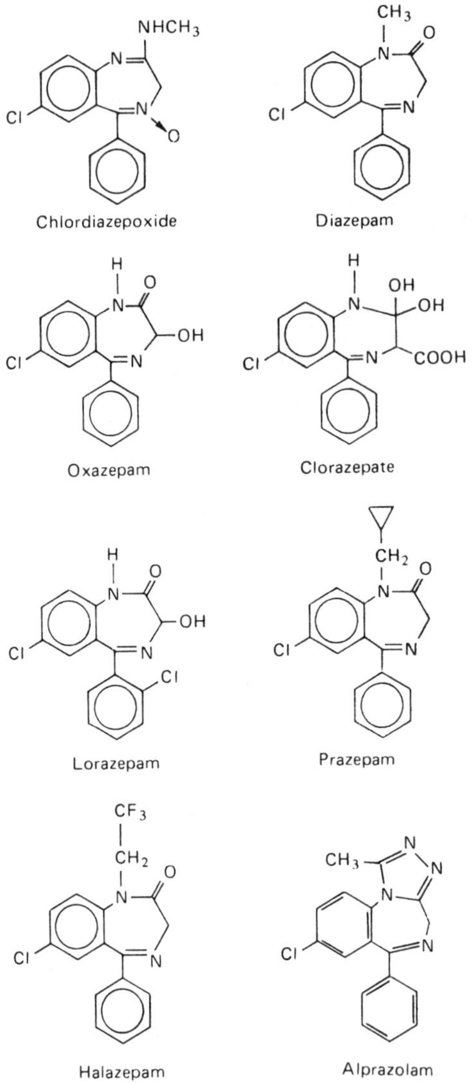

FIGURE 1. Benzodiazepines for the treatment of anxiety.

a benzodiazepine would allow the animal to keep on working despite the anxiety-producing stimulus.

Benzodiazepines block nervous system stimulation that originates in the brain-stem reticular system and diminish activity in areas associated with emotion such as the septal region, amygdala, hippocampus, and hypothalamus. Benzodiazepines

TABLE 2. Benzodiazepine Antianxiety Drugs

Drug	Approximate dose equivalents (mg)	Available dosage forms	Rapidity of absorption	Active metabolites	Half-life (hr)
Alprazolam (Xanax®)	0.5	0.25-, 0.5-, and 1.0-mg tablets	Intermediate	α-Hydroxyalprazolam, des-methylalprazolam, 4-hy-droxyalprazolam	6 to 20
Chlordiazepoxide (Librium® and others)	10	5-, 10-, 25-mg tablets and capsules, 100-mg/2-ml ampule	Intermediate	Desmethylchlordiazepoxide, demoxepam, desmethyl-diazepam, oxazepam	5 to 30
Clorazepate (Tranxene®)	7.5	3.75-, 7.5-, and 15-mg capsules (Tranxene®)	Fast	Desmethyldiazepam,[a] oxazepam	30 to 200
Diazepam (Valium®)	5	2-, 5-, 10-mg tablets, 10-mg/2-ml ampules, 50-mg/10-ml vials, 10-mg/2-ml syringe	Fastest	Desmethyldiazepam, oxazepam	20 to 100
Halazepam (Paxipam®)	20	20-, 40-mg tablets	Intermediate	Desmethyldiazepam, oxazepam	Ca. 14 (parent), 50 to 100 des-methyldiazepam
Lorazepam (Ativan®)	1	0.5-, 1-, 2-mg tablets, 2- and 4-mg/ml syringes, 2- and 4-mg/ml vials[b]	Intermediate	None	10 to 20
Oxazepam (Serax®)	15	10-, 15-, 30-mg cap-sules, 15-mg tablets	Slower	None	5 to 15
Prazepam (Centrax®)	10	5-, 10-mg capsules, 10-mg tablets	Slowest	Desmethyldiazepam,[a] oxazepam	30 to 200

[a] These drugs are pro drugs for desmethyldiazepam: the parent compound is rapidly converted to the active metabolite.
[b] As of this writing, approved by FDA only for preanesthetic use.

raise the seizure threshold and increase the frequency and activity of brain waves—effects that resemble those of other sedative–hypnotic drugs such as barbiturates.

In one experiment, a single dose of diazepam administered to a group of healthy volunteers impaired both immediate and delayed free recall by interfering with the acquisition of new information. Retention remained unaffected. After 3 weeks of diazepam administration, partial tolerance developed to the memory impairment, and 1 week following discontinuation of diazepam, memory returned to normal.[12]

The benzodiazepines' mechanism of action has not been completely elucidated. They interact with specific brain receptors, and they probably resemble an endogenous ligand to these receptors. Benzodiazepine receptors are located in areas of the brain that might explain some of the pharmacological activity of benzodiazepine drugs.[13] Benzodiazepine receptors in the limbic system (hippocampus and olfactory bulb) may mediate antianxiety effects; receptors in the cortex might explain anticonvulsant activity; and receptors in the thalamic nuclei could be involved in sedative effects. The interaction between drug and benzodiazepine receptor allows increased activity of the inhibitory neurotransmitter γ-aminobutyric acid (GABA).

Prolonged administration of benzodiazepine compounds may "turn off" the synthesis of endogenous benzodiazepinelike substances. Various withdrawal effects following drug discontinuation could, in turn, be caused by changes in receptors that follow long-term administration. The time required for recovery from benzodiazepine withdrawal symptoms would then reflect the time it takes either to regenerate an endogenous ligand or to reset receptor sensitivity.

The effects of benzodiazepines on the other organ systems show a strikingly benign pattern. Although decreased respiration, blood pressure, and cardiac output may be observed, they are unlikely to be of clinical consequence. In fact, clinicians have safely administered these drugs to large numbers of patients with disease of the heart, blood vessels, and lungs.

Diazepam, especially, produces significant skeletal muscle relaxation, probably by direct action on spinal neurons. Physicians have administered this drug to many patients with skeletal muscle spasm, spasticity, and other muscle disorders, although its efficacy remains to be proved beyond doubt.

3. Kinetics

The rate at which a drug is absorbed from the gastrointestinal tract determines to a large degree the rapidity of onset and intensity of its acute effects.[14] Among benzodiazepine compounds, the rate of absorption varies considerably (see Table 2). Thus, diazepam and clorazepate reach the blood rapidly and produce prompt and intense effects following a single oral dose: peak concentrations of diazepam may be reached within an hour (even more rapidly in children), which could account for occasional reports of euphoria with diazepam. On the other end of the absorption spectrum, prazepam (and, to a lesser degree, oxazepam) is more slowly absorbed, resulting in a less intense and more delayed onset of action and side effects after single doses. For any drug, absorption is most rapid when medication is taken on an empty stomach. Anything that delays gastric emptying, such as food or aluminum-containing antacids, will slow drug absorption, resulting in delayed onset and diminished intensity of acute clinical effects.

After a single dose of a benzodiazepine drug, its action is terminated primarily by drug distribution, which correlates with its lipid solubility. Thus, diazepam, the most lipid-soluble benzodiazepine, is rapidly and extensively distributed to body tissues following a single oral dose. Even though diazepam has a long elimination half-life, following a single dose its duration of action is relatively brief. Moreover, some degree of central nervous system tolerance also appears to diminish the initial action of the drug, even before blood levels have shown much decline. Less extensively distributed drugs, such as lorazepam, have a longer duration of action after a single dose, even though lorazepam's elimination half-life is much briefer than diazepam's.

Plasma concentrations of benzodiazepine drugs (and their active metabolites) probably correlate to some degree with clinical response, although this relationship is complicated by tolerance that develops within the central nervous system. At present, measurements of benzodiazepine concentrations in blood are not of general clinical use, although studies have correlated unwanted effects of these drugs with blood levels, which may provide clues to problems of overdose and withdrawal.[15]

Benzodiazepines bind tightly to plasma protein and are highly lipophilic, which means they are difficult to remove from the body by dialysis. They probably recirculate enterohepatically, which could account for an occasional secondary rise in blood levels several hours after the initial rise.

Benzodiazepines are metabolized in the hepatic microsomal enzyme system. Many are biotransformed into desmethyldiazepam—a major active metabolite of diazepam and halazepam and a metabolite of chlordiazepoxide. In addition, clorazepate and prazepam have little pharmacological activity themselves but serve as precursors for desmethyldiazepam. The conversion of clorazepate to de-smethyldiazepam occurs rapidly but depends on an acid medium in the stomach, which is compromised by the presence of antacids or food. The transformation of prazepam to desmethyldiazepam occurs more slowly, accounting for the gradual onset of effects with that drug. The metabolic pathways that transform drugs to desmethyldiazepam are influenced by factors such as age, liver disease, or the coadministration of other drugs that may affect hepatic oxidizing activity.

Three benzodiazepines—oxazepam, lorazepam, and the hypnotic temazepam—have no active metabolites and are simply conjugated with glucuronide in the liver. The elimination half-lives of these compounds are in the short to intermediate range. Their metabolism is less affected by the presence of liver disease, the extremes of age, or drug interactions.

Of what use to a clinician is a drug's half-life? First, the half-life determines the time required to achieve a "steady-state" concentration, which occurs in about four half-lives. (At steady state, the amount of drug excreted equals the amount ingested, i.e., a dynamic equilibrium.) Until a drug reaches steady state, it is accumulating, which means that the optimal dosage cannot yet be determined for a given patient. For example, an elderly woman in a nursing home may gradually become more sedated over the first several weeks on a chlordiazepoxide regimen. The drug is continuing to accumulate in her body until steady state is achieved, which may take longer in an elderly patient. For this reason, the clinician must be aware of this lag and adjust dosage accordingly.

A second significance of the half-life is with regard to the frequency of dosing. Drugs with half-lives greater than 24 hr often may be administered in once-daily doses, whereas drugs with short half-lives should be administered more frequently to achieve constant clinical effects. A third implication concerns the other end of the treatment program—termination. Drugs with long half-lives have more built-in tapering: when they are discontinued, their rates of egress from the body are relatively slow. Even so, all benzodiazepines should be discontinued by tapering rather than by abrupt discontinuation, but this is even more of a concern with short-half-life agents.

4. Adverse Effects

The most common unwanted effects of the benzodiazepines (as for other central nervous system depressants) are drowsiness and ataxia, which acutely correlate with elevated blood concentrations of the drugs (and active metabolites). At even higher levels, gross psychomotor impairment and marked sedation occur. The elderly may be more vulnerable to these reactions because they usually achieve higher blood (and tissue) drug levels for a given dose and also because the aging brain is more sensitive to the effects of sedatives.

An increased tendency to express hostility, particularly in group settings, has been reported with diazepam, chlordiazepoxide, and alprazolam but may be less of a problem with oxazepam. Data on this point remain inconclusive,[16] but clinicians may wish to avoid high-potency benzodiazepines in patients who have trouble containing their anger. Occasional patients become depressed during chronic benzodiazepine therapy, but whether this is a causal association remains in doubt.

Aside from the effects noted above, adverse reactions with benzodiazepines tend to be uncommon and are rarely serious. Occasional patients experience increased appetite and weight gain, cutaneous (usually allergic) reactions, nausea, headache, and assorted endocrine changes. Rare reports describe agranulocytosis and cholestatic jaundice, but unusual events in association with drug therapy do not prove an association with the drug. A few investigators have raised the possibility of a relationship between diazepam use and mammary tumors in laboratory animals and in humans; however, a careful review of the data does not support a causal association.[17]

5. Toxicity

Probably the nicest feature about the benzodiazepines is their very high therapeutic index—the ratio of toxic to therapeutic doses. Although frequently ingested in overdose attempts, benzodiazepines alone almost never cause fatalities.[18,19] The combination of a benzodiazepine with alcohol or other sedative–hypnotic agents is more hazardous, although it is unclear whether or not the benzodiazepines actually contribute to the lethality of the combination.[20]

Whenever possible, the use of any drug during pregnancy should be avoided. Some but not all data suggest that benzodiazepines taken during the first trimester may cause an increased incidence of cleft lip/cleft palate. Used late in pregnancy or during nursing, benzodiazepines have been associated with a "floppy infant syndrome." Other observations in infants whose mothers have taken benzodiazepines have included multiple congenital deformities, intrauterine growth retardation, withdrawal symptoms, neonatal depression with poor sucking and hypotonia, hyperbilirubinemia, hypothermia, increased carbon dioxide tension and acidosis, and cardiac arrhythmias (many of which may be coincidental)[21] (see Chapter 10).

6. Tolerance, Dependence, and Withdrawal Reactions

Benzodiazepine drugs have been sought out for "recreational" purposes, i.e., to induce a euphoric state, a "high." Some may have a greater abuse potential than others, conceivably related to their pharmacokinetics (e.g., rapid absorption and onset of action). In general, given the extent of therapeutic use, however, the specific abuse of these drugs has been relatively rare. On the other hand, clinicians typically should not prescribe the benzodiazepines to patients with a history of substance abuse, particularly with sedative–hypnotics. Any suggestion that a patient is taking more of the drug than prescribed should be viewed with concern.

Benzodiazepine withdrawal leads to reactions similar to those observed with other sedative–hypnotic compounds (e.g., barbiturates, alcohol). Cross-tolerance exists among these various classes, and one agent usually can treat a withdrawal syndrome from another. In fact, benzodiazepines are frequently employed in treatment of alcohol-withdrawal syndromes.

We used to believe that withdrawal reactions to benzodiazepines were relatively rare and occurred only when the drugs were used in dosages considerably in excess of therapeutic recommendations. More recently, however, clinicians have observed that patients taking benzodiazepines for long periods of time—even at standard doses—are vulnerable to withdrawal reactions when the drug is abruptly discontinued (possibly because of a "turning off" of endogenous benzodiazepine substances)[22] and to withdrawal symptoms after discontinuing short-term use of some agents, such as alprazolam.

Mild withdrawal symptoms include insomnia, dizziness, headache, anorexia, vertigo, tinnitus, blurred vision, and shakiness. Some of these symptoms may represent a recurrence of anxiety that had been contained by the benzodiazepine. However, if after discontinuing the drug these symptoms increase over several weeks and then begin to wane, a rebound or withdrawal reaction is more likely. Severe signs of benzodiazepine withdrawal include hypotension, hyperthermia, neuromuscular irritability, psychosis, and seizures. Short-acting benzodiazepines, with their relatively rapid egress from the body, might cause a higher incidence of severe withdrawal reactions, including convulsions.

Benzodiazepine Withdrawal Syndrome

Anxiety	Blurred vision
Agitation	Diarrhea
Tremulousness	Hypotension
Insomnia	Hyperthermia
Dizziness	Neuromuscular irritability
Headaches	Psychosis
Tinnitus	Seizures

7. Drug Interactions

Benzodiazepines interact with relatively few other drugs, so they can be coadministered safely with various medical and psychiatric medications. Benzodiazepines have relatively little tendency to induce hepatic microsomal enzymes, especially in comparison with barbiturates and other sedative–hypnotics, which means they usually can be administered to patients taking anticoagulants.

A common pharmacodynamic interaction (i.e., additive effects on target organs) is the **enhancement of central nervous system depression when a benzodiazepine is coadministered with another sedative, including alcohol.** It is doubtful whether coadministration of alcohol with diazepam can increase blood concentrations of diazepam (a pharmacokinetic interaction).

A handful of pharmacokinetic interactions involving benzodiazepines have been reported (see Table 3). In general, benzodiazepines that undergo conjugation as the only metabolic step in biotransformation (i.e., oxazepam, lorazepam, temazepam) have the fewest kinetic interactions. Diazepam increases plasma levels of the anticonvulsant phenytoin (Dilantin® and others) and of the cardiac drug digoxin (Lanoxin® and others). When disulfiram is coadministered with chlordiazepoxide and diazepam, blood levels of the benzodiazepines can rise. Antacids should not be administered at the same time as clorazepate, since they will diminish the biotransformation of clorazepate to the active compound desmethyldiazepam, thus reducing its potency. Antacids and foods slow the absorption of other benzodiazepines but do not diminish the total amount absorbed. Cimetidine (Tagamet®), used for the treatment of acid peptic disease, increases the blood levels of long-

TABLE 3. Benzodiazepine Interactions[a]

Diazepam (Valium®)	↑ Phenytoin (Dilantin® and others) levels
Diazepam	↑ Digoxin (Lanoxin® and others) levels
Disulfiram (Antabuse®)	↑ Chlordiazepoxide (Librium® and others, diazepam levels
Antacids	↓ Clorazepate (Tranxene®) absorption; slower chlordiazepoxide, diazepam absorption
Food	Slower diazepam absorption
Cimetidine (Tagamet®)	↑ Levels of long-acting benzodiazepines
Alcohol	?↑ Diazepam levels; potentiates CNS-depressant effects

[a]From Gelenberg.[23]

acting benzodiazepines (e.g., diazepam, chlordiazepoxide, clorazepate, prazepam) but does not appear to affect the short-acting agents (e.g., lorazepam, oxazepam, temazepam).[23]

8. Nonpsychiatric Uses

Benzodiazepines have a number of uses outside of psychiatry: as anticonvulsants, for premedication before surgery and other procedures (such as cardioversion and endoscopy), in treatment of alcoholic withdrawal states, and as muscle relaxants in such diverse conditions as muscle strain, tetanus, spasticity, and stiff-man syndrome.

B. Buspirone

In light of such evident benzodiazepine drawbacks as sedation, visual motor and memory effects, possible abuse, and discontinuation syndromes, there was considerable enthusiasm and eager anticipation for the new nonbenzodiazepine anxiolytic buspirone (BuSpar®). With a pharmacological profile quite different from other anxiolytic agents, this azaspirodecanedione is a lipophilic heterocyclic compound that lacks sedative, anticonvulsant, or muscle relaxant properties and thus has been deemed "anxioselective." Considerable research and clinical experience documents lack of abuse potential, and the drug has no significant interactions with either alcohol or other central nervous system depressants. The mechanism of action of buspirone does not appear to involve interaction with a benzodiazepine–GABA receptor. It does have a number of complex interactions with dopamine receptors, particularly presynaptic ones, as well as postsynaptic 5-HT$_{1A}$ (serotonin) receptors, for which it is a partial agonist. Despite its effects on dopamine receptors, it does not appear to be a neuroleptic or to have a neuroleptic side-effect and risks profile.

A considerable body of clinical trials indicates that buspirone's effectiveness is comparable with that of benzodiazepines in the treatment of generalized anxiety.[24] There is some evidence that patients with prior treatment with benzodiazepines do rather less well with buspirone than treatment-naive patients.[25] Buspirone is available in 5-mg tablets, and the typical range of effective doses is between 5 mg t.i.d. and 15 mg t.i.d.

Despite reports to the contrary, many clinicians and patients have found buspirone to be a generally disappointing alternative to benzodiazepines. Ineffective in panic attacks, it has proven less than optimal in the treatment of many patients with generalized anxiety as well. Part of the explanation may be that the prescribing of buspirone requires a strategy similar to that of antidepressants, since there is a latency to respond of several days and, for individual patients, a critical dose threshold needed to achieve maximum benefit. Even with optimizing treatment by strategies of dose titration and sufficient duration of treatment, many patients still fail to show improvement. Nonetheless, some successes are achieved (some patients continued on benzodiazepines have improved or required lower doses of the benzodiazepine when buspirone was added). Given its several advantages, and despite uncertainties about general efficacy, a trial of buspirone is to be recommended for persistently anxious patients; for those who do respond to it, their course of treatment, and particularly their eventual discontinuation, may prove less complicated.

C. Other Antianxiety Agents

With the availability of the benzodiazepines, most previously employed antianxiety drugs have become obsolete. However, there are times when a physician may seek an alternative, for instance, in a patient liable to substance abuse. In addition, a doctor may "inherit" a patient who is already taking another type of drug, and thus we should all have some familiarity with other antianxiety compounds.

1. Antihistamines

Because of their sedative effects, drugs that block central and peripheral histamine receptors (primarily H_1) have, largely in the past, been occasionally used to calm anxious patients. Of the antihistamine drugs, clinicians prescribe hydroxyzine (Atarax®, Vistaril®, and others) most often. As shown in Fig. 2, it is chemically unrelated to the antipsychotics, antidepressants, benzodiazepines, or meprobamate.

Data about hydroxyzine metabolism in man are sparse. Apparently the drug is rapidly absorbed, begins to act within 15 to 30 min after a single oral dose, reaches peak plasma levels with in 1 hr, and maintains its effect for at least 24 hr. In addition to its sedative and antianxiety effects, hydroxyzine also possesses antiemetic and antihistaminic properties. It may be more sedating than the benzodiazepines, although in some patients it can cause agitation. Although not as effective an anxiolytic as the benzodiazepines, hydroxyzine does not cause physical dependence or abuse; therefore, it may be used as an alternative for patients prone to drug depen-

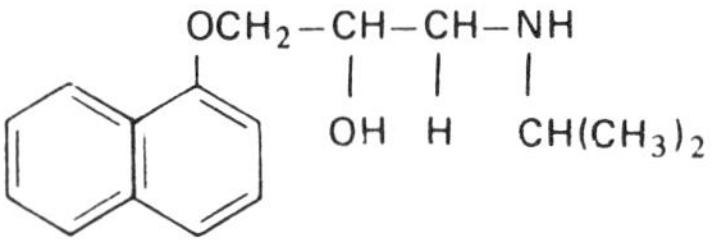

Propranolol

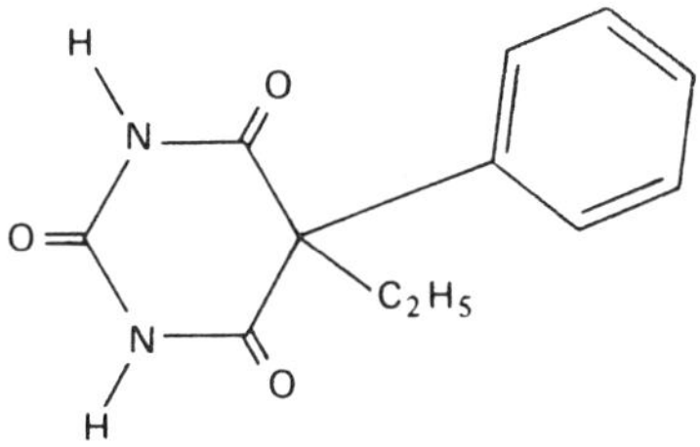

Phenobarbital

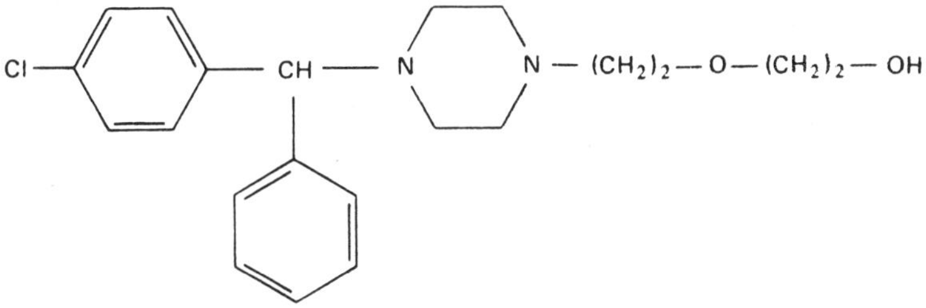

Hydroxyzine

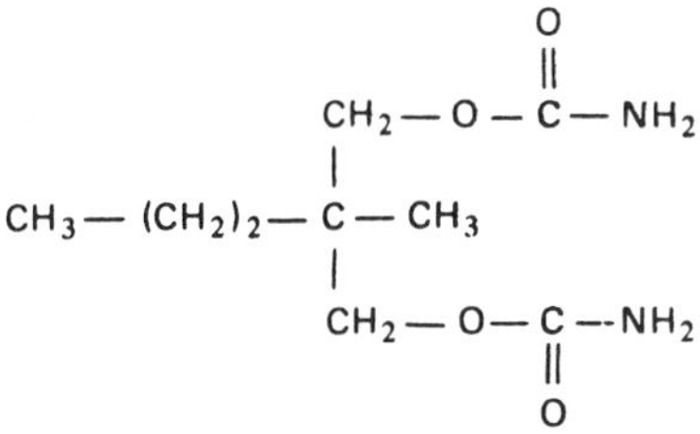

Meprobamate

FIGURE 2. Other drugs used for the treatment of anxiety.

dence. Usual daily doses range from 200 to 400 mg per day in two to four divided doses, which can be administered intramuscularly as well as orally.

In addition to the treatment of anxiety, hydroxyzine has been used for motion sickness, as a preanesthetic medication, and to treat allergic dermatoses. It is not effective for treating alcohol withdrawal syndromes.

Fatal overdose is uncommon with hydroxyzine, and withdrawal reactions have not been observed. Occasionally, cases of abnormal involuntary movements have been reported with antihistamines. Hydroxyzine may be teratogenic when administered to a pregnant woman, but this has not been studied definitively. Whereas benzodiazepines and most other sedative–hypnotic drugs elevate the seizure threshold, antihistamines depress the threshold and thus should be used with caution in patients with seizure disorders. Both peripheral and central anticholinergic toxicity can occur with antihistamine drugs; the elderly tend to be more sensitive to these actions. On a pharmacodynamic basis, hydroxyzine may show additive toxicity with other drugs that possess CNS-depressing antihistaminic or anticholinergic activity.

2. Barbiturates

The benzodiazepines, which have a much more favorable clinical profile, have largely replaced the formerly popular barbiturates for the treatment of anxiety. Barbiturates are classified according to their duration of action, speed of metabolic degradation, and hypnotic potency. Rapidly acting barbiturates are primarily inactivated through metabolism by the hepatic microsomal enzymes of the liver, whereas phenobarbital is eliminated largely unchanged by the kidney.

Barbiturates have a wide variety of effects on the body. They are CNS depressants and, in doses only several times greater than therapeutic, can depress the central respiratory drive. In toxic doses, barbiturates also depress cardiac contractility. Barbiturates interfere with the microsomal drug-metabolizing system in the liver and thus interact with various other drugs.

Barbiturates, most notably phenobarbital (Tedral® and others), probably have some degree of efficacy in the treatment of anxiety, but they seem to be less effective than the benzodiazepines and produce more sedation. It is likely that tolerance develops to antianxiety effects, which most probably result from their sedative activity. Clinical use of these agents is further complicated by the drugs' tendency to induce dependence, to cause serious withdrawal reactions, and to be dangerous in overdoses.

A barbiturate overdose can be extremely serious. Doses in excess of ten times a standard hypnotic dose are generally dangerous. Death occurs in approximately 0.5% to 12% of cases. Short-acting barbiturates are more toxic than the longer-acting agents: 2 to 3 g of amobarbital (Amytal® and others), secobarbital (Seconal®), or pentobarbital (Nembutal® and others) can cause death; in contrast, a lethal dose of the long-acting phenobarbital begins at about 6 to 10 g. Similarly, lethal blood concentrations may be 6 mg/100 ml for phenobarbital and barbital but only 1 mg/100 ml for shorter-acting agents. Obviously, the presence of alcohol or other central nervous system depressants can cause death at lower barbiturate concentrations.

Patients who are moderately intoxicated with barbiturates show central nervous system signs: nystagmus, slurred speech, sedation, and ataxia. More severe levels of intoxication can lead to coma, depressed reflexes, and diminished cardiac contractility. Treatment consists of support of vital functions. Elimination of the drug may be hastened by the use of dialysis or hemoperfusion.

Adverse effects of barbiturates primarily reflect CNS depression. The dose necessary to affect respiration is not much higher than the therapeutic dosage, which gives these drugs a relatively low therapeutic index. Withdrawal reactions are similar to those described for the benzodiazepines but are likely to be much more severe, associated with a high incidence of seizures, and potentially fatal.

Aside from potentiating the sedative and CNS-depressing effects of other similar drugs, barbiturates can interact pharmacokinetically with many other agents, especially oral anticoagulants. **It is best to avoid coadministering barbiturates with anticoagulants,** but if it is unavoidable, the clinician should closely monitor the anticoagulant dosage and prothrombin time.

Barbiturates are contraindicated in patients with acute intermittent porphyria.

3. Propanediols

Meprobamate (Miltown® and others) and tybamate (no longer available in the United States), two propanediols, are labeled for use as antianxiety agents. Meprobamate was an extremely popular drug in the 1950s, but controlled studies have failed to demonstrate its superiority over the barbiturates for treating anxiety. For a vast majority of anxious patients, there is little reason to prescribe one of these drugs. Moreover, adverse effects, interactions, liability to abuse, and withdrawal syndromes resemble those observed with the barbiturates.

The structure of meprobamate is depicted in Fig. 2. The propanediols, similarly to the barbiturates, are metabolized mainly by hepatic microsomal oxidases and also induce these enzymes. The half-life of meprobamate is about 12 hr, and that of tybamate about half as long.

Pharmacological and clinical effects, toxicity, hazards, tolerance, withdrawal reactions, and interactions of meprobamate and tybamate are all similar to those of the barbiturates.

4. β Blockers

Propranolol (Inderal® and others) block β-noradrenergic receptors in the peripheral sympathetic nervous system and probably centrally as well. Although some β blockers discriminate between β_1 (cardiac) and β_2 (pulmonary) receptors, propranolol blocks both receptors competitively and without discrimination. Propranolol's structure is depicted in Fig. 2. Propranolol is almost completely absorbed following oral administration, but only about one-third of the administered drug reaches the systemic circulation, largely because of extensive first-pass metabolism in the liver. As with most drugs, plasma concentrations after a given dose of the drug vary widely among different people. The acute half-life of propranolol is about 3 hr, but with chronic administration it increases to about 4 hr. The drug is largely bound to plasma protein, which may make total blood concentrations less reliable as an index of clinical response. Propranolol is metabolized to inactive products and then excreted in the urine.

Propranolol has extensive effects on the cardiovascular system. It is used to lower blood pressure, prevent attacks of angina pectoris, and control certain cardiac arrhythmias. By blocking β_2 receptors, which are responsible for dilation in pulmonary bronchi, propranolol increases airway resistance and is thus **contraindicated in patients with asthma.** Propranolol has considerable effects on carbohydrate and fat metabolism and is **best avoided or used with great caution in patients with diabetes.** Unlike some β blockers, it readily penetrates the blood–brain barrier, but researchers are uncertain whether or not the effects of propranolol in the brain result from blockade of β receptors. In fact, although its peripheral mechanism of action appears understood, the way in which propranolol acts to treat certain neurological and psychiatric syndromes is unknown.

Propranolol is probably not as effective an antianxiety agent as the benzodiazepine drugs, although it might more specifically alleviate the sympathetic visceral symptoms of anxiety (e.g., tachycardia). It is possible that propranolol has a central antianxiety effect or, alternatively, that the patient experiences a decrease of anxiety because of a lessening of somatic symptoms (as per the James–Lange theory). Perhaps patients with many somatic symptoms or those prone to substance abuse may do best with this agent, but defining a proper role for propranolol in the treatment of anxiety will require much more research.

A β blocker such as propranolol is frequently prescribed for the as-needed use of patients with specific social phobias such as stage fright or public speaking. By diminishing or blocking such peripheral symptoms of anxiety as tachycardia and tremulousness, p.r.n. doses of 10–80 mg appear to be capable of interrupting a vicious cycle of cognitive cuing of anxiety from somatic symptoms that is so typical of these social phobias. Where maintenance treatment with a β blocker is selected for anxiety symptoms, some clinicians favor the use of more long-acting, less centrally acting β blockers such as atenolol (Tenormin® and others), which can be prescribed in a once-daily regimen of 50–100 mg a day. This strategy may reduce the risk of propranolol-induced depression as well. As noted, there is not good evidence, however, that maintenance treatment with β blockers is more effective than placebo in social phobias.

Doses of propranolol vary widely, but for maintenance treatment, initially the physician should administer 10 to 20 mg three or four times daily. Its short half-life makes divided doses necessary, but it also means that propranolol can be administered on an as-needed basis. Since the Food and Drug Administration (FDA) has not labeled propranolol for the treatment of anxiety or any other psychiatric condition, a physician should discuss the risks and benefits with the patient, specify alternative treatments, and document the clinical decision in the record (see Chapter 14).

The major hazards of treatment with propranolol are those of β blockade. In individuals with cardiac disease, propranolol can lead to heart failure or conduction delays. Patients treated for angina or hypertension sometimes experience potentially hazardous rebound phenomena when propranolol is withdrawn suddenly. As noted earlier, asthma is a contraindication to the use of this drug. Patients taking insulin for the control of diabetes may be more susceptible to hypoglycemia when they take propranolol. Other unwanted effects include nausea, vomiting, diarrhea and constipation, depression, delirium, psychosis, and allergic reactions.

5. Antipsychotic Drugs

At times, clinicians use antipsychotic drugs for the control of anxiety. However, phenothiazines and related compounds are generally less effective for treating anxiety than the benzodiazepines, and the former produce a wider range of more serious adverse reactions, particularly long-term neurological effects (namely, tardive dyskinesia). For these reasons, minimize the use of antipsychotic drugs to treat nonpsychotic anxiety. Occasionally, considerations of abuse, failure to achieve satisfactory response with benzodiazepines, or the presence of overwhelming anxiety (as in the borderline personality) may suggest the use of low doses of an antipsychotic drug. In such cases, the physician should administer the lowest effective dose for the shortest period of time, carefully monitor the patient, periodically reevaluate the dose, and use necessary precautions (see Chapter 4).

D. Antidepressants

Although last in our list of available agents for treatment of chronic anxiety, antidepressant agents in general are an underutilized class of drugs for the treatment of persisting anxiety, even when not associated with depression (see Chapter 2). Failure to appreciate the possible efficacy of antidepressants, not just for panic attacks but also for generalized anxiety,[26] may result from the fact that these agents require adequate doses and duration of treatment, as in the treatment of depression. Although such anxiolytics as the benzodiazepines may demonstrate efficacy within the first few hours or days of dosing, antidepressants may require several days or weeks to be efficacious; thus, clinicians may have overlooked their usefulness when compared to the more rapidly acting alternatives. Nonetheless, where benzodiazepines and tricyclic antidepressants have been compared, by the end of 6 weeks, antidepressants are typically comparable in efficacy to the benzodiazepines (and may go on to offer greater improvement). For clinical settings where a clinician wishes to avoid the administration of a benzodiazepine, therapeutic trials of antidepressants may be worth considering.

VI. PHARMACOTHERAPY OF ANXIETY

A. Introduction

Of all the anxiolytic drugs we have just reviewed, the benzodiazepines are the most popular: they are both highly effective and have a broad margin of safety. The

TABLE 4. Standard Pharmocopoeia for Anxiety Disorders

Agent	Usual initial dose (mg)	Dosage range (mg)	Chief limitations	Disorder[a]
TCA				
(e.g., imipramine)	10–25	150–300	Jitteriness TCA side effects	PD, AG, GAD, ?SP, ?PTSD, ?OCD
MAOI				
(e.g., phenelzine)	15–30	15–90	Diet MAOI side effects	PD, AG, SP, ?GAD, ?PTSD, ?OCD
Fluoxetine	10 (½ capsule)	20–80	Jitteriness	OCD, ?PD, ?AG
Benzodiazepines				
(e.g., alprazolam)	0.25 q.i.d.	2–6/day	Sedation	PD, AG, GAD, ?SP, ?PTSD, ?SiP
(e.g., clonazepam)	0.25 h.s.	1–3/day	Discontinuation syndrome	PD, AG, ?GAD, ?SP
(e.g., diazepam)	2.5	5–30/day	Psychomotor and memory impairment	GAD, ?SiP, ?PD, ?SP
Buspirone	5 t.i.d.	15–60/day	Dysphoria	GAD
β blockers				
(e.g., propranolol)	10–20	10–160/day	Depression (maintenance use only)	SP, ?PD, ?GAD

[a]PD, panic disorder; AG, agoraphobia; SP, social phobia; SiP, simple phobia; GAD, generalized anxiety disorder; PTSD, posttraumatic stress disorder; OCD, obsessive compulsive disorder.

physician should also know the standard pharmacopoeia for anxiety disorders (see Table 4). A physician, however, should not use any antianxiety agent as a substitute for listening and talking, developing an ongoing therapuetic relationship, considering environmental, social, and psychological factors, and prescribing nonbiological therapies. At the same time, clinicians should eschew the kind of ideological dogmatism that routinely frowns on the use of substances to alleviate emotional suffering.

B. Evaluating the Patient

Physicians should prescribe an antianxiety agent only after thoroughly assessing each individual and developing a balanced, custom-designed treatment program. They must first rule out other medical and psychiatric syndromes that may respond to specific treatments. Next, they should consider a role for various nonbiological approaches, e.g., counseling, psychodynamic psychotherapy, and behavioral and cognitive treatments.

When panic attacks are frequent or spontaneous, and particularly when associated with such complications as depression or generalized anxiety, pharmacotherapy is likely to be essential to achieve comprehensive benefit. If symptoms are restricted to specific phobic situations and the patient is otherwise comfortable, behavioral and cognitive strategies will possibly be sufficient. Optimal results frequently call for a multimodal approach.

Reviewing alcohol or other substance use history is critical, as is a general review of systems and prescribed medications. Physical examination, laboratory assessments (which may include electrolytes, calcium, magnesium, as well as liver and thyroid function tests), and an electrocardiogram (EKG) are useful baseline measures.

C. Discussing Treatment with the Patient

Once a decision is made to prescribe antianxiety medication, the physician should discuss the treatment plan and advise the patient about potential hazards such as possible problems driving, morning rebound or sedative carryover, and dangers of taking alcohol or other sedative–hypnotic drugs concomitantly. In addition, the clinician will want to suggest that prescription of an antianxiety agent is intended for a limited period of time and that the need for medication will be reevaluated regularly. Also, women who may become pregnant should be advised about potential hazards to the fetus.

D. Panic Attacks

Although treatment of panic attacks early in the course of the disorder may prevent complications, many patients seek treatment after years of symptoms and disability. Even in the face of chronicity, however, most patients achieve substantial if not dramatic benefit with antipanic pharmacotherapy as well as behavioral and cognitive therapies. Given the apparent primacy of the panic attack in the distress and evolution of complications of panic disorder, one approach is to initiate antipanic medications for patients continuing to experience panic attacks, with the expectation of regression and remission of complications once the attacks have ceased. For patients with residual phobic avoidance despite prevention of panic attacks, behavioral and cognitive strategies are employed. (For some patients, behavioral and cognitive strategies are employed initially, especially when symptoms are limited to phobic situations and where the frequency and intensity of unexpected panic are minimal. Pharmacotherapy is subsequently applied if emergence or exacerbation of panic or depression attends the behavioral program.)

Imipramine (Tofranil® and others) has well-established efficacy in panic disorder,[27] usually in the same dosage range as for treatment of depression (150–300 mg/day), but some patients do well at lower doses.

Other tricyclic antidepressants are probably also effective [for example, desipramine (Norpramin® and others) is frequently employed to decrease the anticholinergic burden], but the drawbacks include delayed onset of benefit and treatment emergent adverse effects. In addition to such usual side effects as dry mouth, constipation, and orthostatic hypotension, panic patients are particularly prone to a sudden worsening of their disorder with the first doses. To minimize the impact of this adverse response, treatment can be initiated with very small "test" doses (e.g., 10 mg of imipramine). If this is well tolerated, standard antidepressant dosing can be pursued; for others the adverse response typically fades over a few days, allowing an upward titration of dose. For a small percentage of patients, this initial worsening of the disorder does not subside.

Although clinical experience suggests that most antidepressants are effective, the greatest experience has been registered for imipramine and, to a lesser extent, desipramine. There is some evidence that trazodone (Desyrel® and others), bupropion (Wellbutrin®), and amoxapine (Asendin®) may be less effective in treating panic disorder. The more recently introduced serotonin reuptake-blocking antidepressant fluoxetine (Prozac®) also appears to be very effective for panic attacks except for a propensity to induce jitteriness and augmented anxiety, which will force early discontinuation of treatment for some patients.

The monoamine oxidase inhibitor (MAOI) phenelzine (Nardil®) has stood up well in clinical use and controlled trials,[27] and many clinicians feel that MAOIs work most effectively for panic disorder by blocking panic attacks, relieving depression, and offering a "confidence-enhancing" effect. Except for postural hypotension, MAOIs are usually free of most of the early tricyclic antidepressant side effects including the anxiogenic response. Unfortunately, as treatment proceeds, various challenging problems emerge including insomnia, weight gain, edema, sexual dysfunction, nocturnal myoclonus, and other symptoms. Further, many anxious patients are circumspect about the dietary precautions and instructions about hypertensive crises. Optimal treatment of panic disorder with an MAOI is similar in dose and duration to treatment of depression; all MAOIs are probably effective.

When treatment refusal or treatment failure with antidepressants is considered, the need for better tolerated and effective antipanic treatments becomes apparent. Further, in some cases earlier onset of benefit is desirable. In many respects, the triazolobenzodiazepine alprazolam fills this need because of its therapeutic benefit at initiation of treatment, antipanic efficacy, safety, and high patient acceptance. For most panic disorder patients, the usual dose range of alprazolam is 2–6 mg/day. Clinical response is evident early, but lower doses are necessary to initiate treatment so that the patient can accommodate to sedation. Most patients adapt within a few days to the sedating effects, thus allowing stepwise increase to panic-blocking doses. Adaptation to sedation usually occurs without loss of therapeutic benefit, but some upward adjustment may be required after the first 2 weeks. A very small percentage of patients appear particularly sensitive to the drug and experience persisting sedation despite time and cautious upward dose titration. Alprazolam must be given in divided doses, usually three or four times a day including bedtime, because of its relatively short duration of action.

Despite the ease of administering alprazolam and frequently dramatic results even in the first days of treatment, evident drawbacks include concerns about abuse and dependency, rebound symptoms between doses, withdrawal syndromes, and possibility of early relapse upon discontinuation. The abuse potential of alprazolam may be similar to other benzodiazepines and varies widely among clinical populations. Most well-informed panic and agoraphobic patients who have endured severe distress over time treat their medication with respect and understand the wisdom in maintaining the lowest effective dose. Thus, unless there is evidence that a particular patient is more at risk for abuse, the use of this agent is usually safe for this disorder. As with any benzodiazepine, without controlled prescribing for targeted symptoms, inappropriate use may occur.

Because alprazolam is a relatively short-half-life and high-potency benzodiazepine, discontinuing treatment without a **very gradual taper** tailored to the individual patient's sensitivity to decreasing doses may be followed by rebound symptoms (worsened anxiety) or a withdrawal syndrome.

With these drawbacks in mind, some clinicians initiate treatment with alprazolam and a tricyclic antidepressant with the goal of tapering alprazolam early and continuing treatment with the antidepressant. Although the combination is certainly effective, there is no evidence that the discontinuation of alprazolam is facilitated by this strategy. For patients who require a high-potency benzodiazepine but suffer rebound symptoms, the longer-acting high-potency benzodiazepine clonazepam has proved effective.[28]

With milligram-for-milligram potency about twice that of alprazolam, clonazepam's usually effective dose range for panic patients is between 1 and 3 mg/day given in the morning and at bedtime. Sedation will be the limiting factor in dose titration and is managed by initiating treatment with a low bedtime dose and titrating upward if symptoms persist and as sedation resolves. Initial doses as low as 0.25 mg at bedtime may be used in drug-naive patients or those particularly sensitive to benzodiazepines. Greater doses may be given at bedtime than in the morning to minimize sedation, but many patients function without sedation on equal a.m. and h.s. doses.

The effect of a given daily dose on panic attacks and generalized anxiety will be apparent within a few days. Some patients (fewer than 10%) develop depressive symptoms on clonazepam. Resolution of depressive symptoms typically occurs by lowering the dose or by introducing an antidepressant; clonazepam often can then be tapered with expectation of continued response to the antidepressant; combined treatment can be pursued if anxiety symptoms incompletely resolve with the antidepressant.

Despite the "pharmacological dissection" of anxiety neurosis into antidepressant-responsive panic disorder and benzodiazepine-responsive generalized anxiety disorder, increasing evidence suggests that a number of other benzodiazepines such as diazepam, clorazepate, and lorazepam also improve panic disorder. Whether their efficacy is indeed comparable to alprazolam and clonazepam is insufficiently studied; with a clearly effective shorter-acting alprazolam and longer-acting clonazepam, the need for additional antipanic benzodiazepines is not pressing.

A number of other agents have benefited some patients with panic disorder either alone or adjunctively with standard agents including β blockers, verapamil

(Calan® and others), carbamazepine (Tegretol® and others), valproic acid (Depakene® and others) and, transiently, clonidine (Catapres®).

E. Phobic Disorder

As with panic disorder and agoraphobia, phobic disorders may be treated with an array of interventions ranging from pharmacotherapy to behavioral and cognitive therapies.

The pharmacotherapy of social phobia is less well studied than panic and agoraphobia. Performance anxiety (musical performance, public speaking, stage fright) is often controlled by pretreatment with a β blocker such as propranolol (e.g., 10–80 mg 1 hr before the event), but maintenance treatment with β blockers for frequently encountered or global social phobias may not be better than placebo. The MAOI phenelzine has shown greater promise, however.[29] Benzodiazepines may also relieve social phobic symptoms. Some patients may benefit from rapidly acting agents for specific situations, and others may benefit from maintenance treatment as well.

The most expedient treatments of simple phobia are behavioral therapies, although single-dose pharmacotherapeutic interventions are often undertaken when a rarely encountered simple phobia must be endured, such as an airplane flight.

F. Generalized Anxiety

1. Use of Benzodiazepines for Acute and Chronic Anxiety

Benzodiazepines have been the traditional mainstay of anxiolytic pharmacotherapy,[10] although the clinical decision to prescribe these agents for symptom relief is a difficult one. The attitudes of individual physicians toward prescribing drugs for anxiety fall into a spectrum between "pharmacological Calvinism" and "psychotropic hedonism,"[11] reflecting in part a personal moral stance. The abundant literature on antianxiety agents falls short of providing reliable measures for determining when to prescribe. Given the ubiquitousness of anxiety, the physician must frequently confront the question of whether or not to prescribe.

After considering the differential diagnosis and the relevance of psychological or interpersonal interventions, prescribing a benzodiazepine for the distressed, anxious patient is a therapeutic act analogous to pain relief. It attenuates symptoms as part of a comprehensive diagnostic and treatment plan. Benefits in addition to decreased distress may occur for the medical patient. For example, the drug may provide a cardioprotective effect for the anxious cardiac patient who is vulnerable to arrhythmias.

As compared with barbiturates and nonbarbiturate sedative and hypnotic agents [meprobamate, ethchlorvynol (Placidyl®), glutethimide (Doriden®), methaqualone (no longer available in the United States) and others], the benzodiazepines are more selectively anxiolytic, with less sedation and less morbidity and mortality in overdose and acute withdrawal. Since using a benzodiazepine represents a clinical decision to offer symptomatic relief, it is critical to evaluate the patient's ability to function.

Ideally, benzodiazepine drugs should be administered for short-term use, typically around a period of stress, and always within the context of an ongoing therapeutic relationship. The physician should see the patient for a follow-up visit within a week or two to assess the patient's overall clinical status and response to the medication.

In the real world, however, many patients are on maintenance therapy and take benzodiazepines for months and even years. Is this ever justified, and, if so, what are the risks? To address the latter first, benzodiazepines carry a potential for dependence, abuse, and withdrawal reactions (although all are probably less severe than with other sedatives). Furthermore, the probability of withdrawal reactions (and perhaps the severity) increases with the chronicity of exposure. Also, in occasional patients, long-term use of benzodiazepines may produce chronic sedation and apathy (probably because of excessive levels). To avoid promoting abusive, overly long, and unsupervised use of these substances, physicians should not prescribe large quantities of benzodiazepine drugs or give multiple refills.

Despite our best intentions, however, some patients do take benzodiazepines chronically and function in a relatively stable manner. How does one identify these patients, and how should they be managed? The answer is: empirically. If a patient has been taking a benzodiazepine drug for weeks or a few months, the physician should attempt gradually to taper the agent. Even at low doses and with small decremental decreases, the patient may experience a reemergence of anxiety. The clinician's dilemma is to differentiate between rebound and relapse. If the patient is able to bear the distress, the clinician should recommend that the patient maintain the lower dose or the drug-free state for approximately 2–4 weeks. Some patients will become symptom-free before that time, suggesting that rebound or withdrawal effects accounted for the transient worsening. In other patients, symptoms emerging after tapering persist, suggesting continued need for the drug. If anxiety symptoms (not a withdrawal reaction) regularly reemerge when the dose is tapered, and the patient has not shown a tendency to abuse the drug (e.g., to get "high," to increase the dose beyond the physician's prescription), then the most prudent course may be to allow the patient to continue chronically with the medication. Of course, the clinician should attempt to taper the drug periodically, possibly every 6 to 12 months, to find the lowest effective dose that can control symptoms (see Table 4 for dosage recommendations).

Tolerance does develop to the sedative effects of benzodiazepines, but it is less clear whether tolerance develops to their sleep-maintaining and antianxiety effects. The benzodiazepines probably do retain at least a proportion of their efficacy over many months and possibly longer.

Physicians should advise all patients who are taking benzodiazepines to consult with them before discontinuing the drug. When it is desirable to stop treatment, the drug (especially short-acting agents) should be tapered gradually, by no more than 5% to 10% each day. Even this rate will be intolerable to many, and the smallest dosing decrements possible over long periods of time may be required—in particular for a short-acting, high-potency agent like alprazolam. The β blocker propranolol has been used experimentally as an adjunct to treat benzodiazepine withdrawal reactions, but results have been equivocal.[30,31] Other strategies targeted for use as adjuncts in alprazolam discontinuation are carbamazepine, clonidine, or a switch from alprazolam to clonazepam followed by taper of the longer-acting agent.

Clearly, nonbiological approaches to chronic anxiety if effective are preferable, but in assessing the overall balance between the risks and benefits of the drug and the risks of unchecked anxiety, long-term prescription of a benzodiazepine may be appropriate for some individuals. Perhaps for others, this is merely a "bad habit," but it is equally feasible that some individuals are biologically prone to excessive anxiety.

Is there a role for benzodiazepine drugs in the treatment of anxiety among patients with severe character pathology such as borderline personalities? The answer to this question contains more caveats than recommendations. This diagnostic category probably represents a heterogeneous group of disorders, possibly containing some individuals who will show symptomatic improvement during acute exacerbations of anxiety if treated with standard doses of benzodiazepine drugs. However, for many borderline patients, the dose of a benzodiazepine that would contain the characteristic massive upsurge of anxiety to which they are often prone would be so excessive as to cause unacceptable adverse effects. Moreover, many of these individuals are prone to substance abuse. Additionally, prescription of drugs of this type can vastly complicate therapeutic interactions. Finally, in some individuals, benzodiazepine drugs may disinhibit tendencies to act on socially unacceptable impulses, promoting the expression of hostility, indiscreet sexuality, or self-destructive behavior.

Steps in the Treatment of Generalized Anxiety Disorder

1. Evaluate the nature of the anxiety
   a. Rule out medical and specific psychiatric causes
   b. Define areas of conflict and stress
   c. Note drugs that may interact with anxiolytics

2. Evaluate the patient's character structure and intrapsychic issues to help determine psychological risks and benefits of drug taking

3. Consider nonbiological approaches (e.g., dynamic psychotherapy, ego supportive or crisis intervention therapies, behavioral or cognitive approaches)

4. If medication is indicated, discuss plans with the patient
   a. Outline potential hazards (e.g., problems driving, hangover, additive effects with other drugs)
   b. Propose a short-term trial of drug therapy
   c. Discuss the need for periodic reevaluation
   d. Outline the potential harm to the fetus for women of childbearing age

5. Prescribe short-acting agents in divided doses; long-acting agents may be administered once or twice a day

6. Give benzodiazepines in low doses initially (in small, nonrefillable prescriptions) and increase until
   a. Anxiety is adequately contained
   b. Significant adverse reactions (particularly drowsiness) occur
   c. Recommended dose limits are reached

7. Discontinue drugs slowly (i.e., over several weeks—longer after prolonged therapy or with very high doses)

8. Consider nonbenzodiazepine antianxiety medications for patients who may abuse drugs

9. Only provide ongoing benzodiazepine therapy for patients who
   a. Have significant anxiety without medication
   b. Do not respond to other treatment measures
   c. Do not show signs of abuse

10. Reevaluate drug use
   a. When symptoms do not respond adequately to treatment
   b. When the patient requests increasing doses or requires high doses or prolonged administration
   c. When the period of acute stress passes or after several weeks of acute administration
   d. Every few months in patients on chronic medication

2. Choosing a Specific Benzodiazepine

The choice of a specific benzodiazepine is usually not critical, since all are equally effective for generalized anxiety symptoms. However, the clinician should consider pharmacokinetic differences that may help tailor the drug to the patient and situation. For example, the rapidity with which a specific benzodiazepine is absorbed may affect the speed of onset of therapeutic action after a single dose. Thus, if a patient with a social phobia wished to take a single dose of a benzodiazepine shortly before specific engagements, he might prefer lorazepam, diazepam, or alprazolam because of their rapid onset of action. Similarly, a choice between benzodiazepines with short versus long durations of action during chronic therapy may be affected by the anticipated schedule of drug treatment. For patients who anticipate being under stress for many weeks, a long-acting benzodiazepine (e.g., diazepam, chlordiazepoxide, clorazepate, prazepam) provides the advantages of once- or twice-a-day dosing as well as consistent tissue levels around the clock. (Once-daily dosages are usually administered at night to avoid excessive sedation associated with peak blood concentrations.)

When the long-acting benzodiazepines are discontinued, their long half-lives provide a built-in gradual tapering. Short-acting drugs, by contrast, are excreted more rapidly, and since the time of appearance and severity of symptoms of a withdrawal reaction may correlate with the rapidity of decline of blood levels, short-acting agents may be associated with earlier onset and more severe withdrawal reactions. Consequently, increased attention should be paid to avoiding abrupt discontinuation of these agents. Of course, any benzodiazepine should be discontinued gradually for the patient's safety and comfort.

When short-acting benzodiazepines are used, steady-state concentrations are achieved more rapidly than with long-acting drugs. In the elderly, in whom half-lives of all drugs tend to be prolonged, steady-state levels of long-acting drugs may not be achieved for a number of weeks. During this time, the dose of the drug must be carefully titrated against anxiety symptoms to avoid excessive sedation. Achieving a steady state in a matter of days with a short-acting agent allows determination of the therapeutic dose more rapidly.

The short-acting drugs have fewer kinetic interactions, as with cimetidine (Tagamet®). Finally, because the metabolism of short-acting benzodiazepines is little affected by the presence of liver disease, it may be less complicated to use these agents in patients with hepatic impairment.

Whatever drug is chosen, observe the patient frequently during the early days of treatment to assess his response, monitor unwanted effects, and titrate the dosage until a steady state is reached (i.e., no further accumulation occurs). Dosages will vary widely among patients, based on the fluctuating symptoms of anxiety as well as on individual differences in drug kinetics. If an individual is responding to a transient stress, the drug should be tapered and discontinued soon after the stress has been alleviated.

Chlordiazepoxide, diazepam, and lorazepam are available for parenteral use. Chlordiazepoxide and diazepam produce unreliable blood levels when administered intramuscularly; intravenous use, however, leads to rapid and predictable rises in blood drug levels. Therefore, these two drugs are best administered orally or intravenously. Lorazepam has been approved for intramuscular administration, although at present the labeling is restricted to preanesthetic use. Blood level increases following intramuscular lorazepam administration are generally smooth and reliable.[32]

Diazepam is absorbed by plastic intravenous bags and tubing, which can result in a loss of potency. Moreover, diazepam can precipitate if mixed for infusion with saline or dextrose. To avoid this problem, increase the diazepam concentration, hasten infusion speed, and/or dissolve diazepam in a lipid emulsion.[33] If accidentally administered intraarterially, diazepam may produce severe inflammation with tissue necrosis.[34]

3. Indications for Using Other Antianxiety Drugs

Other antianxiety drugs tend to be less effective, more hazardous, or associated with more unwanted effects than the benzodiazepines. For some patients a β blocker such as propranolol may effectively alleviate some of the somatic manifestations of anxiety. More study will be required, however, before propranolol can be generally recommended or its proper place in therapeutics determined. Clinicians should avoid barbiturates and related sedative–hypnotic compounds (including meprobamate) when treating anxious patients unless there is a specific, compelling reason to employ them.

In patients with a liability to substance abuse, the clinician might want to elect a drug to treat anxiety that does not have potential to produce abuse or dependency. For this purpose, an antidepressant or buspirone may prove efficacious. In some cases, the physician may want to use low doses of a phenothiazine or other antipsychotic drug; if treatment is likely to be prolonged, however, the physician and patient must consider the possibility of producing long-term neurological effects (namely, tardive dyskinesia) even at relatively low doses.

G. Obsessive Compulsive Disorder

Several therapeutic strategies have been used with varying degrees of success in OCD patients, including behavioral treatments, typically consisting of exposure treatments (e.g., in those with contamination fears) and response prevention (e.g., for those with compulsive washing or checking). Reported possible pharmacological interventions include treatments in use for anxiety disorders (e.g., MAOI and non-MAOI antidepressants, antidepressants plus lithium, and, on occasion, benzodiazepines and other anticonvulsants). Recent data suggest particular promise with specific, more serotonergic antidepressants such as clomipramine (Anafranil®) and fluoxetine (Prozac®). Even psychosurgery has proven useful to some patients disabled by treatment refractory OCD.

H. Posttraumatic Stress Disorder

Treatments of PTSD have been less well studied than other anxiety disorders but have included psychotherapies, abreactive techniques, as well as reports of successful psychopharmacological interventions with much of the psychopharmacopoeia. Where specific comorbid syndromes such as depression or panic attacks are featured, the specific treatments for those symptoms or syndromes are indicated.

I. Anxiety Symptoms in Other Syndromes

Benzodiazepine drugs can also serve as ancillary agents in the treatment of depression and psychosis. Although they probably are not effective alone for the treatment of most cases of major depression, particularly of the melancholic type, benzodiazepines may help to alleviate the anxiety that often accompanies depression and thus serve as useful concurrent therapy for other somatic treatments such as antidepressant drugs. For many patients, an antidepressant alone will suffice, but in the early weeks of treatment (before the antidepressant has become effective) a benzodiazepine may provide welcome symptom relief.

Similarly, in the treatment of patients with psychosis, benzodiazepines may at times be used in addition to antipsychotic agents. Benzodiazepines probably have little inherent antipsychotic efficacy (at least in the usual clinical dose range) and, if administered alone, might precipitate hostile or excited outbursts in an occasional psychotic patient. However, coadministered with an antipsychotic drug, benzodiazepines are sometimes useful for the treatment of certain extrapyramidal syndromes such as acute dystonic reactions and akathisia (see Chapter 4) and, in the case of acute psychosis, may permit the use of lower doses of neuroleptics.

VII. SUMMARY

Although anxiety is a common—virtually universal—experience, the recognition of specific anxiety disorders is a critical clinical task requiring careful diagnosis and evaluation. Most patients improve substantially with specific interventions, both pharmacological and psychotherapeutic. When medication treatment of generalized anxiety is indicated, a benzodiazepine may still be the agent of choice, although trials of antidepressants or buspirone should also be considered.

REFERENCES

1. Charney D. S., Heninger G. R.: Noradrenergic function and the mechanism of action of antianxiety treatment. *Arch Gen Psychiatry* 42:458–481, 1985.
2. Insel T. R., Ninan P. T., Aloi J., et al: A benzodiazepine receptor mediated model of anxiety: Studies in non-human primates and clinical implications. *Arch Gen Psychiatry* 41:741–750, 1984.
3. Ballenger J. C.: Biological aspects of panic disorder. *Am J Psychiatry* 143:516–518, 1986.
4. Nemiah J. C.: *Foundations of Psychopathology.* New York, Oxford University Press, 1961, pp 100, 121.
5. Seligman M. E. P.: Phobias and preparedness. *Behav Ther* 2:307–320, 1971.
6. Klein D.: Delineation of two drug responsive anxiety syndromes. *Psychopharmacologia* 5:397–408, 1964.

7. Crowe R. R., Noyes R., Pauls D. L., et al: Family study of panic disorder. *Arch Gen Psychiatry* 40:1065–1069, 1983.

8. Coryell W., Noyes R., Jr., Howe J. D.: Mortality among outpatients with anxiety disorders. *Am J Psychiatry* 143:508–510, 1983.

9. Jenike M. A., Baer L., Minichiello W. E.: *Obsessive/Compulsive Disorders: Theory and Management.* Littleton, MA, PSG Publishing, 1986.

10. Rosenbaum J. F.: The drug treatment of anxiety. *N Engl J Med* 306:401–404, 1982.

11. Klerman G. L.: Psychotropic hedonism versus pharmacological Calvinism. *Hastings Cen Rep* 2:1–3, 1972.

12. Ghoneim M. M., Mewaldt S. P., Beri J. L., et al: Memory and performance effects of single and three-week administration of diazepam. *Psychopharmacology* 73:147–151, 1981.

13. Snyder S. H.: Benzodiazepine receptors. *Psychiatr Ann* 11:19–23, 1981.

14. Greenblatt D. J., Divoll M., Abernethy D. R., et al: Benzodiazepine kinetics: Implications for therapeutic pharmacogeriatrics. *Drug Metab Rev* 14:251–292, 1983.

15. Greenblatt D. J., Shader R. I., Divoll M., et al: Benzodiazepines: A summary of pharmacokinetic properties. *Br J Clin Pharmacol* 11:11S–16S, 1981.

16. Downing R. W., Rickels K.: Hostility conflict and the effect of chlordiazepoxide on change in hostility level. *Compr Psychiatry* 22:362–367, 1981.

17. Jackson M. R., Harris P. A.: Diazepam and tumour promotion. *Lancet* 1:445, 1981.

18. Finkle B. S., McCloskey K. L., Goodman L. S.: Diazepam and drug-associated deaths: A survey in the United States and Canada. *JAMA* 242:429–434, 1971.

19. Greenblatt D. J., Allen M. D., Noel B. J., et al: Acute overdosage with benzodiazepine derivatives. *Clin Pharmacol Ther* 21:497–514, 1977.

20. Divoll M., Greenblatt D. J., Lacasse Y., et al: Benzodiazepine overdosage: Plasma concentrations and clinical outcome. *Psychopharmacology* 73:381–383, 1981.

21. Gelenberg A. J.: Benzodiazepine use during pregnancy. *Mass Gen Hosp Biol Ther Psychiatry Newslett* 3:36, 1980.

22. Gelenberg, A. J.: Benzodiazepine withdrawal. *Mass Gen Hosp Biol Ther Psychiatry Newslett* 3:9–10, 1980.

23. Gelenberg A. J.: Short-acting benzodiazepines and cimetidine (Tagamet): No interaction. *Mass Gen Hosp Biol Ther Psychiatry Newslett* 4:23, 1981.

24. Goa K. L., Ward A.: Buspirone: A preliminary review of its pharmacological properties and therapeutic efficacy as an anxiolytic. *Drugs* 32:114–129, 1986.

25. Schweizer E., Rickels K.: Failure of buspirone to manage benzodiazepine withdrawal. *Am J Psychiatry* 143:1590–1592, 1986.

26. Kahn R. J., McNair D. M., Lipman R. S., et al: Imipramine in chlordiazepoxide in depressive and anxiety disorders. II. Efficacy in anxious outpatients. *Arch Gen Psychiatry* 43:79–85, 1986.

27. Ballenger J. C.: Pharmacotherapy of the panic disorders. *J Clin Psychiatry* 47 (6, Suppl):27–32, 1986.

28. Pollack M. H., Rosenbaum J. F.: Benzodiazepines in panic-related disorders. *J Affect Dis* 2:95–107, 1988.

29. Liebowitz M. R., Gorman J. M., Fyer A. J., et al: Pharmacotherapy of social phobia: An interim report of a placebo-controlled comparison of phenelzine and atenolol. *J Clin Psychiatry* 49:252–257, 1988.

30. Tyrer P., Rutherford D., Huggett T.: Benzodiazepine withdrawal symptoms and propranolol. *Lancet* 1:520–522, 1981.

31. Abernethy D. R., Greenblatt D. J., Shader R. I.: Treatment of diazepam withdrawal syndrome with propranolol. *Ann Intern Med* 94:354–355, 1981.

32. Shader R. I., Greenblatt D. J.: Clinical implications of benzodiazepine pharmacokinetics. *Am J Psychiatry* 134:652–656, 1977.

33. Winsnes M., Jeppsson R., Sjoberg B.: Diazepam adsorption to infusion sets and plastic syringes. *Acta Anesthesiol Scand* 25:95–96, 1981.

34. Rees M., Dormandy J.: Accidental intra-arterial injection of diazepam. *Br Med J* 281:289–290, 1980.

SELECTED READING

1. Rosenbaum J. F., Pollack M. H.: Anxiety, in Hackett T. P., Cassem N. H. (eds): *Massachusetts General Hospital Handbook of General Hospital Psychiatry,* ed 2. Littleton, MA, PSG Publishing, 1987, pp 154–183. Good overall review of anxiety as a syndrome, with particular attention to both psychiatric and medical differential diagnosis and anxiety in the medical setting.
2. Salzman L., Thaler F. H.: Obsessive compulsive disorders: A review of the literature. *Am J Psychiatry* 138:286–296, 1981. Overview of recent developments regarding obsessive compulsive disorder.
3. Baldessarini, R. J.: Drugs and the treatment of psychiatric disorders, in Gilman A. G., Goodman L. S., Gilman A. (eds): *Goodman and Gilman's The Pharmacological Basis of Therapeutics,* ed 6. New York, Macmillan, 1980, pp 391–447. Review of preclinical pharmacology of antianxiety agents.
4. Petursson H., Lader M. H.: Benzodiazepine dependence. *Br J Addict* 76:133–145, 1981. Review of important topic.

6

Insomnia

JEFFREY B. WEILBURG, M.D., and ALAN J. GELENBERG, M.D.

I. INTRODUCTION

Insomnia may be defined as difficulty falling or staying asleep, associated with compromised daytime functioning, that persists for more than 4 weeks. Most patients with insomnia have prolonged (e.g., greater than 30 min) sleep latency (time in bed to time asleep). They often have sleep that is shallow or fragmented by multiple arousals. However, no single disturbance of sleep physiology is found in all insomniacs. Some patients who complain of insomnia appear normal on objective measurements of sleep. Even patients whose complaints are matched by abnormal sleep physiology may have variations in objective sleep measures.[1]

Insomnia is a common problem. Fifteen to 20% of the population complain of insomnia, and 10% of the population report using "sleeping pills." Twenty percent of all adult general medical outpatients and 35% of psychiatric outpatients complain of insomnia. Higher rates are found among women and patients from lower socioeconomic groups. Insomnia is particularly common in elderly patients. The elderly receive 35–40% of sedative–hypnotic prescriptions written each year even though they account for only 12% of the population.[1,2]

II. SLEEP

A. Normal

Sleep laboratories use polysomnography to study sleep physiology. The polysomnogram (PSG) is a continuous all-night recording of the EEG, eye movement,

JEFFREY B. WEILBURG, M.D. • Neuropsychiatry Section, Psychopharmacology Unit, Massachusetts General Hospital, Boston, Massachusetts 02114; Harvard Medical School, Boston, Massachusetts 02115.   ALAN J. GELENBERG, M.D. • Department of Psychiatry, University of Arizona Health Sciences Center, Tucson, Arizona 85724.

muscle activity, respiration, and EKG. Blood oxygen levels, penile tumescence, and video observation may sometimes be included in the evaluation.

The PSG indicates that sleep can be divided into two basic types: rapid eye movement (REM) and non-REM (NREM) sleep. During REM sleep, the EEG becomes activated, pulse and respiration increase and become irregular, muscle tone decreases, and the eyes move in rapid bursts. Dreaming is associated with REM. Non-REM sleep is subdivided into stages 1–4. Stage 1 is a drowsy, transitional state. Stages 3 and 4 are often called "delta" or deep sleep.

REM and NREM reciprocally cycle. The pattern of the cycling is called the sleep architecture. The normal sleeper drowses into stage 1 and then slips into stage 2, where sleep formally begins. After a period of deep or delta sleep, stage 2 returns. The first REM period follows, appearing 90–120 min after sleep onset. NREM and REM then alternate every 90 min throughout the night. Delta sleep is more prominent in the early cycles, and REM is more prominent towards morning.

B. Diagnosis of Insomnia

A wide variety of different pathophysiological processes can present as insomnia. Therefore, insomnia should be considered as a complaint rather than a disease. The clinician should attempt to determine the cause or pathophysiological basis of the complaint before treatment is initiated. Although this task may be difficult, it is not an academic exercise. Sedative–hypnotics can be ineffective and in some cases dangerous if the problem producing the insomnia is not taken into account. For example, the patient who complains of "insomnia" because of underlying sleep apnea may develop significant respiratory and secondary cardiovascular problems if sedatives are given without appropriate caution. The patient with sleep difficulty secondary to the use of or rebound from alcohol or hypnotics may be at risk of overdosage or exacerbation of the substance abuse unless the primary problem is addressed.[3]

The pathophysiological processes that can present as insomnia are summarized in Table 1. Controversy exists regarding the relative importance of these processes. Some investigators believe that in the vast majority of patients insomnia is secondary to psychiatric problems. This group of experts favors the classification of sleep disorders listed in the third revised edition of the *Diagnostic and Statistical Manual of Mental Disorders* (DSM-IIIR) and believes that polysomnography is rarely needed to establish a clinical diagnosis. Other investigators report that a significant proportion of patients presenting with insomnia have a primary disorder of sleep, such as nocturnal myoclonus or sleep apnea, and that these problems may be missed if polysomnogram is not done. This group tends to favor the Association of Sleep Disorders Centers (ASDC) classification of sleep problems over DSM-IIIR. Discussion of this controversy is beyond the scope of this chapter. Rather, this chapter describes some common techniques and clinical pitfalls in diagnosing insomnia.

Attention to accuracy and detail in obtaining the history of the patients' sleep and wake behaviors is important in the diagnosis of insomnia. Using a sleep log,

TABLE 1. Underlying Etiology of Insomnia Complaint

1. Psychiatric disorders
   a. Affective disorders (including dysthymic disorder)
   b. Character disorders
   c. Anxiety disorders
   d. Others: psychosis, adjustment disorders, etc.
2. Drugs
   a. Alcohol (dependence, rebound, or withdrawal phenomenon)
   b. Sedatives (dependence, rebound, or withdrawal phenomenon), barbiturates, benzodiazepines, others
   c. Stimulants (dependence, rebound, or withdrawal phenomenon) amphetamines, methylphenidate; pemoline; cocaine; caffeine and stimulant xanthines in coffee, tea, cola, chocolate
   d. Antidepressants: phenelzine, tranylcypromine, protriptyline, desipramine, imipramine, amoxapine, fluoxetine
   e. Antiasthmatics, decongestants
   f. Tobacco (dependence, rebound, or withdrawal phenomenon)
3. Medical/surgical problems
   a. Pain of any source
   b. Respiratory: chronic obstructive pulmonary disease
   c. Renal: uremia, urinary tract infection
   d. Endocrine: hyper- and hypothyroidism; exogenous steroids
   e. Cardiovascular: nocturnal angina, orthopnea
   f. Delirium: dementia, infection, medication toxicity (including sedative–hypnotics or over-the-counter sleep remedies)
4. Primary sleep disorder
   a. Sleep apnea
   b. Narcolepsy
   c. Periodic movements during sleep: nocturnal myoclonus, restless legs
   d. Circadian rhythm disruption: phase shift
5. Others
   a. Idiopathic primary insomnia
   b. Childhood-onset insomnia
   c. Psychophysiological or conditioned insomnia
   d. Irregular schedule, inadequate sleep time
   e. Unusual polysomnographic pattern

such as that developed by Hobson, and interviewing the spouse, bed partner, or others who can observe the patient while asleep should be part of the work-up.[4] Physical and basic laboratory examinations, including screening of urine and serum for drugs and alcohol if abuse is suspected, are also important diagnostic steps.[5,6]

Some people complain of insomnia but are actually normal. Included in this group are those people who require less than 8 hr of sleep (sometimes only 4 hr) to function normally. These individuals sometimes complain of "insomnia," which to them means having too much time on their hands at night. Likewise, normal elderly persons may complain of being unable to sleep as long as they once did or of lying awake in the early morning. This pattern of sleep may be a function of normal aging. Treatment for these complaints, once the diagnosis is clear, revolves around education and support rather than drugs.[7,8]

Shift workers or long-distance travelers may develop insomnia based on disruptions between their internal pacemakers and ambient day/night conditions.

Many patients with insomnia have underlying psychiatric problems. For example, insomnia may be the primary symptom of a character disorder. Patients may experience significant disruption of sleep initiation or maintenance and develop daytime fatigue and impaired function because of the anxiety caused by conscious or unconscious emotional conflicts. These patients tend to become frustrated and angry about being deprived of the rest to which they feel entitled. They may try to force themselves to sleep, becoming ever more angry and aroused as this fails. Insomnia becomes the main focus, being viewed as the cause of their difficulties. The resultant aroused state can become self-reinforcing.

Patients with insomnia of this sort may become intensely invested in finding "the right sleeping pill." Yet trials of various hypnotics typically fail despite escalating doses. Not surprisingly, these patients are at high risk of becoming drug dependent.

Once this diagnosis is recognized, efforts to help the patient refocus the problem away from insomnia and onto the primary emotional conflicts are the key to managing this type of insomnia. Hypnotic drugs and behavioral sleep strategies may be judiciously employed as adjuncts in this difficult process.

A significant proportion of patients with insomnia have underlying substance abuse problems, such as alcohol, sedative [barbiturates, diazepam (Valium® and others)], stimulant (cocaine, caffeine), or cigarette abuse. These patients often deny, fail to appreciate, or frankly attempt to hide the substance abuse and the adverse effects it has on sleep. Since treating "insomnia" in these patients is futile and potentially dangerous unless the substance abuse problem is addressed first, efforts must be made to recognize patients in this diagnostic group before treatment begins.

Patients with mood and anxiety disorders may have insomnia secondary to the underlying disorder itself or to adverse effects of medications used to treat these problems. Management is discussed below.

Various primary disorders of sleep may present as insomnia; these are outlined below. If a primary sleep disorder is suspected, polysomnography is indicated. A complete description of these and other primary and secondary sleep disorders can be found in Emrich's monograph.[13]

1. Sleep Apnea

The two basic forms of sleep apnea are central and obstructive. Both types may present as insomnia. With central apnea, the diaphragm is not stimulated to contract; thus, no respiration occurs until the patient arouses. In obstructive sleep apnea, the airway is occluded (by excess tissue, failure of tone in the pharyngeal muscles, or both), which prevents air flow. Loud snorts, snores, and partial arousal are noted on the PSG. Many patients have mixed apneas, which contain central and obstructive elements. It can be dangerous to give sedatives to patients with sleep apnea because they raise the arousal threshold and may therefore delay the return of respiration.

When respiration ceases, blood oxygen saturation falls, CO_2 rises, pH falls, and blood pressure rises. Serious cardiorespiratory problems may arise if apnea is untreated. Severe untreated sleep apnea may be fatal.[13]

2. Nocturnal Myoclonus

Nocturnal myoclonus is a stereotypic repeated twitching of the leg and other muscles during sleep. Patients may be unaware that they twitch. The myoclonic activity may partially arouse the patient and thus can produce fragmented, nonrestorative sleep. Nocturnal myoclonus and other sleep-related movement problems are often categorized as disorders involving periodic movements of sleep (PMS). The PMS may be precipitated by psychotropics such as cyclic antidepressants, MAOIs, and antipsychotic agents.[13]

3. Phase Shifts

Some patients have a sleep–wake cycle that is not synchronized with the day–night, light–dark environment. In the case of phase delay, patients will fall asleep several hours after a socially desired bedtime (e.g., 4 a.m.) and will, if undisturbed, awake refreshed at noon. If such patients have to get up for work or school at 7 a.m., they will feel tired. They may thus complain of "insomnia" (e.g., difficulty falling asleep). Behavior treatment, such as phase advance or delay, is the effective treatment for phase-shift problems. It is important to have phase shifts ruled out before initiating drug treatment, since hypnotics may be ineffective or even contraindicated.[9,10]

4. Psychophysiological Insomnia

Although controversial, there are some patients who are overaroused but do not meet criterion for psychiatric disorders. Such patients learn to associate bedtime with anxiety and develop insomnia by classical conditioning. In these individuals, behavioral management (e.g., deconditioning, relaxation, biofeedback) may be judiciously combined with benzodiazepines.[11,12]

5. Primary Idiopathic Insomnia

As the name suggests, some patients have objective (on PSG) and subjective insomnia without an identifiable cause. Some patients with childhood onset insomnia may suffer from some idiopathic problem with sleep.

III. APPROACH TO TREATMENT

Treatment is directed at managing the problem that produces the complaint of insomnia.

Psychotherapeutic modalities are best for those patients whose problems primarily fall along axis II. Simple behavioral strategies may also be helpful for these patients. These include following the basic rules of good "sleep hygiene." Bed should be used only for sleeping or lovemaking. Patients should get out of bed into a chair for reading or other quiet activities (not TV) if they are unable to sleep after 10 min. Moderate exercise early in the day, a regular sleep–wake schedule (including weekends), and a regular bedtime preparation ritual that allows time to "wind down" are useful. A small snack of milk or cheese before bed may help; stimulants and daytime naps should be avoided. Hypnotics may be used adjunctively.

Patients with drug and alcohol abuse (including abuse of hypnotics) must be detoxified, and then be helped to remain abstinent. Dry alcoholics may show objective evidence of disrupted sleep for up to one year after the initiation of sobriety. Supportive psychotherapy sometimes with antidepressants, lithium (Eskalith® and others), antihistamines, or tryptophan (Trofan®, Tryptacin®) may be used to improve insomnia in these cases. Drugs that cross-react with the drug of abuse, such as benzodiazepines with alcohol, should be avoided.

Some patients are very sensitive to the effects of caffeine and must avoid all coffee, tea (including decaffeinated types), colas, chocolate, or stimulants (decongestants, diet pills, etc.) in order to manage their insomnia. Likewise, cigarette smoking can produce insomnia, possibly related to the stimulant and withdrawal effects of nicotine, so some patients may have to stop smoking to treat their insomnia successfully.[14]

Shift workers or those with circadian rhythm disruptions may be helped by schedule manipulation, including incremental realignment of the sleep cycle with the 24-hr day. Bright light, antidepressants, lithium, and hypnotics may also be used adjunctively in some patients.[10]

Insomnia secondary to an affective disorder may respond well to antidepressants and lithium (see Chapter 3). However, if the insomnia is not satisfactorily treated with antidepressants, hypnotics may be used adjunctively. Adjusting the dosage schedule or switching to a more sedating agent may help those who are having difficulty falling asleep because of the stimulating effects of an antidepressant. If the patient has a panic or anxiety disorder and is experiencing insomnia on this basis, optimal management of the anxiety disorder itself is the best strategy for improving sleep.

IV. HYPNOTIC DRUGS

Benzodiazepines are the safest and most effective hypnotic drugs. Tryptophan, antihistamines [e.g., diphenhydramine (Benadryl® and others)], chloral hydrate (Noctec® and others), and paraldehyde may have some utility in selected patients.

Barbiturates, glutethimide (Doriden®), methyprylon (Noludar®), ethchlorvynol (Placidyl®), and ethinamate (Valmid®) are also available but are more toxic, more likely to produce addiction, and generally less effective than the benzodiazepines. The physician occasionally encounters patients, usually older individuals, who have been given one of these agents for a long time. If it is impossible to detoxify the patient or substitute a less toxic regimen, these drugs may be continued with careful supervision and monitoring. Otherwise, their use is rarely indicated.

A. Benzodiazepines

Benzodiazepines tend to decrease sleep latency, diminish the number of nocturnal arousals, and decrease the amount of delta (stage 3 and 4) sleep. The total amount of electroencephalogram (EEG) delta activity probably is not decreased because delta is redistributed throughout other stages. The effects of benzodiazepines on REM sleep are variable, depending on the individual, the underlying illness, and the dose and type of drug.[15]

The four benzodiazepines marketed specifically as hypnotics are flurazepam (Dalmane® and others), temazepam (Restoril® and others), triazolam (Halcion®), and quazepam (Doral®). Many other benzodiazepines, such as clonazepam (Klonopin®), lorazepam (Ativan® and others), and diazepam, are used clinically for managing insomnia.[16,17]

Disruption of mental function is a problem produced by all benzodiazepine hypnotics. Slowing of cognitive and psychomotor function may appear for several days following the use of long-acting benzodiazepines such as flurazepam and diazepam.[18] Memory impairment (anterograde amnesia) may appear, particularly following the use of short-acting agents such as triazolam.[19,20] Depression, excitement, agitation, behavioral disinhibition, and delirium may be induced by benzodiazepine hypnotics and occur with greater frequency and severity in the elderly or already compromised patient. Striking a balance between problems secondary to the insomnia and adverse drug effects is required. The need for such a balance is especially crucial when considering long-term benzodiazepine hypnotic use.

The long-term use of benzodiazepine hypnotics is controversial. There is general agreement that it is best to seek alternatives to long-term benzodiazepine hypnotic use.[21] Periodic reevaluations of treatment with attempts to discontinue hypnotic use are necessary if benzodiazepines are used for more than 1 month. Attempts should be made to use the lowest possible effective doses, and careful records should be kept to ensure that doses do not escalate over time. Attention to problems secondary to drug accumulation, especially if long-acting agents are used, is indicated.

Some studies show that no tolerance to the hypnotic effect of flurazepam develops during 24 weeks of use.[22] Subjective and objective improvement in sleep was seen in insomniacs using flurazepam for periods longer then 4 weeks. However, other investigators report that tolerance to the hypnotic effects of benzodiazepines develops over time and find that chronic insomniacs who have used various hypnotics (including some of the older agents) show very poor quality sleep polysomnographically and continue to complain of disturbed sleep while using drugs.[23]

It is our experience that some patients with otherwise resistant insomnia may require flurazepam 15–30 mg, diazepam 5–10 mg, lorazepam 1–2 mg, or clonazepam 1–3 mg at bedtime two to three times per week for extended periods. However, further study of this subject is required.

Rebound insomnia and rebound anxiety are problems not infrequently caused by benzodiazepine hypnotics, especially with agents that have a rapid or intermediate elimination rate, such as triazolam, midazolam (Versed®), lorazepam, or temazepam.[24,25] Rebound insomnia describes the condition in which the patient experiences subjective and objective disruption of sleep, greater than the base-line difficulty, during night(s) when the hypnotic is not used. Rebound anxiety occurs when anxiety is increased over baseline during the later parts of the night or the day following the use of the drug.

Rebound insomnia and anxiety may foster the need for continued drug use or lead to the use of escalating doses of drug. Therefore, unrecognized or mishandled rebound insomnia and anxiety may foster the development of drug dependence or iatrogenic "treatment resistance."

Rebound insomnia may be managed by tapering the short-acting agent[26] or by switching to long-acting agents such as flurazepam or clonazepam. Patient education is also helpful.

Benzodiazepines may depress respiration and alerting mechanisms. Therefore, they must be used with great care, if at all, in patients with sleep apnea or the obesity–hypoventilation ("Pickwickian") syndrome. Benzodiazepines cross-react with alcohol and are best avoided in patients with a history of alcoholism.

Benzodiazepines are the treatment of first choice for nocturnal myoclonus. Clonazepam 0.5 mg h.s. may be used, although other benzodiazepines may be tried if clonazepam is not effective.

The hypnotic characteristics of flurazepam, temazepam, and triazolam are reviewed below. The reader is referred to Chapter 5 for further details about benzodiazepine use.

1. Flurazepam

Flurazepam (see Fig. 1) is rapidly absorbed. Two metabolites, hydroxyethylflurazepam and flurazepam aldehyde, along with low levels of flurazepam itself, appear in the serum shortly after ingestion. These metabolites, along with flurazepam itself, probably induce sleep. They are eliminated over 8–12 hr and do not accumulate with multiple dosing. Flurazepam's major active component, desalkylflurazepam, appears gradually and is slowly eliminated. Desalkylflurazepam may maintain sleep.[27]

Because elimination half-life is 40–100 hr (up to 300 hr in elderly patients), desalkylflurazepam's effects may be felt the next day. These effects include cognitive and psychomotor slowing, which may become significant in elderly or impaired patients. Tolerance to the cognitive effects may occur in normal volunteers. There are conflicting data regarding the development of tolerance to the hypnotic and

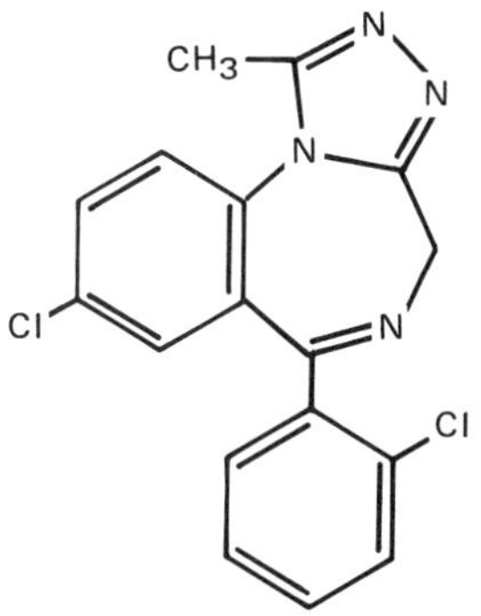

Triazolam

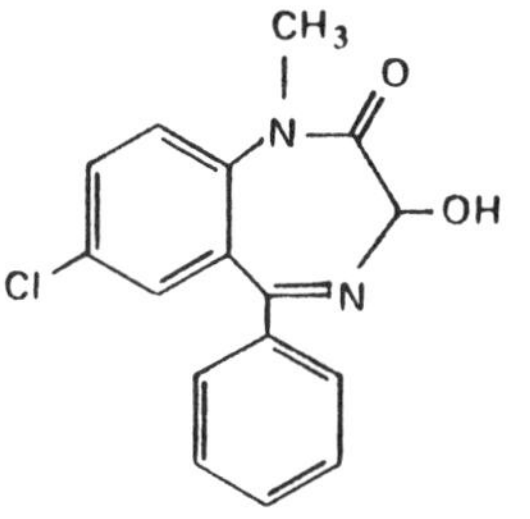

Temazepam

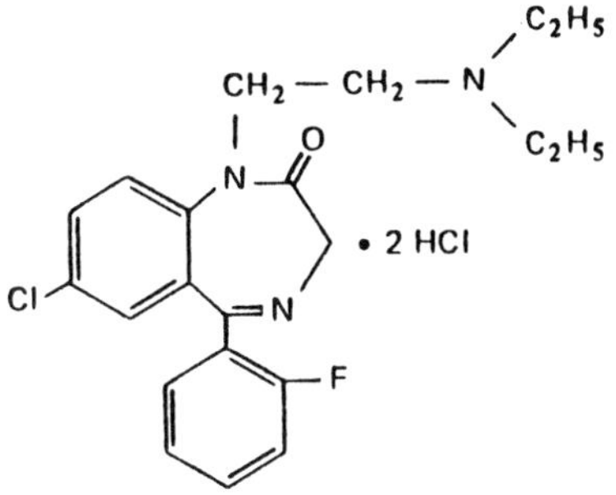

Flurazepam
hydrochloride

FIGURE 1. Benzodiazepines.

psychomotor interference effects. Some patients develop a morning "hangover," but those with generalized chronic anxiety may feel better the day after flurazepam use.[15-17,27]

The long elimination half-life allows flurazepam to accumulate if it is used frequently. Accumulation can cause progressive increases in daytime sedation and cognitive impairment, especially in the elderly. Significant drug accumulation may occur if liver disease is present, since the drug is metabolized by the cytochrome P-450 system to inactive metabolites. Renal insuffiency allows these inactive metabolites to accumulate, but this does not produce excessive sedation. Cimetidine (Tagamet®) may interfere with the metabolism of flurazepam and may produce flurazepam accumulation.[15-17]

Flurazepam generally does not produce rebound insomnia or anxiety, perhaps because accumulation of desalkylflurazepam provides a built-in tapering mechanism.

Some reports suggest that 15 mg may be as effective as 30 mg, so the lower dose should be tried first.

2. Temazepam

With an elimination half-life of 8–20 hr (it may be longer in elderly patients), temazepam has an intermediate duration of action.[15,28] Some accumulation may occur with repeated dosing. The parent drug is glucuronidated in the liver, and there are no active metabolites. Hepatic and renal disease does not lead to drug accumulation or behavioral toxicity. Oral contraceptives may interfere with the clearance of temazepam.

Peak serum concentrations are reached in about 90 min after ingestion. The soft gelatin capsule may be more rapidly absorbed than the hard pill formulation. Temazepam is only moderately powerful inducing sleep, and it may not help patients who have trouble falling (versus staying) asleep. Taking the dose on an empty stomach 1–2 hr before desired sleep may be helpful if sleep initiation is the target symptom.

Temazepam may produce some impairment of cognitive functions, particularly memory. Rebound insomnia and rebound anxiety are not uncommon on the days following termination of use, even if the period of use has been only several days.[28,29]

3. Triazolam

Triazolam (see Fig. 1) is a triazolobenzodiazepine and differs from alprazolam only by the presence of a chlorine atom. Preliminary evidence suggests that it is 5–15 times more potent than alprazolam and 50–150 times more potent than diazepam on a milligram basis.

Triazolam is rapidly absorbed, induces sleep quickly, and is rapidly metabolized. It is virtually gone from the system by the morning (elimination half-life is approximately 2–5 hr). Metabolism occurs on first pass through the liver; hepatic cirrhosis may result in decreased triazolam elimination. Acute ethanol ingestion may impair triazolam clearance and elimination (as well as potentiating its psychomotor effects). Cimetidine may also impair triazolam clearance and elimination.[16,17]

These characteristics make triazolam seem like an ideal hypnotic. Unfortunately, various adverse reactions may occur.[29] Rebound insomnia and rebound anxiety are not uncommon. Tapering the drug by cutting the dose by 50% for 2 days and 50% again for 2 days may ameliorate these problems.[26] Agitation (including delirium, psychotic confusion, and mania) secondary to triazolam use has occurred over a wide dose range. Anterograde amnesia (inability to recall events the morning after drug use) with or without apparent confusion is also not uncommon. Withdrawal seizures may occur with abrupt discontinuation.

Patients should be warned about these potential problems, and the drug discontinued should they appear. If fluctuations in mood appear in depressed patients, another agent should be used. It is best to use as low a dose as possible, beginning with 0.125 mg and not exceeding 0.5 mg; 0.25 mg is considered a usual dose.

4. Quazepam

Quazepam recently has been approved for use as a hypnotic. Maximum quazepam plasma concentrations occur at approximately 1.5 hours following oral ingestion. The drug is metabolized in the liver into two active metabolites, 2-oxoquazepam and N-desalkyl-2-oxoquazepam. N-desalkyl-2-oxoquazepam is identical to N-desalkyl-fluorezapam, the major metabolite of fluorezepam. Quazepam and 2-oxoquazepam have elimination half-lives of around 42 hours, while the desalkyl metabolite has an elimination half-life of 75 hours (longer in the elderly). Accumulation therefore occurs with daily use. Some studies suggest that hydroxyoxoquazepam, an inactive compound, is the major metabolite.

Quazepam appears to be unique because it binds relatively selectively to the benzodiazepine$_1$ (BZD$_1$) receptor. The clinical significance of this interesting observation is not yet firmly established.

Sleep laboratory and clinical studies indicate that a daily dose of 15 mg appears to be effective during short- and long-term (up to four weeks) administration. Dose-dependent daytime sedation (to which tolerance may develop) over several days is not uncommon, but rebound insomnia, rebound anxiety, hyperexcitability, and amnesia have not been observed.[30,31]

Quazepam is available in 7.5-mg and 15-mg tablets. The recommended dosage is 15 mg at bedtime for healthy adults and 7.5 mg at bedtime for the elderly or debilitated.

5. Others

a. Clonazepam

Clonazepam is a nitrobenzodiazepine with structural similarities to nitrazepam, a hypnotic marketed in Europe (not the United States). It has been approved for use as an anticonvulsant and is an effective antianxiety and antipanic agent (see Chapter 5). Because it is very sedating, it is useful as a hypnotic, particularly for patients with insomnia produced by anxiety. If insomnia is produced by nocturnal panic attacks, or if rebound anxiety in the early morning occurs as daytime antipanic agents (e.g., alprazolam) are eliminated, clonazepam may be a useful alternative.

b. Diazepam, Lorazepam

These agents are described in Chapter 5; their moderate hypnotic properties and intermediate elimination half-lives make them suitable for some insomniac patients.[27,32] Patients may prefer these agents if they have been used successfully in the past. These agents may be given at bedtime as a hypnotic in patients who are receiving them during the day for anxiety. Rebound insomnia, cognitive compromise, and drug dependency may occur.

c. Midazolam

Midazolam was developed for use during outpatient medical procedures such as endoscopy, where brief-duration sedation and amnesia are required. It is rapidly absorbed orally and is moderately effective in rapidly inducing sleep. Some studies have demonstrated its effectiveness as a hypnotic,[33] but further data are required to confirm these preliminary findings.

B. Tryptophan

L-Tryptophan is an essential dietary amino acid that is available without prescription, usually in 500-mg tablets. Some authors have reported that 2–5 g of tryptophan shortens sleep latency and improves sleep maintenance in insomniac patients. The maximum beneficial effects of L-tryptophan may appear several nights after this agent has been stopped; a 4-days-on, 3-days-off regimen may be tried. Treatment may begin with 3 g taken one-half hour before bedtime with a sugared drink such as grape juice. No protein should be consumed for 3 hr before tryptophan is taken. The dose may be lowered until a minimum effective dose is found; if 4 g is ineffective, alternative treatment should be sought. Some authors suggest that nicotinamide (100 mg nicotinamide per 500 mg tryptophan) be used adjunctively to steer the metabolism of tryptophan in the direction of serotonin, but the clinical safety and utility of this practice are not established.[32]

Animal studies indicate that L-tryptophan may be hepatotoxic, so caution is advised.

Tryptophan should be used with extreme caution, if at all, in patients using serotonin agonists such as fluoxetine (Prozac®), clomipramine (Anafranil®), and phenelzine (Nardil®), since some may develop a serotonin syndrome.

As this book goes to press, the use of L-tryptophan appears to be associated with over 1300 cases of the "eosinophilia–myalgia syndrome" and has resulted in 15 reported deaths.[34] This syndrome is characterized by serum eosinophil counts > 1000 cell/mm^3 and generalized myalgia with concomitant muscle weakness, cutaneous changes, dyspnea, fever, and other abnormalities. The reason for the association between L-tryptophan and this syndrome remains undetermined,[35] but as of this writing there are suggestions that a contaminant may be the source of tryptophan's toxicity. L-tryptophan has been removed from the market and should not be used until approval is given by the Food and Drug Administration.

C. Antihistamines

Sedating antihistamines such as diphenhydramine or hydroxyzine (Atarax®, Vistaril®, and others) are sometimes used to promote sleep in those who are sensitive to or unable to use benzodiazepines, such as alcoholics or drug abusers, the elderly, or medically ill hospitalized patients. Doxylamine and pyrilamine are the active agents in OTC hypnotics and nighttime cold preparations.

There are conflicting data on the hypnotic efficacy of these drugs, with some studies indicating that they produce no significant change in the objective parameters of sleep. Antihistamines may produce paradoxical excitement, delirium, psychosis, and dystonic reactions. Elderly persons may be especially sensitive to these adverse effects. Antihistamines are therefore not the first choice for treatment of insomnia and must be used with caution.[36]

D. Barbiturates

The barbiturates are discussed in Chapter 5. The short- and intermediate-acting barbiturates have been used as hypnotics [e.g., amobarbital (Amytal® and others), pentobarbital (Nembutal® and others), secobarbital (Seconal®)].

Barbiturates decrease sleep latency, shorten slow-wave sleep, and noticeably decrease the length of REM periods. With short-acting barbiturates, these effects tend to wear off later in the night. Tolerance develops to the hypnotic effects of barbiturates over a matter of days, and rebound phenomena occur following cessation of treatment.

Because of problems of tolerance, abuse, dependence, adverse effects, and the danger of overdose, clinicians should rarely prescribe barbiturates as hypnotics.

E. Chloral Hydrate and Derivatives

Chloral hydrate and two closely related compounds, chloral betaine and triclofos sodium (both no longer available in the United States), are pharmacologically and clinically similar. All three share the same active metabolite, trichloroethanol, which has a plasma half-life of approximately 4 to 12 hr (see Table 2).

Chloral hydrate decreases sleep latency and the number of awakenings and may slightly decrease slow-wave sleep. It has relatively little effect on REM sleep because some degree of tolerance develops. Therapeutic doses of chloral hydrate have little effect on respiration and blood pressure, but toxic doses may impair both.

Chloral hydrate has an irritating and unpleasant taste. Local irritation of the gastrointestinal tract can cause heartburn, nausea, and vomiting. The relatively short half-life of the chloral derivatives makes hangover less of a problem. Allergic reactions are occasionally observed. Because of its potential to cause direct organ

TABLE 2. Chloral Derivatives

Drug	Usual hypnotic dose	Dosage forms	Half-life (hr)
Chloral hydrate (Noctec® and others)	1 to 2 g	250-, 500-mg capsules; 250-mg/5-cc, 500-mg/5-cc, 800-mg/5-cc syrup	4 to 9½
Chloral betain	No longer available in the United States		4 to 9½
Triclofos sodium	No longer available in the United States		4 to 9½

toxicity, chloral hydrate is best avoided in patients with severe hepatic, renal, or cardiac disease. Most adults will require 1 to 2 g of chloral hydrate to induce sleep. Gastric irritation increases with the dose but may be minimized by administering the drug with milk or an antacid.

Chloral betaine is more palatable and less irritating than chloral hydrate; about 870 mg of the former equals approximately 500 mg of the latter. For triclofos sodium, 1.5 g equals approximately 900 mg of chloral hydrate.

Chloral hydrate affects the hepatic drug-metabolizing system and thus may interfere with a number of other drugs that are metabolized in the liver. Another potential problem arises from the fact that trichloroacetic acid, a metabolite of chloral hydrate, displaces acidic drugs from plasma protein binding sites, resulting in increased plasma concentrations of these drugs. Chloral hydrate also may interact with the diuretic furosemide (Lasix® and others) to cause a syndrome of sympathetic instability. The combination of chloral hydrate and alcohol is the legendary "Mickey Finn," reputed to have a "knockout punch." Although the potency of this combination is more legend than fact, the two may enhance each other's blood levels, which adds to the sedative potentiation.

The hypnotic dose of chloral hydrate for adults is approximately 1 to 2 g; a toxic dose is approximately 10 g. The range is considerable, however, with death having followed ingestion of 4 g, and survival after 30 g. Gastric irritation may limit the amount a patient can take at one time, and vomiting diminishes absorption. However, gastric necrosis can produce additional complications. In serious intoxication, support of cardiovascular and respiratory functions is vital. Hemodialysis may enhance excretion.

Tolerance, physical dependence, and addiction are possible with chloral derivatives. A withdrawal syndrome occurs similar to that seen with other sedatives and alcohol.

If a patient has previously used a chloral derivative and prefers it, then careful administration is acceptable. Some centers find chloral hydrate useful for alcoholics who complain of insomnia during inpatient alcohol detoxification. However, there is generally little reason to initiate treatment with one of these agents.

F. Paraldehyde

Paraldehyde, a foul-smelling and irritating liquid, remains in therapeutics largely as a historical curiosity. It is a rapidly acting hypnotic and, as such, rarely causes hangover. Sleep typically occurs within 10 to 15 min (for dosages, see Table 4).

Like the barbiturates, paraldehyde raises the seizure threshold. Before the introduction of the benzodiazepines, it was commonly employed for the treatment of delirium tremens. In the presence of pain, administration of paraldehyde has been associated with paradoxical excitement. Large doses of paraldehyde can suppress respiration and produce hypotension.

Following oral administration, paraldehyde is rapidly absorbed. Most is metabolized in the liver, but a proportion is excreted unchanged via the lungs (imparting a pungent odor to the patient's breath that permeates a hospital room), and a smaller amount in the urine. In patients with liver disease, less is metabolized, and more is excreted by the lungs.

Whether administered orally or parenterally, paraldehyde is extremely irritating to the body's tissues. The lethal dose of paraldehyde is probably very high, but death has been reported after as little as 25 g. Tolerance, dependence, and withdrawal syndromes occur and resemble those observed with alcohol. Paraldehyde is unpleasant and difficult to handle, and it affords no benefits over other medications such as the benzodiazepine hypnotics.

G. Piperidinedione Derivatives

Two piperidinedione derivatives are available—glutethimide and methyprylon. **These drugs have all the drawbacks of the barbiturates without any advantages. There is a narrow range between therapeutic and toxic doses; overdoses may be very difficult to treat and may be lethal. There is no reason to initiate treatment with these agents.** Their use should be reserved for the rare case of a patient who has been using them with good effect for years without problem or dosage escalation and who has been unable to tolerate withdrawal or substitution despite maximal effort.

1. Glutethimide

The half-lives of glutethimide and its sister compound methyprylon are given in Table 3. The pharmacological profile of glutethimide resembles that of the barbiturates except for its anti-motion-sickness effect and anticholinergic activity. Similar to the barbiturates, glutethimide suppresses REM sleep.

Gastrointestinal absorption of glutethimide varies. It is highly lipophilic and about 50% bound to plasma proteins. Almost all the drug is metabolized in the liver. The half-life of glutethimide is about 5 to 22 hr. Similar to the barbiturates, glutethimide stimulates the hepatic microsomal enzyme system, which increases potential for many pharmacokinetic interactions. Glutethimide is contraindicated in patients with acute intermittent porphyria.

Occasionally central nervous system effects occur, including confusion, excite-

TABLE 3. Piperidinedione Derivatives

Drug	Usual hypnotic dose	Dosage forms	Half-life (hr)
Glutethimide (Doriden®)	500 mg	500-mg capsules; 125-, 250-, 500-mg tablets	5 to 22
Methyprylon (Noludar®)	200 to 400 mg	300-mg capsules; 50-, 200- mg tablets	Ca. 4

ment, hangover, drowsiness, and ataxia. Other side effects include anticholinergic effects such as dry mouth, blurred vision, constipation, and difficulty urinating; cutaneous allergic reactions (occasionally serious); gastrointestinal irritation; and rare blood dyscrasias such as thrombocytopenia, aplastic anemia, and leukopenia.

The hypnotic dose of glutethimide is 0.5 to 1 g; as little as 5 g can produce severe intoxication, and 10 g or more may be lethal. In acute glutethimide intoxication, respiratory depression often is less severe than with barbiturate overdoses, but cardiovascular collapse may be worse. Antimuscarinic effects are typically prominent. Neuromuscular irritability can occur with possible seizures. In intoxication, the plasma half-life of glutethimide is prolonged and may exceed 100 hr. Fluctuating levels of consciousness, including lucid intervals, are typical, possibly because of variable absorption and enterohepatic recirculation as well as intermittent release of the drug from lipid stores. The presence of other nervous system depressant compounds obviously will worsen an overdose. Treatment of glutethimide intoxication is aimed at reducing absorption, supporting vital functions, and hastening excretion. The last may be accomplished by hemodialysis, preferably with a lipid solvent such as soybean oil.

Glutethimide can produce tolerance and dependence as well as an abstinence syndrome typical of sedative–hypnotics.

2. Methyprylon

Methyprylon is pharmacologically very similar to its structural relative glutethimide and to the barbiturates. It is a short-acting hypnotic with a plasma half-life of only 4 hr (longer in states of intoxication). Methyprylon suppresses REM sleep, and REM rebound occurs on its withdrawal.

Methyprylon is almost completely metabolized in the liver, where it stimulates the hepatic microsomal enzyme system. It is contraindicated in patients with acute intermittent porphyria. Adverse effects are essentially the same as those described for glutethimide.

Acute intoxication with methyprylon is clinically similar to that with barbiturates, although it may have less of a tendency to depress respiration. A dose of 15 g or 200 mg per kilogram of body weight can cause serious intoxication and possibly death. In fact, death has been reported after as little as 6 g of methyprylon. However, recovery has followed as much as 27 g. Coma after methyprylon intoxication may last for as long as 5 days. Methyprylon is more readily dialyzed than glutethimide because it is more highly water soluble.

H. Ethclorvynol

Ethchlorvynol has a rapid onset of hypnotic action (15 to 30 min), and although its elimination half-life is 10 to 25 hr, rapid distribution terminates its clinical effect within a few hours after an acute dose. Chronic dosing, however, can lead to prolonged effects. Ethchlorvynol resembles the barbiturates in its effect on sleep stages, including suppression of REM. Rebound occurs following drug withdrawal. Adverse effects with ethchlorvynol include confusion, hangover, ataxia, and nausea and vomiting. Other reactions are hypotension, facial numbness, amblyopia, and giddiness. Less common are allergic reactions, blood dyscrasias, and cholestatic jaundice.

Ethchlorvynol seems to interfere with the kinetics of drugs metabolized in the hepatic microsomal system such as anticoagulants. Additive sedative effects can occur with other drugs and alcohol. Delirium has been reported in patients taking ethchlorvynol together with tricyclic antidepressants. Ethchlorvynol is contraindicated in patients with acute intermittent porphyria.

Profound coma can follow an acute overdose with ethchlorvynol. Death has been reported after a dose as low as 2.5 g, but the usual lethal range is given as 10 to 25 g. Treatment includes the usual approaches to diminishing absorption, supporting vital functions, and hastening elimination, possibly using a lipid dialysate with hemodialysis.

There is a high probability that ethchlorovynol use will produce physical dependency. The abuse potential of this agent is likewise high. Tolerance tends to develop to the sedative effects. The lethality of overdosage is high. Therefore, like the piperidinedione derivatives, treatment with this drug should not be initiated. Strenuous efforts should be made to withdraw the drug when patients are already using it or to find a benzodiazepine substitute.

I. Ethinamate

Ethinamate has few differences from the barbiturates. It is a short-acting hypnotic, producing sleep within 30 min and a maximum effect in 45 min (see Table 4). The short duration of action means that hangover is rare. Gastrointestinal irritation

TABLE 4. Other Hypnotic Drugs

Drug	Usual hypnotic dose	Dosage forms	Half-life (hr)
Methaqualone	No longer available in the United States		10 to 42
Ethchlorvynol (Placidyl®)	500 to 1000 mg	100-, 500-, 750-mg capsules	10 to 25
Ethinamate (Valmid®)	500 to 1000 mg	500-mg capsules	Brief
Diphenhydramine (Benadryl® and others)	50 mg	25-, 50-mg capsules, 12.5-mg/5-ml elixir, 50-mg/1-ml injection	Unknown
Paraldehyde	No longer available in the United States		Brief

occurs in some patients, allergic reactions have been reported, and rare cases of thrombocytopenia have been associated with this drug. Paradoxical excitement may be more common in younger patients.

Acute intoxication resembles that with the barbiturates, and death has occurred after as little as 15 g. Treatment of overdoses is the same as for other sedative–hypnotic compounds. Habituation, dependence, and abstinence reactions are also similar. There is no reason to prescribe this drug.

J. Over-the-Counter Drugs

Many over-the-counter preparations are available as "sleeping pills" (see Table 5). Most contain one or a combination of the following: scopolamine, an antihistamine, or a salicylate.

Scopolamine, a centrally active anticholinergic agent, depresses the cerebral cortex as well as the reticular formation and hypothalamus. Scopolamine can cause paradoxical excitement and delirium, particularly in the presence of pain. It decreases REM sleep, and when it is discontinued, REM rebound occurs. Common unwanted effects are anticholinergic, such as dry mouth, blurred vision, urinary hesitancy, and constipation. Because of the anticholinergic effects, overdoses, particularly in young children, can be dangerous. Hazards of anticholinergic drugs and treatment of toxicity have been discussed earlier.

Antihistamines have been previously discussed (see Section III.H). The two contained in over-the-counter sleep preparations are methapyrilene and pyrilamine.

Several salicylates are combined with the above agents in many patent medicines. These are salicylamide, aspirin, acetaminophen, and sodium salicylate. Their effectiveness as sleep aids is in doubt, although it is possible that their analgesic activity may promote sleep by alleviating pain in affected patients. Typical problems with salicylates (other than acetaminophen) include gastric irritation and interference with clotting mechanisms. Intoxication with salicylates can produce tin-

TABLE 5. Some Nonprescription "Sleeping Pills"

Brand name	Contents
Compoz®	Methapyrilene (an antihistamine)
	Pyrilamine (an antihistamine)
Nytol®	Methapyrilene
	Salicylamide (a salicylate)
Sleep-Eze®	Scopolamine hydrobromide
	Methapyrilene
Sominex®	Scopolamine aminoxide
	Methapyrilene
	Salcylamide

nitus and acid–base disturbances and with acetaminophen may result in hepatic necrosis.

There is little evidence that any available over-the-counter sleeping pill is more effective than placebo. Furthermore, they are not innocuous, and some patients may be particularly sensitive to their adverse effects. In addition, when a patient takes an overdose, toxicity (e.g., an anticholinergic syndrome) may occur.

K. Other Psychotropics for Insomnia

1. Antidepressants

Tricyclic antidepressants can be extremely useful when insomnia is the primary symptom of an affective disorder. Some patients respond well to small amounts (e.g., 10–25 mg) of a moderately sedating drug like nortriptyline (Pamelor® and others).[37] Amitriptyline (Elavil® and others) may be useful for patients with childhood-onset insomnia. Amitriptyline and others may be useful for patients with disturbed sleep secondary to fibromyositis.

Trazodone (Desyrel® and others) is very sedating but does not appear to have any particular hypnotic effect or benefit as compared with other antidepressants.[38]

All antidepressants suppress REM sleep; REM rebound with nightmares that disturb sleep may occur when these drugs are stopped. Tapering the drug may obviate this difficulty.

Antidepressants may also induce or exacerbate nocturnal myoclonus, restless leg syndrome, and somnambulism; these problems may lead the patient to complain of "insomnia." Switching to another drug may be helpful.[37]

Stimulating antidepressants may induce difficulty falling asleep. Arranging the dosage schedule so that no drug is used after 4 p.m. may be helpful. Time-limited use of benzodiazepine hypnotics may also be helpful. Recent evidence suggests that trazodone 50 mg h.s. may ameliorate fluoxetine-induced insomnia.

2. Lithium

Lithium can be used to manage the decreased sleep that accompanies mania or hypomania. Patients with dysthymia or cyclothymia who present with insomnia as a main complaint may also find lithium useful alone or as an adjunct to tricyclics.

3. Antipsychotics

Patients whose psychotic disorder disturbs their sleep often benefit from antipsychotic treatment. Some patients who become agitated and ruminative may also benefit from small amounts of antipsychotic at bedtime.

Antipsychotics can induce nocturnal myoclonus, restless leg syndrome, somnambulism, and nightmares. Small amounts of haloperidol (Haldol® and others) probably do not worsen respiration in patients with sleep apnea.

4. Buspirone

Preliminary data show that buspirone (BuSpar®) has no immediate effect on sleep latency or sleep architecture in patients with generalized anxiety disorder but that a decrease in nocturnal time awake, with consequent improvement in sleep efficiency, does appear in these patients after 5 weeks of chronic administration. Buspirone is not a hypnotic but may be useful in managing insomnia secondary to anxiety.

V. CLINICAL APPLICATION SUMMARY

Medication for insomnia should be used only after a diagnosis of the cause of the insomnia is determined, even if this diagnosis is provisional. Specific treatment based on the diagnosis, such as the use of antidepressants for an underlying affective disorder, is essential. Hypnotics are best used as adjuncts to other treatment modalities. Benzodiazepines remain the most effective, safe, and widely used hypnotics. Benzodiazepine hypnotics should be given in the lowest possible effective dose for short time periods. Extra care must be used in elderly patients. Long-term benzodiazepine hypnotic use remains controversial; similar issues are raised by long-term benzodiazepine use in patients with anxiety disorders (see Chapter 5).

VI. THE FUTURE

Various endogenous substances, such as interleukin 1, prostaglandins, and peptides like delta-sleep-inducing peptide (DSIP) have shown to have powerful effects on sleep in animals and humans. Agents that inhibit brain protein synthesis can produce alterations in sleep architecture. Efforts to understand how endogenous proteins and other substances are involved in the physiology of normal sleep suggest that powerful, less toxic, "physiological" agents for treating insomnia may eventually appear. Currently, benzodiazepines like loprazolam and brotizolam and non-benzodiazepines like zolpidem are in development and show some promise.

REFERENCES

1. Weilburg J. B.: Approach to the patient with insomnia, in Goroll A. H., May L. A., Mully A. G. (eds): *Primary Care Medicine,* 2nd ed. Philadelphia, JB Lippincott, 1987.
2. Coleman R. M., Roffwarg H. P., Kennedy S. J., et al: Sleep–wake disorders based on a polysomnographic diagnosis: A national cooperative study. *JAMA* 27(7):997–1003, 1982.
3. Kramer P. D.: Insomnia: Importance of the differential diagnosis. *Psychosomatics* 23:129–137, 1982.
4. Hobson J. A.: *Sleep Charting.* South Norwalk, CT, Medication, Inc, 1983.
5. Tan T. L., Kales J. D., Kales A., et al: Inpatient multidimensional management of treatment-resistant insomnia. *Psychosomatics* 28:266–272, 1987.

6. Zorick F. J., Roth T., Hartze K. M., et al: Evaluation and diagnosis of persistent insomnia. *Am J Psychiat* 38:769–773, 1981.
7. Reynolds C. F., Kupfer D. J., Taska L. S., et al: Sleep of healthy seniors: A revisit. *Sleep* 8:20–30, 1985.
8. Reynolds C. F., Kupfer D. J., Hoch, C. C., et al: Sleeping pills for the elderly: Are they ever justified? *J Clin Psychiatry* 46:9–12, 1985.
9. Weitzman E. D., Czeisler C. A., Coleman R. M., et al: Delayed sleep phase syndrome. *Arch Gen Psychiatry* 38:737–746, 1981.
10. Czeisler C. A., Richardson G. S., Coleman R. M., et al: Chronotherapy: Resetting the circadian clocks of patients with delayed sleep phase insomnia. *Sleep* 4:1–21, 1981.
11. Reynolds C. F., Taska L. S., Sewitch D. E.: Persistent psychophysiologic insomnia: Preliminary research diagnostic criteria and EEG sleep data. *Am J Psychiatry* 141:804–805, 1984.
12. Hauri P.: Treating psychophysiologic insomnia with biofeedback. *Arch Gen Psychiatry* 38:752–758, 1981.
13. Erman M. K. (ed): Sleep disorders. *Psychiatr Clin North Am* 10:4, 1987.
14. Soldatos C. R., Kales J. D., Scharf M. B., et al: Cigarette smoking associated with sleep difficulty. *Science* 207:551–553, 1980.
15. Abernethy D. R., Greenblatt D. J., Shader R. I.: Benzodiazepine hypnotic metabolism: Drug interactions and clinical indications. *Acta Psychiatr Scand* 332(Suppl 74):32–38, 1986.
16. Greenblatt D. J., Abernethy D. R., Shader R. I.: Benzodiazepine hypnotics: Kinetic and therapeutic options. *Sleep* 5:S18–S27, 1982.
17. Greenblatt D. J., Harmatz J. S., Engelhardt N., et al: Pharmacokinetic determinants of dynamic differences among three benzodiazepine hypnotics. *Arch Gen Psychiatry* 46:326–332, 1989.
18. Church M. W., Johnson L. C.: Mood and performance of poor sleepers during repeated use of flurazepam. *Psychopharmacology* 61:309–316, 1979.
19. Morris H. H., III, Estes M. L.: Traveler's amnesia: Transient global amnesia secondary to triazolam. *JAMA* 258:945–946, 1987.
20. Spinweber C. L., Johnson L. C.: Effects of triazolam on sleep performance, memory and arousal threshold. *Psychopharmacology* (Berlin) 76:5–12, 1982.
21. Consensus conference drugs and insomnia. *JAMA* 251:2410–2416, 1984.
22. Dement W. C., Carskadon M. A., Mitler M. M., et al: Prolonged use of flurazepam. *Behav Med* 25–32, 1978.
23. Feinberg I., Fein G., Walker J. M., et al: Flurazepam effects on the sleep EEG. *Arch Gen Psychiatry* 6:95–102, 1979.
24. Kales A., Scharf M. B., Kales J. D., et al: Rebound insomnia. *JAMA* 241:1692–1695, 1979.
25. Kales A., Soldatos C. R., Bixler E. O.: Rebound insomnia and rebound anxiety: A review. *Pharmacology* 26:121–137, 1983.
26. Greenblatt D. J., Harmatz J. S., Zinny M. A., et al: Effect of gradual withdrawal on the rebound sleep disorder after discontinuation of triazolam. *N Engl J Med* 317:722–728, 1987.
27. Greenblatt D. J., Abernethy D. R., Doivoll M. J., et al: Pharmacokinetic properties of benzodiazepine hypnotics. *J Clin Psychopharmacol* 3:129–132, 1983.
28. Mitler M. M., Cascardon M. A., Phillips R. L., et al: Hypnotic efficacy of temazepam: A long-term sleep laboratory study. *Br J Clin Pharmacol* 8:635–689, 1979.
29. Greenblatt D. J., Shader R. I., Divoll M., et al: Adverse reactions to triazolam, flurazepam, and placebo.
30. Kales A., Bixler E. O., Soldatos D. R., *et al:* Quazepam and temazepam: Effects of short- and intermediate-term use and withdrawal. *Clin Pharm Ther* 39:345–352, 1986.
31. Kales A.: Quazepam: Hypnotic efficacy and side effects. *PHarmacother* 10:1–12, 1990.
32. Weilburg J. B., Donaldson S. R.: L-Tryptophan for sleep. *Biol Ther Psychiatry* 11:15–16, 1988.
33. Midazolam. *Med Lett* 28:73–74, 1986.
34. Eidson M., Philen R. M., Sewell C. M., *et al:* L-tryptophan and eosinophilia–myalgia syndrome in New Mexico. *Lancet* 335:645–648, 1990.

35. Clauw D. J., Nashel D. J., Umhau A., *et al:* Tryptophan-associated eosinophilic connective-tissue disease. *JAMA* 263:1502–1506, 1990.
36. Roehers T. A., Tietz E. I., Zoerck F. J., et al: Daytime sleepiness and antihistamines. *Sleep* 7:137–141, 1984.
37. Ware J. C.: Tricyclic antidepressants in the treatment of insomnia. *J Clin Psychiatry* 44:25–28, 1983.
38. Montgomery I., Oswald I., Morgan K., et al: Trazodone enhances sleep and subjective quality but not in objective duration. *Br J Clin Pharmacol* 16:139–144, 1983.

III

Psychoactive Substance Abuse

7

Psychoactive Substance Use Disorders

STEVEN M. MIRIN, M.D., ROGER D. WEISS, M.D., and SHELLY F. GREENFIELD, M.D.

I. INTRODUCTION AND DEFINITIONS

The self-administration of chemical substances that alter mood or thinking has been a widespread practice for centuries. When the use of such substances deviates from societal norms or is deemed harmful to physical or mental health, substance use may merge into substance abuse and dependence.

There are wide cultural variations in acceptable substance use. According to the third revised edition of the *Diagnostic and Statistical Manual of Mental Disorders (DSM-IIIR)*, a substance use disorder is characterized by regular use of psychoactive substances leading to symptoms and maladaptive behavioral changes that would be viewed as undesirable in almost all cultures. These disorders are categorized as either psychoactive substance dependence or psychoactive substance abuse.

Psychoactive substance dependence is described in the DSM-IIIR as a cluster of cognitive, behavioral, and physiological symptoms that indicate that a person has impaired control of substance use and continues this use despite adverse consequences. Dependence is characterized by the presence of at least three of the

STEVEN M. MIRIN, M.D. • McLean Hospital, Belmont, Massachusetts 02178; Department of Psychiatry, Harvard Medical School, Boston, Massachusetts 02115. ROGER D. WEISS, M.D. • Alcohol and Drug Abuse Program, McLean Hospital, Belmont, Massachusetts 02178; Department of Psychiatry, Harvard Medical School, Boston, Massachusetts 02115. SHELLY F. GREENFIELD, M.D. • McLean Hospital, Belmont, Massachusetts 02178; Department of Psychiatry, Harvard Medical School, Boston, Massachusetts 02115.

following symptoms: (1) the substance is often taken in larger amounts or over a longer period than the person intended; (2) a persistent desire or one or more unsuccessful efforts to cut down or control substance use; (3) a great deal of time spent in obtaining, taking, and/or recovering from the effects of the substance; (4) frequent intoxication or withdrawal symptoms when expected to fulfill major role obligations at work, school, or home; (5) important social, occupational, or recreational activities given up or reduced because of substance use; (6) continued substance use despite knowledge of having a persistent or recurrent social, psychological, or physical problem caused or exacerbated by the substance; (7) tolerance, as demonstrated by a 50% increase in amount of substance needed in order to achieve intoxication or a desired effect; (8) characteristic withdrawal symptoms; (9) the substance often taken to relieve or avoid withdrawal. The last two characteristics may not apply to cannabis, hallucinogens, or phencyclidine.

Psychoactive substance abuse is present when a person does not manifest three of the above characteristics but has a maladaptive pattern of psychoactive substance use indicated by (1) continued use despite knowledge of having a persistent or recurrent social, occupational, psychological, or physical problem that is caused or exacerbated by use of the psychoactive substance or (2) recurrent use in situations in which use is physically hazardous. Some of these symptoms must have persisted for at least 1 month or have occurred repeatedly over a longer period of time.

From a pharmacological perspective, drug dependence may result from the direct reinforcing effects of these agents on the central nervous system. The mood-altering effects of many drugs such as relief of anxiety or production of euphoria are mediated through their effects on various brain neurotransmitters. Physical dependence implies a state of altered cellular physiology caused by repetitive drug administration. Repetitive use of some agents may also result in drug tolerance, a state in which increasingly larger doses of the drug must be taken to reproduce its original pharmacological effect. This may result from an increased rate of drug metabolism or a change in the sensitivity of target cells to the drug's pharmacological effect. In addition, inducing tolerance to one member of a particular drug class may be accompanied by developing tolerance to other drugs in the same general class. This phenomenon is called "cross-tolerance."

In individuals in whom both tolerance and dependence have developed, abrupt withdrawal of that agent may precipitate a characteristic abstinence syndrome. The wish to avoid abstinence symptoms can be a powerful force in maintaining drug-taking behavior. The experience of withdrawal, however, may cause some individuals to seek medical treatment. The clinical evaluation of such individuals should ideally include a detailed medical and psychiatric history. The clinician should distinguish emotional problems preceding drug abuse from those that may be drug induced. Since drug dependence and withdrawal often are accompanied by symptoms similar to those seen in patients with other psychiatric disorders (e.g., anxiety, insomnia, depression), a careful developmental and family history, psychological testing, and indicated laboratory tests may be extremely useful. As in other areas of medicine, proper treatment depends on accurate diagnosis. This chapter reviews each major class of commonly abused drugs with respect to basic pharmacology, patterns of abuse, and currently held views about treatment.

II. ABUSE OF OPIOID DRUGS

A. Classification

We can divide the opioids into the following categories based on their derivation:

1. Natural alkaloids of opium derived from the resin of the opium poppy, including opium, morphine, and codeine.
2. Semisynthetic derivatives of morphine, including diacetylmorphine (heroin), hydromorphone (Dilaudid® and others), and oxycodone (Percodan® and others).
3. Purely synthetic opioids, which are not derivatives of morphine, including meperidine (Demerol® and others), methadone (Dolophine® and others), and propoxyphene (Darvon® and others).
4. Opioid-containing preparations, such as elixir of terpin hydrate with codeine and paregoric, having abuse liability.

Among the opioids, the semisynthetic derivatives, like heroin and oxycodone, and pure synthetics, like meperidine and methadone, are most commonly abused. A recent national survey by the National Institute on Drug Abuse found that 2% of all men and 0.5% of all women aged 18–25 had tried heroin; 15% of men and 9% of women at those ages had used other opioids for nonmedical purposes. Estimates of opioid-dependent persons in the United States range from 400,000 to 600,000.

B. Pharmacology

The opioid drugs are readily absorbed from the gastrointestinal tract, nasal mucosa, and lung, but parenteral administration more rapidly elevates blood levels and produces intoxication. For example, following intravenous administration of heroin, the drug is almost immediately hydrolyzed to morphine by the liver, with peak plasma morphine levels attained in approximately 30 min. The drug then rapidly leaves the blood and is concentrated in body tissues. Only small quantities of injected opioids cross the blood–brain barrier, but plasma levels correlate directly with the level of intoxication. Morphine is metabolized primarily by conjugation in the liver. Its half-life (i.e., the amount of time necessary for one-half of a given dose to be cleared from the blood) is approximately $2^1/_2$ hr, and 90% of the total dose administered is excreted in the urine within the first 24 hr. The remainder is excreted through the biliary tract and appears in the feces.

The opioids act by selectively binding to stereospecific neuronal receptor sites, which thus far have been identified in the brain (e.g., the hypothalamus) and gut.

These opioid receptors probably mediate the action of naturally occurring opiate peptides (i.e., the enkephalins and β-endorphin) as well.

C. Acute Effects

1. Central

Acute administration of heroin, especially intravenously, produces an orgasmlike "rush" lasting 30 to 60 sec. This is usually followed by a brief period of euphoria accompanied by a profound sense of tranquility. This state may last for several hours, depending on the dose administered and the plasma level of morphine attained. It is characterized by drowsiness ("nodding"), lability of mood, mental clouding, apathy, and motor retardation. Respiratory depression (i.e., diminished volume and slower rate) secondary to inhibition of the brainstem respiratory center also occurs. Stimulation of the brainstem chemoreceptor trigger zone for emesis may produce nausea and vomiting. However, with repetitive use, tolerance develops to this effect.

2. Peripheral

In the healthy patient, the cardiovascular effects of opioids like morphine are minimal. In some patients, however, peripheral vasodilation may contribute to orthostatic hypotension. These drugs also decrease secretions in the stomach, biliary tract, and pancreas and inhibit the contractility of smooth muscle. As a result of the latter effect, opioid use may be accompanied by constipation; diminished smooth muscle tone in the ureters and bladder may produce urinary hesitancy. Inhibition of the smooth muscle of the iris results in pupillary constriction, an important sign of opioid intoxication.

D. Tolerance and Physical Dependence

Tolerance to some effects of opioid drugs begins to develop after only a week of regular use. Tolerance to other effects develops more slowly. Thus, patients become tolerant rather quickly to the euphoriant, analgesic, sedative, respiratory depressant, and emetic effects while continuing to exhibit pupillary constriction and constipation. In most persons receiving therapeutic doses of morphine four times a day for 2 or 3 days, some degree of physical dependence occurs.

E. Acute Intoxication and Overdose

1. Etiology

Among chronic abusers of opioid drugs, overdose is relatively common and may be accidental or reflect suicidal intent. Unintentional overdose stems from two basic preconditions: (1) the user does not know the dose of opioid actually administered; (2) opioid users vary considerably in their level of drug tolerance. Overdose deaths probably are the result of respiratory depression with subsequent anoxia, although anaphylactic reactions to heroin or its common adulterants may also play a role.

2. Clinical Manifestations

Patients who overdose on illicitly obtained opioid drugs frequently are alone, and the untoward effects are immediate. Thus, diagnosis and medical intervention often are too late. When these patients present for treatment, they may be in stupor or coma with constricted pupils and diminished pulse and respiration. Hypothermia and pulmonary edema (usually noncardiogenic) are seen in severe cases.

The clinician must differentiate opioid overdose from other causes of respiratory depression and coma. Although pupillary constriction usually is a reliable sign, it also occurs in severe barbiturate intoxication. Moreover, in the event of circulatory collapse and cyanosis, the pupils become dilated. In narcotic users, the physician should look for dermatological evidence of repeated intravenous injections (i.e., needle tracks). Signs of other medical illnesses often associated with chronic opioid use (e.g., hepatitis, infectious endocarditis, and multiple abscesses) also may help to make the diagnosis.

3. Management

a. General Life-Support Measures

The first task in treating acute opioid poisoning is to establish an adequate airway. Aspiration can be prevented either by placing the patient on his side or by using a cuffed endotracheal tube. Gastric lavage should be performed if the clinician suspects recent oral intake of opioids or other drugs. If hypoglycemia complicates the clinical picture, establish a reliable intravenous route and slowly infuse a 50% solution of glucose in water. Pulmonary edema, if present, can be treated with positive-pressure respiration.

b. The Use of Naloxone

Narcotic antagonists can dramatically reverse respiratory depression and other symptoms associated with acute opioid toxicity. The current drug of choice is naloxone (Narcan® and others), a pure, potent narcotic antagonist with no agonistic (i.e., opioidlike) effects of its own. Naloxone, 0.4 mg (1 ml) administered intravenously, should reverse manifestations of overdose within 2 min. Increase in respiratory rate and volume, rise in systolic blood pressure, and dilation of the pupils indicate a favorable response. If the initial response is unsatisfactory, the same dose may be given twice more at 5-min intervals. Failure to respond after three doses of naloxone suggests either that the diagnosis of opioid poisoning is erroneous or that another problem (e.g., barbiturate poisoning, head injury) may be complicating the clinical picture. Unfortunately, the duration of action of naloxone is much shorter (1 to 4 hr) than that of most opioid drugs. For this reason, patients with severe overdoses should be hospitalized and monitored closely for at least 24 to 48 hr. A significant decrease in respiratory rate or level of consciousness should prompt additional antagonist treatment to prevent relapse into opioid-induced coma.

Most patients who overdose on drugs like heroin or methadone are also physically dependent. Repeated administration of naloxone, which displaces opioids from their receptor sites in the brain and elsewhere, may precipitate an acute abstinence syndrome. Signs and symptoms of opioid withdrawal may appear within minutes and last for several hours. Moreover, they cannot be easily overcome by giving additional opioid. The severity and duration of abstinence symptoms depend on the degree of opioid dependence and the dose and route of administration of naloxone. The use of a small dose (0.4 mg IV) at 5-min intervals may help to avoid this unwelcome complication.

c. Other Considerations

Patients who use opioids tend to abuse other drugs as well, particularly alcohol, cocaine, and antianxiety drugs. For this reason, the clinician should obtain blood and urine from all patients for toxicological screening. On recovery, the degree of suicidal intent associated with overdose should be assessed; prior to discharge, patients can be referred to an appropriate treatment facility.

F. Opioid Withdrawal

1. Clinical Manifestations

The opioid-dependent patient often comes for treatment when the quality or availability of heroin or other opioids declines on "the street," making it difficult to stave off symptoms of withdrawal. When abstinence symptoms occur, their severity depends on the type of opioid previously used, degree of tolerance, time elapsed since the last dose, and emotional meaning of the symptoms. The symptoms reflect increased excitability in organs previously depressed by chronic opioid use.

In the early stages of withdrawal from heroin (i.e., 6 to 12 hr after the last dose), the patient may yawn and sweat, his nose and eyes may run, and he may experience considerable anxiety. Craving for opioids and drug-seeking behavior intensify. As withdrawal progresses, pupils dilate, and the patient may develop gooseflesh, hot and cold flashes, loss of appetite, muscle cramps, tremor, and insomnia or a restless tossing sleep. Eighteen to 24 hr after the last dose of heroin, nausea, vomiting, and elevations in blood pressure, pulse, respiratory rate, and temperature occur. After 24 to 36 hr, diarrhea and dehydration may develop.

In the case of heroin or morphine, abstinence symptoms generally peak 48 to 72 hr after the last dose. By this time, laboratory values reflect the clinical process. Leukocytosis is common, and ketosis and electrolyte imbalance may develop as a result of dehydration. In untreated cases, most clinical symptoms usually disappear within 7 to 10 days, although physiological disturbances, such as insomnia and malaise, may be detected for weeks to months afterward. Withdrawal from methadone is characterized by slower onset (24 to 48 hr after the last dose) and more gradual resolution of symptoms (3 to 7 weeks). Indeed, patients often complain of fatigue, weakness, and insomnia for several months after stopping the drug.

2. Management

a. Detoxification with Methadone

i. Pharmacology. Withdrawal from opioid drugs may be readily treated by using methadone hydrochloride, a synthetic opioid with pharmacological properties qualitatively similar to those of heroin and morphine. Orally, methadone is approximately half as potent as subcutaneous morphine. Its onset of action is approximately 30 to 60 min, and its duration of action ranges from 24 to 36 hr. Methadone is bound to tissue protein and gradually accumulates. Demethylation in the liver is the principal metabolic pathway. Excretion is primarily through the kidneys and gastrointestinal tract.

The acute effects of methadone are similar to those of other opioid drugs: sedation, analgesia, and respiratory depression. During chronic administration, tolerance develops. Brief periods (7 to 14 days) of use in low (i.e., less than 40 mg/day) and decreasing doses does not result in clinically significant physical dependence.

ii. Principles of Use. The use of methadone in the treatment of opioid withdrawal is based on several factors. As a result of the strong cross-dependence between methadone and the other opioids, sufficient doses **(10 to 40 mg/day) of methadone will prevent abstinence symptoms** in patients who are physically dependent on other opioid drugs. In addition, methadone is well absorbed from the gastrointestinal tract and is effective orally, thus enabling avoidance of the hazards of intravenous use. Since the drug is more slowly metabolized than heroin or morphine, it has a longer duration of action (24 to 36 hr), allowing for once-per-day administration. Finally, oral administration of therapeutic doses to tolerant individuals does not produce the euphoria that follows intravenous use of shorter-acting opioids such as heroin.

iii. Clinical Application. Determining the degree of physical dependence in opioid addicts can be quite difficult because of uncertainty about the percentage of active drug contained in street samples (usually between 3% and 10%) and the variable reliability of these patients. Therefore, the crucial guiding principle in treating opioid withdrawal is to attend to objective clinical signs rather than symptoms. The clinician can use the method of detoxification described in Table 1.

If there is a mixed dependence on both opioids and central nervous system (CNS) depressants, the patient should be temporarily maintained on methadone and withdrawn from the CNS depressants first. Do not prescribe additional methadone

TABLE 1. Method for Detoxifying Opioid Addicts Using Methadone

1. Observe the patient for objective signs of opioid withdrawal. The patient should exhibit two of the following four signs: (a) pulse 10 beats/min over baseline, or (if baseline is unknown) greater than 90 in the absence of a history of tachycardia; (b) systolic blood pressure greater than 10 mm Hg over baseline or greater than 160/95 in the absence of known hypertension; (c) dilated pupils; (d) sweating, gooseflesh, rhinorrhea, or lacrimation. When at least two of the four signs are present, administer methadone 10 mg PO. Signs of withdrawal should abate, at least temporarily, within 90 min. If they do not, the patient's symptoms may not be caused by opioid withdrawal.
2. The evaluation procedure should be repeated every 2–4 hr, and methadone administered as often as necessary for 24 hr. Only rarely should a patient receive more than 40 mg of methadone during this period, unless he or she has been maintained on methadone at higher doses prior to detoxification.
3. The total dose of methadone administered during the first 24 hr (the stabilization dose) should be given the next day in two divided doses.
4. Beginning on the third day, taper the methadone by 5 mg per day, beginning with the morning dose, until the patient is completely withdrawn.
5. All methadone should be dispensed in crushed or liquid form and should be consumed under direct supervision in order to avoid illegal diversion of the drug onto the street.

or CNS depressants for complaints of fatigue, muscle pain, and insomnia. These are common symptoms during and after detoxification. Finally, most outpatient programs monitor patients' urines on a once- or twice-weekly random schedule for the presence of illicit drugs. Naturally, care must be taken to ensure that the patient does not substitute "clean" for potentially drug-contaminated urine. In addition, since toxicology laboratories vary considerably in their degree of accuracy and quality control, laboratory participation in an ongoing proficiency testing program is desirable.

The normal course of outpatient detoxification with methadone is from 3 to 21 days. Although inpatient detoxification can be completed in 3 to 7 days, patients generally prefer a more gradual reduction in dosage. Outpatient programs often have difficulty retaining patients through the entire detoxification process. Patients who drop out may voice dissatisfaction with their current dosage of methadone, fear of its long-term effects, and/or generalized distrust of the treatment program. More pertinent, however, is the patients' ambivalence about giving up opioid use and the fact that tolerance develops rapidly to methadone's euphoriant and analgesic effects. Thus, once the dosage drops to 15 or 20 mg per day, patients may experience a return of abstinence symptoms. Patients should be informed at the outset that they cannot expect a symptom-free withdrawal period.

b. The Use of Clonidine in Opioid Detoxification

In recent years, some investigators have found that clonidine (Catapres®), a nonopioid antihypertensive drug, significantly reduces symptoms of opioid withdrawal. It may act by stimulating α_2-adrenergic receptors in the locus coeruleus of the midbrain, thereby greatly reducing the noradrenergic hyperactivity that usually accompanies opioid withdrawal. Clonidine may be used either instead of or in addition to methadone during detoxification. Following a test dose of 0.025 or 0.05 mg, clonidine is usually initiated at a dose of 0.15–0.3 mg/day in three divided doses. Detoxification from methadone usually requires 2 weeks, and that from heroin may be completed in 5–10 days using relatively low doses. The specific dose is adjusted to reduce withdrawal symptoms as much as possible without causing hypotension. The drug is usually withheld if a patient's blood pressure drops below 90/60 mm Hg. After stabilization on clonidine for between 4 (for short-acting opioids such as heroin or meperidine) and 14 days (for methadone), the drug is tapered over 4–6 days.

Certain medical conditions may contraindicate the use of clonidine, and these may include chronic or acute cardiac disorders, renal or metabolic diseases, and moderate to severe hypertension. Clonidine has been shown to have efficacy equal to methadone in achieving successful short-term detoxification and 6-month abstinence. It has been seen to have a somewhat greater rate of success in inpatient over outpatient settings. The most common side effects of clonidine are hypotension and sedation. In inpatient settings, it is possible to use higher doses of clonidine to achieve greater blockade of withdrawal symptoms such as insomnia and still monitor and support hypotensive side effects. Sedation is also less problematic for the inpatient than the outpatient, who may be trying to maintain routine activities while in the process of detoxification. The advantages of clonidine detoxification are several. It is not a narcotic or controlled substance and thus is usable in a wider range of settings. Second, the induction of opioid antagonists may be achieved more rapidly following clonidine detoxification. The disadvantages outlined above may include insomnia and intolerable hypotension and/or sedation. Furthermore, the use of clonidine for treating opioid withdrawal is not an approved use of the drug at the time of this writing.

c. The Use of Buprenorphine in Opioid Detoxification

Experimental use of buprenorphine for detoxification in inpatient and outpatient settings has shown a high rate of treatment retention and abstinence from illicit opioids. Buprenorphine (Buprenex®) is a partial opioid agonist that produces a feeling of contentment rather than euphoria or a heroinlike rush. At lower doses (2–4 mg/day) it blocks withdrawal symptoms, and at higher doses (greater than 8 mg/day) it can act as an opioid antagonist. Preliminary work has found that 4 mg SL per day can block opioid withdrawal symptoms.[32] Although preliminary studies have reported that buprenorphine does not lead to significant dependence and can be stopped with only minimum withdrawal symptoms, further work with this drug is required before it can be recommended for widespread use.

d. The Use of Lofexidine in Opioid Detoxification

Lofexidine (not available in the United States) is an α-noradrenergic agonist similar to clonidine. It has been used in experimental settings to detoxify patients from long-term methadone dependence and has been found to successfully reduce opioid withdrawal symptoms without the sedative or hypotensive effects seen with clonidine. To determine whether lofexidine may eventually become a safer and more effective alternative to clonidine, more controlled studies are required.

e. Neonatal Addiction and Withdrawal

The infant born to an opioid-dependent mother presents a special problem. As recently as 25 years ago, the mortality rate among passively addicted neonates with signs and symptoms of opioid withdrawal was almost 90%. At present, the mortality rate is negligible, primarily because of increased awareness of this problem.

Like many other drugs of abuse, opioids cross the placenta. Approximately 75% of infants born to opioid-dependent mothers are physically dependent. Most exhibit symptoms of opioid withdrawal within the first 24 hr of life. In infants born to mothers dependent on methadone, the onset of withdrawal symptoms is delayed but usually occurs within the first 3 days. Some infants, however, may first manifest withdrawal symptoms as late as 10 days after birth.

Although there is marked individual variation, the degree of physical dependence developed by the mother correlates well with the severity of the withdrawal syndrome in her infant. These infants are usually tremulous, with a shrill, piercing cry, increased muscle tone, and hyperactive reflexes. Sneezing, frantic sucking of the fists, sleep problems, and regurgitation after feeding are also common. Other symptoms include yawning, sweating, tearing of the eyes, pallor, and diarrhea. High fever, dehydration, and seizures indicate severe withdrawal. Recent studies have also demonstrated a high risk of later neurological and cognitive problems.

The diagnosis of opioid withdrawal in the neonate is often complicated by the fact that a mother may withhold her history of narcotic use or leave the hospital before symptoms develop in her offspring. A maternal history of hepatitis, thrombophlebitis, Acquired Immune Deficiency Syndrome (AIDS), endocarditis, lack of prenatal care, or unexplained neonatal death in the past should raise the clinician's index of suspicion. Other indicators include labor pains that cannot be controlled with the usual doses of analgesics, needle marks along superficial veins, and overt signs of narcotic withdrawal in either mother or infant.

The primary goals in managing neonatal withdrawal are to reduce central nervous system irritability and control diarrhea and dehydration. Although numerous drugs have been used to treat this syndrome, **paregoric,** 0.2 ml PO every 3 to 4 hr, appears to be most helpful. The initial 4-hr dose can be increased in 0.05-ml increments until symptoms are controlled. If symptoms are not alleviated by a dose of 0.75 ml given every 4 hr, the clinician should suspect physical dependence on some other drug class, particularly central nervous system depressants. In such cases, phenobarbital, 5 mg/kg per day in three divided doses, should be added to the regimen.

After a stabilizing dose has been found and maintained for a week, the total daily dose of paregoric should be tapered by 0.06 ml per day. The importance of concomitant supportive therapy (fluids, proper feeding, etc.) cannot be over-emphasized. Indications of therapeutic success include a normal temperature curve, diminished crying and restlessness, increased ability to sleep, weight gain, and vasomotor stability.

Recently, addicted mothers-to-be have been maintained on low doses (40 mg or less) of methadone during pregnancy, tapering the drug a few weeks before delivery. This approach reduces the risk of sepsis from unsterile needle use and makes the mother more available for prenatal care.

G. Treatment of Chronic Opioid Abuse

1. Medical Complications

A wide variety of medical and psychosocial complications accompany chronic opioid use. These, coupled with poor living conditions and lifestyles adapted to support drug habits, produce significant morbidity and mortality. Increasing mortality among urban addicts may be attributable to abuse of heroin in combination with other substances such as cocaine and alcohol and to the rising prevalence of human immunodeficiency virus (HIV) infection in this population.

The majority of health problems associated with chronic opioid abuse originate in either self-neglect or the complications associated with intravenous injections. Injections cause local inflammation of veins (phlebitis) and hyperpigmented scars ("needle tracks"). Contaminants may be injected into the bloodstream and block arteries and cause cardiac and pulmonary complications.

Shared needles spread many infections including HIV, hepatitis B, non-A, non-B hepatitis, malaria, *Staphylococcus aureus* and other agents that cause endocarditis, and syphilis. In the United States and Europe, intravenous drug use plays an important role in the transmission of HIV infection. Twenty-five percent of all cases of AIDS in the United States have occurred in intravenous drug users, with 17% occurring in those for whom intravenous drug use was the only risk factor. High rates of HIV infection in intravenous drug users have geographically clustered, with 82% of all cases of AIDS in intravenous drug users occurring in New York City. There is also a direct association of perinatal and pediatric AIDS with intravenous drug abuse, because women intravenous drug abusers and sexual partners of male intravenous drug abusers make up the largest number of HIV-infected women of childbearing age.

2. Treatment Approaches

a. General Considerations

Chronic opioid users who present themselves for treatment are often both physically ill and psychosocially disabled. The clinician should view opioid dependence as treatable with the constant potential for relapse. There is no known treatment that is completely effective for all individuals addicted to opioids, but a number of treatments are effective for different people. Patients often require a wide range of medical and rehabilitative services. These may include remedial education, job training and placement, legal assistance, and individual, group, and family therapy, as well as pharmacological intervention.

In most cases a multimodal approach will be most helpful, but a treatment program must be individually designed for each patient. Some of the factors that must be assessed when planning treatment are the degree and severity of psychological problems and coexisting psychiatric illness (especially affective disorder). The need is for rehabilitation or habilitation, depending on the individual's previous educational and vocational status, for changing the lifestyle to diminish stimuli that may lead to drug craving, and alertness to coexisting polydrug use that may require additional intervention.

Treatment programs may include a combination of inpatient detoxification and rehabilitation, outpatient drug-free programs, methadone maintenance, support groups such as Narcotics Anonymous, therapeutic communities, narcotic antagonists, and individual, group, and family therapy. Although motivation on the part of the individual is, of course, important, programs can help maintain motivation by helping the patient gain positive rewards from a drug-free life, thus causing more substantial losses if a relapse occurs.

b. Methadone Maintenance

i. Background and Theory. As a treatment modality, methadone maintenance evolved out of the clinical observation that some individuals feel unable to abstain totally from opioid drugs. Its theoretical justification derives from the hypothesis that chronic opioid use induces long-lasting, perhaps permanent, physiological changes at the cellular level. As a result, opioid addicts may experience "narcotic hunger" for months or years after withdrawal; the need to satisfy this hunger results in relapse. In this context, medically supervised maintenance on methadone, in doses high enough to satisfy narcotic hunger, becomes a logical alternative to psychological and physical dependence on heroin or other illicitly obtained opioid drugs.

ii. Clinical Application. See Section II.F.2.a.i for a review of the pharmacology of methadone. Methadone maintenance usually entails administration of 40–120 mg PO per day. Doses exceeding 60 mg per day are generally considered high-dose methadone maintenance; low-dose maintenance ranges between 40 and 60 mg per day. The advantage of low-dose maintenance is the fact that it theoretically limits side effects such as drowsiness while providing sufficient receptor blockade to halt craving. High-dose maintenance theoretically provides receptor blockade that is more difficult to override with heroin and may, therefore, decrease illicit drug use.

After determining that the patient is opioid dependent (i.e., by observing objective signs of withdrawal coupled with a urine specimen positive for the presence of opioids), the clinician should determine a methadone stabilization dose in the manner described in Section II.F.2.a.iii. Thereafter, the daily dose is adjusted to a level (usually greater than 40 mg) at which the patient reports complete blockade of the effects of subsequently administered opioid drugs (e.g., heroin) and random urine samples indicate that the use of opioids other than methadone has ceased.

When it is taken regularly, tolerance develops to methadone's euphoriant effects, and patients generally function without psychomotor impairment or evidence of intoxication. Some patients, however, continue to complain of sedation, excessive sweating, constipation, ankle edema, and decreased interest in sex. The latter is associated with a drug-induced reduction in plasma testosterone, especially at higher doses.

Patients on methadone maintenance generally come to the clinic daily to receive the drug and do not take home any doses until considerable movement toward rehabilitation is noted. Compulsory clinic attendance makes the patient available for participation in individual and/or group counseling, which may be a precondition to receiving methadone. Take-home methadone in locked boxes, which enables the patient to return to the clinic at greater intervals, may be initiated after a successful period of treatment.

Treatment programs are variable in their tolerance of concurrent illicit drug use, level of support (ranging from no ancillary services to a full range of psychotherapy, counseling, medical services, and vocational and rehabilitation programs), and duration of program before at least a trial of detoxification from methadone is required. Other requirements of methadone maintenance programs may include active job seeking and urine specimens to test for illicit drugs.

iii. Detoxification from Methadone. When patients voluntarily or involuntarily leave methadone maintenance treatment, the clinician can decrease the dosage at a 5-mg-a-day rate at a level ranging from 1 mg a week to 25 mg a day; in this way withdrawal generally proceeds gradually. Slow detoxification from methadone maintenance has been associated with a high dropout rate and low abstinence success rate, probably because the patient must tolerate withdrawal symptoms for a longer period of time. Many studies indicate that the use of clonidine in withdrawal from methadone maintenance shortens the detoxification period and is associated with a higher success rate of abstinence. Detoxification from methadone maintenance using clonidine can proceed in the same general manner as described in Section II.F.2.b. Doses used in outpatient detoxification should generally be lower than inpatient doses, however, because of less frequent monitoring and the possibility of hypotension.

iv. Current Issues in Methadone Treatment. Methadone maintenance boasts the highest retention rate of any opioid treatment modality. It is now the primary treatment modality for approximately 75,000 patients. Fifty to 70% of those who begin on methadone maintenance remain in treatment for at least 1 year. The advantages to this treatment approach have been a demonstrated decrease in illegal drug use and arrest rate and an improvement in employment and school records. Problems that persist in methadone maintenance programs are (1) the continued use of illegal drugs including opioids, cocaine, diazepam (Valium® and others), and amphetamines, by patients while on methadone maintenance; (2) alcoholism, which remains a problem for approximately 20% of those on methadone maintenance; (3) diversion of methadone to illicit markets, especially by those using take-home methadone; and (4) the occasional deaths by overdose, especially in children, teen-agers, and heroin addicts who obtain the methadone illicitly or accidentally.

A review of the numerous studies that have measured the outcome of methadone maintenance programs based on eventual detoxification and abstinence have shown that several factors increase success rates. These include (1) screening patients for their potential ability to benefit from rehabilitative treatment; (2) adjunctive psychotherapy, which increased rates of successful detoxification by 15–20%, probably by helping with postdetoxification adjustment; and (3) pharmacotherapy, such as clonidine or lofexidine, which helped with withdrawal and improved rates of successful detoxification by 64%.

c. The Use of Narcotic Antagonists

i. Rationale. The narcotic antagonists are structurally similar to opioids and occupy the same receptor sites in the central nervous system and elsewhere. The first antagonists used were naloxone and cyclazocine (not available in the United States), but their use was discontinued because of poor oral absorption and short duration of action in the former and side effects in the latter. Naltrexone (Trexan®)is a long-acting opioid antagonist. It is rapidly and completely absorbed orally. It has a substantial first-pass extraction and metabolism by the liver, with a half-life of 4–8 hr. In sufficient doses, naltrexone completely blocks the pharmacological effects of subsequently administered opioid drugs through a process of competitive inhibition. As with other narcotic antagonists, if given after opioid dependence has developed, it rapidly displaces opioids from their receptor sites and can precipitate an acute abstinence syndrome.

The rationale for the use of narcotic antagonists in the treatment of opioid dependence is based on the conditioning model originally proposed by Wikler. He noted the development of socially conditioned abstinence symptoms in former addicts who are exposed to the environment in which they have previously experienced both drug taking and withdrawal. In such individuals, increased craving for opioids frequently occurs and often leads to a relapse of drug-taking behavior. Other external stimuli, such as being in the presence of drugs or observing someone else "get high," may produce conditioned abstinence responses. Internal stimuli include feelings of anxiety, depression, and fear of withdrawal. Wikler theorized that if detoxified addicts were allowed to self-administer opioids but were prevented from experiencing opioid reinforcement (i.e., the heroin effect), extinction of conditioned responding would take place, and the patients might eventually give up trying to "get high." The development of narcotic antagonists has provided the pharmacological tool with which to test this theory.

Narcotic antagonists such as naltrexone block the effects of exogenous opioids such as heroin. Unlike agonists such as methadone, however, they are not addicting and, therefore, are not associated with abstinence syndromes when discontinued. These properties have made them clinically useful treatments but also contribute to some of the problems in compliance associated with their use.

ii. Clinical Use. Naltrexone was approved by the Food and Drug Administration in 1984 and is now widely available. Naltrexone should be administered after a person has been off heroin for at least 7 days or off methadone for at least 10 days in order to avoid precipitating acute withdrawal. It is advisable to start with a test dose of naloxone, 0.8 mg subcutaneously, in order to be sure that this short-acting drug does not cause withdrawal symptoms. If no evidence of abstinence symptoms is present, the longer-acting naltrexone can be begun. Fifty milligrams of naltrexone will completely block the euphoric action of narcotics for up to 24 hr, 100 mg is effective for up to 48 hr, and 150 mg is effective for 72 hr. A common dosing schedule, therefore, is to give 100 mg on Monday and Wednesday and 150 mg on Friday. The use of naltrexone blocks the euphoriant action of narcotics and prevents physical addiction even if opioids are used. The problems with compliance to this treatment are related to these same properties. For example, naltrexone has no narcotic effect and, therefore, some patients prefer methadone; there is no withdrawal if naltrexone is stopped, and this may be a disincentive to continue treatment in the face of craving; craving for narcotics can continue in spite of naltrexone treatment, and patients can abuse other classes of psychoactive drugs.

Adverse effects of naltrexone treatment may include gastrointestinal irritation, small increases in blood pressure, lightheadedness, and drowsiness. The most serious side effect is an increase in liver enzymes. It is, therefore, recommended that liver function tests be monitored at base line and then each month for 6 months. If SGOT is twice normal at baseline, do not start naltrexone therapy. Once therapy has begun, if SGOT is elevated to three times base line, discontinue treatment.

The time of most significant dropout from naltrexone therapy is in the induction period; this can be minimized by waiting the recommended time between last opioid use and beginning treatment. Take-home naltrexone has been associated with the highest dropout rates and should only be used in the setting of strong psychosocial supports. The most successful outcomes of naltrexone treatment have been seen in subpopulations that include employed, motivated patients with good vocational stability and less coexisting drug use. Better outcomes are also more likely when naltrexone is administered in highly structured multidisciplinary programs that include family, couple, and individual therapy. Greatest success has been achieved when there are strong family and social supports and in certain groups such as health professionals, occasional heroin users, addicts leaving structured settings and programs, and people from higher socioeconomic groups who do not want methadone maintenance.

Clinical experiments with buprenorphine hydrochloride are under way as described in Section II.F.2.c. Buprenorphine is a mixed opioid agonist and antagonist and results in mild euphoria for 3–4 weeks. This may increase retention in the induction phase but clearly increases the abuse potential. It is 25–40 times more potent than morphine and equipotent to naloxone as an antagonist. Experimentally, a subcutaneous dose of 8 mg a day significantly suppresses heroin self-administration in addicts. Additional clinical trials are needed to assess its long-term efficacy.

d. Outpatient Drug-Free Treatment

The term "outpatient drug-free treatment" (OPDF) actually encompasses various treatment strategies, all of which do not use methadone or narcotic antagonists.

Other psychotropic drugs (e.g., antipsychotics, antidepressants) may be employed as necessary but are usually prescribed temporarily.

Outpatient drug-free treatment often represents the opioid addict's first treatment experience. Many poorly motivated patients choose this type of program, since controls are fewer, and they have to give up little to participate. These programs usually offer individual and/or group counseling. Couple and family work, behavior therapy, and vocational rehabilitation may also be available. Urine screening is used to detect the presence of illicit drug use. Some studies have shown that in highly motivated patients, OPDF treatment can be quite effective in curbing illicit drug use and promoting employment.

e. Therapeutic Communities

Self-help residential treatment programs called "therapeutic communities" or "concept houses" have become one of the primary modalities for the rehabilitation of chronic opioid users, especially those who fail in other modalities. Ex-addicts generally run these programs. Their special perspective has led to the evolution of a confrontational approach and a refusal to accept psychodynamic, biochemical, or environmental justification for continued drug use. Instead, treatment focuses on interpersonal concerns, particularly responsibility toward the community and the competent handling of aggression and hostility. Residents work their way up through a structured hierarchy within the "community" by demonstrating responsibility and honesty and by remaining drug-free. Those who achieve high status in the system, in turn, serve as role models. Some graduates remain as staff members.

The addict's ambivalence about remaining drug-free coupled with direct and highly critical confrontation by peers results in a dropout rate of approximately 80% in the first month. However, courts have more recently begun offering treatment in a therapeutic community as an alternative to jail for addicts convicted of drug-related crimes; this has certainly lowered the overall dropout rate in these programs. For those who stay, the duration of treatment varies, but the average is about 6 months. Graduates from therapeutic communities frequently do quite well in terms of reducing illicit drug use, securing and maintaining employment, and eschewing criminal behavior. Some studies have shown that the length of stay in treatment correlates with treatment outcome.

III. ABUSE OF SEDATIVES, HYPNOTICS, AND ANXIOLYTICS

A. Introduction

The sedatives, hypnotics, and anxiolytics are all included with alcohol in the category of central nervous system (CNS) depressants despite having widely varying chemical structures. These drugs include the barbiturates, the benzodiazepines, and the sedative–hypnotics such as methaqualone (no longer available in the United States), glutethimide (Doriden®), meprobamate (Miltown® and others), chloral hydrate (Noctec® and others), paraldehyde, methyprylon (Noludar®), and ethchlorvynol (Placidyl®). The following sections discuss the various abuse syndromes. (See Chapter 5 for a complete discussion of the pharmacology of these agents.)

B. Tolerance and Physical Dependence

With repetitive administration, three types of tolerance develop to the central nervous system depressants. They include (1) **dispositional tolerance,** in which the enzyme systems of the liver become more capable of rapidly metabolizing these drugs; (2) **pharmacodynamic tolerance,** in which the cells of the CNS adjust themselves to the presence of increasing doses of these drugs; and (3) **cross-tolerance** to the effects of other central nervous system depressants, including alcohol.

Tolerance to the hypnotic effects of a short-acting barbiturate such as pentobarbital (Nembutal® and others) begins to develop within a few days, even when it is administered in therapeutic doses. After 2 weeks, the therapeutic effectiveness of a given dose may be reduced by 50%. Moreover, the simultaneous development of cross-tolerance to other CNS depressants makes substitution of a second drug in this general class therapeutically ineffective, since the patient will likely be tolerant to its effects. On the other hand, since some CNS depressants (including barbiturates and alcohol) compete for hepatic metabolism, concurrent use of two or more of these agents will result in clinical effects far greater than either of the drugs alone. Heightened toxicity and accidental overdose are common sequelae of this practice.

C. Classification

1. Barbiturates

The barbiturates are usually classified by their duration of action into ultra-short-, short-to-intermediate-, and long-acting types. The ultrashort-acting barbiturates, including methohexital (Brevital®) and thiopental (Pentothal® and others), are used primarily as intravenous anesthetics. The short-to-intermediate-acting drugs, including amobarbital (Amytal® and others), pentobarbital, and secobarbital (Seconal®), are employed primarily for their sedative and hypnotic properties. The relatively long-acting barbiturates, including phenobarbital (Tedral® and others), may be employed as sedative–hypnotics but are most useful in the control of some seizure disorders. They may also be used to facilitate withdrawal from other sedative–hypnotics.

2. Benzodiazepines

To a large extent, the benzodiazepines have replaced the barbiturates for the pharmacological treatment of anxiety. In addition, the hypnotic effects of these compounds have contributed to their increasing popularity. Taken for brief periods in therapeutic doses, the benzodiazepines are safe and effective; however, chronic use may result in the development of tolerance and physical dependence.

The benzodiazepines include the anxiolytic benzodiazepines, which are alprazolam (Xanax®), chlordiazepoxide hydrochloride (Librium® and others), clorazepate dipotassium (Tranxene® and others), diazepam, halazepam (Paxipam®), lorazepam (Ativan® and others), oxazepam (Serax® and others), and prazepam (Centrax® and others); the hypnotic benzodiazepines, which are flurazepam hydrochloride (Dalmane® and others), temazepam (Restoril® and others), and triazolam (Halcion®); and the anticonvulsant benzodiazepine clonazepam (Klonopin®).

3. Other Sedative–Hypnotics

In addition to the barbiturates and benzodiazepines, a number of other drugs depress CNS functioning and, therefore, have potential utility (and abuse liability) as sedative–hypnotics. Although they vary considerably in their chemical structure and pharmacological properties, the clinical picture in both intoxication and withdrawal is quite similar. Some of the more commonly used (and abused) agents are described below.

Chloral hydrate, the oldest and best-known drug of this group, will induce sleep at doses of 0.5 to 1.0 g PO. Habitual use may produce tolerance and physical dependence. Abrupt withdrawal may result in delirium, seizures, and death. Patients may occasionally display a "break in tolerance," leading to sudden unexpected death by overdose.

Methaqualone is a drug that achieved wide popularity among drug abusers who enjoyed the sense of well-being, disinhibition, paresthesias, and ataxia that characterize intoxication ("luding out"). In addition, methaqualone has been purported to have aphrodisiac properties, a claim that has not been substantiated. Many users find that methaqualone, unlike the barbiturates, does not make them drowsy. The usual hypnotic dose is 150 to 300 mg PO, but some abusers may take up to 2 g per day. Death has been reported after intravenous injection of 8 g, but most deaths occur in abusers who have also been drinking alcohol. Decrease in the prevalence of abuse of this drug in the last 10 years may be associated with its decreased availability.

Glutethimide in an oral dose of 500 mg rapidly induces sleep in most nontolerant individuals. Doses of greater than 2 g per day for a month produce physical dependence. In these individuals, abrupt discontinuation of the drug may result in a general depressant withdrawal syndrome (see Section III.F). When taken in large doses, glutethimide is stored in the body's fatty tissue and is thus difficult to dialyze. In cases of overdose, episodic release from body stores may occur, and so the patient's degree of intoxication may fluctuate widely. Serum concentrations may not correlate well with the level of consciousness. Unlike other CNS depressants, dilated pupils may accompany glutethimide poisoning. Overdose with this drug has been implicated in many suicides.

Methyprylon in doses of 200 to 400 mg will induce sleep. As is the case with other sedatives, prolonged use of large doses will result in physical dependence. Death has been reported following the ingestion of as little as 6 g, but individuals vary greatly in their degree of tolerance.

Ethchlorvynol is a sedative–hypnotic with a rapid onset and short duration of action. Ingestion of 2 to 4 g per day over several months can produce physical dependence. As with methyprylon, the potentially lethal dose varies widely. Ingestion of more than 10 g at once is usually fatal.

Meprobamate a carbamate derivative, has been widely used as an antianxiety agent and skeletal muscle relaxant. The benzodiazepines have largely supplanted this drug both as a therapeutic compound and as a drug of abuse. The usual therapeutic dose is 400 mg three or four times per day. Chronic administration of slightly higher doses will produce physical dependence with the possibility of seizures on withdrawal. The lethal dose varies between 12 and 40 g. Reports of potential teratogenicity make the ingestion of meprobamate during the first trimester of pregnancy contraindicated (see Chapter 10).

D. Abuse of Sedatives, Hypnotics, and Anxiolytics

According to a recent community study, 1.1% of the adult population in the United States have been dependent on anxiolytics or sedative–hypnotics at some time in their lives. In addition, a 1985 national survey revealed that 15 million Americans had used at least one prescription psychotherapeutic medicine non-medically during that year, with two-thirds of these individuals using the drugs in combination with illicit, nonprescription drugs such as marijuana or cocaine.

Benzodiazepine prescriptions in the United States rose from 80 million in 1983 to 86 million in 1985. In addition, there was also an increase in the number of prescriptions for the newer benzodiazepines such as alprazolam, lorazepam, temazepam, and triazolam. Older drugs in this class, including oxazepam and diazepam, were prescribed less often. This, of course, does not reflect the unknown quantity of benzodiazepines brought into the United States illegally. Significantly, there is evidence of differential abuse liabilities of specific benzodiazepines, with diazepam, lorazepam, and alprazolam generally recognized as having higher abuse liability than other benzodiazepines.

Unlike opioid abusers, the majority of people dependent on benzodiazepines, barbiturates, or other sedative–hypnotics do not belong to a drug-using subculture and an illicit system of drug distribution. Instead, they are primarily individuals who have received legitimate prescriptions from physicians to treat anxiety or insomnia. With prolonged administration, however, tolerance develops to the hypnotic and anxiolytic effects, and this prompts individuals to increase the dose or frequency on their own. The individual may justify this on the basis of continuing to treat symptoms but may then begin to exhibit drug-seeking behavior such as visiting a series of physicians to obtain multiple prescriptions. Extreme tolerance to one medication may develop, as does cross-tolerance to other CNS depressants. Subsequent attempts at withdrawal may thus result in abstinence symptoms.

It is important to note that some people are prescribed long-term benzodiazepine treatment for chronic severe anxiety disorders. These people do not meet criteria for dependence insofar as the drugs do not interfere with their functioning and they do not spend excessive time seeking and taking the medications. Significantly, they are continuing to use the drug for medically prescribed use. However, these individuals may develop tolerance and physical dependence, since abrupt termination of the drug would result in characteristic withdrawal symptoms.

A second group of abusers of anxiolytics and sedative–hypnotics are young people in their teens or early 20s who use illegally obtained medications for nonmedical purposes to induce euphoria. Frequently such use occurs in combination with illicit drugs such as opioids (to enhance euphoria) or cocaine (to counteract the stimulant effects). These individuals are more likely to administer benzodiazepines or barbiturates intravenously, increasing the risk of fatal overdose. People in this group may use these drugs daily or during sprees of intoxication lasting several days, during which they may consume massive doses (in excess of 300 mg/day of diazepam or its equivalent).

E. Acute Intoxication and Overdose

1. Etiology

Not infrequently, depressed patients may make a suicide attempt by overdosing on CNS depressants that they received for the treatment of insomnia and/or anxiety. Occasionally, a patient may become confused after ingesting the therapeutic dose. Ingestion of a second, third, or even fourth dose ("drug automatism") then leads to increased confusion and subsequent overdose.

2. Clinical Manifestations

Mild to moderate intoxication with central nervous system depressants closely resembles alcoholic drunkenness, but patients do not have alcoholic breath. The severity of symptoms depends on the drug(s) used, the route of administration, and the presence or absence of other complicating conditions (e.g., head injury). As summarized in Table 2, patients usually present with drowsiness, slurred speech, motor incoordination, impaired memory, confused thinking, and disorientation. Physical examination may reveal both horizontal and vertical nystagmus, tremor, and ataxia. In addition, patients may exhibit extreme irritability, agitation, inappropriate affect, and paranoia, sometimes accompanied by rage reactions and destructive behavior. The latter is most common when depressant drugs are combined with alcohol.

Severe overdose with CNS depressants is accompanied by signs of cerebrocor-

TABLE 2. Signs and Symptoms of CNS Depressant Intoxication

Drowsiness	Ataxia
Slurred speech	Hypotonia
Motor incoordination	Hyporeflexia
Confusion/agitation	Memory impairment
Disorientation	Respiratory depression
Nystagmus	Inappropriate affect
Tremor	Rage reactions

tical and medullary depression. Patients may be stuporous or comatose with absent corneal, gag, and deep tendon reflexes. Plantar stimulation may produce no response or an extensor response. They may also have impaired cardiopulmonary function with shallow, irregular breathing, hypoxia, respiratory acidosis, and, in the late stages, paralytic dilatation of the pupils. In terminal cases, patients may develop shock, hypothermia, lung complications (e.g., pulmonary edema, pneumonia), and renal failure.

3. Management

The clinical picture described above, coupled with a suspected or confirmed history of sedative intake, depression, and recent psychiatric or medical treatment, suggests the diagnosis of sedative overdose. Additional history from family or friends also can be quite useful. Chromatographic analysis of blood and urine samples will confirm a drug ingestion.

Adequate treatment of sedative overdose requires a well-trained staff in a hospital setting. Generally, if more than ten times the full hypnotic dose of a drug has been ingested, overdose will be severe, particularly if combined with alcohol intake. When the patient arrives in the emergency room, treatment will be dictated by his state of consciousness. In the awake patient who has ingested a drug in the past several hours, vomiting should be induced or gastric lavage instituted. Samples of the vomitus, blood, and urine should be sent for toxicological analysis. If the patient has ingested a fat-soluble drug such as glutethimide, 60 ml of castor oil may be given via nasogastric tube. If bowel sounds are present, a cathartic (e.g., 15 to 30 g of sodium sulfate) should be instilled into the tube to prevent gastric dilatation and regurgitation. If the patient's condition remains stable, he should be continuously observed, with monitoring of respiratory and cardiovascular functioning and the level of consciousness. Following the ingestion of certain fat-soluble drugs including glutethimide, meprobamate, and ethchlorvynol, the patient's level of consciousness may fluctuate widely as the drugs are episodically released from tissues stores.

Patients who present in a stuporous or comatose state or who are initially awake but then lapse into coma clearly need more intensive treatment. The goal is to support vital functions until the patient is able to metabolize and excrete the drug.

Thus, management is directed primarily toward maintaining adequate cardiopulmonary and renal function. If the drug ingestion has occurred within the previous 6 hr, gastric lavage should be attempted, but only after establishing an unobstructed airway by passing a cuffed endotracheal tube. When oxygenation is inadequate, the clinician should ventilate mechanically and administer oxygen. If shock supervenes, transfusions of whole blood, plasma, or plasma expanders should be used to elevate blood pressure and prevent circulatory collapse. The use of vasopressor drugs [e.g., norepinephrine (Levophed® and others)] has been advocated in some situations.

When renal function is satisfactory, forced diuresis should be attempted; enough saline and dextrose in water (approximately 500 ml/hr) should be administered to produce a urine output of 8 to 10 ml/min. With the longer-acting barbiturates such as phenobarbital, alkalinization of the urine with sodium bicarbonate enhances excretion. Some authors advocate the use of diuretics [e.g., furosemide (Lasix® and others)] to promote urinary excretion. In cases of profound intoxication, hemodialysis is often effective when more conservative measures fail. For the more lipid-soluble short-acting barbiturates and some of the nonbarbiturate sedative–hypnotics (e.g., glutethimide), the use of a lipid dialysate or hemoperfusion through charcoal or resins (e.g., the lipid-adsorptive AMBERLITE® XAD-4) may promote drug elimination more effectively than traditional dialysis methods.

In general, survival following an overdose of a CNS depressant depends on the maintenance of adequate respiratory, cardiac, and renal function until drug concentrations drop below potentially lethal levels. On recovery, psychiatric consultation should be obtained, and the patient's suicide potential assessed. Patients who purposefully overdose on CNS depressants are at risk to repeat this action, especially if the underlying problem (e.g., depression) is not effectively managed.

F. Withdrawal from CNS Depressants

Chronic administration of CNS depressants, even in the usual therapeutic doses, may produce tolerance and eventually physical dependence. Although tolerance develops rapidly to the therapeutic dose, tolerance to a potentially lethal dose develops at a somewhat slower rate. Thus, even chronic users of these substances may become quite intoxicated and even comatose when the dose is raised only slightly.

Physical dependence implies the presence of an abstinence syndrome on withdrawal. In severe cases, withdrawal may carry significant risk of mortality and should be carried out in a hospital setting.

1. Clinical Manifestations

Onset of withdrawal symptoms depends on the duration of action of the particular drug. In the case of the short-acting barbiturates, abstinence symptoms may occur within 12 to 16 hr after the last dose. Withdrawal of a longer-acting drug such as diazepam may not result in abstinence symptoms until 7 to 10 days after the last dose. Severity of symptoms depends on the degree of physical dependence and individual variables that are poorly understood.

Early withdrawal symptoms include anxiety, restlessness, nausea, and vomiting. In the case of the short-acting benzodiazepines, such as alprazolam, generalized seizures have been documented within the first 24 hr. As withdrawal progresses, the patient may complain of weakness, abdominal cramps, and may develop postural hypotension, tachycardia, hyperreflexia, and a gross resting tremor that is most prominent in the upper extremities. Although total sleep time is reduced, the percentage of rapid eye movement (REM) sleep increases, and nightmares and insomnia frequently occur.

With the exception of longer-acting benzodiazepines and barbiturates, withdrawal symptoms characteristically peak within 1–3 days after the last dose; generalized seizures may occur within this period, either singly or as status epilepticus. A significant minority of patients who have seizures develop delirium. The latter is characterized by anxiety, disorientation, frightening dreams, and visual hallucinations. Agitation and hyperthermia can lead to exhaustion followed by cardiovascular collapse. Delirium is not easily reversed, even by giving large doses of the abused drug; recovery may be slow. If death does not supervene, the patient, after a lengthy period of sleep, usually clears by the eighth day.

2. Management

a. The Pentobarbital Tolerance Test and Phenobarbital Substitution

The treatment of CNS depressant withdrawal is complicated by the fact that sedative abusers are often inaccurate historians. They may wish to obtain as much drug from the clinician as possible and/or may have impaired memory secondary to chronic drug use. For this reason, the **pentobarbital tolerance test** is a useful tool in the management of withdrawal states. The test takes advantage of the fact that cross-tolerance exists between the various CNS depressants. As illustrated in Table 3, patients showing signs of depressant withdrawal are given 200 mg of pentobarbital orally. One hour later, the clinician should examine the patient for signs of sedative intoxication such as sedation, nystagmus, ataxia, and slurred speech. If mildly intoxicated by this dose, the patient has probably been taking less than the equivalent of 800 mg of pentobarbital a day.

The clinician can stabilize these patients on a dose of 100 to 200 mg of pentobarbital every 6 hr, depending on the degree of intoxication. If, after an initial test dose (200 mg), no signs of intoxication appear, the patient's tolerance is probably greater than that induced by prolonged, daily use of 800 mg of pentobarbital. Consequently, additional increments of 100 mg of pentobarbital may be administered every 2 hr until signs of intoxication become evident or until a total of 500

TABLE 3. Detoxification from CNS Depressants Using the Pentobarbital Tolerance Test

Day 1: 200 mg PO pentobarbital
   If intoxication: 100 to 200 mg PO q.6.h. (nystagmus, ataxia).
   If no intoxication: 100 mg PO q.2.h. until signs of intoxication develop; the total dose required to produce intoxication is then given q.6.h. for the next 24 hr.
Day 2: Give pentobarbital in the same dose as was given in the previous 24 hr.
Day 3 and beyond: Subtract 100 mg/day of pentobarbital from total dose given on previous day until detoxified. If signs of intoxication develop, eliminate a single dose and resume treatment 6 hr later. If signs of withdrawal develop, given 100 to 200 mg pentobarbital PO or IM stat.
Phenobarbital substitution: 30 mg phenobarbital for each 100 mg pentobarbital. Advantages: more constant plasma level and significant anticonvulsant effects.

mg has been given. The clinician should then calculate the total dose required to produce intoxication and give this dose every 6 hr for the next 48 hr. If the patient becomes grossly intoxicated, the next 6-hr dose may be omitted, and the following day's dose reduced by this amount. Once stabilization has been achieved, the total daily dose of pentobarbital is reduced by 100 mg each day until withdrawal is completed. Patients should be free of tremulousness, insomnia, and orthostatic hypotension. If these signs of abstinence recur during the tapering process, additional 100- to 200-mg doses of pentobarbital may be given. However, it is usually sufficient to stop the dosage reduction for 1 day and then cautiously resume tapering the drug.

One disadvantage of pentobarbital-mediated withdrawal is that patients sometimes require intoxicating doses to prevent the development of seizures and perhaps delirium. Thus, practical reasons exist for substituting a long-acting barbiturate such as **phenobarbital** once stabilization on pentobarbital has been achieved. Use of the longer-acting phenobarbital produces a more constant plasma level than can be obtained with the shorter-acting pentobarbital. Phenobarbital also provides a greater degree of anticonvulsant activity relative to its sedative effects, so that the patient need not be intoxicated to avoid the development of seizures. Thirty milligrams of phenobarbital can be substituted for every 100 mg of pentobarbital. The drug can be given in divided doses every 8 hr. Once stabilization on phenobarbital has been achieved, the dose is lowered by 30 mg per day until total withdrawal is achieved. If the patient shows signs of barbiturate toxicity, one or more doses may be omitted. If signs of withdrawal are apparent, an additional 60 to 120 mg of phenobarbital can be given immediately IM, and the total daily dose of phenobarbital increased by 25%. Phenobarbital should not be administered in doses exceeding 500 mg per day.

Treatment with phenobarbital will suppress early symptoms of depressant withdrawal. Once delirium has developed, however, 24 to 72 hr of barbiturate treatment may be required before it clears. Phenytoin (Dilantin® and others) is of questionable efficacy in preventing withdrawal seizures in these patients. The antipsychotic drugs [e.g., chlorpromazine (Thorazine® and others)], which themselves lower seizure threshold, also should be avoided, although some have advocated their use in the treatment of delirium (see Chapter 11).

Although some patients have been withdrawn from CNS depressants as outpatients, this is risky. Thirty milligrams of phenobarbital are substituted for each hypnotic dose of the substance abused, and the total daily requirement of phenobarbital is given in three divided doses. The total daily dose is then reduced by 30 mg per day over 2 to 3 weeks.

b. Gradual Dose Reduction

Although the above-described phenobarbital substitution technique has been used successfully in many centers, some questions have recently emerged regarding the effectiveness of this technique in withdrawing patients from certain drugs, particularly alprazolam. Thus, many clinicians have tapered their patients off alprazolam (when the amount taken was clearly known) by lowering the dose gradually; the *Physicians' Desk Reference* recommends that the daily dose be decreased by no more than 0.5 mg every 3 days. Since even this schedule is very uncomfortable for some long-term, high-dose users, some authors have recommended a gradual switch from alprazolam to clonazepam and tapering the patient off the latter drug.

3. Mixed Opioid–Sedative Dependence

In the last decade, simultaneous abuse of more than one class of psychoactive drug has become commonplace. Heroin addicts and patients on methadone maintenance may also abuse sedatives and alcohol. Although there is no cross-tolerance between opioids and CNS depressants, sedatives may partially alleviate the symptoms of opioid withdrawal. However, even large doses of methadone or heroin will not prevent sedative withdrawal.

When mixed opioid–sedative dependence is suspected, the clinician should first determine the degree of sedative dependence by using the pentobarbital tolerance test. Once this is done, the most prudent course is to maintain the patient on a dose of methadone sufficient to prevent symptoms of opioid withdrawal while gradually tapering CNS depressants. Some have recommended tapering both classes of drugs simultaneously, but this approach complicates the clinical picture.

IV. ALCOHOL ABUSE

A. Introduction

Of all the drugs mentioned in this chapter, alcohol is the most commonly abused. About 70% of the adult population in this country have consumed alcoholic beverages in the last year. Within this group, it is estimated that the number of problem drinkers approaches 10 million, about half of whom are physically dependent on alcohol. Alcoholism is directly or indirectly implicated in 200,000 deaths annually.

Alcohol abuse and dependence are defined by the same criteria used to define all psychoactive substance use disorders (see Section I). Among people who meet these diagnostic criteria, there appear to be several typical patterns of drinking. The first is characterized by drinking large quantities on a daily basis. The second pattern is so-called weekend binge drinking in which people are abstinent during the week but consume large quantities of alcohol on weekends. Another pattern consists of people who have long periods of abstinence punctuated by alcohol binges that may last for weeks or months. In addition to these patterns, there are a number of potential presentations of alcoholism. Indeed, the heterogeneity of the disorder is one of its hallmarks.

B. Risk Factors in Alcoholism

A number of social, environmental, genetic, and biological factors have been shown to correlate with the development of alcoholism. Lower socioeconomic status, ethnic and religious affiliation (e.g., Irish Catholics, American Indians, and Eskimos), marital status (e.g., divorce), and nationality have been reported to be sociodemographic risk factors. The male sex, antisocial personality disorder, younger age, conduct disturbances, and school difficulties have also been associated with increased alcohol abuse and/or dependence. Recent evidence has shown that crude prevalence rates of alcohol dependence and abuse may be higher in rural than in urban areas.

Genetic factors may also play a role in the development of alcoholism. Alcohol dependence tends to cluster in families. Over 100 studies have demonstrated that alcohol dependence/abuse is three to five times more frequent in the parents, siblings, and children of those with alcoholism than in the general population. Monozygotic twins have a higher concordance rate for the disorder than nonidentical twins. Moreover, children of one or more alcoholic parents adopted at birth have a much higher likelihood of developing alcoholism than the offspring of parents without alcoholism, including those adopted by alcoholics. There are no clear biological markers for alcohol dependence or abuse.

Various attempts have been made to correlate "predisposing" personality variables with subsequent drinking behavior. However, recent studies have indicated that personality traits such as passivity, dependency, pessimism, and self-doubt are more likely to result from, rather than cause, alcoholism.

C. Pharmacology

Alcohol is rapidly absorbed from the gastrointestinal tract, but the rate of absorption is modified by food, the volume of liquid ingested, the concentration of alcohol, and its rate of administration. High concentrations limit intestinal absorption by inducing a reflex pylorospasm. As a result, large quantities of alcohol may remain unabsorbed in the stomach before gradually passing into the small intestine. Thus, following an episode of heavy drinking, an individual may experience fluctuating levels of intoxication.

Once absorbed, alcohol appears to be uniformly distributed throughout most body tissues, including the brain, where an equilibrium with the plasma alcohol concentration is quickly reached. In the liver, alcohol is converted to acetaldehyde through the action of alcohol dehydrogenase. Acetaldehyde, in turn, is converted into acetyl coenzyme A, which is then oxidized through the citric acid cycle or used in other anabolic reactions. In addition to alcohol dehydrogenase, liver microsomal oxidases also play a role in alcohol metabolism. The average rate of liver oxidation is 10 ml/hr with a daily maximum of approximately 450 ml. A small unoxidized portion of ingested alcohol is excreted through the kidneys and lungs.

D. Tolerance and Physical Dependence

The degree of tolerance to the effects of alcohol varies considerably. Certainly, people with identical blood alcohol levels may exhibit markedly different behavior responses. In addition, chronic alcohol ingestion results in the development of both metabolic and pharmacodynamic (tissue) tolerance. Metabolic tolerance implies enhanced activity of those enzymes that oxidize alcohol in the liver, e.g., alcohol dehydrogenase. In most individuals, this type of tolerance quickly reaches an upper limit and thereafter is quite uniform. When drinking is steady and does not exceed an individual's metabolic capacity, manifest signs of intoxication may never appear. However, when the rate of consumption rises above this level, blood alcohol concentration increases rapidly, resulting in intoxication. Within 3 weeks after discontinuing drinking, metabolic tolerance is greatly reduced, even in individuals with a long history of heavy use.

Pharmacodynamic or tissue tolerance implies adaptive changes at the cellular level, particularly in the brain, as a result of prolonged exposure to alcohol. Consequently, higher blood alcohol levels are needed to produce the degree of intoxication previously seen at lower blood levels.

Cross-tolerance between alcohol and the other central nervous system depressants results from both pharmacodynamic and metabolic factors. As in the case of other CNS depressants, however, tolerance to the intoxicating effects of alcohol does not imply tolerance to the lethal dose, which is probably not much greater in alcoholics than in nondrinkers.

Chronic ingestion of large amounts of alcohol will result in physical dependence. The time required varies and depends on factors not yet completely defined. In most patients, the presence of physical dependence is confirmed by the development of abstinence symptoms when the drug is withdrawn. Indeed, abstinence symptoms may occur even after a partial drop in the blood alcohol level.

E. Acute Intoxication

1. Simple Type

Although ethyl alcohol affects all organ systems of the body, its most marked and important effect is on the central nervous system (CNS). Physiological signs of intoxication include slurred speech, ataxia, incoordination, nystagmus, and flushed face. There may also be maladaptive behavioral changes such as disinhibition, aggressiveness, poor judgment, impaired attention, poor recent memory, irritability, euphoria, depression, and emotional lability.

Alcohol-induced inhibition of the cerebral cortex releases lower brain centers from the inhibitory and integrating control of the cortex. This accounts for the initial stimulatory effect of alcohol on behavior. In the initial stages of intoxication, the user may become outgoing, loquacious, and emotionally labile. Increased confidence and expansiveness often lead to the perception of enhanced verbal and manual performance. Careful testing, however, reveals that efficiency in both areas is impaired. Recent memory, ability to concentrate, and insight also are diminished.

The acute effects of alcohol on the brain are a function of the blood alcohol level, which itself is determined by dose, speed of absorption, rate of metabolism, and the efficiency of alcohol excretion. In addition, CNS effects are more pronounced when the blood alcohol level is rising than when it is falling. In the majority of individuals, a blood alcohol concentration of 50 to 100 mg per deciliter (dl) results in mild intoxication. Concentrations over 100 mg/dl usually impair an individual's ability to operate a motor vehicle. At 200 mg/dl, most users will be grossly intoxicated, exhibiting ataxia and slurred speech. Blood alcohol levels over 400 mg/dl are potentially lethal. Alcohol usually exerts its lethal effects by leading to respiratory depression or aspiration of vomitus.

There are a number of potential complications of alcohol intoxication. These include highway accidents, household and industrial accidents, and violent acts. Fifty percent of all murders occur when the victim or murderer is intoxicated, and approximately 25% of all suicides occur during a period of intoxication. Falls are common in intoxicated individuals, leading to potentially serious medical sequelae such as fracture, subdural hematoma, and other brain trauma.

2. Alcohol Idiosyncratic Intoxication

In some individuals, many of whom are quiet and shy when sober, relatively small doses of alcohol may induce a transient but profound state of intoxication, which develops either during drinking or shortly thereafter. Such patients appear confused, disoriented, and delusional. Consciousness may be impaired, and auditory or visual hallucinations have been reported. Patients also exhibit increased activity, impulsivity, and aggressiveness with accompanying rage reactions and violence. In those with preexisting depression, suicide attempts may occur. Most episodes are brief and last only 2–3 hr. The episodes generally end with a deep sleep, and the person returns to his or her usual state as the blood alcohol level falls. On awakening, there is usually amnesia for the event. This disorder is apparently uncommon.

F. Chronic Intoxication

1. Peripheral and Central Nervous System Effects

Chronic alcohol abuse may lead to substantial nervous system pathology. The most frequent neurological disorder in alcoholics is **polyneuropathy,** which results primarily from an alcohol-induced vitamin deficiency. A direct toxic effect of alcohol also may be a factor. Most patients are asymptomatic, with absent ankle jerks as the only sign. In more severe cases, burning feet, pain, paresthesias, and distal muscle weakness occur. Foot drop and wrist drop are ultimate complications. Prolonged abstinence, adequate nutrition, and vitamin replacement may gradually reverse this syndrome, but in some, the damage is permanent.

Wernicke's disease, the result of an alcohol-induced thiamine (vitamin B_1) deficiency, occurs in some chronic alcoholics. Nystagmus, bilateral sixth cranial nerve palsies, and paralysis of conjugate gaze are frequently seen along with ataxia and various mental disturbances. Most common is a "quiet delirium" characterized by apathy, lassitude, disorientation, and drowsiness. The symptoms usually have an abrupt onset and may occur singly or in various combinations.

A less common syndrome, **Korsakoff's psychosis** (amnesic or amnestic–confabulatory psychosis, psychosis polyneuritica, or alcohol amnestic disorder), consists of profound anterograde and retrograde amnesia in an otherwise alert individual. Confabulation also may be present. Wernicke's disease and Korsakoff's psychosis are not separate disease entities; rather, the latter is a variably present component of the former. When both are present, the disease is called the **Wernicke–Korsakoff syndrome.** Postmortem brain examination of these patients often reveals structural lesions in the mammillary bodies.

Prompt diagnosis and treatment of Wernicke's disease is imperative. Immediate administration of **thiamine, 50 mg IV and 50 mg IM, followed by 50 mg IM or PO each day** until the patient is eating well should improve the ocular difficulties within hours to days. Ataxia also may respond during this time period. Delirium, when present, is reversible, but as confusional symptoms recede, the symptoms of Korsakoff's psychosis may become more evident. In such patients, recovery takes longer (i.e., a year or more) and may be incomplete.

Other central nervous system sequelae of chronic alcohol abuse include **cerebellar degeneration,** affecting primarily stance and gait, and **vitamin deficiency amblyopia,** causing blurred vision, central scotomata and, if untreated, optic nerve atrophy.

2. Effects on Other Organ Systems

Over time, heavy alcohol consumption may damage other organ systems. Liver disease can take the form of alcoholic fatty liver, alcoholic hepatitis, or cirrhosis. Gastrointestinal disorders include gastritis, pancreatitis, gastric and duodenal ulcers, and malabsorption syndrome. Many alcoholics are anemic, in part because of nutritional (e.g., folic acid) deficiencies that often accompany chronic alcohol abuse. Alcoholic cardiomyopathy also is an important cause of morbidity and mortality in chronic heavy drinkers. The reader is referred to standard textbooks of internal medicine for a more complete description of these clinical entities.

3. Fetal Alcohol Syndrome

The offspring of women who abuse alcohol during pregnancy may manifest one or more congenital abnormalities grouped under the term fetal alcohol syndrome (FAS). These may include low birth weight, microcephaly, mental retardation, short palpebral fissures, epicanthic folds, maxillary hypoplasia, abnormal palmar creases, cardiac anomalies (e.g., septal defects), capillary hemangiomas, and a slowed postnatal growth rate. Generally, these infants sleep and feed poorly and are irritable, tremulous, and hyperactive. Mental retardation or borderline mental retardation is a frequent neurological deficit in children with FAS. More subtle behavioral changes may also result from fetal exposure *in utero* to lower levels of ethanol.

Exposure to alcohol at any time during pregnancy results in significant alterations in neonate reflexive and motoric behavior. The FAS occurs in 33% of infants born to women who drink more than 150 g of alcohol per day during pregnancy; another 33% of infants without clear FAS will be mentally retarded. The perinatal mortality rate is 17%, and seriously affected infants may never achieve normal growth or intellectual performance, even in an optimal postnatal environment.

G. Alcohol Withdrawal

1. Minor Abstinence Syndrome

a. Clinical Manifestations

In physically dependent users, the initial stages of alcohol withdrawal usually begin several hours after the last drink. Tremulousness ("shakes"), muscle tension, blushing, sweating, and a vague sense of anxiety ("jitters") may be followed by nausea, vomiting, anorexia, and abdominal cramps. The patient appears restless and agitated; signs of generalized central nervous system irritability include hyperreflexia and a tendency to startle easily. As the withdrawal syndrome progresses, the patient develops a marked resting tremor and possibly auditory and visual hallucinations accompanied by paranoid delusions. Although initially he may have some insight, it is lost as the syndrome progresses.

In a minority of alcohol-dependent individuals, seizures ("rum fits") develop. They are typically grand mal and begin 7 to 48 hr after cessation of drinking, with over 60% of seizures occurring 17–24 hr after stopping. Seizures may be single or occur in bursts of two to six, but 2% go on to have status epilepticus. This may be particularly true in patients with preexisting seizure disorders, who are at greater risk when intoxicated and withdrawing from alcohol. In one-third of cases, seizures are followed by the development of delirium tremens.

b. Management

Minor abstinence symptoms usually resolve within 3 to 4 days, even in untreated patients. A postwithdrawal syndrome characterized by irritability, hostility, anxiety, insomnia, fatigue, tremor, and headache may last for up to 3 weeks to 3 months. Characteristically, these patients do not come to medical attention unless complications (e.g., seizures) develop. Heavy drinkers soon learn that administration of any central nervous system depressant, including alcohol, will reduce abstinence symptoms. Thus, a drinking bout may be prolonged as the patient attempts to ameliorate abstinence symptoms. If barbiturates or other sedative–hypnotics are used regularly for self-treatment, these patients may develop dependence on those drugs as well.

In patients who seek treatment, the clinician can take advantage of the cross-tolerance between alcohol and other CNS depressants by substituting a longer-acting drug for alcohol and then gradually reducing the dose of the drug. Evidence has demonstrated that benzodiazepines are of value in decreasing the subjective symptoms of withdrawal and in decreasing alcohol withdrawal seizures. Chlordiazepoxide is commonly used for this purpose. A sample regimen would be to administer 50 mg PO immediately and repeat the dose orally every 2 hr as needed for 24 days, followed by 25 mg orally every 2 hr as needed for the second and third day. On the fourth day, the drug can be stopped, and the patient observed for signs of withdrawal. Vital signs must be monitored frequently, and the drug should be given only for objective signs of withdrawal such as hypertension, tachycardia, or tremor. The exact dose will depend on the age, size, general health, liver function, and drinking history of the patient. Within the first 24 hr, up to 400 to 600 mg of chlordiazepoxide may be given. Chlordiazepoxide should not be administered intramuscularly because it is poorly absorbed.

Chlordiazepoxide is metabolized into several active compounds that have slightly longer half-lives than the parent drug. Its onset of action is within 2 hr, with a peak effect in about 4 hr. After a single dose, the half-life of chlordiazepoxide ranges from 6 to 30 hr. The half-life may be longer in the elderly and in patients with liver disease.

The prophylactic use of antiseizure medications in alcohol withdrawal is controversial. Some have recommended that anticonvulsants not be prescribed, even for patients with a past history of withdrawal seizures. Others feel that in patients with a seizure history, combined phenytoin and chlordiazepoxide offer better protection against withdrawal seizures than chlordiazepoxide alone. Further research is needed to establish the efficacy of carbamazepine and phenytoin with or without benzodiazepines in the prevention of alcohol withdrawal seizures.

2. Major Abstinence Syndrome (Delirium Tremens)

a. Clinical Manifestations

In about 5% of patients who develop withdrawal symptoms, the minor abstinence syndrome described above evolves into a state of confusion, disorientation, agitation, and delirium. This severe form of alcohol withdrawal is called delirium tremens (DTs). Symptoms usually begin 24 to 72 hr after the last drink, and 90% of patients who develop DTs manifest symptoms within the first 7 days of abstinence. The syndrome may last for a few hours to a few weeks. Hallucinations (usually visual), tremulousness, disorientation, insomnia, and nightmares are common, as are signs of autonomic nervous system hyperactivity, including fever, sweating, tachycardia, hypertension, and increased respiratory rate. Since other serious medical illnesses (pneumonia, malnutrition, cirrhosis, gastritis, and anemia) often accompany DTs, these should be investigated and promptly treated. Subdural hematomas and meningitis also occur with increased frequency.

The patient with delirium tremens usually has a history of excessive drinking for at least 5 years, along with recent heavy drinking. The severity of the syndrome depends on the degree of prior intoxication, general health, and adequacy of preventive treatment. Once delirium develops, drug treatment and other management efforts may not alter its course. Even in treated cases, delirium tremens is potentially life threatening. Death may occur as the result of either hyperthermia or cardiovascular collapse. Autopsy examinations in patients with delirium tremens have failed to elucidate the exact pathophysiology of the problem.

b. Management

Patients with delirium tremens are seriously ill and require intensive inpatient care. The major objectives of treatment are reduction of CNS irritability, prevention of exhaustion, and correction of potentially fatal fluid and electrolyte imbalance. An intensive care unit is usually preferable for careful monitoring of electrolytes. Intravenous dextrose and saline should be given to replace fluid losses. Hyperthermia should be treated aggressively with acetaminophen for temperature above 101° and a cooling blanket for temperatures above 103.5°. Parenteral thiamine 100 mg per day should be given. Restraints should be used for severe agitation. For sedation, administer diazepam 5–10 mg IV every 15 to 20 min until sedation is achieved. Repeat doses may be given as needed. Infections should be treated with appropriate antibiotics.

3. Alcoholic Hallucinosis

Following a prolonged drinking bout, some patients who are physically dependent on alcohol may experience an organic hallucinosis. This is characterized by vivid and persistent auditory or visual hallucinations that appear within several days of the last drink. These usually take the form of threatening or disturbing voices. Patients may respond by calling the police or arming themselves. Unlike patients with delirium, however, these individuals are oriented to time, place, and person. Alcoholic hallucinations may last from a few hours to a week, although about 10% continue for weeks or months. Some patients will experience a recurrence of hallucinosis following each bout of heavy alcohol use. A few will develop a chronic form of this disorder, which may be mistaken for schizophrenia. **Antipsychotic drugs** such as chlorpromazine or trifluoperazine (Stelazine® and others) are the treatment of choice. On recovery, there is excellent memory for the psychotic episode.

H. Treatment of Chronic Alcohol Abuse

1. Introduction

The alcoholic individual has developed a lifestyle in which he is preoccupied with alcohol and has lost control over its consumption. As a result, drinking usually leads to moderate to severe intoxication. As tolerance develops, the alcoholic tends to increase the dose; if he is physically dependent, he drinks to prevent or relieve abstinence symptoms, which perpetuates the disorder. In the later stages, brief periods of abstinence are consistently followed by relapse, with its accompanying physical, emotional, and social costs.

Even a recovering alcoholic still suffers from a chronic, potentially relapsing illness. Periods of stable abstinence may be interrupted by occasional relapse. Thus, abstinence should not be the sole criterion of successful treatment. Moreover, rehabilitation encompasses not only the physical aspects of the illness but psychological and social recovery as well. To achieve these goals, different treatment approaches must be integrated into a comprehensive program.

2. Biological Treatment

a. Disulfiram

For many years, disulfiram (Antabuse ®) has been used as part of a Pavlovian conditioning model to avoid alcohol abuse. It produces a very unpleasant and potentially serious reaction when a patient taking it ingests alcohol and is a powerful disincentive in cooperative patients.

Disulfiram interferes with the metabolic breakdown of acetaldehyde, an intermediate product in alcohol metabolism. As a result, patients on disulfiram who subsequently ingest alcohol often experience rapid buildup of acetaldehyde in the blood, usually accompanied by signs and symptoms of acetaldehyde poisoning. These include sensations of heat in the face and neck, throbbing headache, flushing, and feelings of constriction in the throat. Nausea, vomiting, hypotension, and profound anxiety are common; chest pain, coughing, and labored breathing may also occur. Some cases may be complicated by seizures, respiratory depression, cardiac arrhythmias, myocardial infarction, acute heart failure, and death.

The severity of the "Antabuse reaction" is usually correlated with the dose of disulfiram and the amount of alcohol subsequently ingested. Symptoms peak 30 min to several hours after alcohol ingestion. Patients with preexisting heart disease, cirrhosis, diabetes, or any debilitating medical disorder are particularly at risk for untoward and dangerous complications, and the drug should therefore be prescribed with great caution if at all. Patients taking disulfiram should be instructed to avoid even disguised forms of alcohol (e.g., aftershave preparations). When reactions occur, they should be treated supportively by maintenance of blood pressure and correcting hypokalemia when necessary.

Disulfiram is rapidly absorbed from the gastrointestinal tract, but it becomes fully effective only after 12 hr. The drug is oxidized by the liver and excreted in the urine. Excretion is slow, with one-fifth of the administered dose remaining in the body after 1 week. As a result, alcohol ingestion may produce an acetaldehyde reaction up to 2 weeks after disulfiram has been discontinued.

Although the drug is largely inert except when combined with alcohol, adverse effects unrelated to alcohol ingestion may include fatigue, gastrointestinal disturbances, acne, tremor, sexual dysfunction, toxic psychosis, restlessness, and persistent garliclike or metallic taste. Although the usual maintenance dose of disulfiram is 250 mg PO per day, a dose of 125 mg per day will generally suffice and will produce fewer undesirable effects. Another rare but potentially serious adverse reaction is hepatotoxicity. Disulfiram-induced hepatitis is an idiosyncratic event that is apparently not dose related and occurs in approximately one out of 25,000 patients treated. Usually liver function tests (LFTs) return to normal when disulfiram is discontinued, but occasionally the patient may develop fulminating (sometimes fatal) hepatitis. A patient with persistently elevated LFTs should not be given disulfiram until the LFTs return to normal. Liver function tests should then be monitored every 2 weeks for the first 2 months and then every 3–6 months thereafter. If LFTs become elevated, disulfiram should be discontinued.

Disulfiram is probably most useful in older, well motivated, socially stable patients whose drinking is precipitated by recurrent psychological stress. The drug should be taken each morning, when the level of stress and the temptation to drink is usually lowest. When taking disulfiram, many patients report that situations previously conducive to drinking generate less craving for alcohol than before. On the other hand, patients with poor impulse control, schizophrenia, or depression with suicidal ideation are at considerable risk of drinking when on disulfiram.

Disulfiram, by itself, is not an extremely effective treatment for alcoholism. A recent study indicated that only 23% of patients treated with disulfiram alone remained alcohol-free for 1 year, compared with 12% of a placebo-treated control group. Taking disulfiram is a daily decision that must be supported by other developments in a patient's life. It should be part of a comprehensive treatment program aimed at helping the patient adjust psychologically and socially to an alcohol-free existence.

b. Psychotropic Drugs

A small subgroup of alcoholic patients may have coexisting psychiatric illness such as major affective illness, panic disorder, or other anxiety disorders. These disorders, if untreated, may exacerbate drinking behavior and may interfere with an individual's efforts at alcohol treatment. Tricyclic antidepressants are useful in individuals with alcoholism and coexisting major depression. Lithium (Eskalith® and others) has proved beneficial in individuals with bipolar illness and alcohol dependence. In addition, one study reported a decrease in alcohol intake in a group of alcoholics without a coexisting mood disorder. This result is controversial, and more research is thus needed in this area. Although alcoholics are frequently anxious, the use of anxiolytic drugs (e.g., diazepam) other than in the treatment of withdrawal is problematic, since alcohol-dependent individuals may abuse other CNS depressants.

3. Behavioral Approaches to Treatment

In the past decade, behavior modification techniques have achieved increasing popularity. The euphoriant effects of alcohol and the ability of the drug to relieve anxiety and tension are powerful reinforcing effects that promote repeated use. Repeated use, in turn, leads to tolerance and physical dependence, at which point drinking to stave off abstinence symptoms constitutes yet another reinforcer. As in the case of opioid use, chronic, heavy use of alcohol is accompanied by dysphoria and anxiety, which is often forgotten when the patient becomes sober. The latter phenomenon has been called state-dependent learning.

a. Aversion Therapy

In aversion therapy or counterconditioning, alcohol ingestion is repeatedly followed by an artificially contrived noxious experience, e.g., a mildly painful electric shock. Over time, a powerful association is forged between drinking and the aversive stimulus. In theory, this association persists beyond the immediate treatment experience so that subsequently even the thought of drinking evokes powerful memories of the aversive stimulus. A number of alcohol treatment programs using chemical aversion (i.e., drugs that produce vomiting) have reported success (i.e., 40 to 60% abstinence for at least 1 year) with this method. The high level of motivation required of patients willing to submit to such treatment may account, in part, for the good results. Outcome results for this therapy are controversial, however, and it is not recommended as an initial form of treatment. If aversion therapy is used, it is recommended in the setting of other treatments aimed at reinforcing sobriety.

b. Other Behavioral Techniques

Contingency contracting is a system in which the patient and therapist agree on a behavior to be encouraged (abstinence) or eliminated (drinking); a system of rewards or punishments is worked out to expedite these goals. In alcoholics who drink in response to anxiety, **systematic desensitization** coupled with **relaxation training** may help to teach alternative coping behavior. **Assertiveness training** attempts to teach the patient social skills to handle interpersonal situations that have made him feel angry, frustrated, or manipulated and have led to drinking.

Although behavioral techniques are successful in some patients, we need carefully controlled studies to define the indications for such treatments. Certainly most successful treatment programs address not only the issue of drinking but also the intrapsychic and environmental cues that trigger this behavior. As with other modes of treatment, behavioral approaches are most successful when tailored to the individual needs of the patient rather than applied on a wholesale basis.

4. The Psychotherapies

A myriad of psychological therapies have been used to treat chronic alcoholism. These include individual, group, and family therapy. Although the alcoholic is the identified patient, problem drinking is frequently accompanied by troubled relationships. Sometimes, a couple or family can be profoundly affected by continued heavy drinking. Organizations like **Alanon** and **Alateen** are often helpful (see Section IV.H.5).

An insight-oriented approach may help some patients; however, most researchers feel that psychotherapy alone does not usually constitute adequate treatment for alcoholism. Exploring sources of external stress and unconscious conflict may be useful but is generally insufficient to maintain sobriety when alcohol is available and the desire to drink is high. For this reason, Alcoholics Anonymous, disulfiram, and/or other measures should be combined with psychotherapy for those patients where this is appropriate.

5. Alcoholics Anonymous and Other Self-Help Groups

Perhaps the most successful treatment for chronic alcoholism is Alcoholics Anonymous (AA), a worldwide organization aiming to help its members achieve lasting sobriety. This is essentially a self-help approach in which members share their recovery experiences and support each other in the struggle to avoid relapse. Unlike the help offered by most professionals and social agencies, AA groups are almost continuously available. Experienced group members (sponsors) are asked to care for new members during periods in which the relapse risk is high. Membership provides an opportunity to give to others, which enhances self-esteem. It also offers consistent reminders of both the rewards of sobriety and the pitfalls of returning to drinking and its disastrous consequences. Of those alcoholics who become long-term active members, about 40–50% remain abstinent for several years, with 60–68% improving to some extent during their participation.

Alanon, a separate organization for significant others, spouses, or other family members and friends, offers support and valuable information about dealing with the alcohol-dependent individual at home. Alateen provides similar help for the children of alcoholic parents. Adult Children of Alcoholics (ACOA) provides education and support for those adults who grew up in a household in which a parent was alcohol dependent. The individual need not be alcohol dependent him/herself to benefit from this group.

6. Adjunctive Services

Halfway houses can be useful for alcoholic patients who, for a variety of reasons, should not live at home. For most people, these houses are transitional facilities between home and hospital, whereas for others they provide long-term custodial care and supervision. Since job failure frequently accompanies chronic alcoholism, vocational rehabilitation programs are also a useful adjunct to treatment.

7. Summary

Concern about the rising cost of health care has naturally prompted questions about the efficacy of various treatments for alcoholism. Some studies have suggested that less intensive treatment, e.g., outpatient group therapy or brief hospitalization for detoxification, does not necessarily result in a poorer outcome compared with other, more intensive treatment regimens. However, as in other addictive disorders, a multimodal treatment approach is clearly the most useful. Reliance on pharmacological intervention alone or on a single psychologically based treatment modality is usually insufficient to disrupt a pattern of repetitive relapse. Finally, although the individual and societal costs of alcoholism are enormous, comparatively little attention has been paid to preventive approaches. More research is clearly needed in this area.

V. ABUSE OF CENTRAL NERVOUS SYSTEM STIMULANTS

A. Cocaine

1. Introduction

Perhaps no other drug has gained as much notoriety during the past decade as cocaine. In 1988, the National Institute on Drug Abuse estimated that nearly a million Americans abused cocaine at least weekly and that 8 million had used cocaine within the past year. Between 1976 and 1986 there was a fivefold increase in emergency room visits because of cocaine-induced medical complications.

Although the chemical structure of cocaine (an ester of benzoic acid) is dissimilar to amphetamine, its actions are similar, though of shorter duration and perhaps greater intensity. Indeed, research subjects who blindly compared the effects of the two drugs in one study were unable to tell them apart.

Cocaine exerts a direct stimulatory effect on the cerebral cortex, producing excitation, restlessness, euphoria, decreased social inhibitions, increased energy, and feelings of enhanced strength and mental ability. Its sympathomimetic effects include vasoconstriction, tachycardia, increased blood pressure and temperature, and dilated pupils.

2. Pharmacology

Cocaine, or benzoylmethylecgonine, is an alkaloid derived from the coca plant, *Erythoxylon coca*. Oral cocaine use is unusual in this country, but Andes Mountains Indians have chewed coca leaves (from which cocaine is derived) for centuries. The usual routes of administration in the United States are intranasal absorption of cocaine hydrochloride, intravenous administration of cocaine hydrochloride dissolved in water, and smoking of alkaloidal cocaine or "freebase." The cocaine alkaloid or freebase is formed when a base (typically baking soda) is added to an aqueous solution of cocaine hydrochloride and allowed to dry into a crystalline chunk or "rock." "Crack" refers to alkaloidal or freebase cocaine that has been prepared by the dealer for sale on the street.

Cocaine is rapidly absorbed when administered intranasally, intravenously, or by smoking. Intranasal administration ("snorting") produces a euphoric effect and activation in seconds to minutes, with a peak at 10 min and effects that last approximately 45 to 60 min. Intravenous cocaine produces an effect within 15 sec, peaks in 5 min, and lasts approximately 15 min. Smoking alkaloidal cocaine ("freebase") produces an effect in 8–12 sec, peaks at 5 min, and lasts 10 to 20 min.

The usual street doses of cocaine are difficult to measure. Approximately 1 g of cocaine produces 30–40 lines, each of which is 15–50 mg and constitutes one "hit," the usual intoxicating dose. Twenty-five to 250 mg may be injected intravenously.

Neurochemically, cocaine is thought to block the presynaptic uptake of dopamine and norepinephrine, with subsequent increases in these neurotransmitters postsynaptically. This alteration is responsible for the activation of the sympathetic nervous system and the resulting sense of activation and sympathomimetic effects. With repeated use, nerve terminals become depleted of dopamine, which is felt to be responsible for postcocaine dysphoria and the desire for more cocaine.

The subjective effects of cocaine are related to the plasma levels achieved, but the rise of the plasma level also is important in determining its effects. The half-life of the drug is approximately 1 hr. After a single dose, plasma levels fall rapidly, and the drug is metabolized by plasma and liver cholinesterases primarily to the water-soluble metabolite benzoylecgonine, which is excreted in the urine. Benzoylecgonine is detectable in the urine for up to 3 days after the last dose and occasionally longer in chronic, high-dose users.

3. Acute Intoxication

Central nervous system stimulants initially produce euphoria and feelings of increased mental ability and self-confidence. Peripheral effects include vasoconstriction and increases in blood pressure, pulse, and temperature. Pupils are often dilated. High doses may lead to repetitive stereotyped behavior, bruxism, formication, irritability, restlessness, tremor, panic, emotional lability, and paranoia.

Acute cocaine poisoning is rare among users of low doses but common in high-doses users and especially among those who use intravenous cocaine or smoke freebase or crack. The symptoms are similar to those of intoxication but lead more frequently to intense anxiety, paranoia, and hallucinations. Elevated blood pressure, tachycardia, ventricular irritability, hyperthermia, and respiratory depression are symptomatic of severe poisoning. Acute heart failure, stroke, and seizures have also been reported.

Initial treatment of acute cocaine poisoning should provide general life support measures, including establishment of an airway, stabilization of the circulatory system, and reduction of severely elevated body temperature. Manifestations of sympathetic nervous system hyperactivity such as hypertension, tachycardia, and tachypnea have been treated in the past with β blockers and calcium channel blockers. However, the use of the β blocker propranolol (Inderal® and others) is controversial and should probably be abandoned because of the increased risk of cardiotoxicity. Patients in hypertensive crisis may be given phentolamine (Regitine®). Intravenous diazepam can be used in patients with repeated seizures. Once the patient is stabilized medically, chlorpromazine or haloperidol (Haldol® and others) may be used to treat any psychotic symptoms that remain.

4. Abstinence Symptoms

The ending of a binge usually produces symptoms of craving and increased cocaine seeking. This period, called the "crash," is initially characterized by agitation, anxiety, depression, and high cocaine craving, followed by fatigue, depression, decreased cocaine craving, exhaustion, hypersomnolence, and hyperphagia. After several days, a withdrawal phase characterized by anergy, decreased interest in the environment, anhedonia, and increased cocaine craving may last 1 to 10 weeks. Finally, during the extinction phase, the individual may be euthymic and have a normal hedonic response. However, conditioned cues (being reminded about cocaine use by environmental stimuli) may activate episodic cocaine craving. The longer the person remains abstinent during this period, the greater is the decrease in the intensity of craving as a result of conditioned cues.

5. Tolerance and Physical Dependence

Cocaine has powerful reinforcing properties. Given the option, rhesus monkeys with prior cocaine experience will consistently choose intravenous cocaine over food, even after several days of food deprivation. Clinical experience in man suggests that some users will pursue the drug experience to the exclusion of almost all other activities. The hypersomnia, hyperphagia, and lassitude seen in the "crash" phase may constitute a discrete abstinence syndrome that may precipitate renewed use.

Although some chronic cocaine users may consume very large amounts of the drug, tolerance to the effects of cocaine has not been consistently demonstrated. Indeed, some regular users may develop a "reverse tolerance" or sensitization to some of the drug's subjective and behavioral effects.

6. Medical Complications

Medical complications of cocaine use affect many organ systems and create significant morbidity and mortality. Cardiovascular effects include hypertension, myocardial infarction, arrhythmias, aortic rupture, and angina pectoris. Central nervous system effects include grand mal seizures, intracranial hemorrhage and other cerebrovascular accidents, ruptured intacranial aneurysms, fungal cerebritis associated with intravenous use, anxiety, paranoia, and psychosis. Because of vasoconstriction and acute increases in blood pressure, gastrointestinal complications such as ischemic or gangrenous bowel are seen. Decreased lung diffusion capacity, bronchitis, pneumothorax, pneumomediastinum, and pulmonary edema are especially associated with smoking freebase; respiratory arrest may be seen as a complication of cocaine use by any route of administration. Other medical complications include rhabdomyolysis with subsequent acute renal failure, hepatic damage, and sometimes death; anosmia and perforation of the nasal septum; HIV infection associated with intravenous administration; hepatitis; and sexual dysfunction.

7. Cocaine Use in Pregnancy

A recent survey of illicit drug use during pregnancy revealed that cocaine was the most commonly used drug. Cocaine is fat soluble and crosses the placenta, where the fetus metabolizes it to water-soluble norcocaine. This is excreted into the amniotic fluid and reswallowed by the fetus. As a result, the fetus may be exposed to a single "hit" of cocaine for 3–5 days.

Cocaine use by pregnant women causes decreased uterine blood flow, increased uterine vascular resistance, and decreased fetal oxygen levels. The increase in maternal blood pressure, heart rate, and vasoconstriction may, therefore, cause fetal hypoxia, preterm labor, abruptio placentae, and perinatal cerebral infarctions and seizures. Cocaine use during pregnancy is also associated with a high rate of intrauterine growth retardation, increased risk of premature birth, and increased risk of congenital malformations such as microcephaly and genitourinary malformations. Intravenous cocaine use may also result in HIV infection passed to the infant. Subtle neurobehavioral impairments are evident for months after birth. Cocaine babies are fragile and irritable with hypersensitive nervous systems. At birth they yawn, sneeze, grimace, and appear overwhelmed by stimuli. This hyperirritability may be observed for months. Cocaine-induced fetal and perinatal complications can occur if cocaine is used even once during any phase of pregnancy.

8. Treatment of Cocaine Abuse

Research on psychopathology in chronic cocaine abusers has shown that a subgroup of this population have concurrent psychiatric disorders, such as mood disorders or attention deficit disorder. These individuals require treatment of these disorders with appropriate therapy in addition to treatment for their cocaine abuse. It is, therefore, necessary to perform a careful psychiatric evaluation on those who present with cocaine abuse or dependence.

Treatment of cocaine use disorders requires the integration of self-help, behavioral, supportive, psychodynamic, family, and pharmacological interventions. Cognitive and behavioral approaches include contingency contracting, frequent contact in the early phase of abstinence, and education about relapse prevention. Self-help groups such as Alcoholics Anonymous, Cocaine Anonymous, and Narcotics Anonymous may provide support, education, and positive reinforcement for abstinence. Individual psychotherapy may help the individual integrate external controls and gain control over impulsive behavior. The use of pharmacotherapy has had some promising preliminary results. Some studies, for instance, have demonstrated decreased craving, dysphoria, and drug use after treatment with the tricyclic antidepressant desipramine (Norpramin® and others). Lithium appears to be effective in patients with coexisting bipolar illness, and stimulants such as magnesium pemoline and methylphenidate (Ritalin® and others) may be useful in individuals who have coexisting attention deficit disorder. Other researchers have had successful early results with bromocriptine (Parlodel®), amantadine (Symmetrel® and others), and imipramine (Tofranil® and others). However, confirming studies must be performed to delineate further which subgroups of cocaine abusers will respond favorably to particular forms of pharmacotherapy.

B. Amphetamines

1. Introduction

A wide variety of drugs stimulate the central nervous system; a number of these have potential for serious abuse. Amphetamine and its related compounds are sympathomimetic agents: their peripheral effects resemble those produced by stimulation of adrenergic (i.e., sympathetic) nerve endings. They also have central stimulatory effects, which contribute to their popularity as drugs of abuse.

Dextroamphetamine, the *d*-isomer of amphetamine, is three to four times as potent as the *l*-isomer. Other amphetamine derivatives include methamphetamine (Desoxyn®), which is available in injectable form, methylphenidate, phenmetrazine (Preludin®), and diethylpropion (Tenuate® and others). This section focuses on amphetamines as examples of this class of agents.

2. Pharmacology

Both amphetamine and methamphetamine are well absorbed from the gastrointestinal tract and are stored in body tissues, including the CNS. Clinical effects generally appear within 30 min after oral ingestion. The drugs are metabolized primarily in the liver via hydroxylation, demethylation, and oxidative deamination. Following oral administration, more than 50% is excreted unchanged in the urine, particularly when urinary acidity is high.

Structurally, the amphetamine molecule strongly resembles the catecholamine neurotransmitters norepinephrine and dopamine. However, amphetamine is more lipid soluble than either and more readily crosses the blood–brain barrier. The central nervous system stimulation, anorexia, insomnia, and enhanced psychomotor activity produced by amphetamines probably result from their facilitation of norepinephrine release from nerve terminals. They also may interfere with the metabolic breakdown of norepinephrine, both by inhibiting reuptake of this neurotransmitter by CNS neurons and by inhibiting monoamine oxidase, an important enzyme in norepinephrine metabolism. Some authors feel that amphetamines may directly stimulate adrenergic receptors.

The amphetamines also exert some effects on dopamine-containing neurons by facilitating dopamine release and perhaps interfering with the metabolic breakdown of this neurotransmitter. The paranoid psychosis that sometimes accompanies chronic use of high doses of amphetamines may result from effects of dopaminergic neurons.

3. Acute Effects

In therapeutic doses, amphetamines produce wakefulness, mood elevation, a sense of initiative and confidence, and increased mental alertness and ability to concentrate on simple tasks. Appetite is diminished, probably through a direct effect on the feeding center in the lateral hypothalamus. Systolic and diastolic blood pressure are increased, and heart rate is reflexly slowed. At higher doses, tremulousness, agitation, insomnia, headache, dizziness, confusion, and dysphoria occur. The electroencephalogram (EEG) is desynchronized, and both total sleep time and REM sleep are markedly reduced.

4. Tolerance and Physical Dependence

With repetitive use, tolerance develops to some of the effects of the amphetamines, especially when high doses are administered frequently. The mechanism is unclear, but some have hypothesized replacement of norepinephrine stores with amphetamine metabolites that act as "false transmitters." Tolerance develops to the mood-elevating, appetite suppressant, and sympathomimetic effects of these drugs. On the other hand, chronic low-dose administration will continue to produce some stimulatory effects.

With sudden drug discontinuation after chronic amphetamine use, abstinence symptoms develop. These include decreased mood and inertia ("crashing"), agitation, nausea, and occasionally severe depression. There is a rebound increase in rapid eye movement sleep. These symptoms frequently set the stage for resumption of use and an alternating pattern of intoxication and withdrawal.

5. Patterns of Abuse

Currently, amphetamine and other CNS stimulants such as methylphenidate are used primarily in the **treatment of narcolepsy, attention deficit disorder, and exogenous obesity.** Until the late 1960s, these drugs were frequently prescribed for the treatment of fatigue and depression as well. However, as their abuse potential became apparent, their prescription was placed under increased regulatory control, which, in turn, created a substantial illicit market. This has led to widespread drug substitution, so that much of what is sold on the street as amphetamine is caffeine or another over-the-counter stimulant.

Many chronic amphetamine users first begin taking the drug in the context of treatment for obesity or depression. Truck drivers, students, and physicians may use the drugs to alleviate fatigue. With the development of tolerance, the user tends to raise the dose. Subsequently, attempts at withdrawal may produce abstinence symptoms (see above) and resumption of use.

A 1981–1983 community study demonstrated that approximately 2% of the adult population abused amphetamines or other similarly acting sympathomimetics. There are two common patterns of abuse. Some amphetamine abusers use the drug intermittently, ingesting large doses (up to several grams) in search of euphoria. Members of this group are often young, tend to abuse other drugs, and are more likely to use stimulants intravenously in the form of methamphetamine ("crystal"). The individuals may also concurrently use CNS depressants or opioids. Intravenous amphetamine produces euphoria more rapidly than oral administration. With this pattern of use, the drug may be injected many times each day in doses as high as 1 g every 3–4 hr. Sprees of intoxication may last days or weeks and are usually followed by abrupt withdrawal ("crashing") and physical exhaustion. The binge is followed by a period of exhaustion and sleep and subsequent nonuse for several days or weeks.

A second pattern of abuse is chronic daily abuse in which amphetamine is used at either high or low doses. Continuous chronic use produces diminished euphorant effects and an increase in dysphoria because of tolerance. Other symptoms are depression, irritability, anergy, anhedonia, social isolation, sexual dysfunction, paranoid ideation, attentional disturbances, and memory problems.

6. Acute Intoxication

a. Clinical Manifestations

Serious toxic reactions from acute ingestion of amphetamine occur primarily in nontolerant (i.e., infrequent) users who consume relatively large doses (i.e., more than 60 mg/day) over a short period of time. The intoxication syndrome is characterized by restlessness, irritability, tremor, confusion, talkativeness, anxiety, and lability of mood. Peripheral effects include headache, chills, vomiting, dry mouth, and sweating. Blood pressure is variably affected, and heartbeat may be irregular. In more severe cases, auditory and/or visual hallucinations, seizures, and hyperpyrexia may occur.

Among intravenous methamphetamine users, prolonged episodes of intoxication ("speed runs") are accompanied by anorexia, weight loss, insomnia, and generalized deterioration in psychomotor abilities. Chronic users, while intoxicated, also exhibit repetitive stereotyped behavior (e.g., taking things apart and putting them back together). Prolonged high-dose use may result in hallucinations, parasitosis (picking at imaginary bugs), lability of mood, and paranoia. These individuals also are prone to episodes of unprovoked violence, especially when amphetamines are combined with CNS depressants.

Tolerance to the toxic effects of amphetamine or methamphetamine varies. Some patients can become quite ill at doses of 30 mg, whereas chronic users may tolerate 1 g of dextroamphetamine or more. Even in chronic users, however, large doses of intravenous methamphetamine may be followed by chest pain, temporary paralysis, or simple inability to function. The conscious, but "overamped" individual may experience racing thoughts coupled with euphoric mood and perhaps some degree of catatonia. Although deaths from amphetamine overdose are relatively rare, hyperpyrexia, seizures, and shock are reported in fatal cases. At autopsy, findings include petechial hemorrhages in the brain and congestion of the lungs, brain, and other organs.

b. Management

Treatment of acute amphetamine toxicity includes **reducing CNS irritability and autonomic nervous system hyperactivity, controlling psychotic symptoms, and promoting rapid excretion of the drug** and its metabolites. Fevers above 102°F should be treated vigorously. Seizures that occur with acute amphetamine toxicity may take the form of status epilepticus, which should be treated with diazepam, 5 to 10 mg IV, injected at a rate not exceeding 5 mg per minute. This procedure may be repeated every 10 to 15 min as necessary to a total dose of 30 mg. Once the patient is no longer in acute distress, CNS irritability can be further reduced by avoiding excessive stimulation. The excretion of unchanged amphetamine can be enhanced by acidification of the urine. Provided there are no signs of liver or kidney failure, ammonium chloride 500 mg PO may be given every 3 to 4 hr.

Psychotic symptoms that accompany acute amphetamine toxicity are best treated by a dopamine-blocking agent such as chlorpromazine or haloperidol. A test dose of chlorpromazine, 25 mg PO, may be followed by up to 50 mg q.i.d. until symptoms disappear, usually within 48 hr. Some patients may require larger doses. If severe agitation or aggressiveness is a problem, intramuscular administration of antipsychotic drugs should be considered. Patients receiving antipsychotics should be monitored for hypotension. In addition, illegally produced amphetamines sometimes are adulterated with anticholinergic substances, which can potentiate anticholinergic effects of antipsychotic agents (see Chapter 4).

7. Amphetamine Withdrawal

Even without treatment, symptoms of acute amphetamine toxicity usually resolve within a week. The withdrawal period is characterized by dysphoria, irritability, anxiety, fatigue, sleep disturbance, hyperphagia, psychomotor agitation, and rebound increase in REM sleep, which lasts for more than 24 hr. As recovery progresses, the clinician should monitor the level of depression, which may be quite severe. The patient's potential for suicide must be evaluated. Treatment with antidepressants should be considered if the depression fails to remit within several weeks.

8. Amphetamine- or Similarly Acting Sympathomimetic-Induced Delusional Disorder

Chronic high-dose amphetamine use may be accompanied by the development of a toxic psychosis. In the early stages, patients are euphoric, loquacious, and overconfident in their abilities. As the syndrome progresses, suspiciousness, fear, and increased aggressiveness are noted, along with delusions of persecution, ideas of reference, and auditory, visual, and tactile hallucinations. Bruxism, parasitosis, distorted time sense, changes in body image, and hyperactivity also are reported. Some patients exhibit compulsive, stereotyped behavior, which may appear purposeful but is characterized by doing and undoing a particular task. Unlike most toxic psychoses, confusion and disorientation are relatively absent.

In most cases of amphetamine psychosis, withdrawal of the drug is followed by slow but complete recovery within days to weeks. In some patients, however, psychotic symptoms may persist for longer. Characteristically, there is hypermnesia for the psychotic episode, allowing the patient to describe his condition in great detail.

Although some have felt that amphetamine psychosis occurs only in patients who are specifically predisposed to psychosis, the syndrome has been produced in normal volunteers given large doses of amphetamine for 5 days. In practice, consistent use of doses over 100 mg per day for weeks or months places the user at greater risk for the development of psychosis. Sudden increases in dosage, even in tolerant individuals, also may precipitate a psychotic episode.

As with acute amphetamine intoxication, withdrawal of the drug and the use of antipsychotics (e.g., haloperidol, chlorpromazine) in doses sufficient to control symptoms is the treatment of choice for amphetamine psychosis. Once the psychosis clears (usually within several days), antipsychotic drugs should be stopped to permit clinical assessment in a drug-free state.

VI. ABUSE OF HALLUCINOGENS

A. Introduction

The hallucinogens are a group of structurally similar agents that produce perceptual distortion (primarily visual illusions and hallucinations) and enhance awareness of internal and external stimuli. They also induce in the user a sense that mundane events are unusually important, a tendency toward introspection and profound emotional lability. Some have referred to this group of substances as "psychedelic" (i.e., "mind expanding"), but this suggested property may be more illusory than real. The term "psychotomimetic" also has been applied to these drugs because they produce a state that mimics functional psychosis, but the resemblance between hallucinogen intoxication and functional psychosis is superficial.

Popular in the late 1960s, hallucinogen abuse has declined in recent years. The drugs have by no means disappeared, however, and the adverse consequences of their use are still apparent in emergency room settings. Moreover, drugs sold on the street as "hallucinogens" often contain other agents (e.g., amphetamine, phencyclidine), so that patients who present with a "bad trip" presumably from mescaline may, in fact, be suffering from a PCP-induced psychosis.

B. Classification

Pharmacologically, the commonly abused hallucinogenic substances may be divided into two major groups. The **indolealkylamines,** including d-lysergic acid diethylamide (LSD), psilocybin, and dimethyltryptamine (DMT), bear a structural resemblance to the neurotransmitter 5-hydroxytryptamine (serotonin). The **phenylethylamines,** including mescaline and the phenylisopropylamines such as 2,5-dimethoxy-4-methylamphetamine (DOM, "STP"), are structurally related to dopamine, norepinephrine, and the amphetamines.

The commonly abused hallucinogens and their important properties are summarized in Table 4. For purposes of this chapter, discussion focuses primarily on LSD, since it is the best known from a pharmacological and clinical standpoint and is the most commonly abused hallucinogen.

C. d-Lysergic Acid Diethylamide

1. Pharmacology

d-Lysergic acid diethylamide is a synthetic hallucinogen derived from an extract of the ergot fungus. It is a potent hallucinogen, with intoxication resulting from doses as low as 50 μg. The drug is colorless, odorless, and tasteless. It is usually ingested as part of a pill or dissolved on a piece of paper. Occasionally, it is administered intravenously. Following oral administration, the drug is well absorbed from the gastrointestinal tract and distributed to body tissues. Only small amounts

TABLE 4. Commonly Abused Hallucinogenic Drugs

Drug	Source	Psychedelic dose	Peak symptoms	Duration of action	Prominent somatic effects	Prominent psychological effects
d-Lysergic acid diethylamide	Synthetic (from fungi)	50 μg	2–3 hr	8–12 hr; undulating activity as effect declines	Increased sympathetic nervous system activity; dilated pupils, increased blood pressure, pulse, tendor reflexes, temperature, blood sugar; tremor	Hypervigilance, illusions, emotional lability, loss of body boundaries, time slowing, increased intensity of all sensations
Psilocybin	Mushroom	10 mg	90 min	4–6 hr	Like LSD, but milder	Like LSD, but less intense, more visual; more euphoria; paranoia
Dimethyltryptamine (DMT)	Synthetic	50 mg	5–20 min	30–60 min	Like LSD, but with more intense sympathomimetic symptoms	Like LSD, but usually more intense, in part because of sudden onset. Must be smoked or injected; cannot be taken orally
Mescaline	Peyote cactus	200 mg	2–3 hr	8–12 hr	Nausea, vomiting; otherwise like LSD; perhaps more intense sympathomimetic effects	Like LSD but perhaps more sensory and perceptual changes; euphoria prominent
Dimethoxymethylamphetamine (DOM, "STP")	Synthetic	5 mg	3–5 hr	6–8 hr at doses below 5 mg; 16–24 hr at high doses (10–30 mg)	Minimal effects at low dose; autonomic effects prominent at doses above 5 mg	May resemble amphetamine combined with LSD, but long-lasting; high incidence of flashbacks, psychosis; chlorpromazine may aggravate symptoms

are detected in the brain, however. Although the mechanism of action is unclear, the drug inhibits the activity of serotonergic neurons in the midbrain dorsal raphe. The plasma half-life of LSD is 2 to 3 hr. It is metabolized into nonhallucinogenic substances, primarily by conjugation in the liver.

2. Tolerance and Physical Dependence

With repeated administration, tolerance develops rapidly (i.e., in 2 to 4 days) to the behavioral effects of LSD. As a result, even chronic users do not typically take the drug more often than twice a week, and the vast majority take it less frequently. Considerable cross-tolerance exists between LSD and other hallucinogens. No physical dependence has been demonstrated in LSD users, and abstinence symptoms are not observed after withdrawal. Only one death has been directly linked to an LSD overdose, although fatal accidents and suicides may occur during periods of intoxication.

3. Acute Intoxication

The effects of LSD begin 20 to 60 min after ingestion, depending on the amount ingested and the degree of tolerance developed. Sympathomimetic effects include tachycardia, increased blood pressure and body temperature, and pupillary dilation. Hyperreflexia, nausea, and muscle weakness also are observed. Peak effects occur between 2 and 3 hr after ingestion. Visual illusions, wavelike perceptual changes, macropsia and micropsia, and extreme emotional lability dominate this period. Perceptions in one sensory modality may affect or overflow into another (synesthesias), so that colors are "heard" and sounds "seen." Subjective time is slowed, and a generalized loss of body and ego boundaries is experienced. Other perceptual changes include depersonalization, derealization, increased fantasy production, illusions, hallucinations, and hyperacusis, which occur in a state of full wakefulness and alertness. Behavioral changes may include impaired judgment, marked anxiety and depression, fear of losing one's mind, paranoia, and referential thinking. In such a state, the lack of a supportive environment or companion can be detrimental, as the individual struggles to control his anxiety and prevent ego disintegration. As the effects of the drug begin to wane (i.e., 4 to 6 hr after ingestion), the patient experiences intermittent "waves of normalcy." After 8 to 12 hr, the intoxication syndrome is mostly cleared, although aftereffects, including a sense of psychic numbness, may last for days.

D. Adverse Reactions following Use of LSD or Other Hallucinogens

1. Panic Reactions

As summarized in Table 5, the major hazards of hallucinogen use are psychological. The most common adverse reaction is a temporary episode of panic (a "bad

TABLE 5. Adverse Reactions to Hallucinogens

Type	Duration	Predisposing factors	Treatment
Acute panic	2–24 hr	Obsessional character, large dose, inexperienced user	Support, reassurance, diazepam if necessary
Toxic delirium	2–24 hr	Large dose, idiosyncratic response	Support, reassurance, antipsychotics if necessary
Drug-precipitated functional psychosis	Indefinite	Vulnerability to schizophrenia or affective disorder	Antipsychotics only after first 48 hr
Flashbacks	Minutes to hours	Recent (<6 months) use of a hallucinogen; use of marijuana or other drug of abuse	Reassurance; usually brief and not as intense as original experience

trip"). The likelihood of panic is determined by the expectations or "set" of the individual taking the drug, the setting in which the drug is taken, the user's current level of psychological health, and the type and dose of the drug administered. Inexperienced users and those who are constricted, schizoid, anxious, and who fear loss of control have difficulty coping with the disorganizing effects of these drugs. Taking the drug alone or in an unfamiliar or hostile setting also seems to predispose to panic.

Fortunately, most panic reactions are limited to the duration of action of the particular drug used. Since there are no known methods for rapidly eliminating hallucinogenic substances from the body, the primary goal of treatment is to support the patient through this period, mostly by reassuring him that he is not "losing his mind" and that he will soon return to his normal state. The presence of calm, supportive friends can be very helpful. Such patients should not be left alone. In the majority of users, symptoms subside as the drug is metabolized, usually within 24 hr. When support and reassurance are ineffective and the patient is severely agitated, diazepam, 10 to 30 mg PO, can be helpful, but in most cases the use of psychotropic medication is not necessary.

2. Toxic Delirium

Some hallucinogen users experience an acute toxic confusional state characterized by hallucinations, delusions, agitation, disorientation, and paranoia. In this state, assaults and inadvertent suicide attempts may occur (e.g., if the patient believes he can fly or stop a speeding train). No prior history of mental disorder, visual rather than auditory hallucinations, and the presence of some degree of observing ego suggest a diagnosis of toxic drug reaction rather than schizophrenia or mania.

As in the case of any toxic delirium, the primary aim is to support the patient through the period of drug effect. In a majority of users, symptoms will subside gradually as the drug is metabolized. Diazepam may be effective in the treatment of associated anxiety but can also cloud the clinical picture and should be prescribed only when support and reassurance are ineffective. Neuroleptics must be given with caution, especially if atropine-type drugs may have been ingested. At times, physical restraint also may be necessary to prevent the patient from harming himself or others. Untreated patients usually become mentally clear within 24 hr (as the drug is metabolized), with no obvious sequelae.

3. Drug-Precipitated Functional Psychosis

Another untoward effect of hallucinogen use is psychosis that fails to clear after the drug is metabolized. This occurs primarily in patients with underlying affective disorder or schizophrenia, some of whom may use these drugs in the hope of achieving psychic reintegration or an anxiety-free state. Such patients often become more disorganized by the drug experience. Treatment may include judicious use of antipsychotic drugs and, when indicated, lithium and/or antidepressants. Following resolution of the acute psychotic episode, attention should be directed toward understanding the precipitants of drug use.

4. Posthallucinogen Perception Disorder ("Flashbacks")

A fourth adverse consequence of hallucinogen use is the spontaneous recurrence of drug effects a few days to several months after the last drug experience. These "flashbacks," lasting a few seconds to several hours, can be precipitated by internal or external stress, cannabis use, emergence into a dark environment, and a number of other stimuli. The symptoms can sometimes be brought on by intention. The most frequent flashbacks are visual hallucinations, which include geometric pseudohallucinations, perceptions in the peripheral field, flashes of color, trailing of color, positive afterimages, halos around objects, macropsia, and micropsia. Other symptoms include feelings of depersonalization, paresthesias, auditory trailing, and emotional lability. Approximately half of the people with this disorder experience remission within months; however, 50% may experience such symptoms for 5 years. Some patients respond to flashbacks with anxiety, depression, and paranoia; others enjoy them. Treatment consists of firm reassurance that the flashback will pass. In some cases, benzodiazepines may be helpful.

E. Sequelae of Chronic Hallucinogen Use

A community study conducted between 1981 and 1983 determined that approximately 0.3% of the adult population had abused hallucinogens at some time in their lives. Today the vast majority of hallucinogen users are experimenters who take the drug several times out of curiosity or to intensify a particular experience. Use is episodic because of the perceptual and cognitive impairments. Regular users are more apt to exhibit serious psychopathology and abuse other psychoactive drugs as well.

Medical sequelae of chronic hallucinogen use are not well defined. Reports of LSD-induced chromosome damage in humans have aroused concern, but the data are equivocal. In very high doses, LSD does produce an increased number of stillbirths and deformities in pregnant laboratory animals; however, the relevance of these findings in humans is unclear. Controlled studies have found no significant increase in the incidence of birth defects among LSD users relative to nonusers. Nevertheless, it is wise to advise pregnant women to avoid the use of this and other drugs, especially during the first trimester.

VII. PHENCYCLIDINE ABUSE

A. Introduction

Phencyclidine (PCP, "angel dust," "hog," "T," "crystal joints," "rocket fuel"), an arylcyclohexylamine, was originally developed as a human anesthetic, but its use in humans was discontinued because it induced psychotic and hallucinatory reactions. It is now legally used only in veterinary medicine as a tranquilizer. Since 1981, PCP abuse has become increasingly prevalent, especially among those over age 20. The number of people reporting known PCP use increased from 7 million in 1979 to 8.3 million in 1982, but more people have taken the drug unknowingly in the form of adulterated marijuana, heroin, or LSD. Phencyclidine-related deaths and emergencies have increased. Street samples of the drug have been shown to contain nearly 30 analogues of PCP, many of which are more potent than PCP itself. Because of its behavioral toxicity, the drug poses a significant health problem, with deaths resulting from drug-induced violence, suicide, and accidents.

B. Pharmacology

Although it is classified as a dissociative anesthetic, PCP has stimulant, depressant, hallucinogenic, or analgesic effects, depending on the dose and route of administration. Phencyclidine is usually administered by dissolving the drug onto marijuana or tobacco and then smoking the mixture. It can also be taken orally, intranasally, or intravenously. The drug is well absorbed via each of these routes, although absorption after oral ingestion may be slow. Phencyclidine is metabolized in the liver and stored in fatty tissues.

Behavioral tolerance to PCP occurs with chronic use, but physical dependence has not been reported. Although the serum half-life is generally about 45 min at low doses, the half-life may be as long as 3 days after an overdose; this is a result of sequestration of PCP in brain and adipose tissue. Phencyclidine may be detected in urine and blood using a gas assay technique. There is some correlation between PCP levels in urine and blood and the degree of intoxication. Blood levels peak 1–4 hr after inhalation, but PCP may be detected in urine for up to a week following high-dose use. Since PCP is a weak base with a pK_a of about 8.5, the drug is increasingly soluble in more acidic aqueous solutions. Thus, acidification of the urine will enhance PCP excretion (see Section VII.D).

C. Acute Intoxication

In low doses (5 mg or less), PCP exerts a depressant effect on the central nervous system, producing a state resembling alcohol intoxication with muscular incoordination, generalized numbness of the extremities, a "blank stare" appearance, and mild ptosis of the eyelids. Ocular findings include vertical or horizontal nystagmus, miotic or normal-sized reactive pupils, and absent corneal reflexes. Impaired perception, mild analgesia, and various forms of motor disturbances are also seen. Higher doses (5 to 10 mg) produce nystagmus, slurred speech, ataxia, hyperreflexia, increased muscle tone, and catalepsy. Severe overdose (greater than 20 mg) may result in hypertensive crisis, muscular rigidity, seizure, respiratory depression, coma, and death.

The diagnosis of PCP intoxication may be difficult because of the wide spectrum of clinical findings that occur with this drug. Phencyclidine intoxication may be mistaken for delirium tremens, acute psychiatric illness (schizophrenia or other psychotic illness), sedative–hypnotic intoxication or withdrawal, or amphetamine intoxication.

In adult patients, there are nine common patterns of intoxication, which may be grouped into major and minor patterns depending on symptom severity and need for hospitalization. The major patterns have an unpredictable course with waxing and waning symptoms and include coma, catatonic syndrome, toxic psychosis, and acute brain syndrome. Minor patterns are associated with mild toxicity and are generally quite brief.

Patients with PCP-induced coma are unresponsive to stimuli and may have autonomic instability. Coma may last from 2 to 24 hr and may be complicated by malignant hyperthermia, status epilepticus, aspiration pneumonia, and rhabdomyolysis. The catatonic syndrome is associated with CNS excitement alternating with stupor and catalepsy. These patients may have hallucinations, delusions, stereotypy, dystonia, and agitation. Toxic psychosis may be diagnosed if a patient is not catatonic but has hallucinations, delusions, paranoia, or other psychiatric manifestations. Acute brain syndrome is characterized by disorientation, lack of judgment, loss of recent memory, and inappropriate affect in the absence of catatonia or other psychiatric symptoms. This may be the most common pattern of intoxication.

Patients with the minor patterns of intoxication are alert and oriented but may have any of the typical motor or autonomic signs, seizures, hypertension and/or nystagmus. Motor signs may include localized dystonia, rigidity, athetosis, tremor, or seizures. Autonomic signs include diaphoresis, hypersalivation, bronchorrhea, and urinary retention. Pupils may be of any size. The most common disturbances of vital signs are tachycardia and hyperthermia. Hypertension is common, but severe hypotension can occur, as can tachypnea or severe respiratory depression or arrest. The subjective effects of PCP include changes in body image, feelings of dissociation, perceptual distortions, auditory and visual hallucinations, and feelings of "nothingness." The behavioral state resembles the effects of sensory deprivation. Amnesia for the period of intoxication (usually 4 to 6 hr) is frequent.

Many PCP users experience some mood elevation, but feelings of anxiety are common as well. Indeed, most regular users report unwanted effects, most commonly perceptual disturbances, restlessness, disorientation, and anxiety, during each period of intoxication, whereas desired experiences (e.g., increased sensitivity to external stimuli, dissociation, stimulation, and elevated mood) occur only 60% of the time. This has led clinicians and researchers to wonder why so many people continue to take a drug that they find largely unpleasant. Some clinicians have proposed that the drug has reinforcing properties; some users enjoy the risks of a PCP experience and the fact that phencyclidine, if nothing else, alters consciousness profoundly. For many polydrug abusers (which most PCP users are), the degree of alteration in affective state is far more important than the type of alteration.

Once the diagnosis is made, treatment efforts should be directed toward calming the patient, ensuring the patient's and the staff's safety, and treating any medical complications that arise. The patient should be placed in isolation (i.e., a "quiet room"), which will decrease external stimulation; restraints may be necessary if the patient is combative or self-destructive. Diazepam (10–20 mg PO or 2.5 mg IV) may be useful in decreasing agitation and muscle hypertonicity. Since the patient may also have taken CNS depressants, diazepam should be used cautiously. To increase PCP excretion, the patient may be given cranberry juice (which contains benzoic acid) or oral ascorbic acid (0.5 to 1.5 g). Urine should be examined for myoglobinuria, and serum chemistries such as uric acid and creatinine kinase (CK) should be sent to rule out rhabdomyolysis. Early signs of rhabdomyolysis are elevations in serum uric acid and CK. Nasogastric suctioning should be reserved for serious cases in which recent oral ingestion is highly suspected, since PCP use may predispose the patient to laryngospasm. Because of the risk of seizures, ipecac should not be used. Hypertensive crisis may be treated with diazoxide (Hyperstat® and others), and status epilepticus, should it occur, is best treated with intravenous diazepam. Metabolic acidosis is treated by administration of sodium bicarbonate. To combat hypoglycemia, 50 ml of 50% dextrose solution IV may be given. Most patients improve within several hours after PCP ingestion. However, the patient should be observed until his sensorium has cleared, his vital signs are stable, and he is no longer combative or agitated.

D. Phencyclidine Overdose

Phencyclidine overdose is differentiated from simple intoxication primarily by the level of consciousness of the patient; the patient who has overdosed will be stuporous or comatose. In severe overdose, coma may last from several hours to 10 days. Hypertension is frequently present and may be severe. Respirations may be decreased, and seizures (sometimes leading to status epilepticus) are frequently seen. Horizontal and vertical nystagmus are always present, and muscle tone is increased.

Treatment of PCP overdose usually follows the procedures described in Table 6.

E. Phencyclidine Psychosis

In some vulnerable individuals, PCP can produce psychotic symptoms that may persist for days or weeks after the last dose. Although this reaction occurs primarily in novice users, it may also occur in chronic users taking moderate to high doses regularly. It is also more likely to occur in patients with a history of prior schizophrenic or manic episodes, even those who are being maintained on antipsychotic drugs.

Generally, the course of PCP psychosis may be divided into three phases, each lasting about 5 days. The initial acute phase of the psychosis, typically the most severe, is often characterized by paranoid delusions, hyperactivity, anorexia, insomnia, agitation or catalepsy, and unpredictable assaultiveness. During this period, patients are extremely sensitive to external stimuli. In the mixed phase of the psychosis, paranoia and restlessness remain, but the patient is usually calmer and intermittently in control of his behavior. In the absence of further drug use, the next 1 or 2 weeks are usually characterized by gradual reintegration and resolution of symptoms. However, in some patients, psychotic symptoms may persist for months. Depression is also a common sequela of PCP psychosis.

The treatment of PCP psychosis is aimed at maintaining the physical safety of the patient and others, decreasing psychotic thinking with the use of antipsychotic

---

TABLE 6. Treatment of PCP Overdose

1. Maintain adequate oxygenation
2. Perform gastric lavage if the drug has been ingested in the previous 4 hr
3. Treat hypertensive crisis with diazoxide or hydralazine
4. Use intravenous diazepam for seizure
5. Acidify the urine in order to facilitate drug excretion and hasten recovery from psychosis or coma
   a. For conscious patients, administer ammonium chloride via nasogastric tube in a dose of 2.75 mEq/kg in 60 cc of saline solution every 6 hr along with intravenous ascorbic acid (2 mg/500 cc IV fluid every 6 hr) until the urine pH drops below 5.5.
   b. For comatose patients, ammonium chloride should be given IV, 2.75 mEq/ kg, as a 1–2% solution in saline. When the urinary pH has reached 5, furosemide, 20–40 mg IV, should be given to promote further drug excretion. Repeat if necessary.
   Renal function should be monitored, since high-dose intoxication may decrease urine output

drugs, and promoting rapid excretion of the drug through acidification of the urine. If drug ingestion has been relatively recent, gastric lavage may also be helpful. Inquiry may be prevented by prompt hospitalization and the use of a "quiet room" to reduce external sensory input. Unlike hallucinogen users on "bad trips," patients intoxicated with PCP do not respond well to "talking down." Use of antipsychotics in conjunction with procedures designed to lower tissue levels of the drug (see Section VII.D) may hasten recovery. However, antipsychotic drugs with significant anticholinergic effects (e.g., chlorpromazine) should be avoided, since they may potentiate the anticholinergic effects of phencyclidine. Haloperidol, 5 mg IM given hourly as needed, may be useful, although response to treatment is often slow. Most patients are substantially better after 2 to 3 weeks of treatment—sooner if the drug can be rapidly eliminated from the body.

F. Sequelae of Chronic Phencyclidine Use

Phencyclidine is a readily available, inexpensive drug with very powerful behavioral effects and is, therefore, difficult to give up. Chronic users rarely seek treatment, although they may develop anxiety, social isolation, severe episodes of depression, and occasional violent outbursts. Some patients exhibit schizophrenia-like symptoms coupled with aggressive behavior, whereas others develop memory and speech impairment. Most PCP users are also active users of other psychoactive drugs, especially marijuana, alcohol, and hallucinogens. Residential drug treatment programs have had some success in the treatment of chronic users, but for many the prognosis is grim.

G. Neonatal Effects of Maternal PCP Use

Infant neurobehavioral effects of *in utero* exposure to PCP have been identi-fied. Clinical characteristics of newborns are similar to those born following *in utero* exposure to heroin or methadone and include tremor, irritability, hypertonia, rest-lessness, and agitation. After 6 months, infants become less jittery and sleep for longer periods of time. In toddlers between the ages of 1 and 2 years, increased activity and impaired attention have been observed. Some studies report that in spite of consistently nurturing environments, infants with a history of *in utero* PCP exposure have borderline abilities in fine motor, adaptive, language, and personal–social development.

VIII. INHALANT ABUSE

A. Introduction

The inhalants comprise a heterogeneous group of volatile organic solvents that have profound toxic effects on the CNS. Some of the commonly inhaled substances and their major active ingredients include glues and paint thinners (toluene), lighter fluid (naphtha), aerosols (fluorinated hydrocarbons, nitrous oxide), cleaning solutions (trichloroethylene, carbon tetrachloride), and gasoline (benzene).

B. Patterns of Abuse

Adults at risk for inhalant abuse include those whose work brings them into contact with these substances, such as shoemakers, cabinetmakers, printers, hair stylists, maintenance workers, painters, and workers in gas stations and dry cleaners. Children and adolescents are also at risk. Young users tend to be 7 to 19 years old, but use has been reported in children as young as 4 to 6 years.

Inhalants tend to be popular drugs of abuse among boys in their early teens, especially in poor and rural areas, where more expensive drugs of abuse are less available. Although the prevalence of inhalant abuse decreases with age, many users go on to abuse other drugs. Most commonly, the volatile gases are inhaled from a handkerchief or rag that has been soaked with the solvent material. Glue squirted into a paper bag gives off a vapor that can then be inhaled. Aerosol spray cans are a source of volatile hydrocarbons and nitrous oxide.

C. Acute Intoxication

High lipid solubility and rapid passage across the blood–brain barrier account for the rapidly intoxicating effects of the inhalants. Drug effects last up to 45 min, depending on the degree of exposure. Central nervous system impairment usually resolves in minutes to hours.

The clinical picture resembles that of alcohol intoxication and includes euphoria, giddiness, and lightheadedness. Irritation of the nasal mucosa, conjunctivitis, and unusual breath odor also suggest the diagnosis. Depending on the solvent used, a variety of toxic effects also may occur. These include induction of cardiac arrhythmias, hypoxia, seizures, and death. Inhalants, especially chlorohydrocarbons, paint and glue, have also produced glomerulonephritis and distal renal acidosis.

Tolerance to the effects of inhalants develops rather quickly, but no cross-tolerance occurs among the various solvents. Although physical dependence has not been documented, a syndrome resembling delirium tremens has been noted in some chronic users who stop using these agents abruptly. Other withdrawal symptoms may include tremulousness, tachycardia, disorientation, hallucinations, seizures, and agitation if inhalant use is abruptly discontinued after chronic use. Abdominal pain, paresthesias, and headaches also can occur during drug-free periods. Strong psychological dependence often develops in inhalant abusers and accounts for much of the difficulty in treating these individuals.

Clinical assessment should include careful individual evaluation of adult and child inhalant abusers and family assessment if the patient is a child or adolescent.

Patients should be assessed for damage to the CNS, liver, kidney, lung, nasal mucosa, heart, and bone marrow. There are no routine biological assays for most inhalants. However, tetraethyl lead gasoline may be detected by basophilic stippling of the red blood cells and increased lead levels in a 24-hr urine. Neuropsychological testing may help identify longer-term effects on attention span, memory, and concentration.

D. Sequelae of Chronic Inhalant Use

Although the exact relationship between inhalant abuse and brain damage is unclear, a number of studies have found a greater than expected prevalence of abnormal electroencephalograms, perceptual motor difficulties, and impaired memory in chronic users. Language skills and the ability to process information and think abstractly also may be affected. Peripheral neuropathies and bone marrow depression are the potential consequences of long-term use. Finally, there is some evidence that as a group, inhalant abusers exhibit more self-destructiveness and antisocial behavior than other types of drug abusers.

E. Treatment Approaches

Chronic solvent abusers are notoriously difficult to treat because of the combined effects of low socioeconomic status, family problems, school failure, antisocial behavior, and brain damage. Therapeutic communities emphasizing work programs, remedial education, and other reality-based therapies may offer the best chance of success with these patients. Behavior modification programs might be useful in treating chronic relapsing behavior.

IX. "DESIGNER" DRUGS: MDMA AND MDEA

MDMA, 3,4-methylenedioxymethamphetamine, also known as "Ecstasy," is a synthetic analogue of MDA, 3,4-methylenedioxyamphetamine, which was developed in 1914 as an appetite suppressant. MDMA came into use in the 1970s as an adjunct to psychotherapy, and in 1983 it became popular as a recreational drug also known as "XTC," "Adam," and "MDM." MDMA causes euphoria, enhanced sociability, and a sense of increased personal insight and empathy. It is not associated with hallucinogenesis. It is ingested as a gelatinous capsule or loose powder and has three phases of action. It causes an initial phase of disorientation. This is followed by a "rush," which may be accompanied by spasmodic jerking of the muscles, and finally a feeling of happy sociability. The effects wear off in 4–6 hr. Some users have reported confusion, anxiety, and depression.

The Federal Drug Enforcement Agency became concerned with the potential for MDMA abuse and possible dangerous side effects based on ill effects of its analogue MDA such as selective damage to serotonin nerve terminals. In 1985, MDMA became a Schedule 1 drug, and there was an increase in its popularity and availability on the black market. The chemically similar MDEA, 3,4-methylenedioxyamphetamine or "Eve," became available as a nonscheduled drug.

The addiction liability, abuse potential, and adverse effects of MDMA and MDEA are controversial. MDMA has been shown to have sympathomimetic effects that have been associated with cardiac arthythmias and deaths, especially in those with underlying cardiovascular disease. Cardiac arrhythmia associated with MDMA has been associated with at least one episode of sudden death in an otherwise healthy young person. As of 1986, reports of MDMA-associated adverse reactions, treatment admissions, or police reports were limited. However, increased street demand has led to anticipation of increasing incidence of adverse reactions associated with adulterated MDMA.

X. MARIJUANA USE AND ABUSE

A. Introduction

Marijuana, a plant used for recreational and medicinal purposes for centuries, is still widely popular today. The term "marijuana" refers to the dried leaves and flowers of the Indian hemp plant *Cannabis sativa,* which grows freely in warm climates and has psychoactive effects when smoked or ingested.

Recent surveys suggest that by 1985, the lifetime prevalence of marijuana use had increased to 62 million. Twenty-three percent of 12- to 17-year-olds and 60% of 18- to 25-year-olds have used marijuana at least once. This represents an increased prevalence of marijuana use in the United States. In addition, the percentage content of the major psychoactive substance in marijuana, Δ^9-tetrahydrocannabinol (THC), has increased from 1–3% in 1970 to 5–15% since 1980. This represents a THC content that is more likely to promote tolerance, dependence, and an opioidlike withdrawal syndrome when abruptly discontinued. The increased potency of marijuana in the last decade has helped make cannabis dependence increasingly common in the United States.

B. Pharmacology

Although marijuana has both stimulant and sedative properties and will produce hallucinations when taken in high enough doses, pharmacologically it appears to be in a class by itself. The plant is extremely complex, containing over 400 identifiable compounds. The principal psychoactive ingredient, however, appears to be THC, which is found in highest concentration in the small upper leaves and flowering tops of the plant. Hashish, a dried, concentrated resinous exudate of the flowers, is several times more potent.

During smoking, the degree of absorption varies, but the average experienced user will absorb approximately half the total dose into his bloodstream. Drug effects are noticeable almost immediately and reach peak intensity within 30 min. Speed of onset is partly determined by the concentration of THC in the preparation. After an hour, plasma levels begin to decline, and most of the subjective effects disappear by 3 hr after the last dose. Oral administration of marijuana is generally 20 to 30% as effective as smoking in delivering THC to the bloodstream. Onset of action also is slower (30 to 60 min after ingestion), but the subjective effects persist for a longer period of time (3 to 5 hr). Predictably, the rate of absorption is influenced by the food content of the stomach.

Following administration by either route, THC leaves the blood rapidly as a result of both hepatic metabolism and efficient uptake by body tissues. It is stored in fat depots, where it may remain for 2 to 3 weeks. Some also remains bound to plasma protein. The drug is metabolized to 11-hydroxy-THC, which is excreted via the gastrointestinal tract and to a lesser extent by the kidneys. Like the parent compound, THC metabolites also bind tightly to plasma proteins and can be sequestered in fat depots for long periods of time, with subsequent rerelease into the bloodstream. As a result, THC can be measured in plasma for up to 6 days after a single episode of marijuana use; THC metabolites can be detected in urine up to a month after a single dose. Evidence of the slow rate of elimination of THC has fostered speculation that repeated use of marijuana may lead to accumulation of THC or its metabolites in body tissues. The experimental data at this point are unclear.

C. Tolerance and Dependence

Daily use of marijuana may result in the development of tolerance. Abrupt discontinuation of marijuana after chronic use may result in withdrawal symptoms, which include insomnia, nausea, vomiting, myalgia, anxiety, restlessness, irritability, chills, anorexia, photophobia, sweating, cannabis craving, depression, yawning, anergy, mental confusion, and diarrhea.

D. Patterns of Use and Abuse

Among the 62 million Americans who have tried marijuana once, there are 5 million youth (24%), 20 million young adults (60%), and 37 million older adults (27%). Among young people, 15 million use the drug monthly, 9 million use it weekly, and 6 million use it daily.

Two types of marijuana dependence have been identified. In one type, individuals will self-administer marijuana every 2–4 hr all day. These individuals frequently present themselves for treatment complaining that their daily dosage has escalated and that they notice a number of withdrawal symptoms on abrupt cessation of the drug. They may or may not notice related memory, motivation, or work impairment. A second type of dependence is most frequently seen in individuals whose marijuana use has been detected by routine employee drug screening. These individuals have a consistent pattern of using marijuana every 24 to 36 hr for several years. They frequently do not report any withdrawal symptoms, but relapse following treatment is common.

Chronic daily use of marijuana is frequently associated with a diminution or loss of the pleasurable effects of the substance, and there may be a corresponding increase in dysphoric effects. Other symptoms associated with chronic use and dependence are lethargy, anhedonia, and attentional and memory problems. Long-term daily use of marijuana has been directly related to subsequent abuse of other drugs, especially alcohol and cocaine.

E. Acute Intoxication

1. Subjective Effects

As with other drugs, the clinical effects following a dose of marijuana depend on the strength of the preparation, route of administration, individual variables (e.g., metabolic rate, prior drug experience, and personal expectations), and the setting in which the drug is consumed.

Shortly after inhalation and absorption through the lungs, users experience a sense of well-being or euphoria, accompanied by feelings of friendliness and relaxation. Intoxicated persons also develop an altered time sense or "temporal disintegration," a state in which the past, present, and future become fused. Although awareness of their environment may be heightened, they may have less ability to communicate: their speech is often disconnected and tangential, and some smokers become remote and withdrawn. Thought processes are slowed, short-term memory is often impaired, and users have difficulty concentrating. Some feel that they have achieved special insights; others find mundane events more humorous or more poignant, and there is accompanying emotional lability.

2. Physiological Changes

Two reliable signs of marijuana intoxication are increased heart rate and conjunctival injection. The rise in heart rate is directly related to the dose of THC. Decreases in salivation, intraocular pressure, and skin temperature also are found, and at high doses, orthostatic hypotension can occur. Bronchodilation occurs acutely, but chronic marijuana smoking often results in obstructive pulmonary disease similar to that seen in tobacco smokers. Marijuana only minimally affects the waking surface EEG; the changes are similar to the drowsy state. Sleep EEG recordings show a significant loss of rapid eye movement (REM) sleep and increases in both stage 4 and total sleep time.

3. Cognitive and Psychomotor Effects

Various authors have studied the effect of marijuana on intellectual and psychomotor performance. Generally, naive and casual users given the drug in a laboratory setting demonstrate deterioration in both areas, probably because of memory disruption. Impairment is dose related and depends on the complexity of the task. More experienced users, however, are able to compensate for the acute intoxicating effects of THC. In most users, driving performance is impaired, with frequent misjudgment of speed and longer time required for braking.

F. Adverse Reactions following Marijuana Use

1. Acute Panic

Considering the extent of usage, the incidence of adverse reactions to marijuana use is quite low. The most common are summarized in Table 7. Panic can occur in inexperienced users; the likelihood is related to the dose of THC, the expectations of the user, and the setting in which the drug is taken. The tachycardia, disconnected thoughts, and paranoid ideation that often accompany marijuana use may precipitate panic attacks. A panic state may simulate acute psychosis, but careful examination reveals that the patient is not disoriented or hallucinating, and his

TABLE 7. Adverse Reactions to Marijuana

Type	Predisposing factors	Symptoms	Treatment
Acute panic	Inexperienced users, hysterical or obsessional characters, oral administration	Anxiety, depression, no psychotic symptoms	Reassurance; occasionally anxiolytics; episode usually short-lived
Toxic delirium	Large dose, oral use	Confusion, disorientation, hallucinations, depersonalization, delusions	Most remit in 12–48 hr; antipsychotics if necessary
Flashbacks	Days or weeks after last dose, prior history of hallucinogen use	Like hallucinogenic experience except brief	Reassurance, anxiolytics if necessary
Chronic psychosis	Prolonged heavy use of very potent marijuana or hashish; rare in U.S.	Paranoia, delusions, hallucinations, panic, bizarre behavior, occasionally violence	Antipsychotics
Amotivational syndrome	Prolonged heavy use; existence of syndrome is controversial	Apathy, decreased attention span, poor judgment, poor interpersonal relationships	Long-term abstinence

ability to test reality is intact; he is aware that his condition is drug related. In some patients, depression may be more prominent than anxiety.

The treatment of panic centers around firm reassurance in a nonthreatening environment. The patient is told that his symptoms have been caused by a strong dose of marijuana and that he will recover within several hours. Physical restraints, seclusion, or administration of antipsychotic drugs can exacerbate the problem. As drug effects subside, panic wanes, and the individual, although shaken, regains control. The persistence of THC metabolites may cause some patients to feel intermittently intoxicated for several days.

2. Toxic Delirium

Following large doses of marijuana, taken either orally or by inhalation, users may experience psychic disorganization accompanied by feelings of depersonalization and changes in perception and body image. Some patients are disoriented, with marked memory impairment. Confusion, derealization, paranoia, visual and auditory hallucinations, and dysphoria are also features of this toxic delirium. Individuals usually remain aware that these effects are drug related. Toxic delirium is more likely after cannabis is ingested, perhaps because of the user's inability to titrate adequately the dose of THC. In most cases, the process is self-limited and lasts from a few hours to a few days; it resolves as plasma THC levels decline. Generally, no pharmacological treatment is indicated, but patients should be carefully observed and prevented from doing harm to themselves or others.

3. Recurrent Reactions ("Flashbacks")

Although more characteristic of LSD, recurrent marijuana experiences occurring days or weeks after the last dose have been reported. Such phenomena are more common in individuals who have previously used hallucinogens. Flashbacks are rare in patients who used the drug more than 6 months previously. When these reactions occur long after the last drug use, a psychiatric or neurological disorder should be suspected, particularly if symptoms continue beyond several hours.

G. Sequelae of Chronic Use

1. Cannabis Psychosis

In Western countries psychosis directly attributable to chronic cannabis use is rare. In Middle Eastern and Asian countries, however, chronic use of more potent marijuana or hashish (the concentrated resin) reportedly may result in a cannabis-induced psychosis. Paranoia, persecutory delusions, visual and auditory hallucinations, panic, bizarre behavior, and occasional aggressive outbursts compose the clinical picture. The role of preexisting psychopathology in these patients is unclear.

Patients with this disorder usually respond quickly to treatment with antipscyhotic drugs, although relapse is common on resumption of cannabis use.

2. Amotivational Syndrome

A number of authors have described an "amotivational syndrome" that develops gradually in chronic heavy marijuana users. It is characterized by decreased drive and ambition, a shortened attention span, poor judgment, a high degree of distractability, impaired communication skills, and diminished effectiveness in interpersonal situations. A tendency toward introversion and magical thinking also has been observed. The individual frequently feels incapacitated and makes few or no plans beyond the present day. Personal habits deteriorate, and there are a progressive loss of insight and feelings of depersonalization. Aggression is profoundly diminished, and, as a result, these patients appear apathetic and withdrawn rather than antisocial. The contribution of predrug psychopathology in this syndrome remains unclear. Indeed, some authors doubt the existence of the amotivational syndrome, since studies of chronic marijuana smokers in Jamaica, Costa Rica, and Greece found no difference in work output as the extent of use increased.

3. Medical Problems

Chronic users of marijuana exhibit an increased frequency of adverse respiratory, cardiovascular, and neuroendocrine effects. They also report an increased frequency of bronchitis and various upper respiratory infections, which improve when drug use is discontinued. Marijuana smoke may impair the ability of alveolar macrophages to inactivate bacteria in the lung. Recent studies demonstrate that daily smoking for only 6 to 8 weeks causes significant airway obstruction and that marijuana smoke, regardless of THC content, results in a greater respiratory burden of carbon monoxide and tar than smoke inhalation from the same quantity of tobacco.

Tachycardia is consistently observed in marijuana smokers, and this may precipitate episodes of angina pectoris in patients with preexisting heart disease. Electrocardiographic studies demonstrate elevations of the S–T segments, flattening of T waves, and increased amplitude of P waves. High doses of marijuana may induce premature ventricular contractions.

Previous reports of gross structural changes in the central nervous system have not been confirmed by computerized tomography (CT). Changes in EEG electrical activity with reduction in alpha rhythm and decrease in beta rhythm activity have been documented, but the relationship to neurophysiological impairment is unknown.

Marijuana affects reproductive hormones in women differently at different phases of the menstrual cycle. Smoking a 1-g marijuana cigarette with 1.8% THC content results in 30% suppression of plasma luteinizing hormone levels during the luteal phase of the menstrual cycle and may have adverse effects on the reproductive function of women. Marijuana smoking suppresses prolactin levels in women and has been reported to decrease serum testosterone, sperm counts, and sperm motility in men.

H. Neonatal Effects of Maternal Marijuana Use

A number of adverse effects of marijuana use on fetal growth and development have been reported. Prenatal maternal marijuana use results in lower-birth-weight infants. This effect appears to be dose related. Neonates also appear to have a higher frequency of tremor and startle responses. Marijuana smoking during pregnancy may also result in higher frequency of congenital malformations and neonatal behavioral disorders.

I. Treatment of Chronic Marijuana Abuse

At present, methods for treating chronic marijuana users are not well developed. Although there is no recognized medical withdrawal regimen for marijuana dependence, discontinuation of the drug is an obvious prerequisite for treatment. Treatment of marijuana abuse may be multimodal and include a combination of therapies. Self-help groups such as Narcotics Anonymous are helpful in providing education and a supportive network that is available at all times and at no cost. Individual therapy may help the individual identify the motivation for drug use. Group therapy is effective in a task-oriented approach to communication and problem-solving skills. Urine monitoring is frequently an important part of the treatment process, as it may help to maintain abstinence. Behavioral approaches including relaxation techniques and contingency contracting are sometimes effective. Generally, treatment may be accomplished on an outpatient basis, but residential treatment should be considered for those individuals with medical or psychiatric illness or without social support systems. As always, a complete medical and psychiatric assessment should be completed to rule out preexisting concurrent affective illness or other psychiatric conditions that may need additional treatment.

XI. SUBSTANCE ABUSE IN PERSPECTIVE

The epidemic spread of substance abuse is a phenomenon best understood through an eclectic perspective. The behavior of individuals who abuse one or more psychoactive substances is partially the result of developmental, sociocultural, and interpersonal factors. In some patients, genetic and/or biological factors may influence vulnerability as well. Although some have suggested that particular personality traits predispose to the development of drug abuse and dependence, some traits (e.g., passivity, manipulativeness, inability to tolerate frustration) may result from repetitive drug use and the drug-using lifestyle. Similarly, although the development of insight and the resolution of intrapsychic conflicts are clearly desirable goals, no hard data suggest that insight, by itself, leads to a decline in drug-using behavior.

Recently, there has been growing interest in the behavioral aspects of substance abuse disorders. Clearly, in some patients phenomena like conditioned abstinence contribute to drug-seeking behavior and relapse. Consequently, patients and their doctors must learn to anticipate the impact of environmental stimuli that may enhance craving and lead to renewed drug use. At the same time, the clinician should attempt to reshape attitudes and behavior through psychotherapy, group pressure, role modeling, and other techniques of social learning. The relative success of self-help organizations like Alcoholics Anonymous is clear testimony to the usefulness of providing concrete guidelines for behavioral change in a supportive setting.

Finally, biological approaches to substance abuse, although appealingly cost effective, also have their pitfalls. In methadone maintenance programs, for example, some opioid addicts become alcoholics, and others spend their time diverting methadone to the street market. The narcotic antagonists, although pharmacologically effective, are not very popular with most opioid users. Disulfiram, an effective deterrent to impulsive drinking, is ineffective by itself in maintaining long-term abstinence.

In summary, substance abuse is an extremely complex and a multidetermined behavior, and the treatment of substance abusers presents a bewildering array of theoretical and practical problems. Those who seek to alter such behavior need to recognize this and devise equally complex and multidisciplinary approaches to treatment. As in other areas of medicine, the doctor's adherence to dogma can be hazardous to the patient's health.

SELECTED READING

General

1. American Psychiatric Association: *Diagnostic and Statistical Manual of Mental Disorders*, ed 3, revised. Washington, American Psychiatric Association, 1987.
2. Baldessarini R. J.: *Chemotherapy in Psychiatry.* Cambridge, Harvard University Press, 1985.
3. Braude M. C.: Perinatal effects of drugs of abuse: Symposium summary. *Fed Proc* 46:2446–2453, 1987.
4. Kozel N. J., Adams E. H.: Epidemiology of drug abuse: An overview. *Science* 234:970–974, 1986.
5. Meyer R. E. (ed): *Psychopathology and Additive Disorders.* New York, Guilford Press, 1986.
6. Mirin S. M. (ed): *Substance Abuse and Psychopathology.* Washington, American Psychiatric Press, 1984.
7. Mirin S. M., Weiss R. D., Michael J.: Psychopathology in substance abusers: Diagnosis and treatment. *Am J Drug Alcohol Abuse* 14:139–157, 1988.

Opioids

8. Bale R. N., Van Stone W. W., Kuldau J. M., et al: Therapeutic communities vs methadone maintenance. A prospective controlled study of narcotic addiction treatment: Design and one-year follow-up. *Arch Gen Psychiatry* 37:179–193, 1980.

9. Batki S. L., Sorensen J. L., Faltz B., et al: Psychiatric aspects of treatment of i.v. drug abusers with AIDS. *Hosp Community Psychiatry* 39:439–441, 1988.

10. Charney D. S., Riordan C. E., Kleber H. D.: Clonidine and naltrexone: A safe, effective and rapid treatment of abrupt withdrawal from methadone therapy. *Arch Gen Psychiatry* 39:1327–1332, 1982.

11. Crabtree B. L.: Review of naltrexone, a long-acting opiate antagonist. *Clin Pharm* 3:273–280, 1984.

12. Crowley T. J., Wagner J. E., Zerbe G.: Naltrexone-induced dysphoria in former opioid addicts. *Am J Psychiatry* 142:1081–1084, 1985.

13. Cushman P.: The major medical sequelae of opioid addiction. *Drug Alcohol Dep* 5:239–254, 1980.

14. Davis D. D., Templer D. I.: Neurobehavioral functioning in children exposed to narcotics *in utero*. *Addict Behav* 13:275–283, 1988.

15. Dole V. P.: Narcotic addiction, physical dependence and relapse. *N Engl J Med* 286:988–992, 1972.

16. Dole V. P.: Implications of methadone maintenance for theories of narcotic addiction. *JAMA* 260:3025–3029, 1988.

17. Dole V. P., Nyswander M. E.: Heroin addiction—a metabolic disease. *Arch Intern Med* 120:19–24, 1967.

18. Edelin K. C., Gurganious L., Golar K.: Methadone maintenance in pregnancy: consequences to care and outcome. *Obstet Gynecol* 71:399–404, 1988.

19. Faltz B. G.: Counseling substance abuse clients infected with human immunodeficiency virus. *J Psychoactive Drugs* 20:217–221, 1988.

20. Friedman A. S., Glickman N. W.: Residential program characteristics for completion of treatment by adolescent drug abusers. *J Nerv Ment Dis* 175:419–424, 1987.

21. Greenstein R. A., Arndt I. C., et al: Naltrexone: A clinical perspective. *J Clin Psychiatry* 45:25–28, 1984.

22. Gritz E. R., Shiffman S. M., Jarvik M. E., McLellan A. T.: Physiological and psychological effects of methadone in man. *Arch Gen Psychiatry* 32:237–242, 1975.

23. Haastrup S., Jepsen P. W.: Eleven year follow-up of 300 young opioid addicts. *Acta Psychiatr Scand* 77:22–26, 1988.

24. Jaffe J. H.: Drug addiction and drug abuse, in Goodman L. S., Gilman A., Gilman A. G. (eds): *Goodman and Gilman's The Pharmacological Basis of Therapeutics*, ed 7. New York, Macmillan, 1985, pp 532–581.

25. Jaffe J. H., Martin W. R.: Opioid analgesics and antagonists, in Goodman L. S., Gilman A., Gilman A. G. (eds): *Goodman and Gilman's The Pharmacological Basis of Therapeutics*, ed 7. New York, Macmillan, 1985, pp 491–531.

26. Jasinski D. R., Pevnick J. S., Griffith J. D.: Human pharmacology and abuse potential of the analgesic buprenorphine: A potential agent for treating narcotic addiction. *Arch Gen Psychiatry* 35:501–514, 1978.

27. Kleber H. D.: Naltrexone. *J Subst Abuse Treat* 2:117–122, 1985.

28. Kleber H. D.: Treatment of narcotic addicts. *Psychiatric Med* 3:389–418, 1987.

29. Kleber H. D., Kosten T. R.: Naltrexone induction: Psychologic and pharmacologic strategies. *J Clin Psychiatry* 45:29–38, 1984.

30. Kleber H. D., Riordan C. E., Rounsaville B.: Clonidine in outpatient detoxification from methadone maintenance. *Arch Gen Psychiatry* 42:391–394, 1985.

31. Kleber H. D., Slobetz F.: Outpatient drug-free treatment, in DuPont R. L., Goldstein A., O'Don-

nell J. (eds): *Handbook on Drug Abuse*. Washington, National Institute on Drug Abuse, 1979, pp 31–38.

32. Kosten T. R., Kleber H. D.: Buprenorphine detoxification from opioid dependence: A pilot study. *Life Sci* 42:635–641, 1988.

33. Kosten T. R., Kleber H. D.: Strategies to improve compliance with narcotic antagonists. *Am J Drug Alcohol Abuse* 11:249–266, 1984.

34. Kosten T. R., Rounsaville B. J.: Psychopathology in opioid addicts. *Psychiatr Clin North Am* 9:515–532, 1986.

35. Kreek M. J.: Medical safety and side effects of methadone in tolerant individuals. *JAMA* 223:665–668, 1973.

36. Martin W. R.: Naloxone. *Ann Intern Med* 85:765–768, 1976.

37. Mello N. K., Mendelson J. H.: Buprenorphine suppresses heroin use by heroin addicts. *Science* 207:657–659, 1980.

38. Meyer R. E., Mirin S. M.: *The Heroin Stimulus: Implications for a Theory of Addiction*. New York, Plenum Press, 1979.

39. Milby J. B.: Methadone maintenance to abstinence—how many make it? *J Nerv Ment Dis* 176:409–422, 1988.

40. Mirin S. M., Meyer R. E., McNamee H. B.: Psychopathology and mood during heroin use. *Arch Gen Psychiatry* 33:1503–1508, 1976.

41. Mirin S. M., Meyer R. E., Mendelson, J. H., et al: Opiate use and sexual function. *Am J Psychiatry* 137:909–915, 1980.

42. National Research Council Committee on Clinical Evaluation of Narcotic Antagonists: Clinical evaluation of naltrexone treatment of opiate-dependent individuals. *Arch Gen Psychiatry* 35:335–340, 1978.

43. Newman R. G.: Methadone treatment—defining and evaluating success. *N Engl J Med* 317:447–450, 1987.

44. O'Brien C. P., Testa T., O'Brien T. J., et al: Conditioned narcotic withdrawal in humans. *Science* 195:1000–1002, 1977.

45. Resnick R. B., Schuyten-Resnick E., Washton A. M.: Treatment of opioid dependence with narcotic antagonists: A review and commentary, in DuPont R. L., Goldstein A., O'Donnell J. (eds): *Handbook on Drug Abuse*. Washington, National Institute on Drug Abuse, 1979, pp 76–104.

46. Rounsaville B. J., Kosten T., Kleber H.: Success and failure at outpatient detoxification. Evaluating the process of clonidine and methadone-assisted withdrawal. *J Nerv Ment Dis* 173:103–110, 1985.

47. Rounsaville B. J., Weissman M. M., Crits-Cristoph K.: Diagnosis and symptoms of depression in opiate addicts. Course and relationship to treatment outcome. *Arch Gen Psychiatry* 39:151–156, 1982.

48. Schuster C. R.: Intravenous drug use and AIDS prevention. *Public Health Rep* 103:261–266, 1988.

49. Sells S. B.: Treatment effectiveness, in DuPont R. L., Goldstein A., O'Donnell J. (eds): *Handbook on Drug Abuse*. Washington, National Institute on Drug Abuse, 1979, pp 105–120.

50. Sells S. B., Simpson D. D.: The case for drug abuse treatment effectiveness, based on the DARP research program. *Br J Addict* 75:117–131, 1980.

51. Senay E. C.: Methadone maintenance treatment. *Int J Addict* 20:803–821, 1985.

52. Simpson D. D., Savage L. J.: Drug abuse treatment readmissions and outcomes: Three-year follow-up of DARP patients. *Arch Gen Psychiatry* 37:896–901, 1980.

53. Snyder S. H.: Opiate receptors in the brain. *N Engl J Med* 296:266–278, 1977.

54. Stimmel B., Adamsons K.: Narcotic dependency in pregnancy: Methadone maintenance compared to use of street drugs. *JAMA* 235:1121–1124, 1976.

55. US Department of Health, Education and Welfare, Public Health Service, Alcohol, Drug Abuse and Mental Health Administration: Clinical management during pregnancy, in: *Drug Dependence in Pregnancy: Clinical Management of Mother and Child*. National Institute on Drug Abuse, Services Research Monograph Series, 1979, Chapter 3.

56. US Department of Health, Education and Welfare, Alcohol, Drug Abuse and Mental Health Administration: Management of labor, delivery and immediate post-partum period, in: *Drug Dependence in Pregnancy: Clinical Management of Mother and Child.* National Institute on Drug Abuse, Services Research Monograph Series, 1979, Chapter 4.

57. Vaillant G. E.: A 20-year follow-up of New York narcotic addicts. *Arch Gen Psychiatry* 29:237–241, 1973.

58. Washton A. M., Resnick R. B., Geyer G.: Opiate withdrawal using lofexidine, a clonidine analogue with fewer side effects. *J Clin Psychiatry* 44:335–337, 1983.

59. Wikler A.: Dynamics of drug dependence. *Arch Gen Psychiatry* 28:611–616, 1973.

60. Zelson C., Lee S. J., Casalino M.: Neonatal narcotic addiction: Comparative effects of maternal intake of heroin and methadone. *N Engl J Med* 289:1216–1220, 1973.

Sedatives, Hypnotics, and Anxiolytics

61. Abuse of benzodiazepines: The problems and the solutions. A report of a Committee of the Institute for Behavior and Health, Inc. *Am J Drug Alcohol Abuse* 14(Suppl 1):1–16, 1988.

62. Braestrup C., Squires R. F.: Brain specific benzodiazepine receptors. *Br J Psychiatry* 133:249–260, 1978.

63. Cohen L. S., Rosenbaum J. F.: Clonazepam: New uses and potential problems. *J Clin Psychiatry* 48(Suppl 10):50–55, 1987.

64. Cronin R. J., Klingler E. L., Jr., Avashti P. S., et al: The treatment of nonbarbiturate sedative overdosage, in Browne, P. G. (ed): *A Treatment Manual for Acute Drug Abuse Emergencies.* Washington, National Institute on Drug Abuse, 1974, pp 58–62.

65. Dorpat T. L.: Drug automatism, barbiturate poisoning and suicide behavior. *Arch Gen Psychiatry* 31:216–220, 1974.

66. Dupont R. L. (ed): Abuse of benzodiazepines: The problems and the solutions. *Am J Drug Alcohol Abuse* 14(Suppl. 1):1–69, 1988.

67. Fleischhacker W. W., Barnas C., Hackenberg B.: Epidemiology of benzodiazepine dependence. *Acta Psychiatr Scand* 74(1):80–83, 1986.

68. Gelenberg A. J.: Benzodiazepine withdrawal. *Mass Gen Hosp Newslett Biol Ther Psychiatry Newslett* 3:9–12, 1980.

69. Herman J. B., Rosenbaum J. F., Brotman A. W.: The alprazolam to clonazepam switch for the treatment of panic disorder. *J Clin Psychopharmacol* 7(3):175–178, 1987.

70. Juergens S. M., Morse R. M.: Alprazolam dependence in seven patients. *Am J Psychiatry* 145(5):625–627, 1988.

71. Mellman T. A., Uhde T. W.: Withdrawal syndrome with gradual tapering of alprazolam. *Am J Psychiatry* 143(11):1464–1466, 1986.

72. Noyes R., Jr., Clancy J., Coryell W. H., et al: A withdrawal syndrome after abrupt discontinuation of alprazolam. *Am J Psychiatry* 142(1):114–116, 1985.

73. Noyes R., Jr., Perry P. J., Crowe R. R.: Seizures following the withdrawal of alprazolam. *J Nerv Ment Dis* 174(1):50–52, 1986.

74. Pevnick J. S., Jasinski D. R., Haertzen C. A.: Abrupt withdrawal from therapeutically administered diazepam. *Arch Gen Psychiatry* 35:995–998, 1978.

75. Piesiur-Strehlow B., Strelow U., Poser W.: Mortality of patients dependent on benzodiazepines. *Acta Psychiatr Scand* 73(3):330–335, 1986.

76. Preskorn S. H., Denner L. J.: Benzodiazepines and withdrawal psychosis. Report of three cases. *JAMA* 237(1):36–38, 1977.

77. Ray W. A., Blazer D. G., Schaffner W.: Reducing long-term diazepam prescribing in office practice. *JAMA* 256 (18):2536–2539, 1986.

78. Setter J. G.: Emergency treatment of acute barbiturate intoxication, in Browne P. G. (ed): *A*

Treatment Manual for Acute Drug Abuse Emergencies. Washington, National Institute on Drug Abuse, 1974, pp 49–53.

79. Smith D. E., Wesson D. R.: Phenobarbital technique for treatment of barbiturate dependence. *Arch Gen Psychiatry* 24:56–60, 1971.

80. Smith D. E., Wesson D. R., Seymour R. B.: The abuse of barbiturates and other sedative–hypnotics, in DuPont R. L., Goldstein A., O'Donnell J. (eds): *Handbook on Drug Abuse.* Washington, National Institute on Drug Abuse, 1979, pp 233–240.

81. Swartzburg M., Lieb J., Schwartz A. H.: Methaqualone withdrawal. *Arch Gen Psychiatry* 29:46–47, 1973.

82. Tennant F. S., Jr., Pumphrey E. A.: Benzodiazepine dependence of several years duration: clinical profile and therapeutic benefits. *Natl Inst Drug Abuse Res Monogr* 55:211–216, 1984.

Alcohol

83. Adams R. D., Victor M.: *Principles of Neurology.* New York, McGraw-Hill, 1985, pp 808–824.

84. Bean M.: Alcoholics Anonymous: Chapter I: Principles and methods. *Psychiatr Ann* 5:7–21, 1975.

85. Blume S. B.: Women and alcohol. A review. *JAMA* 256(11):1467–1470, 1986.

86. Brown C. G.: The alcohol withdrawal syndrome. *Ann Emerg Med* 11(5):276–280, 1982.

87. Ciraulo D. A., Barnhill J. G., Greenblatt D. J.: Abuse liability and clinical pharmacokinetics of alprazolam in alcoholic men. *J Clin Psychiatry* 49(9):333–337, 1988.

88. Cloninger C. R.: Neurogenetic adaptive mechanisms in alcoholism. *Science* 236:410–416, 1987.

89. Cohen S. (ed): The treatment of alcoholism: Does it work? *Drug Abuse Alcoholism Newslett* 7:1–3, 1978.

90. Coles C. D., Smith I., Fernhoff P. M.: Neonatal neurobehavioral characteristics as correlates of maternal alcohol use during gestation. *Alcoholism (NY)* 9(5):454–460, 1985.

91. Council on Scientific Affairs: Aversion Therapy. *JAMA* 258(18):2562–2566, 1987.

92. Edwards G., Orford J., Egert S., et al: Alcoholism: A controlled trial of "treatment" and "advice." *J Stud Alcohol* 38:1004–1031, 1977.

93. Emrick C. E.: A review of psychological oriented treatment of alcoholism: II. The relative effectiveness of different treatment approaches and the effectiveness of treatment versus no treatment. *J Stud Alcohol* 36:88–108, 1975.

94. Emrick C. E.: Alcoholics Anonymous: Affilitation processes and effectiveness as treatment. *Alcoholism (NY)* 11(5):416–423, 1987.

95. Erhart C. B., Wolf A. W., Ernhart C. B., Wolf A. W., Linn P. L.: Alcohol-related birth defects: Syndromal anomalies, intrauterine growth retardation, and neonatal behavioral assessment. *Alcoholism (NY)* 9(5):447–453, 1985.

96. Frances R. J.: Update on alcohol and drug disorder treatment. *J Clin Psychiatry* 49(Suppl):13–17, 1988.

97. Fuller R. K., Roth H. P.: Disulfiram for the treatment of alcoholism. *Ann Intern Med* 90:901–904, 1979.

98. Kwentus M. D., Major L. F.: Disulfiram in the treatment of alcoholism. *J Stud Alcohol* 40:428–446, 1979.

99. Linnoila M., Mefford I., Nutt D. W.: NIH conference. Alcohol withdrawal and noradrenergic function. *Ann Intern Med* 107(6):875–889, 1987.

100. Liskow B. I., Goodwin D. W.: Pharmacological treatment of alcohol intoxication, withdrawal and dependence. A critical review. *J Stud Alcohol* 48(4):356–370, 1987.

101. Liskow B., Mayfield D., Thiele J.: Alcohol and affective disorder: Assessment and treatment. *J Clin Psychiatry* 43(4):144–147, 1982.

102. Mendelson J. H., Babor T. F., Mello N. K., et al: Alcoholism and prevalence of medical and psychiatric disorders. *J Stud Alcohol* 47(5):361–366, 1986.

103. Mendelson J. H., Mello N. K.: Biologic concomitants of alcoholism. *N Engl J Med* 301:912–921, 1979.
104. Mendelson J. H., Mello N. K. (eds): *The Diagnosis and Treatment of Alcoholism*, 2nd ed. New York. McGraw-Hill.
105. Meyer R. E., Kranzier H. R.: Alcoholism: Clinical implications of recent research. *J Clin Psychiatry* 49(Suppl):8–12, 1988.
106. Newsom G., Murray N.: Reversal of dexamethasone suppression test nonsuppression in alcohol abusers. *Am J Psychiatry* 140(3):353–354, 1983.
107. Ng S. K., Hauser W. A., Brust J. C.: Alcohol consumption and withdrawal in new-onset seizures. *N Engl J Med* 319(11):666–673, 1988.
108. Ouelette E. M., Rosett H. L., Rosman N. P., et al: Adverse effects on offspring of maternal alcohol abuse during pregnancy. *N Engl J Med* 297:528–530, 1977.
109. Pomerleau O., Pertschuk M., Stunnett J.: A critical examination of some current assumptions in the treatment of alcoholism. *J Stud Alcohol* 37:849–867, 1976.
110. Schuckit M. A.: A clinical review of alcohol, alcoholism and the elderly patient. *J Clin Psychiatry* 43:396–399, 1982.
111. Sellers E. M., Naranjo C. A., Peachey J. E.: Drug therapy: Drugs to decrease alcohol consumption. *N Engl J Med* 305(21):1255–1262, 1981.
112. Stabenau J. R.: Implications of family history of alcoholism, antisocial personality, and sex differences in alcohol dependence. *Am J Psychiatry* 141(10):1178–1182, 1984.
113. Staiesey N. L., Fried P. A.: Relationships between moderate maternal and alcohol consumption during pregnancy and infant neurological development. *J Stud Alcohol* 44(2):262–270, 1983.
114. Streissguth A. P., Clarren S. K., Jones K. L.: Natural history of the fetal alcohol syndrome: A 10 year follow-up of eleven patients. *Lancet* 2(8446):85–91, 1985.
115. Streissguth A. P., Sampson P. D., Barr H. M.: Studying alcohol teratogenesis from the perspective of the fetal alcohol syndrome: Methodological and statistical issues. *Ann NY Acad Sci* 488:73–86, 1986.
116. Thompson W. L., Johnson A. D., Maddrey W. L., et al: Diazepam and paraldehyde for treatment of severe delirium tremens: A controlled trial. *Ann Intern Med* 82:175–180, 1975.
117. Vaillant G. E.: *The Natural History of Alcoholism: Causes, Patterns, and Paths of Recovery.* Cambridge, MA, Harvard University Press, 1985.
118. Vaillant G. E.: Natural history of male psychological health: VIII. Antecedents of alcoholism and "orality." *Am J Psychiatry* 137:181–186, 1980.
119. Warren K. R., Bast R. J.: Alcohol-related birth defects: An update. *Public Health Rep* 103(6):638–642, 1988.
120. Weissman M. M., Myers J. K.: Clinical depression in alcoholism. *Am J Psychiatry* 137:372–374, 1980.
121. Wilson G. T.: Chemical aversion conditioning as a treatment for alcoholism: A reanalysis. *Behav Res Ther* 25(6):503–516, 1987.
122. Wright C., Vafier J. A., Lake C. R.: Disulfiram-induced fulminating hepatitis: Guidelines for liver panel monitoring. *J Clin Psychiatry* 49:430–434, 1988.

Cocaine

123. Aigner T. G., Balster R. L.: Choice behavior in rhesus monkeys: Cocaine versus food. *Science* 201:534–535, 1978.
124. Byck R., Van Dyke C.: What are the effects of cocaine in man? in Peterson R. C., Stillman R. C. (eds): *Cocaine: 1977.* Washington, National Institute on Drug Abuse, Research Monograph 13, 1977, pp 5–16.
125. Chasnoff I., MacGregor S., Chisum G.: Cocaine use during pregnancy: Adverse perinatal outcome, in: *Problems of Drug Dependence, 1987, Proceedings of the 49th Annual Scientific Meeting,*

Committee on Problems of Drug Dependence, Inc. Washington, National Institute on Drug Abuse Research, Monograph 81, 1988, p 265.

126. Crack. *Med Lett* 28:69–70, 1986.

127. Cregler L. L., Mark H.: Special report: Medical complications of cocaine abuse. *N Engl J Med* 315:1495–1500, 1986.

128. Ellinwood E. H., Jr.: Amphetamine psychosis: Individuals, settings and sequences, in Ellinwood E. H., Cohen S. (eds): *Current Concepts on Amphetamine Abuse.* Washington, National Institute of Mental Health, 1972, pp 143–158.

129. Ellinwood E. H., Jr.: Amphetamines/anorectics, in DuPont R. L., Goldstein A., O'Donnell J. (eds): *Handbook on Drug Abuse.* Washington, National Institute on Drug Abuse, 1979, pp 221–231.

130. Finnegan L.: The dilemma of cocaine exposure in the perinatal period, in: *Problems of Drug Dependence, 1987, Proceedings of the 49th Annual Scientific Meeting, Committee on Problems of Drug Dependence, Inc.* Washington, National Institute on Drug Abuse, Research Monograph 81, 1988, p 379.

131. Fischman M. W., Schuster C. R., Resnekov L., et al: Cardiovascular and subjective effects of intravenous cocaine administration in humans. *Arch Gen Psychiatry* 33:983–989, 1976.

132. Gawin F. H., Ellinwood E. H.: Cocaine and other stimulants: Actions, abuse and treatment. *N Engl J Med* 318:1173–1182, 1988.

133. Gawin F. H., Kleber H. D.: Abstinence symptomatology and psychiatric diagnosis in cocaine abusers: Clinical observations. *Arch Gen Psychiatry* 43:107–113, 1986.

134. Gawin F. H., Kleber H. D.: Pharmacologic treatments of cocaine abuse. *Psychiatr Clin North Am* 9:573–583, 1986.

135. Gawin F. H., Kleber H. D., Byck R., et al: Desipramine facilitation of initial cocaine abstinence. *Arch Gen Psychiatry* 46:117–121, 1989.

136. Giannini A. J., Malone D. A., Giannini M. C.: Treatment of depression in chronic cocaine and phencyclidine abuse with desipramine. *J Clin Pharmacol* 26:211–214, 1986.

137. Grinspoon L., Hedblom P.: *The Speed Culture: Amphetamine Use and Abuse in America.* Cambridge, Harvard University Press, 1975.

138. Isner J. M., Estes M., Thompson P. D., et al: Acute cardiac events temporarily related to cocaine abuse. *N Engl J Med* 315:1438–1443, 1986.

139. Javaid J. I., Fischman M. W., Schuster C. R., et al: Cocaine plasma concentration: Relation to physiological and subjective effects in humans. *Science* 202:227–228, 1978.

140. Kramer J. C.: Introduction to amphetamine abuse, in Ellinwood E. H., Cohen S. (eds): *Current Concepts on Amphetamine Abuse.* Washington, National Institute of Mental Health, 1972, pp 177–184.

141. Kramer J. C., Fischman V. S., Littlefield D. C.: Amphetamine abuse: Pattern and effects of high doses taken intravenously. *JAMA* 201:89–93, 1967.

142. Mody C. K., Miller B. L., McIntyre H. B., et al: Neurologic complications of cocaine abuse. *Neurology* 38:1189–1193, 1988.

143. Pollack M. H., Brotman A. W., Rosenbaum J. F.: Cocaine abuse and treatment. *Compr Psychiatry* 30:31–44, 1989.

144. Post R. M.: Cocaine psychoses: A continuum model. *Am J Psychiatry* 132:225–231, 1975.

145. Post R. M., Kotin J., Goodwin F. K.: The effects of cocaine on depressed patients. *Am J Psychiatry* 131:511–517, 1974.

146. Resnick R. B., Kestenbaum R. S.: Acute systemic effects of cocaine in man: A controlled study by intranasal and intravenous routes. *Science* 195:696–698, 1977.

147. Rosenfeld W., Zabaleta I., Sahdev R., et al: Maternal use of cocaine, methadone, heroin and alcohol: Comparison of neonatal effects, in: *Problems of Drug Dependence, 1987, Proceedings of the 49th Annual Scientific Meeting, Committee on Problems of Drug Dependence, Inc.* Washington, National Institute on Drug Abuse, Research Monograph 81, 1988, p 264.

148. Roth D., Alarcon F. J., Fernandez J. A.: Acute rhabdomyolysis associated with cocaine intoxication. *N Engl J Med* 319:673–677, 1988.

149. Smith D. E., Wesson D. R., Buxton M. E., et al (eds): *Amphetamine Use, Misuse and Abuse: Proceedings of the National Conference.* Boston, G. K. Hall, 1979.
150. Washton A. M.: Nonpharmacologic treatment of cocaine abuse. *Psychiatr Clin North Am* 9:563–571, 1986.
151. Weiss R. D.: Relapse to cocaine abuse after initiating desipramine treatment. *JAMA* 260:2545–2546, 1988.
152. Weiss R. D., Gawin F. H.: Protracted elimination of cocaine metabolites in long term high-dose cocaine abusers. *Am J Med* 85:879–880, 1988.
153. Weiss R. D., Mirin S. M.: *Cocaine.* Washington, American Psychiatric Press, 1987.
154. Weiss R. D., Mirin S. M.: Subtypes of cocaine abusers. *Psychiatr Clin North Am* 9:491–501, 1986.
155. Weiss R. D., Mirin S. M., Griffin M. L., et al: Psychopathology in cocaine abusers: Changing trends. *J Nerv Ment Dis* 176:719–725, 1988.
156. Weiss R. D., Mirin S. M., Michael J. L., et al: Psychopathology in chronic cocaine abusers. *Am J Drug Alcohol Abuse* 12:17–29, 1986.
157. Woods, J. R., Plessinger M. A., Clark K. E.: Effect of cocaine on uterine blood flow and fetal oxygenation. *JAMA* 257:957–961, 1987.

Hallucinogens

158. Abraham H. D.: Visual phenomenology of the LSD flashback. *Arch Gen Psychiatry* 40:884–889, 1983.
159. Blacker K. H., Reese T. J., Stone G. C., et al: Chronic users of LSD: The "acidheads." *Am J Psychiatry* 125:341–351, 1968.
160. Bowers M. B., Jr.: Acute psychosis induced by psychotomimetic drug abuse. I. Clinical findings. *Arch Gen Psychiatry* 27:437–439, 1972.
161. Bowers M. B., Jr.: Acute psychosis induced by psychotomimetic drug abuse. II. Neurochemical findings. *Arch Gen Psychiatry* 27:440–442, 1972.
162. Bowers M. B., Jr., Chipman A., Schwartz A., et al: Dynamics of psychedelic drug abuse: A clinical study. *Arch Gen Psychiatry* 16:560–566, 1967.
163. Brawley P., Duffield J. C.: The pharmacology of hallucinogens. *Pharmacol Rev* 24:31–66, 1972.
164. Cohen S.: Flashbacks. *Drug Abuse Alcoholism Newslett* 6:1–3, 1977.
165. Faillace L. A., Snyder S. H., Weingartner H.: 2,5-Dimethoxy-1-methylamphetamine: Clinical evaluation of a new hallucinogenic drug. *J Nerv Ment Dis* 150:119–126, 1970.
166. Freedman D. X.: The use and abuse of LSD. *Arch Gen Psychiatry* 18:330–347, 1968.
167. Gilmour D. G., Bloom A. D., Lele K. P., et al: Chromosomal aberrations in users of psychoactive drugs. *Arch Gen Psychiatry* 24:268–272, 1971.
168. Glass G. S., Bowers M. B., Jr.: Chronic psychosis associated with long-term psychotomimetic drug abuse. *Arch Gen Psychiatry* 23:97–102, 1970.
169. Grinspoon L., Bakalar J. B.: *Psychedelic Drugs Reconsidered.* New York, Basic Books, 1979.
170. Jacobson C. B., Berlin C. M.: Possible reproductive detriment in LSD users. *JAMA* 222:1367–1373, 1972.
171. McGlothlin W. H., Arnold D. O.: LSD revisited. *Arch Gen Psychiatry* 24:35–49, 1971.
172. McGlothlin W. H., Arnold D. O., Freedman D. X.: Organicity measures following repeated LSD ingestion. *Arch Gen Psychiatry* 21:704–709, 1969.
173. Snyder S. H., Faillace L. A., Weingartner H.: DOM (STP), a new hallucinogenic drug, and DOET: Effects in normal subjects. *Am J Psychiatry* 125:113–120, 1968.
174. Snyder S. H., Weingartner H., Faillace L. A.: DOET (2,5-dimethoxy-4-ethylamphetamine), a new psychotropic drug: Effects of varying doses in man. *Arch Gen Psychiatry* 24:50–55, 1971.
175. Tucker G. J., Quinlan D., Harrow M.: Chronic hallucinogenic drug use and thought disturbance. *Arch Gen Psychiatry* 27:443–447, 1972.

Phencyclidine

176. Clouet D. H. (ed): *Phencyclidine: An Update.* Washington, National Institute on Drug Abuse, 1986.
177. Domino E. F.: Treatment of phencyclidine intoxication. *Psychopharmacol Bull* 16:83–85, 1980.
178. Fauman M. A., Fauman B. J.: The psychiatric aspects of chronic phencyclidine use: A study of chronic PCP users, in Peterson R. C., Stillman R. C. (eds): *PCP: Phencyclidine Abuse: An Appraisal.* Washington, National Institute on Drug Abuse, Research Monograph 21, DHEW Publ. No. 78-728, 1978, pp 183–200.
179. Gelenberg A. J.: Psychopharmacology update: Phencyclidine. *McLean Hosp J* 2:89–96, 1977.
180. Graeven D. B.: Patterns of phencyclidine use, in Peterson R. C., Stillman R. C. (eds): *PCP: Phencyclidine Abuse: An Appraisal.* Washington, National Institute on Drug Abuse, Research Monograph 21, DHEW Publ. No. 78-728, 1978, pp 176–182.
181. Luisada P. V.: The phencyclidine psychosis: Phenomenology and treatment, in Peterson R. C., Stillman R. C. (eds): *PCP: Phencyclidine Abuse: An Appraisal.* Washington, National Institute on Drug Abuse, Research Monograph 21, DHEW Publ. No. 78-728, 1978, pp 241–253.
182. Peterson R. C., Stillman R. C.: Phencyclidine: An overview, in Peterson R. C., Stillman R. C. (eds): *PCP: Phencyclidine Abuse: An Appraisal.* Washington, National Institute on Drug Abuse, Research Monograph 21, DHEW Publ. No. 78-728, 1978, pp 1–17.
183. Smith D. R., Wesson D. R., Buxton M. E., et al: The diagnosis and treatment of the PCP abuse syndrome, in Peterson R. C., Stillman R. C. (eds): *PCP: Phencyclidine Abuse: An Appraisal.* Washington, National Institute on Drug Abuse, Research Monograph 21, DHEW Publ. No. 78-728, 1978, pp 229–240.

Inhalants

184. Cohen S.: Inhalants, in DuPont R. L., Goldstein A., O'Donnell J. (eds): *Handbook on Drug Abuse,* Washington, National Institute on Drug Abuse, 1979, pp 213–220.
185. Glaser F. B.: Inhalation psychosis and related states, in Browne P. G. (ed): *A Treatment Manual for Acute Drug Abuse Emergencies.* Washington, National Institute on Drug Abuse, 1975, pp 95–104.
186. Lewis J. D., Moritz D., Mellis L. P.: Long-term toluene abuse. *Am J Psychiatry* 138:368–370, 1981.
187. Westermeyer J.: The psychiatrist and solvent–inhalant abuse: Recognition, assessment, and treatment. *Am J Psychiatry* 144:903–907, 1987.

Marijuana

188. Benedikt R. A., Cristofaro P., Mendelson J. H., et al: Effects of acute marijuana smoking in postmenopausal women. *Psychopharmacology* 90:14–17, 1986.
189. Bernstein J. G.: Marijuana—new potential, new problems. *Drug Ther* 10:38–48, 1980.
190. Chopra G. S., Smith J. W.: Psychotic reactions following cannabis use in East Indians. *Arch Gen Psychiatry* 30:24–27, 1974.
191. Cohen S.: Cannabis: Impact on motivation. Part I. *Drug Abuse Alcoholism Newslett* 9:1–3, 1980.
192. Cohen S.: Cannabis: Impact on motivation. Part II. *Drug Abuse Alcoholism Newslett* 10:1–3, 1981.
193. Harding T., Knight F.: Marijuana-modified mania. *Arch Gen Psychiatry* 29:635–637, 1973.
194. Jones R. T.: Marijuana-induced "high": Influence of expectation, setting and previous drug experience. *Pharmacol Rev* 23:359–369, 1971.

195. Kupfer D. J., Detre T., Koral J., et al: A comment on the "amotivational syndrome" in marijuana smokers. *Am J Psychiatry* 130:1319–1322, 1973.
196. Melges F. T., Tinklenberg J. R., Hollister L. E., et al: Temporal disintegration and depersonalization during marijuana intoxication. *Arch Gen Psychiatry* 23:204–210, 1970.
197. Mendelson J. H.: Marijuana, in Meltzer H. Y. (ed): *Psychopharmacology: The Third Generation of Progress*. New York, Raven Press, 1987, pp 1565–1571.
198. Mendelson J. H., Ellingboe J., Kuehnle J. C., et al: Effects of chronic marijuana use on integrated plasma testosterone and luteinizing hormone levels. *J Pharmacol Exp Ther* 207:611–617, 1978.
199. Mendelson J. H., Mello N. K., Ellingboe J., et al: Marijuana smoking suppresses luteinizing hormone in women. *J Pharmacol Exp Ther* 237:862–886, 1986.
200. Mendelson J. H., Rossi A. M., Meyer R. E. (eds): *The Use of Marijuana: A Psychological and Physiological Inquiry.* New York, Plenum Press, 1974.
201. Meyer R. E.: Psychiatric consequences of marijuana use: The state of the evidence, in Tinklenberg J. R. (ed): *Marijuana and Health Hazards. Methodological Issues in Current Research*. New York, Academic Press, 1975, pp 133–152.
202. Millman R. B., Sbriglio R.: Patterns of use and psychopathology in chronic marijuana users. *Psychiatr Clin North Am* 9:533–545, 1986.
203. Mirin S. M., Shapiro L. M., Meyer R. E., et al: Casual versus heavy use of marijuana: A redefinition of the marijuana problem. *Am J Psychiatry* 127:54–60, 1971.
204. Peterson R. C. (ed): *Marijuana Research Findings: 1980.* Washington, National Institute on Drug Abuse, Research Monograph 31, 1980.
205. Schnoll S. H., Daghestani A. N.: Treatment of marijuana abuse. *Psychiatr Ann* 16:249–257, 1986.
206. Tashkin D. P., Shapiro B. J., Lee Y. E.: Subacute effects of heavy marijuana smoking on pulmonary function in healthy men. *N Engl J Med* 294:125–129, 1976.
207. Tennant F. S.: The clinical syndrome of marijuana dependence. *Psychiatr Ann* 16:225–234, 1986.
208. Tennant F. S., Jr., Groesbeck C. J.: Psychiatric effects of hashish. *Arch Gen Psychiatry* 27:133–136, 1972.
209. Thacore V. R., Shukla S. R. P.: Cannabis psychosis and paranoid schizophrenia. *Arch Gen Psychiatry* 33:383–386, 1976.
210. Treffert D. A.: Marijuana use in schizophrenia: A clear hazard. *Am J Psychiatry* 135:1213–1220, 1978.
211. Weil A. T.: Adverse reactions to marijuana: Classification and suggested treatment. *N Engl J Med* 282:997–1000, 1970.
212. Wu T., Tashkin D. P., Djahed B., et al: Pulmonary hazards of smoking marijuana as compared with tobacco. *N Engl J Med* 318:347–351, 1988.
213. Zuckerman B., Frank D. A., Hingson R., et al: Effects of maternal marijuana and cocaine use on fetal growth. *N Engl J Med* 320:762–768, 1989.

"Designer" Drugs: MDMA and MDEA

214. Dowing G. P., McDonough E. T., Bost R. O.: 'Eve' and 'ecstasy': A report of five deaths associated with the use of MDEA and MDMA. *JAMA* 257:1615–1617, 1987.
215. Greer G., Strassman R. J.: Information on "ecstasy." *Am J Psychiatry* 142:1391, 1985.
216. Newmeyer J. A.: Some considerations on the prevalence of MDMA use. *J Psychoactive Drugs* 18:361–362, 1986.

IV

Special Topics

8

Geriatric
Psychopharmacology

CARL SALZMAN, M.D.

I. INTRODUCTION

Older persons, representing more than 10% of the population in the United States, take all forms of medication more frequently than younger individuals. Estimates of psychotropic drug use in the aged range from 7% to 92% in institutional settings and up to 30% in medical settings. A greater incidence of polypharmacy accompanies increased utilization of drugs. A recent survey in general hospitals showed that older patients received an average of 5–12 medications per day. These statistics reflect many factors: a longer life span accompanied by a greater incidence of chronic illnesses; increased medical sophistication; and drug therapy as a substitute for nonbiological interventions.

To care for the geriatric patient, the clinician must not only master the usual skills necessary for effective drug administration but also understand the unique characteristics of older individuals. In this age group, medical disorders commonly mimic or contribute to psychiatric presentations (for example, see Table 1 in Chapter 1). Physical illnesses, medical drugs, and significant organ system alterations increase the number and severity of adverse reactions from psychotropic agents. Age-related physiological changes as well as drug–drug interactions may alter psychotropic drug pharmacokinetics; medications often take longer to work, last longer in the body, and produce greater clinical effects per milligram dosage. In addition, older individuals frequently have complicated psychosocial problems that may decrease drug compliance (see Table 1).

CARL SALZMAN, M.D. • Department of Psychiatry, Harvard Medical School, Boston, Massachusetts 02115; Department of Psychopharmacology, Massachusetts Mental Health Center, Boston, Massachusetts 02115.

TABLE 1. Problems of Drug Taking in the Elderly

Problem	Causes
Forgetting medication	1. Organic, affective, or psychotic disorders may cause the patient to take too much or too little medication.
Poor compliance	1. Drugs are too expensive.
	2. Adverse effects may be prohibitive.
	3. Arthritic patients cannot open bottles (particularly with safety caps).
	4. Elderly cannot get to the pharmacy (particularly in winter months).
Unusual reactions	1. Unknown polypharmacy; patient taking over-the-counter medications or other drugs without doctor's knowledge; more than one doctor is prescribing drugs without knowledge of the other.
	2. Unknown or underestimated use of alcohol, nicotine or caffeine.
	3. Undiagnosed illness that will alter absorption, distribution, metabolism, or elimination.
	4. Undiagnosed decline in nutritional status, which also may alter drug metabolism.
Confusion over medicine taking	1. Pills that look alike (shape, color, size) may be interchanged; pills taken concurrently but on different schedules (e.g., q.i.d. and t.i.d.) may be confused.
Capricious noncompliance	1. May take either excessive medicines or none at all because of irrational ideas or struggles over control.

II. PHARMACOKINETIC CHANGES

Because of structural and functional changes within the aging body, the clinical effect(s) of psychotropic drugs may be altered. In most individuals, age-related changes result in reduced clinical effect, increased toxicity, and prolonged clinical and toxic effects. Pharmacokinetic studies of psychotropic drugs and drug–drug interactions have focused on age-related changes in (1) drug absorption and distribution, (2) protein binding, (3) metabolism, and (4) excretion.

A. Absorption and Distribution

In the absence of gastrointestinal pathology, the amount of gastrointestinal (GI) absorption is not significantly altered by the aging process. Because of decreased gastric motility, however, the **rate** of gastric absorption sometimes may be lowered. Although this may delay peak blood levels, actual age-related changes in absorption have little clinical or practical relevance.

The use of other drugs, however, commonly alters psychotropic drug absorption in the elderly (see Table 2). Antipsychotic agents may be precipitated in the stomach by coffee or tea. Aluminum-, magnesium-, or calcium-containing antacids, kaolin–pectin preparations (Kaopectate® and others) and activated charcoal can decrease absorption of benzodiazepines and chlorpromazine (Thorazine® and others) and should be given at least 2 hr before or after the psychotropic drug to prevent alterations in the plasma level. Milk of magnesia, which is frequently given to older medical and surgical patients who are taking psychotropic drugs, also may delay psychotropic drug absorption. Similarly, drugs with anticholinergic activity such as atropine (Antrocol® and others), antipsychotics, and cyclic antidepressants may decrease gut motility and interfere with absorption. This may result in inadequate blood levels and compromised clinical effects. By delaying gastric emptying and permitting more degradation of the drug in the gut, anticholinergic agents lower chlorpromazine blood levels.

B. Protein Binding

All psychotropic drugs [except lithium (Eskalith® and others)] bind extensively to plasma albumin. Since albumin levels decrease with age, older patients may be more susceptible to toxic responses and thus require lower doses of medication. In addition, other drugs given to elderly persons may displace psychoactive agents from protein binding sites, resulting in a higher proportion of unbound drug and increased adverse reactions (see Table 2).

Cyclic antidepressants bind to α_1 acid glycoprotein (AAGP) as well as to albumin. This protein may be increased during illness; the effect of age on this protein in an otherwise healthy person is uncertain. Decreased binding of psychotropic drugs by albumin may be offset by an increased binding by AAGP.

Protein-bound drugs are in a constant equilibrium with the unbound fraction. Unless this equilibrium is upset by an acute disturbance such as illness or alteration of binding by the presence of other drugs, then age will not affect protein binding in any given **individual elderly patient,** although as a class of people, protein binding may be somewhat reduced in the elderly. Thus, because of this equilibrium as well as the presence of AAGP binding, the overall impact of alterations in protein binding for any individual older patient cannot be predicted. However, if an older patient is given drugs that will alter the number of available binding sites or will displace other drugs, or if the older person becomes acutely ill, then there may be a change in binding of psychotropic drugs, and greater toxicity or lesser clinical effect may result.

C. Hepatic Metabolism

The first-pass effect (i.e., degration of drugs as they initially traverse the liver) removes more than 80% of most psychotropic drugs taken orally and absorbed from

TABLE 2. Pharmacokinetic Effects of Drug Interactions

Psychotropic drug	Interacting agent and effect
Absorption	
Antipsychotics	Coffee, tea may cause precipitation in the stomach (although some authorities now doubt this effect).
Benzodiazepines	Aluminum-, magnesium-, or calcium-containing antacids delay hydrolysis.
	Kaopectate and milk of magnesia decrease absorption.
All drugs	Anticholinergic agents [e.g., atropine (Antrocol® and others), scopolamine, meperidine (Demerol® and others)] decrease absorption.
Protein binding	
Diazepam (Valium® and others)	Increases binding of digoxin (Lanoxin® and others)
Cyclic antidepressants	TSH increases binding.
Excretion	
All drugs	Propranolol (Inderal® and others) increases serum levels from reduced renal blood flow.
Cyclic antidepressants	Tetracycline (Surmycin® and others) and spectinomycin (Trobicin®) increase serum levels by reducing clearance.
	Sodium bicarbonate increases serum levels by increasing tubular reabsorption secondary to alkalinization of the urine.
	Ascorbic acid and ammonium chloride decrease serum levels by a decrease in reabsorption secondary to acidification of the urine.
Lithium (Eskalith® and others)	Ibuprofen (Motrin® and others), indomethacin (Indocin® and others) (and perhaps all prostaglandin synthetase inhibitors) increase serum levels by reducing renal clearance.
	Some Na-losing diuretics increase serum levels by increasing reabsorption.
	K^+-sparing diuretics may decrease serum levels by decreasing reabsorption.

Hepatic metabolism

Decreases metabolism of psychotropic drugs (raises blood level)
   Isoniazid (INH® and others)
   Chloramphenicol (Chloromycetin® and others)
   Methylphenidate (Ritalin® and others)
   Disulfuram (Antabuse®)
   Antipsychotics (raises heterocyclics)
   Cimetidine (Tagamet®) [affects chlordiazepoxide (Librium® and others), diazepam, prazepam (Centrax® and others), and clorazepate (Tranxene® and others)]
   Propranolol (Inderal® and others)
   Norethindrone (Loestrin® and others)
   Barbiturates (when toxic doses of heterocyclics are present)
Increases metabolism of psychotropic drugs (lowers blood level)
   Alcohol
   Barbiturates
   Nonbarbiturate sedative–hypnotics [except perhaps flurazepam (Dalmane® and others)]
   Caffeine

(continued)

TABLE 2. *(Continued)*

Increases metabolism of psychotropic drugs (lowers blood level) *(cont.)*
Smoking
Carbamazepine (Tegretol® and others)
Phenytoin (Dilantin® and others)
Chloral hydrate (Noctec® and others)
Antiparkinson agents
Lithium [affects chlorpromazine (Thorazine® and others); also decreases chlorpromazine levels by delaying gastric emptying, inhibiting absorption, and increasing renal excretion]

the stomach and intestine. This hepatic function depends on liver blood flow, which decreases as cardiac output decreases. Thus, the first-pass drug extraction decreases with age and may, therefore, increase the fraction of metabolized psychotropic drug that reaches the general circulation and enhance the potential for toxicity. Propranolol (Inderal® and others) and cimetidine (Tagamet®), which diminish hepatic blood flow, may also contribute to a decreased first-pass effect.

All psychotropic drugs (except lithium) are primarily metabolized in the liver microsomal enzyme system. With age, hepatic enzyme metabolism may be reduced, producing higher blood levels of psychotropic drugs and increased adverse reactions. Medical drugs combined with psychotropic drugs also may inhibit or induce liver microsomal enzymes, further altering psychotropic drug metabolism (see Table 2). In addition, protein–caloric malnutrition and vitamin deficiency (e.g., vitamin A, folic acid, C, and B_{12}) may decrease hepatic microsomal enzyme activity.

Since hepatic metabolism proceeds in a stepwise fashion, the metabolism of some psychotropic drugs results in the production of metabolites that are active and may produce side effects. This is true for neuroleptics, cyclic antidepressants, and some benzodiazepines. Furthermore, some cyclic antidepressant metabolites are water soluble, so that the duration of their effect will depend on renal function (see below) as well as on hepatic metabolic processes.

D. Excretion

Psychotropic drugs are excreted primarily through the kidney. Reduced renal blood flow, glomerular filtration, and tubular excretory capacity all have been reported in the elderly. Delayed excretion (as well as impaired hepatic metabolism) is reflected in a prolonged elimination half-life ($t_{1/2\beta}$) of many psychotropic drugs.

The clinical effect of age-related decreased renal function and prolonged half-life may prove hazardous in the case of lipid-soluble psychotropic drugs (e.g., long-acting benzodiazepines, tertiary tricyclic antidepressants, and some antipsychotic drugs) that form active metabolites. Accumulation of these drugs and their metabolites in the increased fatty tissue as well as in plasma may predispose the older patient to substantially increased risk of toxicity. Other coadministered drugs also can affect excretion of psychotropic agents (see Table 2).

Cyclic antidepressants may produce a water-soluble hydroxy metabolite in addition to lipid-soluble active metabolites. The excretion of these water-soluble metabolites, like excretion of lithium, depends on renal function. Because renal excretion decreases with age, blood levels of these water-soluble hydroxy metabolites may increase, and the duration of the effect may be prolonged (see antidepressants).

III. ADVERSE REACTIONS

Adverse reactions occur much more frequently in older patients. In addition to altered pharmacokinetics, the elderly exhibit a generally increased sensitivity to psychoactive medication. The most frequent and troublesome effects involve the central nervous system (CNS) (sedation, confusion, and extrapyramidal effects) and heart (orthostatic hypotension and alterations of heart rate, rhythm, and contractility) (see Tables 3 and 4). In addition, drug combinations, so prevalent in elderly patients, increase the incidence and severity of both central and peripheral adverse reactions (see Table 5).

TABLE 3. Adverse Reactions from Psychotropic Drugs That May Be More Common and Severe in the Elderly

Effect	Cause	Clinical examples
Sedation	Most psychotropic drugs either produce or enhance sedation of other medication	Increased daytime sedation and napping; increased confusion and irritability at night
Confusion	Anticholinergic properties of cyclic antidepressants and antipsychotics Lithium Secondary to increased sedation	Disorientation, visual hallucinations, agitation, assaultiveness, loss of memory
Orthostatic hypotension	Results from cyclic antidepressants and antipsychotics	Falling when getting out of bed or a chair
Cardiac toxicity	Primarily from cyclic antidepressants	Increase in heart rate and in frequency of irregular heart beat rhythms; altered EKG patterns
Extrapyramidal symptoms	Antipsychotics	Akathisia—restlessness, which is confused with agitation; akinesia—decreased activity and interest, which is confused with depression

TABLE 4. Pharmacological Effects of Psychotropic Drugs That Cause Adverse Reactions in the Elderly

Drug effect	Symptoms
Decreased central nervous system arousal level	Sedation, apathy, withdrawal, depressed mood, disinhibition, confusion
Peripheral anticholinergic blockade	Dry mouth, constipation, atonic bladder, aggravation of narrow-angle glaucoma and prostatic hypertrophy
Central anticholinergic blockade	Confusion, disorientation, agitation, assaultiveness, visual hallucinosis
α-Adrenergic blockade and central pressor blockade	Orthostatic hypotension
Quinidine effect, anticholinergic effect, decreased myocardial contractility	Tachycardia, cardiac arrhythmia, heart block, increased PR interval and widening of QRS complex, decreased inotropic effect, heart failure
Dopaminergic blockade	Extrapyramidal symptoms, (?) tardive dyskinesia

A. Central Nervous System

1. Sedation

Excessive sedation in the elderly may be caused by many psychiatric medications, including benzodiazepines, the sedating cyclic antidepressants [e.g., doxepin (Sinequan® and others), amitriptyline (Elavil® and others)], and the sedating antipsychotic agents [e.g., chlorpromazine and thioridazine (Mellaril® and others)]. Oversedation not only can be mistaken for depression but also can decrease the elderly patient's contact with his surroundings, impair his cognitive capacities, and reduce his self-esteem.

Oversedation is also one of the most common and problematic effects resulting from inappropriate drug combinations. In medically ill elderly patients, sedating psychotropic medications are often administered together with narcotics, analgesics, alcohol, or sedatives. For example, two recent studies showed that one or more narcotics were given to many elderly patients who received a hypnotic drug, a tricyclic antidepressant, diazepam (Valium® and others), or an antipsychotic agent. The CNS depression (oversedation) that results from such drug combinations may lead to a lowered mood, a sense of helplessness, an attitude of "resignation," and withdrawn or disordered behavior. There may be a decline in self-care and social activities, as though the older patient had given up and was quietly awaiting death.

2. Confusion

Confusion in the elderly most often results from three etiologies: CNS depression or oversedation (see above), CNS anticholinergic effects, or toxic effects of

TABLE 5. Unwanted Clinical Effects of Psychotropic–Psychotropic/
Medical Drug Combinations

Unwanted effects	Psychotropic drug used	Combination drug
Oversedation: disorientation, confusion, agitation, irritability, hallucinations	Antipsychotics Benzodiazepines Sedative–hypnotics Amitriptyline (Elavil® and others) Doxepin (Sinequan® and others)	Narcotics Analgesics Alcohol Sedatives
Central anticholinergic: confusion, disorientation, assaultiveness, toxic hallucinosis	Antipsychotics Cyclic antidepressants	Narcotics Atropine-containing drugs (often over-the-counter antihistamines)
Stimulation: excitation, hyperpyrexia, convulsions	MAOIs	Cyclic antidepressants
Neurotoxic responses:		
Confusion, extrapyramidal symptoms, seizures, ataxia, hyperreflexia, abnormal EEGs	Lithium (Eskalith® and others)	Antipsychotics
Cerebellar ataxia, polyuria, decreased thyroxine index, decreased libido	Lithium	Phenytoin (Dilantin® and others)
Transient delirium	Cyclic antidepressants	Ethylclorvynol (Placidyl®)
Inhibition of response to L-Dopa (Larodopa® and others)	Antipsychotics	L-Dopa
Cardiac:		
Prolonged conduction	Cyclic antidepressants	Quinidine Procainamide (Pronestyl® and others)
Nodal bradycardia with slow atrial fibrillation	Chlordiazepoxide (Librium® and others)	Digoxin (Lanoxin® and others) Thiazide diuretics
Inhibition of antihypertensive effects	Cyclic antidepressants Antipsychotics (MAOIs, amphetamines?)	Guanethidine (Ismelin® and others) Debrisoquine, clonidine (Catapres®)
Exacerbation of orthostatic hypotension	Cyclic antidepressants	Diuretics
Peripheral anticholinergic:		
Dry mouth, urinary stasis, constipation	Antipsychotics Cyclic antidepressants	Narcotics Atropine-containing drugs (often over-the-counter) Antihistamines
Respiratory depression	Heterocyclic antidepressants Phenothiazines	Meperidine (Demerol® and others)

(*continued*)

TABLE 5. (*Continued*)

Unwanted effects	Psychotropic drug used	Combination drug
Lithium-related:		
Prolongation of neuromuscular blockade	Lithium	Succinylcholine (Anectine® and others)
Exacerbation of lithium side effects		Pancuronium (Pavulon®)
Increase or decrease in glucose tolerance (?)		Methyldopa (Aldomet® and others)
		Insulin

lithium therapy. When significant sedation occurs, older individuals often develop difficulties with concentration and recall.

Anticholinergic effects can lead to CNS confusion as well as peripheral toxicity. Central anticholinergic syndromes commonly appear with the use of various antidepressants [e.g., amitriptyline, imipramine (Tofranil® and others), trimipramine (Surmontil®), doxepin, and nortriptyline (Pamelor® and others)] and some antipsychotics [e.g., chlorpromazine, thioridazine, chlorprothixene (Taractan®)]. More often, however, confusion results from administration of multiple drugs with anticholinergic properties. For instance, the combination of an antipsychotic, an antidepressant, and an antiparkinson agent may cause significant toxicity, since the elderly have increased CNS sensitivity to cholinergic blockade. The clinician should remember that even medications with few anticholinergic effects can cause CNS impairment when given alone to sensitive individuals, when given in higher doses, and when administered in combination with other psychiatric or medical drugs.

The elderly patient with organic brain dysfunction is especially susceptible to central anticholinergic toxicity. Antipsychotics and cyclics may be prescribed with narcotics [e.g., meperidine (Demerol® and others)], atropine-containing drugs (which are often available as over-the-counter drugs), and antihistamines that can produce either central and/or peripheral effects. In mild forms, it may appear as disorientation, confusion, agitation, and irritability, particularly at night (i.e., the "sundowner's syndrome"). In more severe cases, it may result in a toxic delirium consisting of disorientation, hallucinations, other psychotic symptoms, and even assaultiveness. This condition can be life-threatening because of self-destructiveness, combativeness, or poor self-care. Typically, in general hospital or nursing home settings, such patients attempt to climb out of bed, mistake nurses for family members, have frightening periods of amnesia, or may even hallucinate. A common medical response is to add more sedating drugs, particularly antipsychotics; this, of course, only further aggravates the condition or puts the patient to sleep.

Lithium alone commonly causes confusion and disorientation in the elderly. Confusion secondary to neurotoxicity also may result from the combined use of an antipsychotic drug and lithium. Other symptoms include extrapyramidal features, seizures, ataxia, hyperreflexia, and abnormal electroencephalograms (EEGs). If lithium and an antipsychotic are used together, we recommend that the patient's status be monitored with EEGs and that the plasma lithium concentrations be kept below 1.0 mEq/liter.

The combination of lithium and phenytoin (Dilantin® and others) also should be used with caution, since patients who overdose with both drugs have developed persisting cerebellar ataxia. In addition, lithium-type toxicity, including polyuria, polydipsia, decreased free thyroxine index, and decreased libido, has been noted despite normal serum levels of both drugs. In one case, substitution of carbamazepine (Tegretol®) for phenytoin resulted in remission of symptoms.

3. Extrapyramidal

Older patients develop frequent and severe extrapyramidal symptoms. This may result from altered pharmacokinetics or decreased dopamine levels in the nigrostriatal pathways of the elderly. The most common effects include akathisia, parkinsonism, akinesia, and the "Pisa syndrome." As in younger patients, akathisia is often misdiagnosed as agitation. Further medication often increases the severity of the symptoms. Akinesia, consisting of decreased speech and energy and masklike facies, may be mistaken for depression. The "Pisa syndrome" is a form of dystonia in which the trunk is flexed to one side. The clinician can treat these extrapyramidal symptoms with antiparkinson drugs. However, he must weigh the added risk of anticholinergic effects against the possible advantages and disadvantages of switching the class of antipsychotic drug.

Tardive dyskinesia is much more common in the elderly (especially women) and in those with neuropathological disorders. No effective treatment exists, although the condition may remit if medication is stopped shortly after the appearance of symptoms (remission occurs less often in older patients; see Chapter 4).

Parkinson's disease itself is more common in old age. Antipsychotic drugs may worsen a patient's condition by inhibiting his response to L-Dopa (Larodopa® and others) (i.e., by blocking dopaminergic receptors in the CNS). When the older patient is treated with L-Dopa and antipsychotics, a change in dosage may be necessary.

Tardive dyskinesia is more likely in the older chronic patient because of the greater cumulative neuroleptic treatment. It is also more likely in older patients because changes in the aging central nervous system predispose them to developing tardive dyskinesia. Advanced age is now known to be the most consistent and most likely predictor of the development of tardive dyskinesia following use of neuroleptic drugs.

4. Other Central Nervous System Symptoms

Sometimes, undesirable CNS effects occur when drug combinations cause excessive stimulation of the noradrenergic system. Cyclics or monoamine oxidase inhibitors (MAOIs), for example, increase the central effect of amphetamines, and MAOIs increase the effect of a large number of agents (see Table 5). In the elderly, this stimulation may increase agitation, excitation, restlessness, and insomnia.

The combination of cyclic antidepressants and MAOIs may cause severe reactions with symptoms that include excitation, hyperpyrexia, convulsions, and hypertension; fatalities have occurred. However, in most of these patients one or more of the following three conditions was present:

1. Excessive doses of one or more drugs were used.
2. The cyclic was given parenterally.
3. Other psychotropic drugs were also administered.

Generally, the clinician should avoid combining MAOIs and cyclics in the elderly unless they do not respond to the usual therapies or have had a previous good response to combination therapy.

Coadministration of ethclorvynol (Placidyl®) and cyclic antidepressants may cause a transient delirium and also should be avoided.

B. Cardiac

Unwanted cardiovascular effects from psychotropic drugs and from the combination of psychoactive and medical agents include changes in cardiac conduction, rhythm, and blood pressure. Although less frequent than CNS sedation or anticholinergic effects, cardiovascular changes are particularly hazardous in the elderly, especially if they have preexisting heart disease.

1. Effects on Blood Pressure

Orthostatic hypotension in the elderly probably results from several causes: blockade of central vasomotor centers, negative inotropic effects (decreased cardiac contractility), and peripheral α-adrenergic blockade. This adverse reaction occurs most often with the use of cyclic antidepressants (particularly amitriptyline and imipramine) and antipsychotics (particularly chlorpromazine and thioridazine).

In clinical practice, transient hypotension may cause dizziness, falls and fractures, myocardial infarcts, or strokes. In one study, about 40% of geriatric patients reported dizziness and falling. In the elderly, hypotension may occur at night when patients get up to urinate. Therefore, each patient should be checked for orthostatic changes and told to change positions slowly. Sometimes support stockings can also help.

Cyclic antidepressants decrease the antihypertensive properties of many drugs, particularly guanethidine (Ismelin® and others), bethandine, and debrisoquine. Cyclics also inhibit the effects of clonidine (Catapres®), but probably not reserpine (Serpasil® and others) or α-methyldopa (Aldomet® and others). Antipsychotic agents, like cyclic antidepressants, can also inhibit the effect of antihypertensive agents by blocking their uptake. Low-potency antipsychotics may be more inhibitory than their high-potency counterparts.

2. Effects on Rate, Rhythm, and Contractility

In the elderly patient, there is an increased risk of cardiac effects from antipsychotics and cyclic antidepressants. Because of their anticholinergic properties and effects on intracardiac conduction, these agents cause tachycardia. They also may produce an increased incidence of premature ventricular contractions (although at some doses they decrease), heart block, atrial and ventricular arrhythmias, heart failure, or worsening of congestive heart failure [see Chapters 2 and 4 for electrocardiogram (EKG) changes].

Spontaneously occurring atrial and ventricular arrhythmias are commonly seen in the geriatric population. Since cyclic antidepressants retard cardiac conduction, their combination with drugs that prolong conduction, such as quinidine, procainamide (Pronestyl® and others), or disopyramide (Norpace® and others), may result in toxic cardiac effects. Digoxin (Lanoxin® and others), when combined with lithium, may cause severe nodal bradycardia with slow atrial fibrillation or sinoatrial block. The combination of thiazide diuretics (which deplete sodium and potassium), lithium, and digoxin is particularly hazardous; sodium depletion leads to lithium retention, which, in turn, worsens potassium depletion and increases digoxin toxicity. Permanent ventricular pacemakers are not affected by the tendency of cyclic antidepressants to interfere with intraventricular conduction.

C. Other Organ Systems

1. Respiratory depression is a rare effect in all patients; it appears more frequently in the elderly with lung disease who are taking sedative–hypnotics. Moreover, meperidine-induced respiratory depression is enhanced by both cyclic antidepressants and phenothiazines.
2. Ocular and dermatologic effects from long-term antipsychotic drug use usually occur only in older patients after large cumulative doses of medication (most often seen with low-potency phenothiazines).
3. Peripheral anticholinergic effects such as dry mouth may cause oral infections, loss of teeth, and a poor fit for dentures; decreased gastrointestinal and urinary bladder motility appear more often in the elderly.
4. Lithium prolongs the neuromuscular blockade produced by succinylcholine (Anectine® and others) and pancuronium (Pavulon® and others). Therefore, in lithium-treated patients requiring elective surgery and a muscle-relaxant anesthetic, lithium should be discontinued 48 to 72 hr preoperatively and resumed on return of bowel function. Depressed patients taking lithium who also require ECT should be closely monitored after anesthesia to ensure the return of adequate respiration.

5. The interaction of lithium with insulin and carbohydrate metabolism is complex; controversy exists about whether lithium increases or decreases glucose tolerance. Therefore, glucose control of insulin-dependent diabetics on lithium therapy should be closely monitored.

IV. CLINICAL APPLICATION

A. Principles of Medication Use

To Treat Elderly Patients Safely

1. Know the properties of each drug, including pharmacokinetics, possible adverse reactions, and clinical effects.
2. Take a careful medical history and complete a physical examination to rule out an organic etiology and to document physical symptoms and illnesses that may interfere with psychotropic medication (e.g., bowel obstruction or constipation, kidney disease or urinary tract disease, glaucoma).
3. Be aware of other medications the patient is taking (even over-the-counter preparations).
4. When medication is indicated, do not withhold it because of age.
5. Offer a simple plan for taking medication. For example, show the patient a picture of the pill to be taken, outline an exact dosage schedule, and use family members to help if necessary.
6. Start medication in small amounts and increase slowly while monitoring adverse reactions. Rely on both verbal reports and physical signs.
7. Continue the trial of medication until the patient receives adequate doses or has significant clinical effects or prohibitive adverse reactions (clinicians have a tendency to stop drug trials too early).
8. Avoid polypharmacy by giving only one psychotropic agent whenever possible and by intermittently reviewing all medications.
9. Always provide psychological support.
10. Assume that any older patient taking psychotropic medication who is agitated, confused, restless, forgetful, or depressed may be drug toxic. Consider reducing the dose or discontinuing medication before adding another drug.

B. Clinical Use

1. Agitated Behavior and Psychotic Thinking

In older as well as younger people, antipsychotic medication may be used to treat psychosis. Although a late emergence of schizophrenia is rare, other types of psychoses (i.e., depressive, organic, schizoaffective, and manic) may appear in older patients. Most commonly, however, antipsychotic medication is used in elderly patients to control severe behavioral disturbances such as agitation, wandering, self-mutilation, and assaultiveness. A number of nonneuroleptic drug treatments have also been developed to control severe agitation and disruptive behavior.

a. Neuroleptic Treatment of Acute Psychosis or Behavior Problems

No studies show that one antipsychotic medication is more effective than another in controlling acute psychosis or disordered behavior in elderly patients. However, adverse reactions differ markedly with different antipsychotic medications. The antipsychotic effects that are frequent and particularly hazardous in older people include sedation, orthostatic hypotension, and extrapyramidal and anticholinergic symptoms. High-potency antipsychotic medications such as haloperidol (Haldol® and others), thiothixene (Navane® and others), or piperazine phenothiazine [trifluoperazine (Stelazine® and others), perphenazine (Trilafon® and others), fluphenazine (Prolixin® and others)] are less likely to produce sedation and hypotension although more likely to produce extrapyramidal symptoms. A low-potency antipsychotic such as thioridazine is strongly sedative and anticholinergic but produces fewer and less severe extrapyramidal reactions in the elderly. The selection of an appropriate antipsychotic medication, therefore, rests on a consideration of unwanted effects. For example, nonsedating antipsychotic drugs are more likely to produce extrapyramidal symptoms; conversely, sedating antipsychotics are less likely to produce extrapyramidal symptoms.

In addition, the clinician can administer other antipsychotics if the older individual has previously responded favorably and safely and if no new physical conditions preclude their use. Antipsychotics should be started in low oral doses (e.g., 1 to 2 mg of a high-potency drug or 10 to 25 mg of thioridazine) two to three times per day and increased slowly.

b. Nonneuroleptic Treatment of Acute Psychosis or Behavior Problems

A number of nonneuroleptic treatments are now available to the clinician to control severe disruptive behavior in elderly patients. Although research studies in this area are virtually nonexistent, there is growing clinical experience that suggests that some of these may be quite useful:

1. β Blockers. The use of doses of β-blocking drugs such as propranolol (10–100 mg/day) may dramatically reduce agitation, wandering, and screaming

behavior associated with dementing illness. In most cases, adverse effects are minimal. The older patient must have an adequate cardiovascular evaluation before β-blocking drugs are used and must not have a disorder such as asthma that would contraindicate the use of these drugs.

2. Trazodone (Desyrel® and others). This antidepressant drug in low doses (75–150 mg) is remarkably effective in controlling severe agitation and disruptive behavior. Side effects other than sedation are unusual at this low dose.

3. Anticonvulsants. Carbamazepine and anticonvulsants, when used in very low doses (e.g., carbamazepine 100–200 mg/day), are sometimes effective in controlling disruptive behavior that has not been managed by other medications.

4. Lithium. Low doses of lithium carbonate (e.g., 75–300 mg/day) given daily on a maintenance basis have also been found effective in controlling or preventing outbursts of yelling, agitation, and disruptive behavior. Low-dose lithium must be used with extreme caution in older people (see below), especially since symptoms of lithium central nervous system toxicity may be worsening of confusion, agitation, and disruptive behavior.

5. Buspirone (BuSpar®). This new antianxiety agent (5–15 mg/day) effectively controls disruptive behavior. Experience with buspirone at this time is considerably less than with other drugs, but reports suggest that toxicity is very low and the therapeutic benefit substantial.

c. Treatment of Extrapyramidal Symptoms

Extrapyramidal effects caused by the antipsychotics and occasionally by cyclic antidepressants are treated with low doses of antiparkinson agents [e.g., trihexyphenidyl hydrochloride (Artane® and others), 0.5 mg]. The clinician should increase dosage only after careful assessment of the patient's mental and physical status, since these drugs often combine with antipsychotic agents to produce serious anticholinergic syndromes. In the treatment of dystonias, the antiparkinson medications should be discontinued after 3 to 7 days. In the treatment of parkinsonism and akathisia, antiparkinson agents should be stopped after 6 weeks to 3 months. If symptoms reemerge, the clinician should consider changing the antipsychotic medication.

2. Dementia

Chronic brain syndromes may result from many etiologies. Dementias caused by tumors or increased intracranial pressure may be reversible. However, the majority of the elderly with dementia have progressive, diffuse loss of brain cells (Alzheimer's disease accounts for 50–75% of cases) and ongoing deterioration of mental capacities and functioning (see Section III.A.2). Less often, older patients may develop dementia from arteriosclerosis. Multiple infarcts often cause a clinical picture in the elderly marked by sudden onset and stepwise progression with a deteriorating mental status and focal neurological signs. About one-third of demented patients have both Alzheimer's dementia and multiple infarcts. Older patients also may present with "pseudodementia." Although not a specific disorder, this term describes a decline in cognitive processes associated with depression. The patient appears somewhat confused and disoriented and has impaired recent memory. In severe cases, there are behavioral disturbances and a deterioration of self-care. Adequate antidepressant treatment often brings about a surprising restoration of thinking, memory, and behavior. Factors associated with this condition include:

1. History of affective disorder.
2. Symptoms of major depressive illness accompanying the cognitive disturbance.
3. Poor immediate recall but adequate delayed verbal recall.
4. Intact recognition memory.
5. Few objective deficits on mental status.
6. Improved intellectual functioning when the clinician structures tests, prompts the patient, or allows extra time for answers.

Older patients with well-established organic dementia also may have significant depressive symptoms that compromise thinking and behavior. Administering an antidepressant often improves the patient's depression, memory, orientation, and behavior.

Patients with senile dementia of the Alzheimer's type may have a selective deterioration of acetylcholine neurons. Precursors of acetylcholine such as lecithin and choline have been administered to these patients with equivocal results. Some late-life dementias have been thought to result from decreased central nervous system oxygenation. Clinical and research experience with cerebral vasodilators as well as drugs that increase cerebral blood flow [heparin (Panheprin® and others), papaverine (Cerebid® and others), cyclandelate (Cyclospasmol® and others)] and hyperbaric oxygen have also produced questionable results. The ergot alkaloid combination ergoloid mesylates (Hydergine® and others) has been shown to have slightly positive effects in the mildly demented older patient, but these results have not been striking. Central nervous system stimulants have had no positive therapeutic effect on thinking or memory. Recent research has also focused on new protein complexes, steroids, and metabolic agents such as piracetam, $ACTH_{4-10}$, procaine (Novocain® and others), and RNA, but the effects of these agents have been either negative or equivocal.

3. Anxiety

The treatment of anxiety in the elderly is similar to that in younger patients. Benzodiazepines are the safest and most effective agents for the short-term management of moderately severe anxiety states that do not respond to nonpharmacological methods. However, the clinician should carefully monitor the drugs, since they often cause sedation, uncoordination, and confusion. Combining benzodiazepines with drugs that have sedative or disinhibiting properties often can produce central nervous system depression, lowered mood, disinhibited behavior, confusion, disorientation, memory loss, and agitation.

Benzodiazepines in clinical use can be divided into two groups on the basis of their pharmacokinetic properties. In one, the drugs are long-acting, have active metabolites, and tend to accumulate, lasting about twice as long in the elderly as in young or middle-aged adults. These drugs include chlordiazepoxide (Librium® and others), diazepam, clorazepate (Tranxene® and others), prazepam (Centrax® and others), and flurazepam (Dalmane® and others). These drugs may be useful for the older patient who is forgetful or is taking many drugs on different daily schedules and inadvertently omits the benzodiazepine; if a dose is missed, clinical effect will not be compromised. Since these agents tend to accumulate, however, they should be given only once a day or perhaps every other day. Unfortunately, the long-acting benzodiazepines have several disadvantages for the elderly. They take longer to reach steady-state plasma concentrations. Should toxicity occur, elimination of these drugs may take days to weeks, thus subjecting the older patient to prolonged unwanted effects.

The other group of benzodiazepines has a much shorter duration of activity, has no active metabolites, and tends not to accumulate. These agents include oxazepam (Serax® and others), lorazepam (Ativan® and others), and alprazolam (Xanax®). They must be administered throughout the day, since their duration of action is measured in hours. Symptoms of rebound anxiety and insomnia are common several hours or days after the last dose of a short-half-life benzodiazepine. These rebound symptoms may be interpreted as a return of anxiety, prompting the continuation of drug treatment. Should toxicity occur, however, the drugs will be more rapidly eliminated than the long-acting benzodiazepines. Since the older patient is more sensitive to the benzodiazepines' unwanted effects and to the problems of polypharmacy, the use of short-acting drugs may be preferred.

Benzodiazepines should **not** be prescribed to older patients on a regular maintenance basis for the treatment of anxiety or of insomnia. Long-term benzodiazepine use, even at therapeutic doses, can be associated with subtle but progressive increase in confusion, disorientation, and forgetfulness. Older patients may become dependent on benzodiazepines, especially short half-life drugs, and have considerable difficulty in giving up their use.

When an older patient reacts adversely to benzodiazepines or exhibits persistent, severe, or disorganizing anxiety, the practitioner should avoid other sedative–hypnotic drugs. These medications, particularly the barbiturates, may cause respiratory depression, CNS impairment, and frequent paradoxical responses. Instead, he might consider using low doses of an antipsychotic agent.

4. Insomnia

The elderly patient commonly suffers from insomnia. The clinician should carefully evaluate the patient, since sleep disorders frequently are secondary to various other conditions (e.g., depression, anxiety states, pain, dyspnea, daytime napping, urinary urgency, recent emotional trauma, or caffeine). If medication is appropriate, the practitioner should only administer a hypnotic for a short period. Flurazepam, 15 mg at bedtime, is effective; with longer use, this drug may accumulate and cause sundowning and disinhibition. The shorter-acting benzodiazepines (e.g., oxazepam, 5 to 10 mg, or lorazepam, 0.5 to 1.0 mg) offer other alternatives. Temazepam (Restoril® and others), 15 mg at bedtime, has strong hypnotic properties and a short half-life. Although there is little experience with this drug in the elderly, it may have significant advantages.

Triazolam (Halcion®) is a benzodiazepine with an extremely short elimination half-life (4 hr) that is also rapidly absorbed. Thus, for short-term treatment of sleep disturbance, triazolam has been found to help older patients. Doses must be kept low (0.125–0.25 mg/night), since the older patient is extremely sensitive to higher doses of this drug. Increased confusion, agitation, and forgetfulness have been observed at higher doses. Because triazolam is so rapidly eliminated, rebound insomnia may develop 4 or 5 hr after the last dose is taken. It is not unusual for older patients who take triazolam at bedtime to reawaken at 2:00 or 3:00 in the morning. It is difficult to distinguish between such awakening on the basis of endogenous sleep disorder problems versus triazolam rebound insomnia. Nevertheless, some clinicians prescribe a second dose of triazolam for such awakenings. This practice leads to long-term higher-dose triazolam use, dependence, and in some older patients progressive cognitive impairment. Triazolam treatment of sleep disturbance, even at very low doses, is recommended only for very brief periods of time.

5. Depression

Cyclic antidepressants are the drugs of choice for the severely depressed elderly patient. However, older patients are more susceptible to adverse effects than younger individuals. In particular, orthostatic hypotension, cardiac irritability, and anticholinergic toxicity occur more often in the elderly. Since no data suggest clinical superiority of one cyclic antidepressant over another, the clinician should choose a cyclic on the basis of its adverse reactions. In general, tertiary tricyclic amines such as imipramine and amitriptyline produce more severe untoward effects than their demethylated secondary amine metabolites, desipramine (Norpramin® and others) and nortriptyline, respectively. Tertiary amines also last longer and, because of altered hepatic metabolism with age, tend to accumulate more than secondary amines. Desipramine or nortriptyline may be recommended for the depressed elderly patient who has psychomotor retardation or hypersomnia. In those with anxious or agitated depressions, doxepin, which is more sedating, should be used.

Therapeutic blood levels are not different in older people compared with younger adults. However, these therapeutic blood levels are often achieved with doses that are substantially lower than those given to younger adult patients. For example, desipramine may be therapeutic at doses as low as 30–75 mg/day; nortriptyline 20–

50 mg/day; doxepin 10–75 mg/day. It must be remembered, however, that some older patients require a full therapeutic dose to achieve full antidepressant effect. Therefore, as a general rule, cyclic antidepressants should be prescribed in lower doses for older patients. The actual dose necessary to treat an individual elderly patient has to be based on therapeutic and toxic response.

In addition to desipramine, nortriptyline, and doxepin, two other cyclic antidepressants may be recommended for the older patient. Trazodone is a sedating antidepressant with no anticholinergic activity. Although response to this drug is less predictable and reliable than the aforementioned antidepressants, it is sometimes therapeutic for patients who have not responded to other antidepressants. Fluoxetine (Prozac®) is a stimulating antidepressant without anticholinergic effects. At the present time, it has not been used extensively in older patients, but recent experience suggests that it has therapeutic benefit with relatively limited toxicity.

Monoamine oxidase inhibitors [phenelzine (Nardil®), tranylcypromine (Parnate®)] are also useful in treating depressive symptoms in older patients. These drugs are more commonly recommended for less severe depressions that are characterized by psychomotor slowing, decreased motivation, decreased energy, and a reduced enthusiasm and pleasure in life. Because of dietary restrictions and numerous drug interactions, they can only be given to older patients who are able to follow prescribing instructions or can only be given in settings with strict medical supervision. The doses of monoamine oxidase inhibitors, like those of cyclic antidepressants, tend to be one-third to one-half lower than doses used in younger adult patients, although occasionally full therapeutic doses are necessary. Adverse effects tend to be more problematic for older patients and more frequent. They include orthostatic hypotension, agitation, and aggravation of dementing symptoms.

Psychomotor stimulants such as methylphenidate (Ritalin® and others) are also sometimes used to treat older patients who describe themselves as depressed. Symptoms that tend to respond to stimulants include withdrawal of enthusiasm, interest, and energy, fatigue, and lassitude. These symptoms may sometimes be interpreted by the older person as "giving up." Low doses of methylphenidate (5–20 mg/day) in a patient with a healthy cardiovascular system dramatically increase an older person's interest in participating in daily life. Adverse effects of these doses are relatively mild and are chiefly limited to sleeplessness and hyperstimulation, so the drug should not be prescribed during the afternoon and evening hours. Psychomotor stimulants should not be given for prolonged periods of time, prescribed in high doses, or used for patients with major depressive disorder. Psychomotor stimulants, like MAO inhibitors, may induce a manic episode in bipolar patients.

Every elderly patient must have a physical examination and a pretreatment EKG prior to the initiation of cyclic antidepressant therapy. During the course of treatment, blood pressure, urine output, and EKG should be monitored periodically. Should the patient become toxic or incompletely respond to treatment, plasma levels of the drug should be obtained. Occasional elderly patients will require high doses of medication.

For the severely depressed elderly patient whose life may be jeopardized by the depression (either through inanition or suicidal impulses), electroconvulsive therapy (ECT) should be administered. ECT is also preferred for elderly patients who have responded positively in the past, who have nihilistic or somatic delusions, or who have moderately to severely impaired cardiovascular status.

6. Bipolar Affective Disorder

The manic elderly patient presents a difficult therapeutic challenge. In acute manic states, antipsychotic medication may be necessary to control hyperactivity, which can rapidly lead to exhaustion. Most commonly, high-potency agents such as haloperidol, fluphenazine, or thiothixene are used.

Since lithium may produce serious adverse effects in the elderly, its use in the treatment of mania requires special care. The elderly are sometimes more sensitive to toxicity resulting from the interaction of lithium and an antipsychotic drug. This toxicity is manifested by a confusional state with memory loss, disorientation, fear, and agitation.

Lithium excretion and its elimination half-life tend to be delayed in older patients, leading to increased likelihood of toxicity. For this reason, adverse effects are more common among the elderly than in younger patients. Symptoms of lithium toxicity may be confused with common symptoms of older patients, such as nausea, vomiting, restlessness, tremor, and confusion.

Lithium, therefore, should be restricted to the treatment of the acute manic episode and for prophylaxis only in patients with well-documented recurrent bipolar illness. Acute mania may be treated with an initial dose of 150 mg of lithium b.i.d.; dosages should be increased by 150-mg increments. Blood levels above 0.8 mEq/liter should **not** be routine. If high levels are necessary, the clinician should carefully monitor the patient's electrolytes, fluid status, and mental status. For prophylaxis, a blood level of 0.4 to 0.8 mEq/liter is often adequate. Cardiac and renal function should be carefully checked, and serum lithium levels, EKGs, thyroid, and kidney function should be reviewed frequently.

V. CONCLUSION

Medicating the elderly presents many difficulties for the clinician. The marked heterogeneity of this population and their special problems (particularly polypharmacy, altered pharmocokinetics, and CNS sensitivity) must be confronted with each patient. Often, serious adverse reactions develop before the clinician can give adequate therapeutic doses of medication. Therefore, the practitioner should carefully monitor each drug and change the regimen if the initial trial fails. Even with skilled pharmacological and nonbiological interventions, older patients may have limited responses. Despite these problems, caretakers who provide comprehensive treatment, including psychosocial, medical, and psychopharmacological approaches, often see gratifying results.

SELECTED READING

1. Salzman C.: *Clinical Geriatric Psychopharmacology.* McGraw-Hill, 1984.
2. Salzman C., Lebowitz B. (eds): *Anxiety in the Elderly.* New York, Springer, 1990.
3. Salzman C.: Practical considerations on the pharmacologic treatment of depression and anxiety in the elderly. *J Clin Psychiatry* 51(Suppl):40–43, 1990.
4. Glassman R., Salzman C.: Interactions between psychotropic and other drugs: An update. *Hospital and Community Psychiatry* 38:236–242, 1987.
5. Salzman C.: Treatment of the elderly agitated patient. *J Clin Psychiatry* 48(Suppl):19–22, 1987.
6. Salzman C.: Treatment of agitation in the elderly, in Meltzer H.Y. (ed): *Psychopharmacology: The Third Generation of Progress.* New York, Raven Press, 1987, pp. 1167–1176.
7. Salzman C.: Treatment of the agitated demented elderly patient. *Hospital and Community Psychiatry* 39:1143–1144, 1988.
8. Rosebush P., Salzman C.: Memory disturbance and cognitive impairment in the elderly, in Tupin J. P., Shader R. I., Harnett D. S. (eds): *Handbook of Clinical Psychopharmacology,* ed 2. New Northvale, New Jersey, London, Jason Aronson, 1988, pp. 159–210.
9. Salzman C.: Treatment of the depressed elderly patient, in Altman H. J. (ed): *Alzheimer's Disease and Dementia: Problems, Prospects and Perspectives.* New York, Plenum Press, 1987, pp. 171–182.
10. Liptzin B., Salzman C.: Psychiatric aspects of aging, in Rowe J. R., Besdene R. (eds): *Geriatric Medicine,* ed 2. Boston, Little, Brown, 1988.
11. Berezin M., Liptzin B., Salzman C.: Geriatric psychiatry, in Nicholi A. (ed): *Harvard Guide to Modern Psychiatry,* ed 2. Cambridge, Harvard University Press, 1988.
12. Salzman C.: Treatment of agitation, anxiety, and depression in dementia. *Psychopharmacology Bulletin* 24:139–142, 1988.

9

Pediatric
Psychopharmacology

JOSEPH BIEDERMAN, M.D., and
RONALD STEINGARD, M.D.

I. INTRODUCTION

The origins of pediatric psychopharmacology can be traced back to 1937, when
Bradley published his findings on the use of the stimulant D-amphetamine (Dex-
edrine®, and others) in agitated children with various diagnoses. Despite its long
history, the field of pediatric psychopharmacology has shown relatively little progress over the last 40 years beyond the use of stimulants. Most available information
regarding use of psychotropics in the pediatric population has been anecdotal, and
very little of it grounded on empirically based studies. Because of the lack of
research data, the Food and Drug Administration (FDA) has not recommended the
use of most psychotropics in children. This, in turn, has led to the clinical use of
limited numbers of psychotropics for few indications and usually at conservative
doses.

   The use of psychotropics has not been taught until recently in psychiatric
training programs (adult or child), and few psychiatrists have been willing (or
interested) to use medications as part of their treatment strategy. Despite these
limitations, they are commonly prescribed by nonpsychiatric physicians such as
pediatricians, family doctors, and neurologists—practitioners who lack adequate
training or expertise in psychopathology or psychopharmacology. This situation has
led to inappropriate uses and even abuses of psychotropics in juvenile patients,
particularly in the administration of antipsychotics in institutionalized retarded pa-

JOSEPH BIEDERMAN, M.D. • Department of Psychiatry, Harvard Medical School, Boston, Massachusetts 02115; Pediatric Psychopharmacology Unit, Massachusetts General Hospital, Boston, Massachusetts 02114.   RONALD STEINGARD, M.D. • Department of Psychiatry, Harvard Medical
School, Boston, Massachusetts 02115; Pediatric Psychopharmacology Unit, Massachusetts General
Hospital, Boston, Massachusetts 02114.

tients. Although deviant behavior in children may be the result of social disadvantage, some may be rooted in biological and family/genetic factors. Physicians lack the tools to change societal problems, but they can make treatment tools such as psychotropics available to the patients and their families. When psychotropics are appropriately used and properly administered to children, they can often produce small but clinically significant benefits.

II. GENERAL CONSIDERATIONS

Various guiding principles should be considered when using psychotropics in children and adolescents. Psychotropics ameliorate symptoms and lack curative properties. Although target symptoms may be similar in different disorders, treatment will vary according to the underlying condition. For instance, the treatment of insomnia will differ if it occurs in the context of a psychotic disorder, a mood disorder, attention-deficit hyperactivity disorder, an anxiety disorder, or psychosocial stresses. It is essential to determine the diagnosis and to consider alternative treatment approaches in conjunction with the use of psychotropics. The use of pharmacotherapy should be part of a treatment plan in which consideration is given to all aspects of the child's life, and medications should not be used instead of other interventions or only after other interventions have failed. Realistic expectations of pharmacotherapy based on knowledge of what it can and cannot do, as well as careful definition of target symptoms, are essential to a successful pharmacological intervention.

We review pretreatment, treatment, and posttreatment considerations in the use of medications in children.

A. Pretreatment Considerations

The use of psychotropics should always follow a careful assessment of the child and family, including psychiatric, social, cognitive, and educational evaluations. Diagnostic information should be gathered from the child, parents or caretakers, and, whenever possible, from teachers. The clinician should complete a multiaxial diagnostic evaluation based on the third revised edition of the *Diagnostic and Statistical Manual of Mental Disorders* (DSM-IIIR). As in any other psychiatric evaluation, careful attention should be paid to differential diagnosis, including medical, neurological, and psychosocial factors contributing to the clinical presentation.

Psychiatric disorders of children and adolescents can be associated with cognitive deficits and learning difficulties that may not respond to psychotropics. Therefore, it is imperative that the child undergo a careful neuropsychological evaluation aimed at pinpointing deficits and defining remedial interventions. This evaluation may help in designing and implementing an educational plan tailored to the child's needs.

Psychopharmacologic evaluation of the child should address the basic question of whether the patient has a psychiatric disorder (or disorders) that may respond to psychotropics. If that is the case, the clinician should decide which is the best psychotropic to use, taking into consideration the age of the child and the severity and constellation of the symptomatic picture. Target symptoms should always be defined before pharmacotherapy is introduced.

Once a psychotropic has been chosen as the most likely to be efficacious, the family and child should be familiarized with its risks and benefits, availability of alternative treatments, and likely adverse effects. Adverse effects can be divided into those emerging at the beginning of pharmacotherapy (short term), those associated with chronic administration (long term), and those associated with abrupt discontinuation of the drug (withdrawal). Certain adverse effects can be anticipated based on known pharmacological properties of the drug (e.g., anticholinergic adverse effects of tricyclic antidepressants), whereas others, generally rare, are unexpected (idiosyncratic) and are difficult to anticipate based on the properties of the drug. Short-term adverse effects can be minimized by introducing the medication at low initial doses and titrating slowly. Long-term side effects require monitoring of anticipated adverse effects (e.g., growth when using stimulants, kidney and thyroid function when using lithium carbonate). Idiosyncratic adverse effects require drug discontinuation and selection of alternative treatment modalities.

B. Treatment Considerations

Treatment should be started at the lowest possible dose; this usually is the lowest manufactured dose. This initial low dose is recommended as a test dose, since it can alert the clinician to unexpected or unusual adverse effects. By slowly titrating the dose, several additional goals can be attained. First, it diminishes the chances for adverse effects; second, it permits the family and patient to adjust to this new intervention, and it enhances the opportunity to develop rapport; third, it permits the use of the lowest effective dose, since increasing the amount will be dictated by the clinical picture rather than by a set dose range.

Once pharmacotherapy is initiated, frequent contact with the patient and family is recommended to carefully monitor response to the intervention and adverse effects. Active liaison with the child's school should be initiated either directly or with the parents' help. Each visit should include a careful assessment of potential adverse effects and an adjustment of dosage and timing of administration as clinically indicated to maximize beneficial effects and avoid adverse effects. Evaluation of adverse effects should include both subjective reports from the patient and family (e.g., stomach aches, appetite changes) as well as appropriate evaluation of objective measures (e.g., heart rate, blood pressure changes).

C. Long-Term Treatment Considerations

Given the current understanding of childhood psychiatric disorders and concerns regarding long-term exposure to any extrinsic agent, it seems prudent to reevaluate the need for continued psychopharmacological intervention over time. A general rule is to discontinue the medications following a sufficient period of clinical stabilization for approximately 6 months. A trial off medication should occur during a period of relative psychosocial stabilization. Discontinuing a medication during a stressful period can confuse the clinical picture. During the withdrawal phase, the clinician should observe and monitor clear target symptoms. The patient and family should be fully apprised of possible withdrawal symptoms in advance. Withdrawal symptoms should be distinguished from exacerbation of the disorder for which the psychotropic was prescribed. To minimize reactions, it is important to withdraw medications gradually and, if clinically indicated, be prepared to slow the withdrawal of the drug.

Since most psychiatric disorders are chronic or recurrent, a mechanism for timely follow-up after drug discontinuation is necessary. In general, patients should be seen again at 6 months and 1 year. The patient and family should be educated about "warning signs" that may signal an early recurrence. Patients and families should be instructed to alert the treating clinician at the earliest sign of a possible recurrence.

III. PSYCHOTROPICS COMMONLY USED IN PEDIATRIC PSYCHIATRY

A. Stimulants

Stimulants (see Table 1) were the first class of compounds to have been reported as effective in treating behavioral disturbances in children with attention-deficit hyperactivity disorder (ADHD). Stimulants are sympathomimetic drugs structurally similar to endogenous catecholamines. The most commonly used compounds in this class include methylphenidate (Ritalin® and others), D-amphetamine, and magnesium pemoline (Cylert®). These compounds are thought to act both in the central and peripheral nervous system, by preventing reuptake of catecholamines into presynaptic nerve endings and by preventing the degradation of the nerve endings by monoamine oxidase. Methylphenidate and D-amphetamine are both short-acting compounds with an onset of action within 30 to 60 min and a peak clinical effect usually seen between 1 and 3 hr after administration. Therefore, multiple daily administrations are required for a consistent daytime response. A slow-release spansule is available for both methylphenidate and D-amphetamine with a peak clinical effect between 1 and 5 hr; it allows a single dose to be administered in the morning that will last during the school day. Typically these compounds have a rapid onset of action so that clinical response will be evident

TABLE 1. Common Psychotropics Used in Pediatric Psychiatry

Drug	Daily dose	Dosage schedule	Common adverse effects
Stimulants			
Dextroamphetamine Methylphenidate	0.3–2.0 mg/kg	Twice	Insomnia, decreased appetite, weight loss, depression, psychosis (rare, with very high doses), increase in heart rate and blood
Magnesium pemoline	1.0–2.5 mg/kg	Once	pressure (mild), abnormal liver function tests (with pemoline); possible reduction in growth velocity with long-term use; withdrawal effects include possible rebound phenomena
Antidepressants			
(Tricyclics, i.e., imipramine, desipramine, amitriptyline, nortriptyline)	2.0–5.0 mg/kg	Once or twice	Anticholinergic (dry mouth, constipation, blurred vision), weight loss, cardiovascular (mild increase in diastolic blood pressure and delays in electrocardiograph conduction parameters) with
	1.0–3.0 mg/kg (dose adjusted with serum levels)	Once or twice	daily doses <3.5 mg/kg; treatment requires serum levels and electrocardiograph monitoring; no known long-term side effects; withdrawal effects can occur (severe gastrointestinal symptoms, malaise); overdoses can be fatal
Monoamine oxidase inhibitors (MAOIs, i.e., phenelzine, tranylcypromine)	0.5–1.0 mg/kg	Once or twice	Severe dietary restrictions (high-tyramine foods); hypertensive crisis with dietetic transgression or with certain drugs; weight gain, drowsiness, changes in blood pressure, insomnia, liver toxicity (remote)
Antipsychotics			
Low potency (i.e., chlorpromazine, thioridazine)	3–6mg/kg	Once or twice	Anticholinergic (dry mouth, constipation, blurred vision, more common with low potency agents); weight gain; extrapyramidal reactions (dystonia, rigidity, tremor, akathisia, more com-
High potency (i.e., haloperidol, fluphenazine, thiothixene)	0.1–0.5 mg/kg	Once or twice	mon with high potency); drowsiness; risk for tardive dyskinesia with chronic administration; withdrawal dyskinesia with drug discontinuation

(continued)

TABLE 1. (Continued)

Drug	Daily dose	Dosage schedule	Common adverse effects
Lithium	10–30 mg/kg (dose adjusted with serum levels)	Once or twice	Polyuria, polydipsia, tremor, nausea, diarrhea, weight gain, drowsiness, skin abnormalities; possible effects on thyroid and renal functioning with chronic administration; therapy requires monitoring of lithium levels, thyroid and renal tests
Antianxiety			
Benzodiazepines			
Clonazepam	0.01–0.04 mg/kg	Once or twice	Drowsiness, disinhibition, agitation, confusion, depression
Alprazolam	0.02–0.07 mg/kg	Three times	
Other drugs			
Clonidine	3–10 μg/kg	Twice / Three times	Sedation, hypotension, dry mouth, confusion, depression; localized irritation with transdermal preparation
Propranolol	2–8 mg/kg	Twice / Three times	Sedation, hypotension, bradycardia, dry mouth, confusion, depression, rebound
Fenfluramine	1–2 mg/kg	Twice	Anorexia, irritability, drowsiness, weight loss, insomnia (when discontinued)
Anticonvulsants			
Carbamezapine	10–20 mg/kg	Twice	Bone marrow suppression (requires baseline and close monitoring of blood counts), dizziness, drowsiness, rashes, nausea, abnormal liver function tests
Phenytoin	5–10 mg/kg	Once or twice	Nystagmus, slurred speech, confusion, nausea, rash, megaloblastic anemia (treated with folic acid), gingival hyperplasia, hypertrichosis

when a therapeutic dose has been obtained. Magnesium pemoline is a longer-acting compound with a duration of action that is greater than 24 hr and thus can be given as a single daily dose. The onset of action can often be delayed for as long as 2–8 weeks, and it is considered to be a less reliable medication.

The primary indication for the use of these compounds is to treat ADHD. This disorder is manifest by poor attention, impulsive behavior, and often significant motor restlessness and overactivity (see Section IV.A). The stimulant medications diminish the motor overactivity and impulsive behaviors and allow the patient to sustain attention. Stimulants can also be effective in children and adolescents with attention-deficit disorder without hyperactivity and also in treating mentally retarded patients with ADHD. Finally, as in adults, stimulants can be helpful as adjunct treatment for refractory mood disorders (depressive disorders).

Because of their short half-life, the short-acting stimulants should be given in divided doses throughout the day, typically 4 hr apart. The total daily dose ranges from 0.3 mg/kg per day to 2 mg/kg per day. Starting dose is generally 2.5 to 5 mg/day given in the morning with the dose increased if necessary every few days by 2.5 to 5 mg in a divided dose schedule. In some cases, a useful strategy can be gradually to titrate a single morning dose to assess an effective dose and subsequently to repeat this process for the second and, if necessary, third daily dose. Because of the anorexogenic effects of the stimulants, it may be beneficial to administer the medicine after meals. Magnesium pemoline is typically given as a single daily dose in the morning. The dose ranges from 1.0 to 2.5 mg/kg per day. The typical starting dose is 18.75 to 37.5 mg, with increments of 18.75 mg given every few days thereafter until the appearance of desired effects or of side effects that preclude further increments. Both D-amphetamine and methylphenidate are available in tablet and slow-release preparations. Pemoline is available in tablet form.

The most commonly reported side effects associated with stimulant medication are appetite suppression and sleep disturbances. Appetite suppression correlates with the clinically active phase of drug action. It is not necessary to administer the medication prior to meals, as the presence of food in the stomach does not appear to alter the pharmacokinetics or the behavioral effects of the medication. Therefore, it is possible to administer the medication in a scheduled way so that the anorexogenic effects are minimized at meal time. If the anorexogenic effects are pronounced and multiple doses are required for behavioral stabilization, then weight loss can occur. Commonly the child experiences a delay of sleep onset, a side effect that usually accompanies late afternoon or early evening administration of the stimulant medications. Although less commonly reported, mood disturbances ranging from increased tearfulness or emotional sensitivity to a full-blown major depression-like syndrome can be associated with stimulant treatment. Other infrequent side effects include headaches, abdominal discomfort, and increased lethargy and fatigue. Mild increases in baseline pulse and blood pressure (of little clinical significance) can also be observed. A stimulant-associated toxic psychosis has been observed, usually in the context of either a rapid rise in the dosage or very high doses. In children the psychosis resembles a toxic phenomenon (visual hallucinosis and formication) and is dissimilar from the exacerbation of schizophrenic symptoms. However, a preexisting psychotic disorder should be carefully ruled out. Administration of magnesium pemoline has been associated with hypersensitivity reactions involving the liver and accompanied by elevations in liver function studies (SGOT and SGPT). This reaction has been reported after several months of treatment. Thus, baseline liver function studies and repeat studies are recommended.

Sometimes a chronic tic disorder is precipitated or exacerbated by stimulant administration. Children with a personal or family history of tic disorders are at greater risk for developing a chronic tic disorder, which can persist after discontinuation of the stimulant. It is unknown whether the child would have spontaneously developed a tic disorder. The administration of stimulants should be avoided in children suspected of having a risk for tic disorders.

Administration of stimulants for long periods can affect growth. Initial reports, based on a small number of cases, suggested that stimulants decrease the growth rate in children. However, more recent studies do not support this contention. In general, the consensus is that stimulants can produce a small negative (deficit) impact on growth velocity that can be easily offset by drug holidays. It would be unsound practice to attribute serious growth retardation to stimulants. Careful monitoring of growth is certainly indicated. If a decrease in growth rate occurs, either a drug holiday or alternative treatment options should be considered.

Transient behavioral deterioration can occur on the abrupt discontinuation of stimulant medications in some children. The prevalence of this phenomenon and the etiology are unclear. Symptoms usually last for less than a day but can persist in some cases for several days. Thus, it would be advisable to taper off stimulants. Rebound phenomena can also occur in some children between doses, creating an uneven, often disturbing clinical course. In those cases, alternative treatments should be considered.

In addition, stimulants are contraindicated for patients with marked anxiety, since they can aggravate these symptoms.

Stimulants can adversely interact with some antihypertensive agents by decreasing their antihypertensive effect. They should be used with caution in combination with monoamine oxidase inhibitors. Since the stimulants can inhibit the metabolism of anticoagulants, anticonvulsants, and tricyclic antidepressants, the combined use of stimulants with these medications requires appropriate monitoring (serum levels) and dose adjustments when necessary.

In summary, this class of compounds has been extensively studied in treating ADHD and has been shown to be very effective. The most effective stimulants are the short-acting compounds, which require multiple daily doses to maintain their clinical effect.

B. Antidepressant Drugs

Antidepressant medications comprise two main families of compounds: the imipramine-like drugs, usually referred to as tricyclic antidepressants or TCAs, and the monoamine oxidase inhibitors or MAOIs. Although well established in the psychopharmacological armamentarium of adult psychiatric disorders, their use in pediatric psychiatry has been limited and almost exclusively focused on the use of imipramine (Tofranil® and others). There is no adequate information regarding the efficacy and toxicity of either the monoamine oxidase inhibitors [phenelzine (Nardil®), tranylcypromine (Parnate®)] or the newer antidepressants such as maprotiline (Ludiomil® and others), trazodone (Desyrel® and others), and amoxapine (Asendin®) in treating childhood psychiatric disorders.

1. Tricyclic Antidepressants

The TCAs are structurally similar to the phenothiazines but have a different spectrum of clinical and adverse effects. Several TCAs are currently available. They include the tertiary amines amitriptyline (Elavil® and others), imipramine, doxepin (Sinequan® and others), and trimipramine (Surmontil®), and the secondary amines desipramine (Norpramin® and others), nortriptyline (Pamelor® and others), and proptriptyline (Vivactil®). Imipramine is the only TCA currently approved by the FDA for use in pediatric populations, in daily doses of up to 2.5 mg/kg, and only in treating enuresis.

The postulated mechanism of action of the tricyclic antidepressants in treating depression has focused on the capacity of these drugs to prevent the reuptake of brain neurotransmitters, especially norepinephrine (NE) and serotonin (5-HT). Although, in general, antidepressants have similar spectra of action, their inhibitory effects on transmitter reuptake and their anticholinergic effects vary. Since the response to antidepressants is seen more rapidly in children with enuresis or ADHD than in children treated for mood disorders, a different mechanism of action has been postulated for these disorders.

Early reports in the 1970s suggested that TCAs may be more toxic in children than in adults. Children were thought to be more susceptible for two reasons: first, children have a smaller adipose compartment than adults, which could affect the uptake, redistribution, and storage of these drugs in the body; and second, children have decreased binding of TCAs to plasma albumin, which may result in greater amount of circulating free drug. Although these concerns may apply to very young children, available evidence, as well as clinical experience, supports similarities rather than differences in efficacy and toxicity with the use of TCAs in children and adolescents when compared with the clinical experience with adults.

Antidepressants are long-acting drugs with half-lives in children that may range from 10 to 17 hr and therefore can be administered once or twice daily with good effect.

Available information from controlled and open studies as well as case reports suggests that TCAs may be effective in various childhood psychiatric disorders. Established indications for TCAs in pediatric psychiatry include the treatment of enuresis, attention-deficit disorder, and major depression. Other possible uses for TCAs include the treatment of Tourette's disorder and childhood anxiety disorders.

As for any other psychotropic medication, dosage should be individualized and the drugs administered should be the lowest effective dose. For imipramine, an upper dose limit of 5 mg/kg has been suggested for children. However, this absolute dosage limit has little meaning, since a substantial interindividual variability in the metabolism and elimination of TCAs has been demonstrated in children. Therefore, some children tolerate only lower doses, and others may require doses above such a ceiling. High doses (up to 5 mg/kg) of imipramine have been used to treat school phobia and major depression in children. The use of such high doses of imipramine or desipramine in treating childhood psychiatric disorders reflects the emerging impression that children are relatively efficient in metabolizing and eliminating TCAs.

Treatment with a TCA should be initiated with 10 mg or 25 mg depending on the size of the child (approximately 1 mg/kg) and titrated slowly every 4 to 5 days by 20% to 30% increments. When a daily dose of 3 mg/kg (or a lower effective dose) is reached, steady-state serum levels and an electrocardiogram (EKG) should be obtained. Most TCAs are available in tablet or capsule forms. Nortriptyline is also available in a liquid preparation.

Widely varying drug concentrations may occur in different patients on the same dose, potentially leading to markedly different outcomes ranging from no response to improvement or to iatrogenic toxicity. Prolongation of intracardiac conduction, as well as increases in diastolic blood pressure and heart rate were found in children with serum levels higher than 250 ng/ml but not at lower levels. In contrast, subjective adverse effects such as dry mouth or dizziness were unrelated to drug serum levels. Since the clinical effectiveness and toxicity of TCAs almost certainly are dose related, and since there is a wide interindividual variability in serum levels attained at a given dose, monitoring serum TCA levels may be more pertinent to maximizing benefits and avoiding toxicity than the daily dose per se, even if based on body size. Monitoring of serum levels can facilitate treatment response by providing objective guidelines for dosage adjustment. Serum TCA monitoring can also assure compliance.

Common short-term adverse effects of the tricyclic antidepressants include anticholinergic effects, such as dry mouth, blurred vision, and constipation. However, there are no known deleterious effects associated with chronic administration of these drugs. Gastrointestinal symptoms and vomiting are manifested when these drugs are discontinued abruptly. Since the anticholinergic effects of TCAs limit salivary flow, they may promote tooth decay in some children. More frequent dental evaluation may be advisable in children receiving long-term TCA treatment.

Concerns have been raised regarding the possible cardiac toxicity of TCAs in young children, especially at daily doses above 3.5 mg/kg. For example, it has been reported consistently that children tend to have small but statistically significant elevations in diastolic blood pressure induced by imipramine and desipramine. In contrast, in studies of adult patients, orthostatic hypotension is the most frequent potentially dangerous cardiovascular effect of TCA treatment. Sinus tachycardia (>100 bpm) was also reported in studies of TCAs in children. In young children, tachycardia is not in itself abnormal or of hemodynamic significance. In fact, sinus tachycardia is the rule rather than the exception in pediatric patients at the time of medical examination. In older children or adolescents, however, a persistent heart rate above 130 bpm is of greater concern and may require noninvasive evaluation of cardiac function such as echocardiography with Doppler, which permits measurement of ventricular ejection fraction and cardiac output.

Electrocardiographic abnormalities also have been consistently reported in children receiving TCA doses higher than 3.5 mg/kg. Preskorn et al.[1] reported that, at daily doses of imipramine of up to 5 mg/kg, EKG changes (reduction in conduction efficiency) were observed uniformly when "total" TCA (imipramine plus desipramine) serum levels exceeded 250 ng/ml but not at lower levels. Biederman et al.[2,3] reported similar findings associated with desipramine.

A widened PR interval during TCA treatment, in the absence of the development of A–V block (PR interval > 200 msec), is of no hemodynamic consequence. Incomplete right intraventricular conduction defect is a normal EKG finding in the first decade of life, and in the absence of underlying heart disease there is no documented evidence that it is of any hemodynamic consequence to the patient. Nevertheless, the development of incomplete right intraventricular conduction defects in patients receiving TCA treatment deserves closer EKG and clinical monitoring, especially when relatively high doses of TCA (especially doses above 3.5 mg/kg) are used. In the context of a healthy heart, a QRS duration above 120 msec (complete right bundle branch block) does not necessarily imply impaired cardiac function. However, since complete right bundle branch block delays the electricomechanical function of the right ventricle, its development warrants further (noninvasive) assessment of the quality of myocardial contraction (ejection fraction and cardiac output). In the context of cardiac disease, however, complete right intraventricular conduction defect has potentially more serious implications. Thus, patients with documented congenital or acquired cardiac disease, cardiac murmurs, pathological rhythm disturbances (i.e., A–V block, supraventricular tachycardia, ventricular tachycardia, Wolf–Parkinson–White syndrome), family history of sudden cardiac death or cardiomyopathy, diastolic hypertension (>90 mm Hg), or when in doubt about the cardiovascular state of the patient, a complete (noninvasive) cardiac evaluation is indicated before initiating treatment with a TCA to help determine the risk/benefit ratio of such intervention. Cardiac evaluation should include cardiologic consultation, 24-hr Holter monitoring, and echocardiographic evaluation.

The possible clinical significance of the cardiovascular changes observed, especially of small delays in cardiac conduction in generally healthy children, is not clear. To some extent, they reflect pharmacological actions of TCAs that may not be harbingers of dangerous cardiac malfunction. TCAs can increase heart rate and can delay intracardiac conduction, but these effects, if small, seem rarely to be pathophysiologically significant in noncardiac patients (adults and children) with normal baseline EKGs, or even in adults with heart disease. Although toxic concentrations of TCA drugs resulting from overdoses regularly produce significant depressant effects on myocardial conduction and contractile efficiency, therapeutically effective doses or serum levels in adults free of cardiac disease are reasonably safe and have no adverse effects on left ventricular ejection fraction. Nevertheless, there still may be serious cardiovascular reactions, possibly idiosyncratic, to unusually high doses or tissue levels of TCAs in pediatric patients, as suggested by the sudden death of one 6-year-old girl receiving a daily dose of imipramine of 14.7 mg/kg. Although the reversibility of TCA-induced adverse cardiovascular effects in children has not been investigated, clinical experience indicates that imipramine- and desipramine-induced cardiovascular changes are reversible with decreased dose or drug discontinuation.

Biederman et al. reported no significant differences in cardiovascular parameters between children and adolescents. Thus, the available data do not support the frequently held notion that prepubertal children have an increased risk for TCA-associated cardiac toxicity and suggest a similar risk for these adverse effects, especially when daily doses above 3.5 mg/kg are used.

Patients with cardiovascular disease should receive antidepressants under close medical supervision. Tricyclics should be used with caution in patients with narrow-angle glaucoma or history of urinary retention. They should also be used with caution in patients with a history of seizures, since tricyclics may lower the seizure threshold. Care is required when antidepressants are given to hyperthyroid patients or to those receiving thyroid medication, since cardiac arrhythmias can develop. Excessive consumption of alcohol in combination with tricyclics should be avoided because of possible potentiating effects. The use of tricyclics in schizophrenic and manic patients requires caution, since these drugs can exacerbate psychotic and manic symptoms.

Concomitant use of a tricyclic antidepressant and an MAOI antidepressant is currently contraindicated. It is generally recommended that the tricyclic or MAOI be discontinued at least 2 weeks before switching to the other family of antidepressants. The antihypertensive effects of some antihypertensive drugs may be blocked when used concomitantly with tricyclics. Tricyclics can decrease the metabolization of anticonvulsants and antipsychotics with resulting possible increased toxicity of both drugs. The combined use of tricyclics with sympathomimetic amines can result in hypertension and a hypertensive crisis.

In summary, TCAs are long-acting compounds found effective in treating enuresis, attention-deficit hyperactivity disorder, childhood depression, and juvenile anxiety disorders. Treatment with TCAs, especially at daily doses of above 3.5 mg/kg or at serum levels higher than 150 ng/ml, may increase the risk of asymptomatic electrocardiographic changes (especially of the PR and QRS intervals) indicative of delayed cardiac conduction as well as minor increments in diastolic blood pressure and heart rate. Although daily doses of 5 mg/kg may be needed clinically and to attain serum TCA levels above 100 ng/ml, in some cases, doses over 3.5 mg/kg may be excessive for some children who develop excessively high serum levels or cardiac conduction defects. Optimal doses probably range between 2.5 and 5 mg/kg for most children, but TCA therapy in children requires optimization clinically and monitoring of serum TCA levels and EKG. Despite the effectiveness of TCAs in several childhood psychiatric disorders, the use of these compounds requires special caution because of the potential risk of overdosage, whether accidental, suicidal, or iatrogenic.

2. Monoamine Oxidase Inhibitors

The available information on clinical uses of MAOIs in pediatric psychiatry is extremely limited. This class of drugs includes hydrazine (i.e., phenelzine) and nonhydrazine (i.e., tranylcypromine) compounds. There are no known established indications for the use of MAOIs in pediatric psychiatry. Anecdotal reports have suggested that the MAOIs may be useful in major depressive disorders, conduct disorders, schizophrenic and autistic children, as well as phobic anxiety states and obsessive compulsive disorder. A recent controlled study in a small group of patients suggested that MAOIs are useful in treating ADHD. Pediatric dose ranges have not been established. In two studies using phenelzine in children, the dose started at 15 mg q.i.d. and increased up to 15 mg t.i.d. in older children. Short-term adverse effects include the potential for hypertensive crisis [treatable with phentolamine (Regitine®)] associated with dietetic transgressions or drug interactions, orthostatic hypotension, weight gain, drowsiness, and dizziness. Long-term adverse effects are not known in children. Extrapolating from the adult literature, those may include hypomania, hallucination, confusion, and hepatotoxicity (rare). Major general limitations for the use of this class of drugs in children and adolescents are the dietetic restrictions of tyramine-containing foods (i.e., most cheeses), pressor amines (i.e., sympathomimetic substances), or drug interactions (i.e., most cold medicines, amphetamines). Drug interactions are one of the most serious risks with MAOIs.

C. Antipsychotics

The major classes of antipsychotic drugs used clinically are (1) phenothiazines, which include the low-potency compounds (requiring high dosages) such as chlorpromazine (Thorazine® and others) and thioridazine (Mellaril® and others) and the high-potency compounds such as trifluoperazine (Stelazine® and others) and perphenazine (Trilafon® and others); (2) butyrophenones [i.e., haloperidol (Haldol® and others) and pimozide (Orap®)]; (3) the thioxanthenes [i.e., thiothixene (Navane® and others)]; (4) indolone derivatives [i.e., molindone (Moban®)]; and (5) dibenzazepines [loxapine (Loxitane® and others)]. These chemically varied drugs are remarkably similar pharmacologically and generally yield comparable benefit when given in equivalent doses and induce a similar variety of adverse effects. In addition, those with low potency (e.g., chlorpromazine and thioridazine) are particularly likely to induce unwanted effects such as hypotension and sedation. All antipsychotic agents are about equally effective.

The best postulated mechanism of action of these drugs in treating schizophrenia is thought to be inhibition of dopaminergic receptor sites in the brain. However, clozapine (Clozaril®), a new and atypical but effective antipsychotic agent, has a relatively strong antagonistic interaction with central α-adrenergic receptors, lacks acute extrapyramidal adverse effects, and exerts only weak antagonism of dopaminergic transmission in the basal ganglia and the limbic forebrain. Antipsychotic medications have a relatively long half-life (24 to 38 hr in adults), and, therefore, they need not be administered more frequently than twice daily.

Antipsychotics are only indicated in treating childhood psychotic disorders. However, they are prescribed for a variety of other disorders. Antipsychotics are commonly used in treating **complications** of pervasive developmental disorders and mental retardation such as severe agitation, aggression, self-abuse, and insomnia. This class of drugs is also used in treating Tourette's disorder and attention-deficit hyperactivity disorder as well as in the management of juveniles with aggressive forms of conduct disorders. The target symptoms that most commonly respond to antipsychotics include excessive motor activity (agitation), aggressiveness, tics, stereotypies, delusions, and hallucinations. Antipsychotic agents should not be used for treating anxiety or for sedation, conditions for which antianxiety medication can be highly effective.

The usual oral dosage of antipsychotic drugs ranges between 3 and 6 mg/kg per day for the low-potency phenothiazines and between 0.5 and 1.0 mg/kg per day for the high-potency phenothiazines, butyrophenones, thioxanthenes, and indole derivatives. Most antipsychotic preparations are available in either tablet or capsule form. In addition, at least one compound from each class of antipsychotic is available in a liquid concentrate form. Several compounds including chlorpromazine, thiothixene, loxapine, haloperidol, and fluphenazine are available in injectable form for intramuscular administration. Fluphenazine and haloperidol are also available in "depot" preparations for intramuscular use; these can be administered every 2 to 4 weeks and are used primarily in the adult population to help assure treatment compliance in severely disturbed patients.

Common short-term adverse effects of antipsychotic drugs are drowsiness, increased appetite, and weight gain. Anticholinergic effects such as dry mouth, nasal congestion, and blurred vision are more commonly seen with the low-potency phenothiazines. Extrapyramidal effects such as acute dystonia, akathisia (motor restlessness), and parkinsonism (bradykinesia, tremor, facial inexpressiveness) are more commonly seen with the high-potency compounds (phenothiazines, butyrophenones, and thioxanthenes).

As in adults, the long-term administration of antipsychotic drugs may be associated with tardive dyskinesia. Although children appear generally less vulnerable than adults for developing tardive dyskinesia, there is an emerging consensus that this potentially worrisome adverse effect affects children and adolescents. The potentially irreversible tardive dyskinesia should be distinguished from the more common, generally benign withdrawal dyskinesia associated with the abrupt cessation of these drugs, which tends to subside after several months of drug discontinuation. A behavioral syndrome of deteriorating behavior ("tardive dysbehavior," "supersensitivity psychosis") has also been reported to be part of the withdrawal phenomena associated with antipsychotic drug discontinuation. This behavioral syndrome seems to be qualitatively different from the target symptoms for which the drug was initially prescribed and usually ceases spontaneously after a few weeks. To minimize withdrawal reactions, taper antipsychotic drugs **very slowly** over several months. Little is known about the potentially lethal neuroleptic malignant syndrome in juveniles.

Prevention (appropriate use, clear indication, clear target symptoms, periodic drug discontinuation to assess need for drug use) and early detection (with regular monitoring) is the only available treatment for tardive dyskinesia. Reduction of dose or drug discontinuation (if clinically feasible) should be undertaken rapidly whenever clinically feasible once tardive dyskinesia has been detected. Other short-term adverse effects of antipsychotics are more easily managed. Excessive sedation can be avoided by using less sedating antipsychotics and managed by prescribing most of the daily dose at night. Drowsiness should not be confused with impaired cognition and can usually be corrected by adjusting the dose and timing of administration. In fact, there is no evidence that antipsychotics adversely affect cognition when used in low doses. Anticholinergic adverse effects can be minimized by choosing a medium- or high-potency compound. Extrapyramidal reaction can be avoided in most cases by slow titration of an antipsychotic dose. Antiparkinsonian agents [i.e., anticholinergic drugs, antihistamines, amantadine (Symmetrel® and others)] should be avoided unless strictly necessary because of the added adverse effects that these drugs may produce. Akathisia may be particularly problematic in young patients because of underrecognition. It should be considered in the differential diagnosis of agitation and anxiety in a patient receiving antipsychotic therapy. The centrally acting β-adrenergic antagonist propranolol (Inderal® and others) is often very helpful in treating this bothersome adverse effect.

Antipsychotics are contraindicated in patients with a seriously compromised sensorium and in the presence of blood dyscrasias or bone marrow depression. They are also contraindicated in patients with suspected or established subcortical brain damage because of the risk for hyperthermic reaction. Antipsychotics should be used with great caution in patients with liver disease and in patients with seizure disorders (because of the potential for lowering seizure threshold). Antipsychotic drugs may elevate prolactin levels and result in associated amenorrhea and galactorrhea.

When concomitantly used with a tricyclic antidepressant, they can increase antidepressant toxicity by decreasing metabolism of the TCA.

In summary, antipsychotic drugs are only indicated in treating psychotic symptomatology. However, they are widely used in pediatric psychiatry for the treatment of a variety of disorders and target symptoms for which alternative effective pharmacological and nonpharmacological treatments may be available. Since long-term (more than 3 months) antipsychotic treatment may be associated with potentially irreversible tardive dyskinesia, the risk/benefit ratio of such intervention in non-psychotic disorders should be carefully evaluated, and these drugs possibly reserved for conditions in which other treatments have failed. Antipsychotics are long-acting drugs that can (and should) be given not more often than twice daily. Antiparkinsonian agents should not be used prophylactically, since they tend to aggravate the already taxed adverse effect profile of antipsychotic therapy.

D. Lithium Carbonate

Lithium (Eskalith® and others) is a simple solid element, and it bears chemical similarities to sodium, potassium, calcium, and magnesium. As with other psychotropic agents, the precise cellular mechanism of action is unknown. Lithium has diverse cellular actions that alter hormonal, metabolic, and neuronal systems. It is unknown if lithium-induced changes in these systems are the cause or the effect of the illness. Based on its putative biological and biochemical actions, proposed theories of lithium's mechanism of action include neurotransmission (i.e., interaction with catecholamine, indoleamine, cholinergic, and endorphin systems, inhibition of β-adrenoreceptors), effects on the endocrine system (i.e., blocking release of thyroid hormone and the synthesis of testosterone), circadian rhythm (i.e., normalization of altered sleep–wake cycles), and cellular processes (i.e., ionic substitution, inhibition of adenylate cyclase). However, until more is known about the pathophysiology of the psychiatric disorders for which lithium is used, it is not possible to determine which of lithium's effects is responsible for its therapeutic actions.

In adults, the elimination half-life of lithium is approximately 24 hr, and it takes 5 to 7 days to reach steady state. Thus, there is no need to obtain serum lithium levels more often than once a week unless there is concern about toxicity. Since work done to determine lithium therapeutic ranges has been based on a 12-hr sampling interval, it is imperative that blood samples for serum lithium determination always be drawn 12 hr after the last dose. Micromethods for lithium serum level determinations are available and permit samples to be obtained using fingerstick technique. Recent data suggest that lithium determination in saliva correlates with serum lithium levels. When this technique becomes more commonly available, it may facilitate the monitoring of therapy in young children in whom venipuncture may be problematic.

Despite its wide and clearly documented use in adult psychiatry, there is very limited empirical evidence documenting the uses of lithium in pediatric psychiatry. Based on the experience in adult psychiatry, the main use of lithium is in the acute and prophylactic treatment of bipolar disorder, manic type. Lithium can also be effective in the treatment and prophylaxis of major depression (bipolar and nonbipolar), schizoaffective disorders, and as an adjunct treatment of schizophrenia. Although lithium has been shown in controlled studies to have an antiaggressive effect in certain populations, its usefulness in treating aggressive or explosive behavior has not been established. Nevertheless, lithium may be a viable treatment consideration for patients with episodic aggressive behavior, particularly those who have not been responsive to other treatments. Lithium has been used with some success in adults in treating certain types of headaches (chronic cluster and cyclic migraine). It has been proposed that lithium may be efficacious in treating tardive dyskinesia; however, the evidence from clinical trials has been inconclusive. It has also been suggested that lithium may be effective in treating a wide range of behavioral disorders in children of lithium-responsive bipolar disorder adults.

Because of the complex relationship between lithium and sodium excretion, dose–serum level associations can be variable and require careful monitoring. The usual lithium starting dosage ranges from 150 to 300 mg/day in divided doses twice or three times a day. There is no standard therapeutic serum lithium level range for children. Based on the adult literature, suggested guidelines include serum levels of 0.8 to 1.5 mEq/liter for acute episodes and levels of 0.6 to 0.8 mEq/liter for maintenance therapy. Nevertheless, as with any other intervention, the lowest effective dose or serum level should always be used.

Lithium preparations include standard lithium carbonate 300 mg (tablets or capsules). Scored tablets permit the introduction and titration of smaller doses (by permitting the breaking of the tablets into quarters or halves). Slow- or controlled-release lithium carbonate preparations are available. The slow-release preparations are more gradually absorbed, reaching peak serum levels about 3 hr later than the regular preparations. Although there does not appear to be a clear advantage to using the standard or the slow-release preparations, some individuals may tolerate one better than another. Lithium is also available in a liquid preparation (lithium citrate) containing 8 mEq of lithium per 5 cc (the same amount of lithium contained in 300 mg of lithium carbonate), which may facilitate its intake in young patients unable or unwilling to take pills.

Common short-term adverse effects include gastrointestinal symptoms such as nausea and vomiting, renal symptoms such as polyuria and polydipsia, and central nervous symptoms such as tremor, sleepiness, and memory impairment. Chronic administration of lithium may be associated with metabolic (decreased calcium metabolism, weight gain), endocrine (decreased thyroid functioning), and possible renal damage. Thus, it is imperative that children be screened for renal (BUN, creatinine) and thyroid function (T_3, T_4, TSH) and calcium metabolism (Ca,P) before lithium treatment is started and that these tests be repeated at least every 6 months. Particular caution should be exercised when lithium is used in patients with neurological, renal, and cardiovascular disorders.

Although unsubstantiated, concerns have been raised regarding adverse interactions of lithium with antipsychotics (particularly haloperidol) leading to encephalopathy. Nevertheless, in clinical practice the judicious use of lithium with antipsychotics is not only safe but often necessary. Caution should be exercised when lithium is used with drugs that reduce renal lithium clearance such as thiazide diuretics and nonsteroidal antiinflammatory drugs, which may result in an increase in the serum concentration of lithium.

In summary, although well established in the acute and prophylactic treatment of adults with mood disorders, lithium's use in pediatric psychiatry is promising but still experimental. The use of lithium salts requires pretreatment and periodic follow-up evaluations of renal and thyroid functioning as well as close monitoring of blood levels. New technologies may permit monitoring of lithium levels in saliva.

E. Antianxiety Drugs

There is little information available regarding the efficacy and toxicology of antianxiety agents in pediatric psychiatry. The most important agents in this class are the benzodiazepines. Other available compounds include several barbiturates, several compounds structurally related to alcohol [i.e., chloral hydrate (Noctec® and others), paraldehyde, meprobamate (Miltown® and others)], and sedative antihistamines [i.e., diphenhydramine (Benadryl® and others), hydroxyzine (Atarax®, Vistaril®, and others), and promethazine (Phenergan® and others)]. No information is available for the use of the new atypical nonbenzodiazepine antianxiety drug buspirone (BuSpar®) in pediatric psychiatry.

Antianxiety drugs decrease spontaneous and conditioned behavioral responses in laboratory animals, elevate seizure threshold, produce ataxia, and potentiate the sedative effects of other sedatives such as barbiturates. Benzodiazepines also have high-affinity binding sites, as well as interactions with membrane components of γ-aminobutyric acid (GABA) receptors and chlorine ion transport systems.

Because of their pharmacological (clinical effects) and toxicological (comfortable margin of safety) properties, the benzodiazepines are the most widely used antianxiety drugs in adults. In adults, most benzodiazepines are absorbed at an intermediate rate, with peak plasma levels appearing in 1 to 3 hr. Most benzodiazepines are lipophilic and highly bound to plasma membranes. Many of these agents have active metabolites that dominate their course of activity. Benzodiazepines tend to have minimal pharmacokinetic interactions with other drugs. In adults, the benzodiazepines can produce tolerance (and cross-tolerance with other benzodiazepines) and dependence [physiological (addiction) and psychological (habituation)].

There are no established indications for the use of antianxiety agents in pediatric psychiatry. However, because of their low toxic profile and comfortable margin of safety, antianxiety agents, particularly antihistamines, are commonly prescribed to young children to treat poorly diagnosed symptoms of agitation and insomnia. Since the pharmacological profile of antianxiety drugs includes behavioral disinhibition, and since many childhood psychiatric disorders are characterized by behavioral disinhibition, the use of these agents in the absence of a specific indication carries the risk of worsening the clinical picture.

Possible indications for the use of antianxiety agents in pediatric psychiatry include the treatment of childhood anxiety symptoms and disorders. In recent years, the high-potency benzodiazepines alprazolam (Xanax®) and clonazepam (Klonopin®) have received increasing attention as effective and safe treatment for adult panic disorder with and without agoraphobia. Childhood anxiety disorders include separation anxiety disorder, overanxious disorder, and avoidant disorder. Recent reports suggest that children also can manifest adult-type anxiety disorders such as panic disorder and agoraphobia, which may also respond to high-potency benzodiazepines. Additional possible uses of benzodiazepines include adjunct treatment in acute psychotic episodes and in treatment-refractory schizophrenia. The chlorinated benzodiazepines clonazepam and clorazepate (Tranxene® and others) may be particularly helpful in treating complex partial seizures. In adults, the rapid activity of diazepam (Valium® and others) and its tendency to produce euphoria have made it a popular drug for abuse. Lorazepam (Ativan® and others) and oxazepam (Serax® and others) do not have active metabolites and do not tend to accumulate in tissue, making them preferable for short-term symptomatic use. When long-term use is anticipated, longer-acting benzodiazepines are preferable.

The potency of these compounds is variable. In general, as previously discussed, treatment should be initiated with the lowest available dose and titrated according to clinical response and side effects.

In general, the clinical toxicity of the benzodiazepines is low. The most commonly encountered short-term adverse effects are sedation, drowsiness, and decreased mental acuity. In high doses they can produce a confusional picture. In adults, benzodiazepines have been reportedly associated with depressogenic effects. With the exception of the potential risk for tolerance and dependence (risk suspected but not well studied in adults, risk unknown in children), there are no known long-term adverse effects associated with benzodiazepines. Adverse withdrawal effects can occur, and benzodiazepines should always be tapered slowly.

Since treatment with benzodiazepines can produce central nervous system depression, patients should be cautioned about possible decreased alertness and the potentiation of this effect with the concomitant use of alcohol, barbiturates, narcotics, other antianxiety medications, antipsychotics, antidepressants, and anticonvulsants.

Use in patients with acute narrow-angle glaucoma is contraindicated, and patients with open-angle glaucoma should receive these compounds only if they are receiving appropriate treatment for their glaucoma.

These agents represent a potentially safe and effective treatment of childhood anxiety disorders. They are a potential alternative to the use of antidepressant medications in the management of these disorders. Given the low side-effect profile and high efficacy in adult patients, further clinical investigation is warranted.

F. Other Drugs

1. Fenfluramine

Fenfluramine hydrochloride (Pondimin®) is a sympathomimetic amine related to the amphetamines and utilized as an appetite suppressant. Fenfluramine has been shown to decrease blood and brain serotonin concentrations in animals. The finding that 30–40% of patients with the syndrome of autism have elevated blood serotonin levels prompted investigation of this agent in treating autism. Early studies suggested improvement in social awareness, communication, and cognitive functioning and a decrease in hyperactivity and sensorimotor symptoms. Follow-up studies have been unable to support this earlier optimism. However, clinical studies still suggest improvement in a select population of autistic children, particularly in children with an IQ over 40 and pronounced agitation. The recommended dose is 1.5 mg/kg per day, but a range of optimal doses has been reported as 1.1 to 1.8 mg/kg per day. Flexible dose schedules are suggested, but generally a b.i.d. dose schedule has been employed. Onset of response has been reported as early as the second day of treatment. Therapeutic effects that have been reported include increased relatedness and animated facial expression, decreases in irritability, temper tantrums, aggressiveness, self-mutilation, hyperactivity, and improved sleep pattern. Side effects reported have included drowsiness, irritability, weight loss, and appetite suppression. Transient increases in irritability, restlessness, and sleep interruption have been reported on discontinuation of the drug. In general, however, side effects appear to be dose related and minimal at therapeutic doses, or they respond rapidly to dose reduction or discontinuation.

Fenfluramine is contraindicated in patients with glaucoma. It should not be administered concomitantly or within 14 days following the administration of an MAOI. Caution should be exercised when concomitantly using potent anesthetic agents because of the catecholamine-depleting effect associated with prolonged use of fenfluramine. Other drugs with stimulant effects on the central nervous system should be used with caution in combination with fenfluramine because of potential additive effects.

In conclusion, fenfluramine seems to have limited amphetamine-like therapeutic effects in some children with autism. No definitive conclusions regarding relationships among initial serotonin levels, clinical effects, and symptom responses can yet be drawn with the available information.

2. Clonidine

Clonidine (Catapres®) is an imidazoline derivative with α-adrenergic agonist properties that has been primarily used in treating hypertension. At low doses, it appears to stimulate inhibitory presynaptic autoreceptors in the central nervous system. Established indications in psychiatry include the treatment of withdrawal syndromes, Tourette's disorder, and, more recently, ADHD. Possible additional indications include the treatment of self-injurious and aggressive behaviors. Anecdotal reports suggest that clonidine may be efficacious in affective disorders, obsessive compulsive disorder, anxiety disorders, tardive dyskinesia, and psychosis. Clonidine is a relatively short-acting compound with a plasma half-life ranging from approximately 5.5 hr in children to 8.5 hr in adults.

Daily doses should be titrated and individualized. Usual daily dose ranges from 3 to 8 μg/kg given generally in divided doses, b.i.d. or t.i.d. Therapy is usually initiated at the lowest manufactured dose of a full or half tablet of 0.1 mg depending on the size of the child (approximately 1 to 2 μg/kg) and increased depending on clinical response and adverse effects. Initial dosage can more easily be given in the evening hours or before bedtime because of sedation. The most common short-term adverse effect of clonidine is sedation. It can also produce, in some cases, hypotension, dry mouth, depression, and confusion. Clonidine is not known to be associated with long-term adverse effects. In hypertensive adults, abrupt withdrawal of clonidine has been associated with rebound hypertension. Thus, it requires slow tapering when discontinued. Clonidine should not be administered concomitantly with β blockers since adverse interactions have been reported with this combination.

In summary, clonidine is a new and welcome addition to the pharmacotherapy of juvenile psychiatric disorders for conditions and symptoms in which antipsychotics have been previously used. Since it has a relatively low side-effect profile compared with the antipsychotic compounds, it should be considered first for those indications in which both drug classes are recommended. Clonidine offers a treatment alternative in Tourette's disorder and perhaps in attention-deficit hyperactivity disorder, especially when the other agents are either ineffective or poorly tolerated.

3. β Blockers

β-Adrenergic blocking agents have been studied extensively in the management of various medical disorders, including hypertension and cardiac arrhythmias. Propranolol, a nonselective β_1 and β_2 adrenergic receptor antagonist, has recently received considerable attention with regard to its potential use for psychiatric disorders including drug-induced akathisia, anxiety disorders, schizophrenia, as well as aggressive and self-abusive behavioral disorders.

Propranolol effects are mediated through its ability to block β-adrenergic receptors at multiple sites in the body. It also crosses the blood–brain barrier, and this probably accounts for some of its efficacy in psychiatric disorders but also contributes to concerns regarding potential CNS toxicity. At this time, it is unclear whether the benefits obtained from propranolol are primarily related to peripheral or central effects of the drug.

Propranolol is available in both oral and injectable forms. It is almost completely absorbed following oral administration. Much of the medication is metabolized during the first pass through the liver, and as little as one-third may reach systemic circulation. However, as much as a 20-fold variability in interindividual hepatic elimination can occur. Hepatic elimination is lessened as the dose is increased, which suggests saturation. The half-life after chronic administration is about 4 hr. The drug is 90–95% bound to plasma proteins, which can contribute to variability of the drug in the plasma and affect plasma concentrations as a guide to therapeutic dosing. The drug is virtually completely metabolized prior to excretion in the urine.

There are no current established indications for the use of β blockade in treating child and adolescent psychiatric disorders, but there are possible indications currently under investigation. At present, this investigation centers around the use of propranolol in the management of severe aggressive and self-injurious behaviors. It has been suggested that propranolol may also be efficacious in treating schizophrenia, anxiety disorders, lithium-induced tremors, neuroleptic-induced akathisia, and perhaps Tourette's disorder.

Several investigators have reported success using propanolol in managing severe aggressive behavior in all age ranges of patients with mental retardation, adults with schizophrenia, and explosive behavior disorders in young patients. In the child and adolescent populations, reduction in the intensity and frequency of aggressive episodes has been witnessed in patients who had failed to respond to other psychiatric interventions. Some investigators have reported the need to use high doses of propranolol for efficacy, but more recent studies suggest a dose range of 40–280 mg/day given in divided doses two to four times a day. Patients have generally been started at doses of 10–20 mg b.i.d. The dose range used in pediatric disorders requiring propranolol is approximately 1–8 mg/kg per day.

Short-term adverse effects of propranolol are usually not serious and generally abate on cessation of drug administration. Nausea, vomiting, constipation, and mild diarrhea have been reported. Psychiatric side effects appear to be relatively infrequent but can occur and include vivid dreams, depression, and hallucinations. Allergic reactions manifested by rash, fever, and purpura are infrequent but have been reported and warrant discontinuation of the drug. There are no known long-term effects associated with chronic administration of propranolol. Since abrupt cessation of this drug may be associated with rebound hypertension, gradual taper is recommended.

Propranolol's capacity to induce β-adrenergic blockade warrants caution in select patient populations. Propranolol can cause bradycardia and hypotension as well as increase airway resistance. Thus, it is contraindicated in asthmatic and certain cardiac patients. Propranolol can also augment the hypoglycemic action of insulin and mask tachycardia that may alert the patient and clinician of a pending hypoglycemic reaction. Therefore, propranolol should be used with caution in the diabetic population.

In summary, propranolol represents an alternative in the pharmacological management of aggressive behavioral disorders and in the alleviation of some adverse effects of antipsychotics (akathisia) and lithium (tremor). It offers the benefit of having a relatively low side-effect profile and should be tried before antipsychotics for indications other than psychosis (i.e., aggression, dyscontrol).

G. Anticonvulsants

Anticonvulsants are indicated for the treatment of seizure disorders. Their use in treating nonseizure childhood psychiatric disorders is currently not recommended. Chronic seizure disorders in children and symptoms of varied psychopathological states have been linked, including deficits in cognitive skills, mood variation, and behavioral problems. In these circumstances, it is the diagnosis and appropriate management of the underlying seizure disorder that is important. The discussion of nonpsychiatric uses of anticonvulsants is beyond the scope of the current presentation. However, in clinical practice, despite lack of adequate scientific evidence, anticonvulsants, mainly phenytoin (Dilantin® and others) and carbamazepine (Tegretol® and others), have been used in treating nonspecific behavioral disturbances, particularly those associated with aggressive outbursts, impulsiveness, and restlessness in the absence of clinical or electroencephalographic evidence for seizure activity.

Phenytoin is the most prominently used hydantoin anticonvulsant compound. It saturates its metabolic pathway within therapeutic ranges. Therefore, the relationship between dose and plasma level is not linear, and the half-life varies with plasma levels. About 90% of the compound is irreversibly bound to plasma protein, so that conditions or agents that can affect plasma protein levels may alter the free plasma concentration of phenytoin. The drug is metabolized by the liver and excreted in the urine. Rates of liver metabolism vary inversely with age (phenytoin metabolism decreases as age increases). The mechanism of action of seizure control appears to be membrane stabilization. There is wide interindividual variation in plasma levels at a given dose, and this necessitates careful monitoring of clinical response and plasma levels when using phenytoin. Initial recommended daily doses range from 5 to 10 mg/kg per day, and therapeutic plasma levels range from 10 to 29 μg/ml.

Side effects of phenytoin are dependent on duration and route of administration and dosage or plasma levels of the drug. Gastrointestinal symptoms including nausea, vomiting, and epigastric discomfort can occur but are usually ameliorated by giving the drug either in divided doses or with meals. Rare idiosyncratic responses to phenytoin have been reported, including liver dysfunction, immunologic disorders, bone marrow depression, and skin reactions of varying severity. Acne, hirsutism, and coarsening of the skin and facial features have been reported. Acute toxicity is primarily manifested by central nervous system symptoms. These include nystagmus, ataxia, dysarthria, lethargy, diplopia, vertigo, blurred vision, mydriasis, and hyperactive tendon reflexes. Behavioral symptoms of toxicity include hyperactivity, silliness, drowsiness, hallucinations, and confusion. The presence of these symptoms even within the therapeutic range necessitates, and often responds to, dose reduction. Chronic administration of the drug has been associated with gingival hyperplasia and mild peripheral neuropathies. In addition, a toxic syndrome characterized by disturbances of behavior and cognition associated with chronic use of phenytoin, even at therapeutic doses, has been reported. Phenytoin should be discontinued if a skin rash appears and should not be resumed if the rash is exfoliative but can be resumed if the rash is of a milder, measles-like type after it completely subsides. Serum levels sustained above the optimal range should be avoided, since they may produce a confusional state. Phenytoin serum level determination is necessary for achieving optimal dosing. Determination of serum levels is particularly important when phenytoin is used concomitantly with other drugs, since many drugs interfere with its pharmacokinetics, resulting in either increase or decrease of phenytoin levels. Phenytoin should be used with caution in patients with compromised liver function, since the liver is the chief site of biotransformation of phenytoin.

Carbamazepine is an anticonvulsant structurally related to the tricyclic anti-depressants. The drug effectively limits the spread of seizure activity, but its site and mechanism of action remain unclear. It is the drug of choice in treating temporal lobe epilepsy (complex partial seizures). When given orally, carbamazepine is rapidly absorbed. It is metabolized in the liver prior to urinary excretion and has several active metabolites. It is an autoinducer, and therefore, after chronic administration, its half-life varies from that of a single dose. The plasma half-life after chronic administration is between 13 and 17 hr, and that of its active metabolites is probably 5 to 8 hr. The short half-life necessitates multiple daily doses, often 6 to 8 hr apart. The therapeutic plasma concentration is variably reported at 4–8 μg/ml, and recommended daily doses in children range from 20 to 30 mg/kg. Since the relationship between dose and plasma level is variable and uncertain, with marked interindividual variability, close plasma level monitoring is recommended.

Although there are no specific psychiatric indications, carbamazepine has been used in treating children with impulsive and aggressive behaviors without evidence about its effectiveness. Recent studies with adults have suggested efficacy of carbamazepine in treating bipolar disorders, manic type, as an alternative to lithium treatment.

Frequent short-term side effects include dizziness, drowsiness, nausea, vomiting, and blurred vision. Tolerance to these side effects tends to develop over time and can often be managed by gradual dose increments or by reducing the dose. Acute symptoms of toxicity include ataxia and diploplia and are associated with high plasma levels. Idiosyncratic reactions such as bone marrow suppression, liver toxicity, and skin disorder have been reported but appear to be rare. However, given the seriousness of these reactions, careful monitoring of blood counts and liver and renal function are warranted during treatment.

Carbamazepine is contraindicated in patients with a history of bone marrow depression. The drug should be discontinued if evidence of significant bone marrow depression develops. The concomitant use of more than one anticonvulsant can result in reduction of serum levels of individual drugs requiring dose adjustments. The concomitant use of anticonvulsants with lithium may increase the risk for neurotoxicity. Serum level determinations are necessary for achieving optimal dosing. Determination of serum levels is particularly important when anticonvulsants are used concomitantly with other drugs, since many drugs interfere with anticonvulsants' pharmacokinetics, resulting in either increase or decrease of serum levels.

Reports of behavioral disinhibition as a result of phenobarbital (Tedral® and others) or primidone (Mysoline® and others, which have phenobarbital as their major metabolite) therapy have called into question the utility of these medications in children. It has been suggested that this phenomenon may be more common in children with preexisting behavioral disorders. However, there are children who are successfully treated for seizure disorders without this complication. In general, caution should be used with these compounds in treating children with behavioral disorders.

In summary, there are no clear indications for the use of anticonvulsant agents in treating childhood psychiatric disorders in the absence of clear seizure activity. Carbamazepine may be useful as a second-line treatment for juvenile bipolar disorder, manic type.

IV. DIAGNOSTIC CATEGORIES AND CLINICAL CONSIDERATIONS

A. Attention-Deficit Hyperactivity Disorder

Attention-deficit hyperactivity disorder (ADHD) is one of the major clinical and public health problems in the United States in terms of morbidity and disability of children, adolescents, and adults. Its impact on society is enormous in terms of financial cost, stress to families, impact on schools, and potentially damaging effects on self-esteem. Data from cross-sectional, retrospective, and follow-up studies indicate that children with ADHD are at risk for developing other psychiatric disorders in childhood, adolescence, and adulthood such as antisocial behaviors, psychosis, and substance abuse as well as depressive symptoms and depressive disorders.

ADHD, formerly called hyperkinesis or minimal brain dysfunction and more recently attention deficit disorder (ADD), is a disorder characterized by impaired attention, impulsivity, and excessive motor activity. These symptoms can be shared with other childhood psychiatric disorders such as childhood depression, conduct disorders, and anxiety states. Eliciting a consistent pattern of ADHD symptoms from two or more settings (home, school, physician's office) adds to the certainty of the diagnosis. With developmental variations, ADHD affects children of all ages, as early as age 3, and not uncommonly the disorder persists throughout adolescence and into adulthood. The diagnosis of ADHD requires careful attention to differential diagnosis, assessment of psychological and learning difficulties, and comprehensive treatment planning that addresses family, education, and social factors. Parents need guidance throughout childhood in setting reasonable expectations and providing opportunities for positive experiences that enhance a child's often damaged self-esteem. The pharmacological component of treatment, although not curative, and limited in terms of helping educational achievement, can decrease some of the most disturbing symptoms of this disorder.

Stimulants have been used widely in treating ADHD (Table 2). Yet as many as 30% of children do not improve. Since stimulants are short-acting drugs, their use is

TABLE 2. Indications, Contraindications, and Interactions of Common Psychotropics Used in Pediatric Psychiatry

Drug	Indications	Possible indications	Contraindications	Interactions
Stimulants				
Dextroamphetamine	Attention deficit	Adjunct treatment for	Marked anxiety, psycho-	Decreased effects of some
Methylphenidate	Hyperactivity	refractory depres-	sis, agitation	antihypertensives, anti-
Magnesium pemoline	Narcolepsy	sion	Known hypersensitivity	coagulants, and anticon-
				vulsants
Antidepressants				
Tricyclics	Enuresis	Nonspecific disinhibi-	Concomitant use of mono-	Decreased effects of some
		tion	amine oxidase inhibitors	antihypertensives, anti-
	Depression	Pain syndromes	Use immediately after a	psychotics, and anticon-
	Attention deficit	Severe headaches	myocardial infarction	vulsants
	Hyperactivity		Known hypersensitivity	
	Anxiety disorders	Tourette's Disorder	Acute narrow-angle	
		Sleep disorders	glaucoma	
MAOIs	Depression		Concomitant use of mono-	Sympathomimetic or re-
	Attention deficit		amine oxidase inhibitors	lated drugs, high-
	Hyperactivity		and other antidepres-	tyramine-containing
	Anxiety disorders		sants	foods, narcotics
				(meperidine), most cold
				medicines
Antipsychotics	Psychotic	Mania	Known hypersensitivity	When used with tricyclics
	disorders	Extreme agita-	Acute narrow-angle	can increase antidepres-
		tion/aggression	glaucoma; patients on	sant toxicity by increas-
		nonresponsive to	large doses of central	ing metabolization
		other treatments	nervous system depres-	
		Nonresponsive atten-	sants; comatose patients	
		tion deficit disorder	Great caution in patients	
		Nonresponsive tic dis-	with bone marrow sup-	
		order	pression or liver	
			damage	
			Known hypersensitivity	

Drug				
Lithium	Manic–depressive illness	Severe aggressive pictures Adjunct treatment in refractory depression, severe dysphoria in other syndromes, cyclothymia, dysthymia/depression	No absolute contraindication Caution in patients with renal insufficiency	Drugs that reduce renal lithium clearance (thiazide diuretics nonsteroidal antiinflammatory) may increase serum concentration of lithium
Antianxiety drugs	Anxiety disorders Seizure disorders (clonazepam)	Severe situational anxiety/insomnia (short-term)	Known hypersensitivity Acute narrow-angle glaucoma (untreated)	Potentiation of effect of other central nervous system depressant drugs and alcohol
Clonidine	Withdrawal syndromes	Attention-deficit hyperactivity Tourette's disorder Nonspecific aggression	Hypersensitivity Caution in diabetes mellitus	Interaction with β blockers
Propranolol		Nonspecific aggression Tourette's disorder Anxiety disorders Akathisia	Asthmatic and certain cardiac patients Hypersensitivity Bronchial asthma Caution in diabetes	Interaction with clonidine β blockers
Fenfluramine		Pervasive developmental disorders	Known hypersensitivity Acute narrow-angle glaucoma Concomitant use of MAOIs	Potentiation of effect of other central nervous system stimulant drugs

(continued)

TABLE 2. (*Continued*)

Drug	Indications	Possible indications	Contraindications	Interactions
Anticonvulsants Carbamezapine	Seizure disorders	Mania	Bone marrow depression Hypersensitivity	Reduction of serum levels requiring dose adjustments with other anticonvulsants, antidepressants, and antipsychotics
Phenytoin	Seizure disorders		Hypersensitivity	Increased metabolization with decreased effects of other anticonvulsants, antidepressants, and antipsychotics

complicated by the need to take medicine in school and the troublesome rebound of symptoms in the evening hours at home. In addition, insomnia as well as dysphoric mood and some slowing of growth in weight and height during development may occur. These problems encourage the search for appropriate pharmacological alternatives in treating ADHD. An important aspect of the search for alternatives to stimulant treatment in general is the need to define an effective and acceptable treatment for adolescents with ADHD. Although it is clear that ADHD syndrome persists into adolescence and adulthood in at least 30% to 50% of patients who manifested the disorder as children, there is very limited information regarding appropriate pharmacotherapy in the older age groups. Several open and three controlled studies have reported beneficial effects of stimulants in adolescents with ADHD-like symptoms. However, concerns about treating adolescents with stimulant medication include the hypothetical risk of abuse and dependence by the patient or his/her associates, concerns about growth suppression, and the common dislike by adolescents of the subjective effects of stimulant medication.

Tricyclic antidepressants (TCAs), mainly imipramine, have been proposed as an alternative treatment for ADHD. Possible advantages of TCAs over stimulants include a longer duration of action and the feasibility of once-daily dosing without symptom rebound or insomnia, greater flexibility in dosage, the option of monitoring plasma drug levels, and minimal risk of abuse or dependence. Initial open trials were followed by controlled studies and showed that, in general, TCAs are superior to a placebo, although not always superior to methylphenidate.

Desipramine, a major active metabolite of imipramine, has relatively selective neuropharmacological actions on noradrenergic neurotransmission and, like other TCAs, eventually enhances the functional availability and activity at α-adrenergic receptors. Because of its pharmacological properties, it is associated with somewhat smaller risks of adverse effects such as sedation, dry mouth, and impairment in cognition, commonly encountered with imipramine and other tertiary amine TCAs. Two controlled studies have investigated the efficacy of desipramine in children with attention-deficit hyperactivity disorder using daily doses of up to 3.5 mg/kg. Our group reported favorable results in open, dose-ranging, long-term trials of desipramine in children and adolescents using daily doses of up to 5 mg/kg. In our series, these relatively high doses of desipramine were well tolerated without clinically significant cardiovascular effects.

In addition to TCAs, other compounds have been recently evaluated in treating prepubertal children with ADHD that may also prove to be effective in treating adolescents with this disorder. In a 12-week double-blind crossover trial, Zametkin et al.[4] reported significant and immediate reduction in ADHD symptoms with minimal adverse effects associated with MAOIs in 14 children. Using a similar design, Hunt et al.[5] recently reported the beneficial effects of the α-adrenergic stimulating agent clonidine in treating children with ADHD with daily doses of up to 4 to 5 μg/kg. Clonidine was well tolerated, and the main adverse effect was sleepiness, which tended to subside after 3 weeks of continued treatment.

B. Childhood Anxiety Disorders

The DSM-IIIR includes a subclass of three disorders of childhood or adolescence in which anxiety that is not caused by psychosocial stressors is the predominant feature. These are separation anxiety disorder (SAD), avoidant disorder, and overanxious disorder. Childhood anxiety disorders are relatively common disorders that bear striking similarities to the adult anxiety disorders and in many cases persist into adult life.

Separation anxiety disorder may represent a childhood equivalent to adult agoraphobia. It is a relatively common disorder affecting children of both sexes. The predominant disturbance is excessive anxiety on separation from familial surroundings. It is called separation anxiety because it is assumed that the main disturbance is the inability of the child to separate from the parent or from major attachment figures. When separation occurs or is anticipated, the child may experience severe anxiety to the point of panic. Children with this disorder are uncomfortable when away from the house or from familiar areas. They may refuse to leave the house or attend school or camp. Because of that, they are sometimes called "school refusers" or "school phobics." However, not all school refusal is related to separation anxiety. It can be caused by a variety of reasons such as other psychiatric problems (i.e., depression, psychosis), family stresses, or school difficulties. Children with SAD may be unable to stay by themselves and often display clinging behaviors. The onset may be as early as preschool age. However, it appears more commonly in older children. Typically there are periods of exacerbations and remissions over several years, and the disorder persists in some cases into adulthood. In its severe form it can be highly incapacitating in that the child is unable to attend school and function independently. Children with physical symptoms may undergo complex medical evaluations. When school refusal develops, common complications include academic failure and social avoidance. Demoralization and overt depression are often seen. In most cases the disorder develops after some type of life stressor, typically a loss of a dear one (real or perceived), an illness of the child, or a change in the child's environment. The disorder seems to be familial, being more common in family members than in the general population.

Avoidant disorder resembles adult social phobia or adult avoidant personality disorder. The predominant feature is a persistent and severe shrinking from contact with strangers that interferes with psychosocial functioning. Affected children are interested in social relationships but fearful of them. In social settings they may appear to be inarticulate or mute. Usually children with this disorder are unassertive and lack self-confidence. Age-appropriate socialization skills may not develop, and in severe cases the child fails to form social bonds beyond the immediate family. The disorder may develop in early childhood after stranger anxiety should have disappeared. The course is unknown but probably is chronic.

Overanxious disorder could be considered the childhood equivalent to adult generalized anxiety disorder. It is a common disorder more frequently seen in boys than in girls. Similar to the adult syndrome, the essential feature is excessive worrying and fearful behavior that is not focused on a specific situation or object and not related to psychosocial stressors. The child worries about future events such as examination, injury or disease, or about meeting expectations, such as deadlines, appointments, chores. The anxiety is commonly expressed as concerns with self-competence and self-performance. Because of these concerns, children may be erroneously seen as "hypermature." In some cases physical concomitants of anxiety are apparent, such as headaches, stomach aches, shortness of breath, dizziness, or other somatic complaints. These physical symptoms may result in unnecessary medical evaluations and interventions. The child may frequently report "feeling nervous." The onset may be acute or gradual, with exacerbations usually associated with stress. In severe cases, children may show poor academic performance and failure to engage in age-appropriate activities.

With the exception of SAD, there are no systematic evaluations of treatment modalities for childhood anxiety disorders. It is, however, possible that children and adolescents with anxiety disorders will respond to the same pharmacological approaches as adult patients. In addition to tricyclic antidepressants, high-potency benzodiazepines such as clonazepam and alprazolam can be effective, as can the MAOI (i.e., phenelzine) drugs. In a single study, high doses of the antidepressant drug imipramine were effective after 2 to 8 weeks of treatment. In affected children, in addition to pharmacotherapy, treatment requires a variety of methods including psychotherapy, family therapy, and behavioral therapy. In severe cases hospitalization may be required.

C. Obsessive Compulsive Disorder

Many people experience obsessive thoughts and compulsive urges or actions in their everyday life. However, when these obsessive thoughts and compulsive actions become so frequent or intense that they interfere with an individual's functioning, then the diagnosis of obsessive compulsive disorder (OCD) is given. OCD can be severely incapacitating for its victims. The symptoms of the disorder can spread to interfere with social and occupational functioning and often involve the person's entire family. OCD can develop early in life; nearly one-third of adult obsessional patients report the onset of their symptoms before the age of 15 years, and cases of the disorder have been described as early as the age of 3. Clinically, childhood-onset OCD closely resembles the adult syndrome. Because the disease is likely to run a chronic course, early treatment is desirable. Symptoms generally consist of checking or cleaning rituals and obsessive thoughts. Individuals with cleaning rituals will attempt to avoid coming in contact with what they believe to be contaminated objects. Individuals with checking rituals usually have terrible feelings that they have done something wrong. These individuals usually ask others for reassurance and engage in less checking behavior if there is someone else present to share responsibility for their actions. They repeatedly engage in checking behaviors to ensure that they performed an action correctly, did not harm someone in the process, or did not create a hazard that might produce harm in the future. Virtually all individuals with cleaning or checking rituals also have obsessive thoughts. However, there are some individuals with obsessions alone. These obsessive thoughts are usually of an aggressive, religious, or sexual nature.

OCD is considered to have one of the poorest prognoses of all psychiatric disorders; it is one of the most chronic conditions known to psychiatry. Many drug treatments have been tried with little success. The most promising recent development in the drug treatment of OCD has been the use of clomipramine (Anafranil®), a tricyclic antidepressant that has been used for 15 years in Europe, Canada, and South America and has recently become available in the United States. Clomipramine was marketed in Switzerland for the treatment of depression in 1966 and in several other countries shortly thereafter. At present it is marketed in a total of 77 countries for depression and in 11 of these countries for the treatment of OCD. Since 1976, there have been six double-blind studies of the treatment of adult OCD patients with clomipramine. All of these studies have indicated the effectiveness of clomipramine in adult OCD patients; discontinuation of treatment was generally followed by relapse. Clomipramine is thought to be effective in this disorder because of its effects in blocking the reuptake of the neurotransmitter serotonin in the brain. A deficit in serotonin neurotransmission has been suggested as a possible etiology of this disorder. Other newer antidepressants such as fluoxetine (Prozac®) that have a similar biochemical property of enhancing serotonin effects in the central nervous system also appear to be effective in treating OCD.

In contrast to the relatively large number of reports concerning adult OCD patients, little attention has been paid to the treatment of children with this condition. This may be because of underrecognition of the disorder in children, as these children frequently conceal their symptoms. OCD in children, as in adults, generally resists psychological and drug treatments. In a double-blind study, Flament and colleagues[6] from the National Institutes of Health recently reported the results of clomipramine treatment of 19 children and adolescents with severe OCD. Half of these children had not responded to previous treatment with other tricyclic antidepressants. They reported a significant improvement in the children receiving clomipramine when compared with placebo. However, treatment with clomipramine was not innocuous. A fair number of side effects were reported by the patients—mainly anticholinergic symptoms such as dry mouth, constipation, and blurred vision, which are commonly encountered with tricyclic antidepressants. Some unusual and more serious complications also occurred, such as abnormal movements in two children, a grand mal seizure in one child, and psychotic symptoms in one child. However, the severity of the disorder and the absence of other effective therapies for this disorder in the United States seem to justify further investigation of this drug in children despite these possible side effects. Therapeutic intervention in the early form of the disorder is important before complications, chronicity, and social incapacitation occur, which can make treatment and restabilization of functional life habits more difficult.

Recommended daily doses of clomipramine after titration and individualization of dose are similar to dose ranges recommended for other tricyclic antidepressants, 3 to 4 mg/kg to be taken in two divided doses and not to exceed 250 mg/kg. Daily doses of fluoxetine are not yet clearly established. Initial experience using this drug suggests a daily dose range of 0.5 to 1 mg/kg that can be administered once daily, in the morning.

D. Mood Disorders

Despite controversy over its existence, childhood depressive disorder is recognized by the DSM-IIIR as a disorder with core symptoms similar to those found in adult depression and with developmentally specific associated features such as school difficulties, school refusal, negativism, aggression, and antisocial behavior. The depressive disorders should be carefully differentiated from feeling states such as unhappiness or disappointment that commonly occur during childhood. Further, recent evidence suggests the existence of a childhood symptom complex that closely resembles adult mania.

There is little doubt that affective disorders occur in adolescence. Kraepelin[7] reported that among 903 manic–depressive patients, 18.9% had onset of symptoms between the ages of 10 and 20 years. Based on retrospective accounts from adults with bipolar disorder, the rate of adolescent onset varies between 11% and 35%. More recent studies have demonstrated the occurrence in adolescents of bipolar and nonbipolar major depression as well as cyclothymia, dysthymia, and adjustment disorder with depressed mood. Although the prevalence of adolescent affective disorders is not well known, it has been estimated to be as high as 15% of adolescent intakes. Recent studies have shown that adolescent affective disorders are frequently chronic and recurrent. In addition, these disorders can have a severe impact on psychosocial functioning and are associated with an increased risk for suicidal behavior.

The DSM-IIIR defines the same diagnostic criteria for affective disorders in all age groups with consideration to age-specific developmental variations in the clinical picture, such as somatic complaints, negativistic behavior, restlessness, aggression, and school difficulties. Since not all forms of affective disorders are typical, and most affective symptoms are subjective, the use of multiple sources of information and of structured interview techniques is crucial for the assessment of adolescent affective disorders. No major differences in the clinical picture have been identified between adult and adolescent patients with nonbipolar major depression. However, psychotic symptoms are more frequently encountered in adolescent than in adult mania, and mixed presentations associated with severe behavioral disturbances as well as rapid cycling often occur in adolescent bipolar disorder. In addition, as shown by Strober and Carlson,[8] adolescents presenting initially with nonbipolar major depression often develop mania at follow-up. Because of these presentations, adolescent bipolar disorder can often be misdiagnosed as schizophrenia, nonbipolar major depression, or conduct disorder.

Currently, intensive investigational efforts are under way to define the role of family history, the natural course of the disorder, and the efficacy of psychosocial and pharmacological interventions. Preliminary evidence suggests that a positive family history places some children at risk for recurrent depressions of varying intensities. The limited pharmacological evidence available tends to suggest that children with well-defined major depressive disorders can respond favorably to antidepressant therapy and that the degree of improvement relates to plasma levels of the medication. The use of lithium carbonate in these disorders appears promising but remains experimental. A recent investigation suggests that lithium may be useful in treating childhood disorders in the offspring of manic–depressive, lithium-responsive parents.

E. Psychotic Disorders

The term psychosis is used in DSM-III to describe abnormal behaviors of individuals with grossly impaired reality testing. Impaired reality testing implies the inability of the individual to differentiate between internal and external reality. It can cause the individual to evaluate incorrectly the accuracy of his or her perceptions and thoughts and to make incorrect inferences about external reality. The term psychosis is also used when the individual's behavior is grossly disorganized and it can be inferred that reality testing is impaired. The diagnosis of psychosis requires the presence of either delusions or hallucinations. Psychotic disorders in children, as in adults, can be functional or organic. Functional psychotic syndromes include schizophrenia and related disorders and the psychotic forms of mood disorders. Organic psychosis can develop secondary to lesions to the central nervous system as a consequence of medical illnesses, trauma, or drug use, both licit and illicit.

The mainstays of treatment of psychoses are antipsychotic medications. In cases where the psychotic process occurs in the context of a mood disorder, the concomitant use of specific treatments for the mood disorder is crucial for clinical stabilization. In cases where the clinical picture of psychosis is associated with severe agitation, the adjunct use of benzodiazepines such as lorazepam or clonazepam can facilitate the management of the patient and may permit the use of lower doses of antipsychotics. The extent to which antiparkinsonian agents should be used prophylactically when antipsychotics are introduced is controversial. Whenever possible, our approach is to use antiparkinsonian agents only when extrapyramidal symptoms emerge. Extrapyramidal reactions can be prevented in many cases by avoiding rapid neuroleptization and high-potency antipsychotics. When a child or adolescent started on antipsychotics develops an acutely agitated clinical picture with associated inability to sit still and aggressive outbursts, the possibility of akathisia should be rapidly considered in the differential diagnosis. If suspected, the dose of the antipsychotic may need to be lowered, and β blockers should be added. In recent years, the syndrome of postpsychotic depression has received increasing attention. Initial trials with antidepressant drugs added to the antipsychotic treatment appear to be promising in relieving patients from associated depression, thus fostering rehabilitation efforts. Postpsychotic depression should be distinguished from akinesia, which is an adverse extrapyramidal effect that can respond to antiparkinsonian agents.

F. Enuresis

Enuresis in children is characterized by episodes of either daytime or nighttime wetting beyond an age when this type of behavior might be developmentally appropriate. Enuresis is either primary or secondary (when it occurs in the context of a history of prior adequate toilet training experience). An initial medical evaluation is indicated when this occurs to rule out the presence of a treatable medical illness (i.e., urinary tract infection, structural anomalies, etc.).

Children with functional enuresis, which implies the absence of a definable medical etiology, usually respond to nonpharmacological therapies (e.g., behavior modification, psychotherapy), and these treatments should be considered first. When an immediate therapeutic effect is necessary, an antidepressant drug, commonly imipramine, may be used. In most cases, symptoms reappear after the drug is withdrawn. Antidepressant therapy should not be continued for a period of more than 6 months, as enuresis may remit spontaneously.

G. Tic Disorders

Tic disorders include three diagnostic entities. Simple tic disorder is the presence of a single vocal or motor tic. Chronic motor tic disorder or chronic vocal tic disorder is the presence of various motor or vocal tics. Tourette's disorder is the presence of various motor and vocal tics. Simple tic disorder appears to be a benign clinical disorder that rarely requires pharmacological intervention and is genetically distinct from the other syndromes. Chronic tic disorders and Tourette's disorder, however, can be a severe neuropsychiatric syndrome of childhood onset and lifelong duration. These disorders can often be associated with other behavioral and psychological symptoms. When a chronic tic disorder or Tourette's is present and causes the patient significant dysfunction, pharmacological intervention can be very beneficial. Antipsychotic drugs, particularly haloperidol, have been very beneficial in low doses. More recently, clonidine has proven effective in some children with this disorder. In addition, clonazepam, β blockers, desipramine, and calcium-channel blockers were reported as helpful in some children with Tourette's disorder. Stimulant use has been detrimental.

H. Developmental Disorders

This class of disorders is coded in axis II and includes mental retardation, pervasive developmental disorders (autistic and autistic-like disorders), and the specific developmental disorders, previously called learning disabilities. It is estimated that at any given time approximately 1% to 3% of the population meets the diagnostic criteria for mental retardation. The syndrome of autism is a pervasive developmental disorder diagnosed by the presence, before 30 months of age, of disturbances in the rate of development and coordination of physical, social, and language skills, abnormal responses to sensory inputs (hyperreactivity alternates with hyporeactivity), and disturbances in the capacity to relate appropriately to people, events, and objects. No specific treatment alters the natural history of the syndrome (Table 3). Test results of the vast majority of autistic children fall in the ranges for the mentally retarded. In all cases, however, certain symptoms and developmental delays remain present throughout life. The specific developmental disorders represent a mixed group of cognitive dysfunctions in the context of an overall average or above-average IQ and adequate educational opportunities.

Psychotropic agents, particularly antipsychotics, continue to be widely used and abused in treating mentally retarded individuals, particularly in the care of institutionalized patients, despite the lack of controlled studies regarding their usefulness. Although psychotropics can temporarily control behavioral and psychiatric complications in some developmentally disordered children, they do not affect the cardinal symptoms of the underlying disorder, and even those children who respond to therapy may continue to display the abnormal behaviors and impaired communication skills that characterize the disorder.

Developmentally disordered children often have psychiatric disorders and behavioral problems including hyperactive, aggressive, distractible, and self-abusive behaviors. They also often manifest multiple neurological abnormalities. Psychotropics are used in this population primarily for the treatment of agitation, aggression, and self-injurious behaviors. Antipsychotics have been traditionally used to control these symptoms. Few studies support the use of one type of antipsychotic drug over another. Although a more sedating phenothiazine (e.g., chlorpromazine, thioridazine) may be beneficial for the more agitated patient, a more potent phenothiazine (e.g., perphenazine, trifluoperazine) or the butyrophenone haloperidol may be helpful in the withdrawn, inactive child. Increasingly, β blockers and clonidine have been reported to be very useful in populations of developmentally disordered patients for managing agitation, aggression, and self-abusive behaviors. Considering the relatively low toxic profile of these drugs compared with the antipsychotics, they are the preferred treatment for the management of these complications and should always be tried first.

Antidepressant drugs and lithium could conceivably be effective in controlling affective symptoms and disorders, and stimulants may improve symptoms of ADHD. Antianxiety agents should be used with caution in the management of developmentally disordered children because they tend to produce disinhibition, which may result in increased restlessness and more disturbed behavior. Fenfluramine had initially been reported to have a beneficial effect in some children with autism; however, further investigation has not replicated these initial reports.

The treatment of specific developmental disorders is largely remedial and supportive. There are no medications that are effective in altering the basic course of

TABLE 3. Pharmacotherapy of Common Disorders

Disorder	Pharmacotherapy

Developmental disorders
  Mental retardation
  Pervasive developmental
    disorder
	Main characteristics: disturbance in the acquisition of cognitive, language, and motor skills; it can be global or in specific or multiple areas; coded as axis II diagnoses
	No specific treatment for core disorder
	Pharmacotherapy of complications (aggression, self-abuse, insomnia, agitation):
	  β Blockers
	  Clonidine
	  Antipsychotics
	  Lithium
	  ?Fenfluramine (stimulantlike effects)

Specific developmental disorders (learning disabilities)	Remedial help; no specific pharmacotherapy; in association with attention-deficit hyperactivity, treat the associated disorder

Disruptive behavioral disorders
  Attention-deficit hyperactivity
    disorder (ADHD)
	Main characteristics: inattentiveness, impulsivity, with or without hyperactivity; 30% will not respond to stimulants; 50% will continue to manifest the disorder into adulthood
	Pharmacotherapy:
	  Stimulants
	  Tricyclics
	  Clonidine
	  Antipsychotics
	  Combined pharmacotherapy for treatment-resistant cases

Conduct disorder Oppositional disorder	Main features: persistent and pervasive patterns of antisocial behaviors (conduct) and negativistic defiant behaviors (oppositional) Frequently associated with attention-deficit hyperactivity and depression No specific pharmacotherapy available for core disorder For aggression and self abuse, follow guidelines for complications of pervasive developmental disorders In combination with attention-deficit hyperactivity or depression, treat the associated disorder

Tic disorders (Tourette's
  Disorder)
	Main characteristics: multiple motor and one or more vocal tics
	Frequently associated with ADHD and obsessive compulsive disorder
	Pharmacotherapy:
	  Clonidine
	  Antipsychotics
	  Tricyclics
	  Clonazepam
	  β Blockers

Elimination disorders Enuresis/encopresis	Main features: repeated involuntary age-inappropriate incontinence

(continued)

TABLE 3. (*Continued*)

Disorder	Pharmacotherapy
	Main treatment approach: psychoeducational and behavioral Pharmacotherapy of enuresis: Tricyclics
Schizophrenia	Main features: delusions and hallucinations Pharmacotherapy: similar to adult disorder Antipsychotics For treatment-resistant cases: Antipsychotics + lithium Antipsychotics + carbamezapine Antipsychotics + β blockers Antipsychotics + benzodiazepines
Mood disorders Bipolar disorder (BPD) Depressive disorder	Similar to the adult disorders with age-specific associated features Main features: disturbance of mood (manic or depressive) and associated symptoms occuring together for a period of time; more frequent psychotic symptoms in juvenile mania Pharmacotherapy: similar to adult disorders For BPD: lithium or carbamezapine For BPD, manic: add benzodiazepines or antipsychotics For BPD, depressed: add antidepressants
Anxiety disorders Separation anxiety disorder Overanxious disorder Avoidant disorder Adultlike anxiety syndromes	Nosologic uncertainty; more similar than dissimilar to the adult disorders Main features: generalized or focused excessive anxiety Pharmacotherapy: High-potency benzodiazepines (clonazepam, alprazolam) Antidepressants (tricyclics and monoamine oxidase inhibitors) Combined pharmacotherapy for treatment-resistant cases

the disorder. If the child also meets diagnostic criteria for ADHD, major depression, or any other axis I diagnosis, then the guidelines described for the treatment of these disorders are applicable.

I. Sleep Disorders

Sleep disorders can be characterized by deviation from age-appropriate patterns of sleep. Sleep patterns change throughout development, and therefore the presentation of a sleep disorder must be evaluated in this context. Disorders of sleep can consist of difficulties with the initiation and maintenance of the sleep cycle or insomnias, the presence of excessive sleep or hypersomnias, variations in the sleep–wake schedule or phase-lag syndromes, and alteration of sleep stages or parasomnias.

In adolescents, a syndrome of altered sleep–wake cycle called delayed sleep phase syndrome has been described. It is characterized by insomnia with subsequent undisturbed sleep associated with difficulty on arising in the morning. This syndrome represents the presence of an altered sleep–wake cycle and appears to be alleviated by chronotherapy, which involves gradual advancement of the onset of sleep until sleep normalizes. Other related treatment approaches include partial or total sleep deprivation. These techniques are assumed to "reset" the individual's circadian clock.

Narcolepsy is characterized by the spontaneous initiation of sleep during wakeful periods. Affected individuals present with frequent and inappropriate daytime naps that interfere with school or work performance. Stimulants and the "activating" tricyclic antidepressants, such as desipramine and protryptiline, have been described as useful in the control of narcolepsy. Using the lowest effective dose is always recommended, and drug holidays are suggested, as patients appear to develop tolerance to these interventions. The drug holiday may forestall the development of tolerance. The MAOIs may be of use in those patients with narcolepsy who have not responded to the stimulants or tricyclics.

Cataplexy is characterized by the spontaneous initiation of sudden inability to perform voluntary movements during wakeful periods. It is described as a more episodic phenomenon, and treatment may be tailored to active phases of the illness. Imipramine has been described as beneficial in treating cataplexy.

Sleep apnea is a phenomenon of either upper airway obstruction (obstructive sleep apnea syndrome) or upper airway occlusion resulting from failure of respiratory effort during sleep (central sleep apnea syndrome). This results in frequent brief awakenings occurring during normal nighttime sleep. As a result the patient often complains of excessive daytime fatigue and can exhibit a decline in performance. The presence of this disorder requires a careful medical evaluation to assess the nature of the respiratory difficulty and to rule out the etiology of the airway obstruction when present. Treatment is generally aimed at improving airway function during sleep. In adults with the central sleep apnea syndrome, treatments have included low-flow nasal oxygen, respiratory stimulants, and diaphragmatic pacing. Tricyclic antidepressants have also been used with some benefit.

Parasomnias are disturbances of sleep stages that can occur in prepubertal children during the stage of sleep characterized by the transition from deep sleep (stages 3 to 4) to rapid eye movement (REM) sleep. Stage 3 to 4 sleep typically occurs in the earlier part of the sleep cycle, and therefore, these disorders more frequently manifest themselves in the earlier parts of the night (1 to 3 hr after falling asleep). Typically, the patient cannot recall the episode the next morning. The parasomnias include night terrors (pavor nocturnus), sleepwalking, and sleeptalking.

Night terrors are characterized by episodes in which the child seemingly awakens in a very frightened agitated state. The child is in a state of autonomic arousal and may manifest excessive sweating, tachycardia, heart palpitations, and dilated pupils. The child is typically unresponsive to the environment, is inconsolable, and appears to be responding to "nightmares." The child may demonstrate complex motor movements, engage in sleepwalking and sleeptalking during this episode, and be very difficult to arouse. Once aroused, the child is described as confused and disoriented and frequently has no recollection of the event in the morning. Sleepwalking and sleeptalking are episodes of complex motor behavior and speech that occur during the sleep cycle. These events are non-goal-directed in nature and generally benign, but the presence of sleepwalking does obviously pose some risk, and precautions should be taken by parents to protect the patient during these episodes.

Both benzodiazepines and tricyclics have been shown to be effective in curtailing these episodes. However, pharmacotherapy should be limited to those patients in whom these episodes cause some persistent dysfunction or whose episodes are so severe that they pose some risk to the patient or his family.

The presence of insomnia in children and adolescents requires a careful history to elucidate the origin of the disturbance and the presence of comorbid factors such as depression, anxiety, ADHD, or psychosocial stressors. Nonpharmacological approaches to the treatment of insomnia are generally preferred, particularly when it is caused by psychosocial stressors. These approaches include behavioral treatment, reassurance, and support. However, when the insomnia is secondary to an existing psychiatric disorder such as ADHD, anxiety disorder, or major depression, the treatment program should address the underlying disorder. For example, the presence of major depression would argue for treatment with an antidepressant medication. If the sleep disorder is part of a mood disorder, then the sleep disturbance should improve as the depression resolves. For more prolonged (several weeks) states of nonspecific insomnias associated with next-day tiredness and dysfunction, the judicious short-term use of low doses of antihistamines or short-acting benzodiazepines may be very helpful.

The approach to evaluating hypersomnia is similar to that described earlier for insomnia. That is, the presence of associated disorders that interfere with the nighttime sleep pattern or cause daytime lassitude needs to be carefully assessed in the differential diagnosis. Hypersomnia often manifests initially and more frequently in adolescence. This phenomenon may be related to hormonal changes.

V. CONCLUSION

Although the origins of pediatric psychopharmacology began over 50 years ago, its longterm outlook is very dependent on careful clinical applications and future research. The lack of FDA approval for many of the medications for the specific indications we have discussed does not prohibit the careful introduction of innovative uses of available compounds. As we have stated, it is essential to apply careful differential diagnostics, assessing psychiatric, social, cognitive, educational, and medical/neurological factors that may contribute to the child's clinical presentation and, therefore, consider the use of pharmacotherapy as part of a broader treatment plan that encompasses all aspects of a child's life. Pharmacotherapy should be integrated into this treatment plan as an adjunct to individual psychotherapy, family therapy, educational interventions, behavioral interventions, and careful medical management rather than as an alternative to these other interventions or only when these other interventions have failed. In defining the role of pharmacotherapy in the treatment plan, realistic expectations of pharmacotherapeutic interventions, careful definition of target symptoms, and careful assessment of the potential risks and benefits of this type of intervention for behaviorally disordered children are major ingredients for a successful pharmacological intervention.

REFERENCES

1. Preskorn S. H., Weller E. B., Weller R. A.: Depression in children: Relationship between plasma imipramine levels and response. *J Clin Psychiatry* 43:450–453, 1982.
2. Biederman J., Baldessarini R., Wright V., et al: Efficacy and safety of desipramine in the treatment of children and adolescents with attention deficit disorder: A double-blind placebo-controlled study. *J Am Acad Child Adol Psychiatry* 28:777–785, 1989.
3. Biederman J., Baldessarini R., Wright V., et al: Serum drug levels and cardiovascular findings in children and adolescents with attention deficit disorder treated with high doses of desipramine. *J Am Acad Child Adol Psychiatry* 28:903–911, 1989.
4. Zametkin A., Rapoport J. L., L. M. D., et al: Treatment of hyperactive children with monoamine oxidase inhibitors: I. Clinical efficacy. *Arch Gen Psychiatry* 962–966, 1985.
5. Hunt, R. D., Ruud B., Minderaa M. D., et al: The therapeutic effect of clonidine in attention deficit disorder: A comparison with placebo and methylphenidate. *Psychopharmacol Bull* 22:229–236, 1986.
6. Flament M. F., Rapoport J. L., Berg C. J., Sceery W., Kilts C., Mellstrom B., Linnoila M.: Clomipramine treatment of childhood obsessive compulsive disorder: A double-blind controlled study. *Arch Gen Psychiatry* 42:977–983, 1985.
7. Kraepelin E.: *Manic–Depressive Insanity and Paranoia.* Edinburgh, E. and S. Livingstone, 1921.
8. Stroeber M., Carlson G.: Predictors of bipolar illness in adolescents with major depression: A follow-up investigation. *Adolesc Psychiatry* 10:299–319, 1982.

SELECTED READING

General

1. Biederman J., Jellinek M. S.: Psychopharmacology in children. *N Engl J Med* 310:968–972, 1984.
2. Campbell M., Green W., Deutsch S.: *Child and Adolescent Psychopharmacology.* Beverly Hills, Sage Publications, 1985.
3. Cantwell D. P., Carlson G. A.: *Affective Disorders in Childhood and Adolescence—An Update.* New York, SP Medical and Scientific Books, 1983.
4. Carlson G. A.: Classification issues of bipolar disorders in childhood. *Psychiatr Dev* 2:273–285, 1984.

5. Carlson G. A., Strober M.: Affective disorder in adolescence: Issues in misdiagnosis. *J Clin Psychiatry* 39:59–66, 1978.

6. Coffey B. J.: Therapeutics III: Pharmacotherapy, in Robson K. (ed): *Manual of Clinical Child Psychiatry.* Washington, American Psychiatric Press, 1986, pp 149–184.

7. Davis, R. E.: Manic–depressive variant syndrome of childhood: A preliminary report. *Am J Psychiatry* 136:702–706, 1979.

8. *Diagnostic and Statistical Manual of Mental Disorders* (DSM-IIIR), ed. 3, revised. Washington, American Psychiatric Press, 1987.

9. Evans R. W., Clay T. H., Gualtieri C. T.: Carbamazepine in pediatric psychiatry. *J Am Acad Child Adol Psychiatry* 26:2–8, 1987.

10. Gittelman R. (ed): *Anxiety Disorders of Childhood.* New York, The Guilford Press, 1986.

11. Gualtieri C. T., Golden R. N., Fahs J. J.: New developments in pediatric psychopharmacology. *J Dev Behav Pediatr* 4:202–209, 1983.

12. Gualtieri C. T., Golden R., Evans R. W., et al: Blood level measurement of psychoactive drugs in pediatric psychiatry. *Ther Drug Monit* 6:127–141, 1984.

13. Gualtieri C. T., Ondrusek M. G., Finley C.: Attention deficit disorders in adults. *Clin Neuropharmacol* 8:343–356, 1985.

14. Hastings J. E., Barkley R. A.: A review of psychophysiological research with hyperkinetic children. *J Abnorm Child Psychol* 6:413–417, 1978.

15. Hechtman L.: Adolescent outcome of hyperactive children treated with stimulants in childhood: A review. *Psychopharmacol Bull* 21:178–191, 1985.

16. Herskowitz J.: Neurologic presentations of panic disorder in childhood and adolescence. *Dev Med Child Neurol* 28:617–623, 1986.

17. Kuyler P. L., Rosenthal L., Igel G., et al: Psychopathology among children of manic–depressive patients. *Biol Psychiatry* 15:589–597, 1980.

18. Mannuzza S., Gittleman Klein R., Bonagura N., et al: Hyperactive boys almost grown up: II. Status of subjects without a mental disorder. *Arch Gen Psychiatry* 45:13–18, 1988.

19. Michels R., Cavenar J. O. (eds): *Psychiatry,* Vol 6: *Child Psychiatry* (Solnit A. J., Cohen D. J., Schowalter J. E., eds). New York, Basic Books, 1986.

20. Popper C. (ed): *Psychiatric Pharmacosciences of Children and Adolescents.* Washington, American Psychiatric Press, 1987.

21. Rapoport J. L.: DSM-III-R and pediatric psychopharmacology. *Psychopharmacol Bull* 21:803–806, 1985.

22. Reiss A. L.: Developmental manifestations in a boy with prepubertal bipolar disorder. *J Clin Psychiatry* 46:441–443, 1985.

23. Rutter M., Hersov L. (eds): *Child and Adolescent Psychiatry: Modern Approaches* (ed 2). Boston, Blackwell Scientific Publications, 1985.

24. Rutter M., Tuma H. A., Lann I. S.: *Assessment and Diagnosis in Child Psychopathology.* New York, The Guilford Press, 1988.

25. Ryan N. D., Puig-Antich J.: Pharmacological treatment of adolescent psychiatric disorders. *J Adol Health Care* 8:137–142, 1987.

26. Solnit A. J., Cohen D. J., Schowalter J. E. (eds): *Psychiatry,* Vol. 6: *Child Psychiatry.* Philadelphia, J. P. Lippincott, 1986.

27. Sovner R., Hurley A. D.: Do the mentally retarded suffer from affective illness? *Arch Gen Psychiatry* 40:61–67, 1983.

28. Weiss G., Hechtman L.: *Hyperactive Children Grown Up.* New York, The Guilford Press, 1986.

29. Weller R. A., Weller E. B., Tucker S. G., et al: Mania in prepubertal children: Has it been underdiagnosed? *J Affect Disord* 11:151–154, 1986.

30. Wender P. H.: *The Hyperactive Child, Adolescent, and Adult.* New York, Oxford University Press, 1987.

31. Werry J. S. (ed): *Pediatric Psychopharmacology: The Use of Behavior Modifying Drugs in Childhood.* New York, Brunner-Mazel, 1978.

32. White J. H., O'Shanick G.: Juvenile manic–depressive illness. *Am J Psychiatry* 134:1035–1036, 1977.
33. Wiener J. M.: Psychopharmacology in childhood disorders. *Psychiatr Clin North Am* 7:831–843, 1984.
34. Zametkin A. J., Rapoport J. L.: Neurobiology of attention deficit disorder with hyperactivity: Where have we come in 50 years? *J Am Acad Child Adol Psychiatry* 26:676–686, 1987.

Stimulants

35. Barkley R. A.: The use of psychopharmacology to study reciprocal influences in parent–child interaction. *J Abnorm Child Psychol* 9:303–310, 1981.
36. Barkley R. A., Karlsson J., Pollard S., et al: Developmental changes in the mother–child interactions of hyperactive boys: Effects of two dose levels of Ritalin. *J Child Psychol Psychiatry* 26:705–715, 1985.
37. Bond W. S.: Recognition and treatment of attention deficit disorder. *Clin Pharmacol* 6:617–624, 1987.
38. Cantwell D. P.: A clinician's guide to the use of stimulant medication for the psychiatric disorders of children. *J Dev Behav Pediatr* 1:133–140, 1980.
39. Charles L., Schain R.: A four-year follow-up study of the effects of methylphenidate on the behavior and academic achievement of hyperactive children. *J Abnorm Child Psychol* 9:495–505, 1981.
40. Cohen N. J., Thompson L.: Perceptions and attitudes of hyperactive children and their mothers regarding treatment with methylphenidate. *Can J Psychiatry* 27:40–42, 1982.
41. Comings D. E., Comings B. G.: A controlled study of Tourette syndrome. I. Attention-deficit disorder, learning disorders, and school problems. *Am J Hum Genet* 41:701–741, 1987.
42. Evans R. W., Gualtieri C. T., Amara I.: Methylphenidate and memory: Dissociated effects in hyperactive children. *Psychopharmacology (Berl)* 90:211–216, 1986.
43. Friedmann N., Thomas J., Carr R., et al: Effect on growth in pemoline-treated children with attention deficit disorder. *Am J Dis Child* 135:329–332, 1981.
44. Gadow K. D.: Effects of stimulant drugs on academic performance in hyperactive and learning disabled children. *J Learn Disabil* 16:290–299, 1983.
45. Gadow K. D.: Prevalence and efficacy of stimulant drug use with mentally retarded children and youth. *Psychopharmacol Bull* 21:291–303, 1985.
46. Gauthier M.: Stimulant medications in adults with attention deficit disorder. *Can J Psychiatry* 29:435–440, 1984.
47. Golinko B. E.: Side effects of dextroamphetamine and methylphenidate in hyperactive children—a brief review. *Prog Neuropsychopharmacol Biol Psychiatry* 8:1–8, 1984.
48. Greenhill L. L., Puig-Antich J., Novacenko H., et al: Prolactin, growth hormone and growth responses in boys with attention deficit disorder and hyperactivity treated with methylphenidate. *J Am Acad Child Psychiatry* 23:58–67, 1984.
49. Gualtieri C. T., Wargin W., Kanoy R., et al: Clinical studies of methylphenidate serum levels in children and adults. *J Am Acad Child Psychiatry* 21:19–26, 1982.
50. Hechtman L., Weiss G., Perlman T.: Young adult outcome of hyperactive children who received long-term stimulant treatment. *J Am Acad Child Psychiatry* 23:261–269, 1984.
51. Kalachnik J. E., Sprague R. L., Sleator E. K., et al: Effect of methylphenidate hydrochloride on stature of hyperactive children. *Dev Med Child Neurol* 24:586–595, 1982.
52. Lowe T. L., Cohen D. J., Detlor J., et al: Stimulant medications precipitate Tourette's syndrome. *JAMA* 247:1168–1169, 1982.
53. Mattes J. A., Gittelman R.: Growth of hyperactive children on maintenance regimen of methylphenidate. *Arch Gen Psychiatry* 40:317–321, 1983.

54. Porrino L. J., Rapoport J. L., Behar D., et al: A naturalistic assessment of the motor activity of hyperactive boys: II. Stimulant drug effects. *Arch Gen Psychiatry* 40:688–693, 1983.
55. Price R. A., Leckman J. F., Pauls D. L., et al: Gilles de la Tourette's syndrome: Tics and central nervous system stimulants in twins and nontwins. *Neurology (NY)* 36:232–237, 1986.
56. Rapoport J. L., Buchsbaum M. S., Zahn T. P., et al: Dextroamphetamine: Cognitive and behavioral effects in normal prepubertal boys. *Science* 199:560–563, 1978.
57. Rapport M. D., DuPaul G. J.: Hyperactivity and methylphenidate: Rate-dependent effects on attention. *Int Clin Psychopharmacol* 1:45–52, 1986.
58. Satterfield J. H., Cantwell D. P., Schell A., et al: Growth of hyperactive children treated with methylphenidate. *Arch Gen Psychiatry* 36:212–217, 1979.
59. Satterfield J. H., Schell A. M., Barb S. D.: Potential risk of prolonged administration of stimulant medication for hyperactive children. *J Dev Behav Pediatr* 1:102–107, 1980.
60. Schmidt K., Solanto M. V., Sanchez-Kappraff M., et al: The effect of stimulant medication on academic performance, in the context of multimodal treatment, in attention deficit disorders with hyperactivity: Two case reports. *J Clin Psychopharmacol* 4:100–103, 1984.
61. Schmidt K.: The effect of stimulant medication in childhood-onset pervasive developmental disorder—a case report. *J Dev Behav Pediatr* 3:244–246, 1982.
62. Shapiro A. K., Shapiro E.: Do stimulants provoke, cause, or exacerbate tics and Tourette's syndrome? *Compr Psychiatry* 22:265–273, 1981.
63. Steinhausen H. C., Kreuzer E. M.: Learning in hyperactive children: Are there stimulant-related and state-dependent effects? *Psychopharmacology (Berl)* 74:389–390, 1981.
64. Swanson J. M., Lerner M., Cantwell D.: Blood levels and tolerance to stimulants in ADDH children. *Clin Neuropharmacol* 9(Suppl):523–525, 1986.
65. Varley C. K., Trupin E. W.: Double-blind administration of methylphenidate to mentally retarded children with attention deficit disorder: A preliminary study. *Am J Ment Defic* 86:560–566, 1982.
66. Volkmar F. R., Hoder E. L., Cohen D. J.: Inappropriate uses of stimulant medications. *Clin Pediatr (Phila)* 24:127–130, 1985.
67. Wallander J. L., Schroeder S. R., Michelli J. A., et al: Classroom social interactions of attention deficit disorder with hyperactivity children as a function of stimulant medication. *J. Pediatr Psychol* 12:61–76, 1987.
68. Weiss G.: Controversial issues of the pharmacotherapy of the hyperactive child. *Can J Psychiatry* 26:385–392, 1981.
69. Weizman A., Weitz R., Szekely G. A., et al: Combination of neuroleptic and stimulant treatment in attention deficit disorder with hyperactivity. *J Am Acad Child Psychiatry* 23:295–298, 1984.
70. Wender P. H., Reimherr F. W., Wood D. R.: Stimulant therapy of 'adult hyperactivity' (letter). *Arch Gen Psychiatry* 42:840, 1985.
71. Zahn T. P., Rapoport J. L., Thompson C. L.: Autonomic and behavioral effects of dextroamphetamine and placebo in normal and hyperactive prepubertal boys. *J Abnorm Child Psychol* 8:145–160, 1980.
72. Zametkin A. J., Karoum F., Linnoila M., et al: Stimulants, urinary catecholamines, and indoleamines in hyperactivity. A comparison of methylphenidate and dextroamphetamine. *Arch Gen Psychiatry* 42:251–255, 1985.

Antidepressants

73. Biederman, J., Gastfriend D., Jellinek M. S., et al: Cardiovascular effects of desipramine in children and adolescents with attention deficit disorder. *J Pediatr* 106:1017–1020, 1985.
74. Biederman J., Gastfriend D. R., Jellinek M. S.: Desipramine in treating children with attention deficit disorder. *J Clin Psychopharmacol* 6:359–363, 1986.
75. Biederman J., Gonzalez E., Bronstein B., et al: Despiramine and cutaneous reactions in pediatric outpatients. *J Clin Psychiatry* 49:178–183, 1988.

76. Dillon D. C., Salzman I. J., Schulsinger D. A.: The use of imipramine in Tourette's syndrome and attention deficit disorder: Case report. *J Clin Psychiatry* 46:348–349, 1985.

77. Donnelly M., Zametkin A. J., Rapoport J. L., et al: Treatment of childhood hyperactivity with desipramine: Plasma drug concentration, cardiovascular effects, plasma and urinary catecholamine levels, and clinical response. *Clin Pharmacol Ther* 39:72–81, 1986.

78. Gastfriend D. R., Biederman J., Jellinek M. S.: Desipramine in treating adolescents with attention deficit disorder. *Am J Psychiatry* 141:906–908, 1984.

79. Gittelman-Klein R., Klein D.: Controlled imipramine of school phobia. *Arch Gen Psychiatry* 25:204–207, 1971.

80. Greenberg L. M., Yellin A. M.: Blood pressure and pulse changes in hyperactive children with imipramine and methylphenidate. *Am J Psychiatry* 132:1325–1326, 1975.

81. Hoge S. K., Biederman J.: A case of Tourette's syndrome with symptoms of attention deficit disorder treated with desipramine. *J Clin Psychiatry* 47:478–479, 1986.

82. Hoge S. K., Biederman J.: Liver function tests during treatment with desipramine in children and adolescents. *J Clin Psychopharmacol* 7:87–89, 1987.

83. Lake C. R., Mikkelsen E. J., Rapoport J. L., et al: Effect of imipramine on norepinephrine and blood pressure in enuretic boys. *Clin Pharmacol Ther* 26:647–653, 1979.

84. Mikkelsen E. J., Rapoport J. L.: Enuresis: Psychopathology, sleep stage, and drug response. *Urol Clin North Am* 7:361–377, 1980.

85. Muller U. R., Goodman N., Bellet S.: The hypotensive effect imipramine hydrochloride in patients with cardiovascular disease. *Clin Pharmacol Ther* 2:300–307, 1961.

86. Rapoport J. L.: Antidepressants in childhood attention deficit disorder and obsessive compulsive disorder. *Psychosomatics* 27(Suppl):30–36, 1986.

87. Rapoport J. L., Mikkelsen E. J., Zavadil A. P.: Plasma imipramine and desmethylimipramine concentration and clinical response in childhood enuresis. *Psychopharmacol Bull* 14:60–61, 1978.

88. Rapoport J. L., Mikkelsen E. J., Zavadil A., et al: Childhood enuresis. II. Psychopathology tricyclic concentration in plasma, and antienuretic effect. *Arch Gen Psychiatry* 37:1146–1152, 1980.

89. Ryan N. D., Puig-Antich J., Cooper T., et al: Imipramine in adolescent major depression: Plasma level and clinical response. *Acta Psychiatr Scand* 73:275–288, 1986.

90. Ryan N. D., Puig-Antich J., Cooper T., et al: Relative safety of single versus divided dose imipramine in adolescent major depression. *J Am Acad Child Adol Psychiatry* 26:400–406, 1987.

91. Saraf K. R., Klein D. F., Gittelman-Klein R., et al: Imipramine side effects in children. *Psychopharmacologia (Berlin)* 37:265–274, 1974.

92. Saraf K. R., Klein D. F., Gittelman-Klein R., et al: EKG effects of imipramine treatment in children. *J Am Acad Child Psychiatry* 17:60–69, 1978.

93. Weller E. B., Weller R. A., Preskorn S. H., et al: Steady-state plasma imipramine levels in prepubertal children. *Am J Psychiatry* 139:506–508, 1982.

94. Zametkin A., Rapoport J. L., Murphy D. L., et al: Treatment of hyperactive children with monoamine oxidase inhibitors. I. Clinical efficacy. *Arch Gen Psychiatry* 42:962–966, 1985.

95. Zametkin A., Rapoport J. L., Murphy D. L., et al: Treatment of hyperactive children with monoamine oxidase inhibitors. II. Plasma and urinary monoamine findings after treatment. *Arch Gen Psychiatry* 42:969–973, 1985.

Lithium

96. Campbell M., Schulman D., Rapoport J. L.: The current status of lithium therapy in child and adolescent psychiatry. A report of the Committee on Biological Aspects of Child Psychiatry of the American Academy of Child Psychiatry, December 1977. *J Am Acad Child Psychiatry* 17:717–720, 1978.

97. Campbell M., Small A. M., Green W. H., et al: Behavioral efficacy of haloperidol and lithium

carbonate. A comparison in hospitalized aggressive children with conduct disorder. *Arch Gen Psychiatry* 41:650–656, 1984.

98. Carroll J. A., Jefferson J. W., Greist J. H.: Psychiatric uses of lithium for children and adolescents. *Hosp Community Psychiatry* 38:927–928, 1987.

99. DeLong G. R., Aldersh A. L.: Long-term experience with lithium treatment in childhood: Correlation with clinical diagnosis. *J Am Acad Child Adol Psychiatry* 26:389–394, 1987.

100. Hsu L. K.: Lithium-resistant adolescent mania. *J Am Acad Child Psychiatry* 25:280–283, 1986.

101. Jefferson J. W.: The use of lithium in childhood and adolescence: An overview. *J Clin Psychiatry* 43:174–177, 1982.

102. Khandelwal S. K., Varma V. K., Srinivasa-Murthy R.: Renal function in children receiving long-term lithium prophylaxis. *Am J Psychiatry* 141:278–279, 1984.

103. Lena B.: Lithium in child and adolescent psychiatry. *Arch Gen Psychiatry* 36(Spec No):854–855, 1979.

104. Marini J. L., Sheard M. H.: Antiaggressive effect of lithium ion in man. *Acta Psychiatr Scand* 55:269–286, 1977.

105. McKnew D. H., Cytryn L., Buchsbaum M. S., et al: Lithium in children of lithium-responding parents. *Psychiatry Res* 4:171–180, 1981.

106. Perry R., Campbell M., Grega D. M., et al: Saliva lithium levels in children: Their use in monitoring serum lithium levels and lithium side effects. *J Clin Psychopharmacol* 4:199–202, 1984.

107. Platt J. E., Campbell M., Green W. H., et al: Cognitive effects of lithium carbonate and haloperidol in treatment-resistant aggressive children. *Arch Gen Psychiatry* 41:657–662, 1984.

108. Platt J. E., Campbell M., Green W. H., et al: Effects of lithium carbonate and haloperidol on cognition in aggressive hospitalized school-age children. *J Clin Psychopharmacol* 1:8–13, 1981.

109. Rogeness G. A., Riester A. E., Wicf J. S.: Unusual presentation of manic depressive disorder in adolescence. *J Clin Psychiatry* 43:37–39, 1982.

110. Siassi I.: Lithium treatment of impulsive behavior in children. *J Clin Psychiatry* 43:482–484, 1982.

111. Steinberg D.: The use of lithium carbonate in adolescence. *J Child Psychol Psychiatry* 21:263–271, 1980.

112. Steingard R., Biederman J.: Lithium responsive manic-like symptoms in two individuals with autism and mental retardation. *Am Acad Child Adol Psychiatry* 26:932–935, 1987.

113. Thorneloe W. F., Crews E. L.: Manic depressive illness concomitant with antisocial personality disorder: Six case reports and review of the literature. *J Clin Psychiatry* 42:5–9, 1981.

114. Vetro A., Szentistvanyi I., Pallag L., et al: Therapeutic experience with lithium in childhood aggressivity. *Neuropsychobiology* 14:121–127, 1985.

115. Vitiello B., Behar D., Ryan P., et al: Saliva lithium monitoring [letter]. *J Am Acad Child Adol Psychiatry* 26:812–813, 1987.

116. Weller E. B., Weller R. A., Fristad M. A., et al: Saliva lithium monitoring in prepubertal children. *J Am Acad Child Adol Psychiatry* 26:173–175, 1987.

117. Weller E. B., Weller R. A., Fristad M. A.: Lithium dosage guide for prepubertal children: A preliminary report. *J Am Acad Child Psychiatry* 25:92–95, 1986.

118. Youngerman J., Canino I. A.: Lithium carbonate use in children and adolescents. A survey of the literature. *Arch Gen Psychiatry* 35:216–224, 1978.

Antipsychotics

119. Biederman J., Lerner Y., Belmaker R. H.: Combination of lithium carbonate and haloperidol in schizo-affective disorder: A controlled study. *Arch Gen Psychiatry* 36:327–333, 1979.

120. Campbell M., Fish B., Shapiro T., et al: Acute responses of schizophrenic children to a sedative and a "stimulating" neuroleptic: A pharmacologic yardstick. *Curr Ther Res* 14:759–766, 1972.
121. Campbell M., Grega D. M., Green W. H., et al.: Neuroleptic-induced dyskinesias in children. *Clin Neuropharmacol* 6:207–222, 1983.
122. Gualtieri C. T., Patterson D. R.: Neuroleptic-induced tics in two hyperactive children. *Am J Psychiatry* 143:1176–1177, 1983.
123. Gualtieri C. T., Guimond M.: Tardive dyskinesia and the behavioral consequences of chronic neuroleptic treatment. *Dev Med Child Neurol* 23:255–259, 1981.
124. Gualtieri C. T., Hawk B.: Tardive dyskinesia and other drug-induced movement disorders among handicapped children and youth. *Appl Res Ment Retard* 1:55–69, 1980.
125. Gualtier C. T., Quade D., Hicks R. E., et al: Tardive dyskinesia and other clinical consequences of neuroleptic treatment in children and adolescents. *Am J Psychiatry* 141:20–23, 1984.
126. Gualtieri C. T., Sprague R. L.: Preventing tardive dyskinesia and preventing tardive dyskinesia litigation. *Psychopharmacol Bull* 20:346–348, 1984.
127. Perry R., Campbell M., Green W. H., et al: Neuroleptic-related dyskinesias in autistic children: A prospective study. *Psychopharmacol Bull* 21:140–143, 1985.

Antianxiety Agents

128. Biederman J.: Clonazepam in treating prepubertal children with panic-like symptoms. *J Clin Psychiatry* 48(Suppl):38–42, 1987.
129. Emde R. N.: Early development and opportunities for research on anxiety, in Tuma A. H., Maser J. D. (eds): *Anxiety and the Anxiety Disorders.* Hillsdale, NJ, Lawrence Erlbaum, 1986, pp 413–420.
130. Gittelman R., Koplewicz H. S.: Pharmacotherapy of childhood anxiety disorders, in Gittelman R. (ed): *Anxiety Disorders in Children.* New York, Guilford Press, 1987, pp 188–203.
131. Simeon J. G., Ferguson H. B.: Alprazolam effects in children with anxiety disorders. *Can J Psychiatry* 32:570–574, 1987.

Fenfluramine

132. August G. J., Naftali R., Papanicolaou A. C., et al: Fenfluramine treatment in infantile autism: Neurochemical, electrophysiological, and behavioral effects. *J Nerv Ment Dis* 172:604–612, 1984.
133. Campbell M., Small A. M., Palij M., et al: Efficacy and safety of fenfluramine in autistic children: Preliminary analysis of a double-blind study. *Psychopharmacol Bull* 23:123–128, 1987.
134. Donnelly M., Rapoport J. L., Ismond D. R.: Fenfluramine treatment of childhood attention deficit disorder with hyperactivity: A preliminary report. *Psychopharmacol Bull* 22:152–154, 1986.
135. Geller E., Ritvo E. R., Freeman B. J., et al: Fenfluramine in autism. *N Engl J Med* 307:1450–1451, 1982.
136. Geller E., Ritvo E. R., Freeman B. J., et al: Preliminary observations on the effect of fenfluramine on blood serotonin and symptoms in three autistic boys. *N Engl J Med* 307:165–168, 1982.
137. Ritvo E. R., Freeman B. J., Geller E., et al: Effects of fenfluramine on 14 outpatients with the syndrome autism. *J Am Acad Child Psychiatry* 22:549–558, 1983.
138. Ritvo E. R., Freeman B. J., Yuwiler A., et al: Study of fenfluramine in outpatients with the syndrome autism. *J Pediatr* 105:823–828, 1984.
139. Volkmar F. R., Paul R., Cohen D. J., et al: Irritability in autistic children treated with fenfluramine (letter). *N Engl J Med* 309:187, 1983.

Clonidine

140. Cohen D. J., Detlor J., Young J. G., et al: Clonidine ameliorates of Gilles de la Tourette's syndrome. *Arch Gen Psychiatry* 37:1350–1357, 1980.
141. Goetz C. G., Tanner C. M., Wilson R. S., et al: Clonidine and Gilles de la Tourette's syndrome: Double-blind study using objective rating methods. *Ann Neurol* 21:307–310, 1987.
142. Hunt R. D.: Treatment effects of oral and transdermal clonidine in relation to methylphenidate: An open pilot study in ADD-H. *Psychopharmacol Bull* 1:111–114, 1987.
143. Hunt R. D., Ruud B., Minderaa M. D., et al: Clonidine benefits children with attention deficit disorder and hyperactivity: Report of a double-blind placebo–crossover therapeutic trial. *J Am Acad Child Psychiatry* 5:617–629, 1985.
144. Hunt R. D., Ruud B., Minderaa M. D., et al: The therapeutic effect of clonidine in attention deficit disorder: A comparison with placebo and methylphenidate. *Psychopharmacol Bull* 1:229–236, 1986.
145. Leckman J. F., Detlor J., Harcherik D. F., et al: Short- and long-term treatment of Tourette's syndrome with clonidine: A clinical perspective. *Neurology* 35:343–351, 1985.

β Blockers

146. Dominguez R. A., Goldstein B. J.: Beta-blockers in psychiatry. *Hosp Comm Psychiatry* 6:565–566, 568, 1984.
147. Jenkins S. C., Maruta T.: Therapeutic use of propanolol for intermittent explosive disorder. *Mayo Clin Proc* 62:204–214, 1987.
148. Lipinski J. F., Zubenko G. S., Cohen B. M., et al: Propranolol in treating neuroleptic-induced akathisia. *Am J Psychiatry* 3:412–415, 1984.
149. Ratey J. J., Mikkelsen E. J., Smith G. B., et al: Beta-blockers in the severely and profoundly mentally retarded. *J Clin Psychopharmacol* 2:103–107, 1984.
150. Sorgi P. J., Ratey J. J., Polakf S.: Beta-adrenergic blockers for the control of aggressive behaviors in patients with chronic schizophrenia. *Am J Psychiatry* 6:775–776, 1986.

10

Psychotropic Drug Use in Pregnancy

LEE S. COHEN, M.D., JERROLD F. ROSENBAUM, M.D., and VICKI L. HELLER, M.D.

I. INTRODUCTION

The pregnant psychiatric patient on or requiring psychotropic medication presents an unusual clinical dilemma, often placing the physician between a teratologic rock and a clinical hard place.[1] Although ethical considerations have precluded adequately controlled studies with pregnant woman, the rate of psychotropic use in pregnancy[2,3] together with the prevalence of certain psychiatric conditions in women during pregnancy[4] and in women of childbearing potential[5] in general underscore the need for treatment guidelines to help clinicians manage pregnant patients with psychiatric disorders.

Several reviews over the last decade[6,7] describing the outcome of offspring exposed to psychotropics in utero suggest that agents including antipsychotics,[8–11] antidepressants,[12,13] and benzodiazepines[14–16] have low teratogenic potential; yet one cannot, however, assert their safety. Data are insufficient to determine the absolute risk of organ dysgenesis or long-term neurobehavioral sequelae that may result from drug exposure on the one hand or the morbidity and potential mortality of untreated psychiatric symptoms during pregnancy on the other. This dilemma—

LEE S. COHEN, M.D. • Pregnancy Consultation Service, Clinical Psychopharmacology Unit, Massachusetts General Hospital, Boston, Massachusetts 02114; Department of Psychiatry, Harvard Medical School, Boston, Massachusetts 02115. **JERROLD F. ROSENBAUM, M.D.** • Clinical Psychopharmacology Unit, Massachusetts General Hospital, Boston, Massachusetts 02114; Department of Psychiatry, Harvard Medical School, Boston, Massachusetts 02115. **VICKI L. HELLER, M.D.** • Department of Obstetrics and Gynecology, Beth Israel Hospital, Boston, Massachusetts 02215; Department of Obstetrics and Gynecology, Harvard Medical School, Boston, Massachusetts 02115.

the uncertain risk of pharmacotherapeutic intervention versus that of untreated psychiatric symptoms in pregnancy—makes the evaluation of the psychiatrically ill pregnant patient a distressing challenge. The purpose of this chapter is (1) to review the course of certain psychiatric illnesses in pregnancy, (2) to describe the putative risks of specific psychotropic drug use in pregnancy, and (3) to provide the clinician with an approach to assessing the risks and benefits of treating psychiatrically ill pregnant patients and for choosing and instituting therapy with these agents.

II. ASSESSMENT OF THE PREGNANT PSYCHIATRIC PATIENT

Concern about psychotropic drug use in pregnancy is typically raised in one of three scenarios. First, some women experience onset of psychiatric symptoms during pregnancy. Second, and more frequently, patients with known psychiatric illness already managed on drugs register concern about the risks of in utero psychotropic exposure versus risk of relapse off drug should they conceive, or the risk of untreated symptoms. Third, some women inadvertently conceive while taking psychotropics and request urgent consultation regarding discontinuation or changes in pharmacotherapeutic regimens. Decisions regarding institution, alteration, or discontinuation of treatment are made only after the risks and benefits of pharmacotherapeutic intervention are weighed. In general, psychotropic drug use in pregnancy should be reserved for those clinical situations in which the risk to mother and fetus from the disorder is felt to outweigh the risk of drug treatment.

A. Risks of Pharmacotherapy

Risks associated with psychotropic use in pregnancy may be divided into five types: (1) teratogenic risk; (2) risk of long-term behavioral sequelae, so-called "behavioral teratogenesis"; (3) risk of direct toxic effects on fetus; (4) impact of drug on labor and delivery; and (5) effect of drug on the breast-feeding infant. Clearly, these factors may combine to cause a net effect following maternal medication use.

Teratogenic risk refers to the risk of gross organ malformation. Risk of organ dysgenesis appears greatest when drug exposure occurs between weeks 2 and 10 of gestation. Between weeks 2 and 4, corresponding to the first 14 days of development (and that time prior to the first missed menstrual period), toxic injury to the conceptus is more likely to result in a nonviable, blighted ovum.[17,18]

Behavioral teratogenesis refers to the potential for long-lasting behavioral sequelae associated with in utero exposure to a given drug. Animal studies evaluating the neurobehavioral effects of antipsychotics, antidepressants, and benzodiazepines have not revealed consistent results.[19–21] Neurotransmitter changes in the dopaminergic, cholinergic, and noradrenergic systems have been noted in animals after in utero exposure to psychoactive agents,[22] though the relevance of these findings insofar as they might be extrapolatable to humans is unclear.

The direct toxic effects of a drug on a developing fetus both during pregnancy and also at the time of labor and delivery have been cited as other risks associated with psychotropic drug use during pregnancy. Transient neonatal distress syndromes following antidepressant[23,24] and benzodiazepine[25] administration in pregnancy, and extrapyramidal symptoms noted with antipsychotics,[26–28] respectively, have also been described. These phenomena may result from a heightened inherent vulnerability of the immature central nervous system to exogenous agents[29,30] or from prolonged or exaggerated drug effects possibly secondary to decreased neonatal hepatic microsomal activity.[31] Decreased plasma protein and protein binding affinity noted in neonates may also lead to increased amounts of free drug and may contribute to some of the transient toxicity seen in newborns exposed to psychoactive agents during the latter portions of pregnancy and during labor and delivery.[31]

Nearly all drugs that cross the placenta appear in breast milk.[32] Since psychotropics cross the placenta, it is not surprising that they have also been regularly measured in breast milk. The concentration, however, and the resulting effect of these agents on the newborn are extremely variable. Toxicity in infants of breast-feeding mothers on psychotropic agents is rare; accumulating cases of severe neonatal distress syndromes, however, in babies breast fed by such women have been reported,[33] as have significant levels of these agents in their breast milk.[34] Less clear, and perhaps of greater significance too, is the potential effect of prolonged drug exposure on the brain of the developing infant.

B. Risks Associated with Psychiatric Illness

Though difficult to quantify, attendant morbidity (and potential mortality) associated with a particular psychiatric illness must be weighed against potential risk of in utero psychotropic drug exposure. Impaired self-care including the inability to follow appropriate prenatal instructions as well as suicidal, impulsive, or potentially dangerous manic behavior are examples of clinical risks that may argue for the use of pharmacotherapy in pregnancy. Of uncertain clinical effect are the neuroendocrine changes associated with major psychiatric disorders[35,36] or the physiological consequences of untreated symptoms of anxiety, for example,[1,37] on the fetoplacental unit. The relevance, too, of untreated psychosis, anxiety, or depression for the maternal attachment process and subsequent neuropsychological development of the child is controversial[38–41] and at this time is a matter of speculation[42] but also of concern.

C. Assessing Risk versus Benefit

Several studies[2,3] estimate that up to 35% of pregnant women ingest psychotropics during pregnancy. Other data suggest that approximately 25% of inquiries to drug information centers are requests for assistance in counseling women or clinicians about the attendant risk of in utero drug exposure.[3] These statistics may at first glance be reassuring of the relatively low risk of drug exposure. However, the difficulty in evaluating such patients springs from our inability to quantify relative risk of in utero psychotropic drug exposure versus that of psychiatric disorder. In this state of uncertainty, the physician must design the most thoughtful treatment plan tailored to the individual patient's overall clinical situation. Uncertainty, however, does not imply a lack of clinical options.

III. COURSE AND MANAGEMENT OF PSYCHIATRIC ILLNESS IN PREGNANCY

A. Psychosis

Although few studies evaluate the natural course of schizophrenia during pregnancy, up to 20% of pregnant schizophrenic patients may decompensate during the postpartum period.[43] The patient with onset of psychotic symptoms during pregnancy requires a thorough evaluation to rule out organic causes of mental status change.[1] It is a potentially serious error to assume that changes in mental functioning in pregnancy are merely a "reaction" to the pregnancy. A vigorous diagnostic workup should be pursued to rule out metabolic derangement, infection, intracranial masses, the presence of toxic substances, and other etiologies of thought disorder.

B. Antipsychotics

Although case reports have cited a possible relationship between first-trimester exposure to certain neuroleptics such as haloperidol (Haldol® and others) and resultant limb deformities,[44] several large prospective studies have failed to demonstrate significant teratogenic risk associated with this and other neuroleptic agents.[8,10] Rumeau-Rouguette and colleagues,[9] for example, retrospectively studied 12,764 women including 315 cases of first-trimester exposure to neuroleptic; 3.5% of women exposed gave birth to malformed infants. An increase in congenital malformations compared to nonexposed offspring was reported for patients exposed to phenothiazines with three-carbon aliphatic side chains such as chlorpromazine (Thorazine® and others) but not for other antipsychotics.

Results from a multicenter study by Slone et al., including 5282 pregnant women with 1309 cases of phenothiazine exposure, failed to document an increased rate of malformations in the group of women exposed.[8] These findings are consistent with those from the California Child Health and Development Project. Citing results from this prospective study of 20,504 pregnant women, Milkovich and Van den Berg noted no increased rate of malformations in women to whom the neuroleptic prochlorperazine (Compazine and others) had been administered during the first trimester.[10] More recently, however, Edlund and Craig have reanalyzed the data from this study and have reported a trend toward increased numbers of anomalies in women who received this agent between the sixth and tenth week of gestation (3.2% incidence of anomalies in controls versus 5.4% in exposed offspring).[11]

Treatment-associated adverse effects evident during the perinatal period have been described in neonates exposed to neuroleptics in utero. Extrapyramidal signs including dystonic movements, parkinsonian-like effects, and hypertonicity have been noted.[26–28] Although usually of short duration, these effects have been reported to last up to 6 months of age in some babies.[28] Neonatal jaundice has also occurred in babies exposed to neuroleptics in utero, possibly related to immature hepatic microsomal activity in the newborn.[45] Functional bowel obstruction,[46] a possible result of heightened sensitivity of the neonate to anticholinergic activity of neuroleptic agents, has also been cited in a case of a baby exposed in utero to these agents.

Despite multiple efforts to document neurobehavioral sequelae of in utero exposure to antipsychotics in animals, results of these studies have not been consistent.[20,47] Even less is known about the neurobehavioral consequences for humans who have been exposed to these agents in utero. Transient agitation, restlessness, and hypertonicity during the immediate postnatal period have been observed in some antipsychotic-exposed neonates.[48] At least one study has demonstrated increased stature and weight in offspring of neuroleptic-exposed women.[49] Evidence for long-term behavioral sequelae has been lacking, nonetheless, when children have been evaluated in follow-up between 4 and 5 years of age.[50]

Despite prospective studies that fail to support a teratogenic potential of antipsychotic agents, it remains wise to avoid in utero antipsychotic exposure in the first trimester, a recommendation that must be weighed against the risk and nature of relapsing psychosis for the individual patient and the possible requirement of compensatory treatment with high doses of neuroleptics. When the clinician has decided either to institute or to continue antipsychotic therapy during pregnancy, tapering and, if possible, discontinuation of antipsychotic medication 10 days to 2 weeks before the estimated date of confinement is recommended to minimize the chance of neonatal extrapyramidal side effects or other transient neonatal treatment-related effects. Such decisions need also to be weighed against the risk of relapsing psychosis, which sometimes dictates the maintenance of low-dose antipsychotics, particularly in severe chronically ill patients who have already demonstrated repeated decompensation when not taking these agents.

IV. UNIPOLAR AND BIPOLAR MOOD DISORDERS

Although pregnancy has been described as a period of relative emotional stability,[51] recent work suggests that as many as 9–10% of pregnant women may meet research diagnostic criteria (RDC) criteria for major or minor depression during pregnancy and that women with pregravid histories of depression may be at greater risk for prepartum episodes of major mood disorder.[4,52,53] During the postpartum period, furthermore, as many as two-thirds of women experience "postpartum blues."[54] These symptoms are most commonly transient. In a prospective study, however, O'Hara and colleagues reported that 12% of a group of 99 women met RDC criteria for major or minor depression during the immediate postpartum period. Others have also suggested that women with a history of bipolar affective disorder may be at particular risk for developing postpartum mood episodes.[55,56]

A. Antidepressants and Monoamine Oxidase Inhibitors

Although depression may not be uncommon during pregnancy, use of antidepressants should be reserved for the more severe cases, where symptoms of neurovegetative dysfunction (decreased appetite, sleep disturbance, psychomotor retardation, and suicidal ideation) begin to compromise the well-being of mother and therefore fetus. The therapeutic goal need not be complete remission; for example, supportive psychotherapy and stress management strategies may sufficiently attenuate suffering to defer use of a chemotherapeutic agent until after delivery.

Early reports of a possible association between in utero exposure to tricyclic antidepressants and limb dysgenesis[57] have been followed by other large studies that do not support such an association. For example, in two British reviews, 25,000 pregnancies were evaluated, and 85 cases of either imipramine (Tofranil® and others) or amitriptyline (Elavil® and others) exposure were examined.[12,58] Increased risk of malformation was not described in tricyclic-exposed offspring. In a report from the Finnish Register of Congenital Malformations during 1964–1972, an analysis of 2784 cases of birth defects was also undertaken and was compared with a group of matched controls. This study did not reveal evidence of a statistical association between first-trimester exposure to tricyclic antidepressants and congenital malformations.[13] To date, no studies have addressed the neurobehavioral sequelae of in utero exposure to antidepressants in humans. Despite reports of behavioral changes in animals exposed to these agents (including decreased exploratory responses and impaired reflex development),[59] the relevance of these findings for humans is unclear.

The efficacy of monoamine oxidase inhibitors (MAOIs) in the treatment of depression is well described.[60,61] There are few accumulated data, however, regarding the use of these agents in pregnancy. Phenelzine (Nardil®) has been presumed to be teratogenic based on animal testing.[62]

Should the decision be made to continue or to initiate antidepressant therapy during pregnancy, tricyclics are the treatment of choice, given the paucity of data regarding teratogenicity of MAOIs as well as the "newer antidepressants" such as trazodone (Desyrel® and others), maprotiline (Ludiomil® and others), fluoxetine (Prozac®), or bupropion (Wellbutrin®). In the case of fluoxetine, an additional concern is the extended half life of its major metabolite, norfluoxetine. Women who might conceive on fluoxetine are subject to at least 4 weeks of continued in utero exposure to the drug given the pharmacokinetics of this agent. When clinicians choose to institute antidepressant therapy, secondary amines such as desipramine (Norpramin® and others) and nortriptyline (Pamelor® and others) with fewer anticholinergic effects are preferable to tertiary amines. Plasma tricyclic levels should also be followed at least once per trimester given changes in total extracellular volume during pregnancy and labor and the potential effects of these changes on serum drug concentration. Although it has been suggested that adequacy of dosing may be an important factor in determining treatment response in depressed patients,[63] for the pregnant patient, treatment with the lowest effective dose is a more suitable interim goal and may avoid such observed neonatal side effects as urinary retention, transient breathlessness, and irritability noted at delivery.[23,24]

B. Lithium

Although it has been suggested that bipolar women are particularly vulnerable to postpartum episodes of depression, less is known about the actual course of bipolar illness during pregnancy. The teratogenic potential of lithium carbonate (Eskalith® and others) is more clearly established based on a review of the Lithium Register of Babies begun in 1968.[64] Babies exposed to lithium for at least the first trimester of pregnancy are eligible for inclusion in the Register, though underreporting of normal outcomes following in utero exposure to lithium likely results in an overestimation of the risk of malformation.[65] Weinstein and Goldfield,[66] reporting on 166 cases of lithium exposure, describe 18 cases of malformations in exposed infants; 12 of these included the heart and great vessels, and four were noted to have Ebstein's anomaly. This malformation is characterized by right ventricular hypoplasia, patent ductus arteriosus, and tricuspid valve insufficiency. In one series of 36 patients evaluated with Ebstein's anomaly, slightly less than half presented in the first week of life with cyanosis and congestive heart failure. Approximately half of these died within days of presentation, though the majority of the others went on to have surgery for correctable lesions and had a good prognosis for long-term survival.[67] As of 1980, 225 cases of in utero exposure to lithium have been reported. A recent review has indicated for infants exposed to lithium during at least the first trimester a 0.1% (one in 1000 cases of exposure) risk of having Ebstein's anomaly, approximately 20 times the risk of the general population.[68]

Family planning is important for women being treated with psychoactive agents and is particularly important for the bipolar woman managed on lithium. Discussion among patient, husband/partner, obstetrician, and psychiatrist allows for thoughtful selection from the variety of treatment options both before attempts at conception and during pregnancy itself. Choice among such strategies as discontinuation, maintainance, or "as-needed" institution of pharmacotherapy for bipolar women wishing to conceive depends largely on previous history. The greater the number of prior affective episodes, the greater the risk of subsequent relapse. Thus, a patient with one circumscribed episode of mania or infrequent recurrent episodes may benefit from lithium discontinuation with close follow-up.

Pharmacotherapeutic treatment decisions for more brittle bipolar patients with clear-cut need for antimanic prophylaxis are more complicated. Patients with histories of decompensation off lithium treatment may have the drug reinstituted in the second trimester, as this has not been associated with adverse effects as long as careful monitoring of plasma lithium level is performed.[1] The use of low-dose neuroleptic prophylactically for this population has also been proposed by some investigators.[69] Though frequently used as an alternative or adjunct treatment for many bipolar patients,[70,71] the putative safety of carbamazepine (Tegretol® and others) use in pregnancy[72–75] as an alternative to lithium[1] for brittle bipolar women has recently been questioned. Reports from the California Teratogen Registry describing prospective evaluation of offspring with documented in utero exposure to carbamazepine suggest increased rates of craniofacial anomalies, postnatal growth deficiency, and developmental delay compared to a sample of nonexposed children.[76] Though the drug is widely used as a first-line drug for pregnant women with seizure disorders requiring anticonvulsant prophylaxis,[77] given the attendant morbidity associated with uncontrolled seizures in pregnancy, its use as an alternative to lithium for pregnant bipolar women is best deferred.[78] Similarly, the use of valproic acid, another anticonvulsant with putative antimanic activity that is increasingly used in psychiatry,[79] is also best avoided during pregnancy; increased rates of spina bifida in offspring exposed to this drug in utero[80] have been reported.

Thus, treatment planning for the bipolar woman who wishes to conceive but who describes a history of severe decompensation off antimanic therapy may include continuation of lithium treatment until documentation of pregnancy. The rationale for this approach is twofold. First, continuation of treatment minimizes the risk of decompensation during a period of time of variable duration when the woman is attempting conception. Second, in the case of conception while taking lithium, injury to the conceptus during the earliest period of development (corresponding to days 14–28 of the menstrual cycle) and prior to the first missed menstrual period tends to result in either total repair or a nonviable blighted ovum.[17,18]

Brittle bipolar women who conceive on lithium should discontinue the drug with documentation of pregnancy with reintroduction of the agent if necessary during the second or third trimester. When exposure to lithium has occurred during the first 12 weeks of gestation, clinicians need to apprise patients of the increased risk to offspring of cardiovascular malformations as well as the reassuring absence of other neurobehavioral difficulties in follow-up reports of children exposed to lithium in utero.[81] Fetal cardiac ultrasonagraphy as early as 18 weeks of gestation, as well, is advised to rule out cardiovascular anomalies. If documentation of pregnancy occurs in the second or third trimester, continuation of lithium therapy is not absolutely contraindicated and may be a more reasonable course than introduction of other medications as some have suggested.[82]

The pregnant patient on lithium requires particularly close observation of plasma levels given the changes in renal function noted with pregnancy,[1] including increased glomerular filtration rate (GFR) and total body water. Lithium dosing schedules should feature small divided doses. Plasma levels should be monitored as least monthly.

During the third trimester, the patient's need for lithium should be reevaluated to allow for the possibility of tapering and discontinuation of the drug around the estimated date of confinement (EDC). For women who require continuation of the agent, lithium dose should be reduced at that time by 50%, given the dehydration and fluid shifts associated with labor and delivery, which may result in marked elevations of lithium levels and lithium toxicity. Lithium levels of 1.0 mEq/ml have also been associated with neonatal toxicity, including cyanotic states, bradycardia, impaired respiratory function, T-wave abnormalities, "floppy babies," and nephrogenic diabetes insipidus.[83] Lithium doses can be readjusted or reinstituted after delivery.

C. Electroconvulsive Therapy

Although not widely used currently in the treatment of acute mania, electroconvulsive therapy (ECT) has proven efficacy for this condition. One review of 318 pregnant women treated with ECT did not reveal a higher complication rate than in psychotic women treated without psychotropic agents or electroconvulsive therapy.[84] Recent cases of ECT administration in pregnant women have been reported employing real-time ultrasonagraphy and monitoring of arterial oxygenation and have not revealed adverse effects to mother and fetus.[85,86] Remick and Maurice[87] and others[1,86] have offered guidelines for ECT administration in pregnancy, including (1) performing ECT with a team comprised of anesthesiologist, psychiatrist, and obstetrician, (2) low-voltage, nondominant ECT with EEG monitoring, (3) use of tocodynamometry for evaluation of uterine tone, (4) arterial blood gas monitoring before and after ECT, and (5) Doppler ultrasonagraphy for monitoring fetal heart rate.

Clinical situations favoring use of ECT are those in which mother and fetus are at greatest risk, as with a pregnant bipolar woman in acute decompensation. Although the risks of mania are well described,[88] potential impulsivity, for example, associated with mania in a pregnant woman poses added risk to fetus. Hospitalization and treatment with neuroleptic agents alone or with adjunctive benzodiazepines[1] may suffice, though ECT should be considered. Similarly, in the depressed patient whose psychotic symptoms impair ability to care for herself (as evidenced by decreased food intake, guilty ruminations, and self deprecatory hallucinations), ECT may be the treatment of choice.[1] Withholding treatment in this setting is unacceptable.

V. ANXIETY DISORDERS AND BENZODIAZEPINES

Normal pregnancy is associated with anxiety and anticipation. Normative anxiety must, however, be distinguished from anxiety disorder. Little is known about the course of anxiety disorders in pregnancy or the impact of untreated anxiety and panic attacks on the developing fetus. An association of high levels of anxiety and obstetrical complications with poor neonatal outcome has been reported by some though not by others.[89] Although a case has been reported of panic-attack-associated abruptio placenta and fetal distress in a pregnant woman with panic disorder,[90] George and colleagues[91] have described three cases of women whose symptoms of panic diminished during pregnancy. Cohen et al. however, describe persistence, worsening, or emergence of panic attacks in a series of 22 women seeking consultation for panic disorder.[92]

Panic disorder afflicts two to three times more women than men and aggregates in women of childbearing potential. In this light, and given the high prevalence of anxiety disorders in general, it is particularly important to consider the issue of anxiolytic use during pregnancy.

Benzodiazepines and tricyclic antidepressants remain the most frequently used agents for the treatment of anxiety disorders.[93,94] Tricyclic antidepressants, as discussed, have an apparently low teratogenic potential, though risks associated with benzodiazepine use remain controversial. Retrospective reports evaluating the use of diazepam (Valium® and others) during pregnancy describe an increased risk of oral cleft abnormalities. Safra and Oakley[95] reported a fourfold use of diazepam in mothers of children with cleft lip with or without cleft palate compared to mothers of children with other malformations. A significant association between oral clefts and first-trimester diazepam exposure was also noted in a study by Saxen[96] following a retrospective study from the Finnish Register of Congenital Malformations of 599 cases of oral clefts.

Rosenberg and colleagues[15] have presented more recently the first of a series of reports that suggest an absence of increased risk of anomalies associated with first-trimester benzodiazepine use. These authors described a case-controlled study of 445 infants with cleft lip with or without cleft palate and 166 infants with cleft palate alone and compared them with 2498 control infants with other birth defects. First-trimester diazepam exposure did not differ among the groups. In a prospective study evaluating 249 women conducted by the National Institute of Child Health and Development, Shiono and Mills[16] also report an absence of increased rates of oral clefts in first-trimester diazepam users.

Laegreid and colleagues[97] present data describing a range of malformations (epicanthal folds, slanted eyes, uptilted short nose, dysplastic auricles, high arched palate, wide-spaced nipples, and webbed neck) and central nervous system abnormalities (mental retardation, seizures) in seven of 37 infants of women followed prospectively who were exposed to oxazepam (Serax® and others) or diazepam throughout pregnancy. The report has been criticized since some patients used antidepressants and antipsychotics concurrently, and since this so-called "benzodiazepine embryofetopathy" resembled an unrelated genetically transmitted malformation.[98] Abnormalities in the 30 unexposed infants were also not described. The report differs from findings of the Hungarian case-controlled surveillance system; despite a 15.5% exposure rate to diazepam in a large population of pregnant

Hungarian women obtained from a large population-based data base, increased rates of malformations were not observed.

Thus, considerable data, although contradictory, exist regarding the potential teratogenic effects of diazepam; fewer reports, however, address the risks of in utero exposure associated with other benzodiazepines such as alprazolam (Xanax®) and clonazepam (Klonopin). Information regarding the teratogenicity of these high-potency benzodiazepines frequently used in the treatment of panic disorder[99] and other anxiety disorders[100] is limited to the manufacturer's case register of clinician reports of offspring exposed to the drug in utero. Over 400 cases of alprazolam exposure during pregnancy have been compiled.[101] Although there have been individual cases of malformations noted, accurate incidence data are not available since the total population exposed is unknown.

Prior to attempts at conception, patients with anxiety disorders require careful reevaluation of their pharmacotherapeutic regimens. Patients with panic disorder maintained on benzodiazepines who wish to conceive should be tapered on a schedule adjusted to the patient's ability to tolerate decreases in dose with close regard for breakthrough symptoms of panic and/or withdrawal. Even with gradual tapering of a benzodiazepine, some patients may be unable to tolerate discontinuation of the drug without severe recurrent symptoms of anxiety. This may be particularly evident with short-acting benzodiazepines such as alprazolam. For these patients, addition of a tricyclic may be helpful to treat reemerging symptoms of panic. Clonazepam has putative antipanic activity based on multiple reports[102–104] and apparently has low teratogenic potential.[75] Because of its longer duration of action than alprazolam, a switch to clonazepam may afford the alprazolam-dependent patient who wishes to conceive a more gentle weaning from a benzodiazepine with less "rebound" anxiety.[105]

Benzodiazepine use during pregnancy has been associated with a variety of effects during the perinatal period. Decreased Apgar scores, hypothermia, hypotonia, decreased respirations, and difficulty feeding have all been described during this period.[6] Neonatal abstinence syndromes have also been described with diazepam, including symptoms of tremor, hypertonicity, and irritability;[90] a similar abstinence syndrome has also been described in patients maintained on alprazolam.[101]

VI. BREAST FEEDING

Pharmacotherapeutic treatment during pregnancy necessarily entails drug exposure for the infant. Following delivery, however, treatment of mother does not imply a need to include treatment of infant. During the postpartum period the clinician is freed from the constraint of having to treat fetus in order to treat mother.

All psychotropics including antipsychotics, antidepressants, lithium carbonate, and benzodiazepines are secreted into milk.[32] Concentrations of these agents in the milk, however, vary greatly and depend on the specific characteristics of the drug as

TABLE 1. Psychotropic Drug Use in Pregnancy[a]

Psychosis
  Maintenance low-dose antipsychotic therapy for chronically psychotic patients
    may offset risk of relapse and need for higher doses
  Medical differential diagnosis and workup for new-onset psychotic states
  ? High-potency neuroleptics safer
  Medication present in breast milk
Mania
  Medical differential diagnosis and work-up for new onset manic symptoms
  Careful contraceptive history
  Evaluate need for prophylaxis
  Trimester I
    Avoid lithium carbonate
    Exposure to lithium before week 12: consider cardiac ultrasound at week 20
    Clear need for antimanic prophylaxis: discontinue lithium at documentation
      of pregnancy
    Consider prophylactic antipsychotics in brittle bipolar women
  Trimester II and III
    After week 12 and with need for treatment, consider lithium carbonate
      re(introduction)
    Discontinue lithium before estimated date of confinement or decrease dose by
      50% prior to delivery
    Lithium in small divided doses
  Mania in pregnancy
    Hospitalization
    Neuroleptics, ?adjunctive clonazepam, electroconvulsive therapy
  Lithium in breast milk
Depression
  Medical differential diagnosis
  Withold medication in trimester I if possible
  Inability to care for self or to provide prenatal care indicates need for somatic
    therapy
  Secondary- over tertiary-amine tricyclics or "newer antidepressants"
  ECT for delusional depression
  Antidepressants in breast milk
Anxiety disorders
  Medical differential diagnosis
  Tricyclics are treatment of choice.
  Adjunctive behavioral/cognitive and supportive psychotherapy
  Withhold medication in first trimester if possible
  Taper benzodiazepine therapy prior to conception: slow taper vs. change to
    tricyclic antidepressant
  Unable to taper short-acting benzodiazepine, consider change to clonazepam
  Patient on anxiolytic when pregnancy confirmed, attempt taper
  Avoid additional drug introduction, particularly in trimester I
  Benzodiazepines in breast milk

[a]Adapted from Cohen et al.[1]

well as on maternal metabolism. The effect of these agents on the neonate is variable and depends on the drug, its bioavailability, and the maturity of the infant's metabolism and central nervous system. The variability of these factors among newborns may account for the range of consequences of individual drugs from nondetectable effects to symptoms of severe neonatal distress.[34] Frequency of severe complications after neonatal exposure to psychotropics from breast feeding is small, although appreciable concentrations of these agents can be found in human breast milk. The effect of drug exposure on the developing infant's brain is uncertain; hence, patients who require psychotropics should be cautious about breast feeding until the potential risks to infant are better understood.

VII. CONCLUSION

Although antipsychotics, antidepressants, and benzodiazepines have low known teratogenic potential, their safety cannot be asserted; thus, they are best avoided in pregnancy if possible. Pharmacotherapeutic intervention may be called for, nonetheless, when risk to mother and fetus outweighs the risk of using pharmacotherapy. Absolute quantification of these risks is difficult, but this does not imply a lack of clinical options. A summary of proposed guidelines for treating psychiatrically ill pregnant patients is outlined in Table 1.

Given the prevalence of psychotropic use during pregnancy, it is critical that the clinician have a prepared approach to administration of these agents. Coordinated care among patient, husband or partner, obstetrician, and psychiatrist is essential, as is careful medical record documentation. Psychotropics are generally used in pregnancy without discernible adverse consequences. Pending good controlled prospective data on the impact of drugs on fetal and later development, the clinician will continue to have to act in a state of potential uncertainty, weighing partially calculated risk to manage individual clinical dilemmas.[1]

REFERENCES

1. Cohen L. S., Heller V. L., Rosenbaum J. F.: Treatment guidelines for psychotropic drug use in pregnancy. *Psychosomatics* 30:25–33, 1989.
2. Doering J. C., Stewart R. B.: The extent and character of drug consumption during pregnancy. *JAMA* 239:843–846, 1978.
3. Kasilo O., Romero M., Bonati M., et al: Information on drug use in pregnancy from the Viewpoint Regional Drug Information Center. *Eur J Clin Pharmacol* 35:447–453, 1988.
4. O'Hara M. W., Zekoski E. M., Phillips L. H., et al: A controlled prospective study of postpartum mood disorders: Comparison of childbearing and nonchildbearing women. *J Abnorm Psychol* (in press).
5. Frank E., Kupfer D. J., Jacob M., et al: Pregnancy related affective episodes among women with recurrent depression. *Am J Psychiatry* 144:288–293, 1987.
6. Calabrese J. R., Gulledge A. D.: Psychotropics during pregnancy and lactation: A review. *Psychosomatics* 26:413–426, 1985.

7. Robinson G. E., Stewart D. E., Flak E.: The rational use of psychotropic drugs in pregnancy and postpartum. *Can J Psychiatry* 31:183–190, 1986.

8. Slone D., Siskind V., Heinonen O. P., et al: Antenatal exposure to the phenothiazines in relation to congenital malformations, perinatal mortality age, birth weight and intelligence quotient score. *Am J Obstet Gynecol* 128:486–488, 1977.

9. Rumeau-Rouguette C., Goujard J., Huel G.: Possible teratogenic effect of phenothiazines in human beings. *Teratology* 15:57–64, 1977.

10. Milkovich L., Van den Berg B. J.: An evaluation of the teratogenicity of certain antinauseant drugs. *Am J Obstet Gynecol* 125:244–248, 1976.

11. Edlund M. J., Craig T. J.: Antipsychotic drug use and birth defects: An epidemiologic reassessment. *Compr Psychiatry* 25:32–37, 1984.

12. Kuenssberg E. V., Knox J. D.: Imipramine in pregnancy. *Br Med J* 2:292, 1972.

13. Idanpaan-Heikkila J., Saxen L.: Possible teratogenicity of imipramine-chloropyramine. *Lancet* 2:282–284, 1973.

14. Czeizel A., Lendvay A.: In-utero exposure to benzodiazepines (letter). *Lancet* 1:628, 1987.

15. Rosenberg L., Mitchell A., Parsells J., et al: Lack of relation of oral cleft to diazepam use during pregnancy. *N Engl J Med* 309:1282–1285, 1983.

16. Shiono B. H., Mills J. L.: Oral clefts and diazepam use during pregnancy (letter). *N Engl J Med* 311:919–920, 1984.

17. Langman J.: Human development—normal and abnormals, in Langman J. (ed.): *Medical Embryology* (ed. 5). Baltimore, Williams & Wilkins, 1985, p 123.

18. Dicke J. M.: Teratology: Principles and practice. *Med Clin North Am* 73:567–581, 1989.

19. Coyle I. R.: Changes in developing behavior following prenatal administration of imipramine. *Pharmacol Biochem Behav* 3:799–807, 1975.

20. Robertson R. T., Majka J. A., Peter C. P., et al: Effects of prenatal exposure to chlorpromazine on postnatal development and behavior of rats. *Toxicol Appl Pharmacol* 53:541–549, 1980.

21. Kellogg C., Ison J., Miller R.: Prenatal diazepam exposure: Effects on auditory temporal resolution in rats. *Psychopharmacology* 79:332–337, 1983.

22. Miller G. C., Friedhoff A. G.: Prenatal neurotransmitter programming of postnatal receptor function. *Prog Brain Res* 2:509–522, 1987.

23. Webster P. A. C.: Withdrawal symptoms in neonates associated with maternal antidepressant therapy. *Lancet* 2:318–319, 1973.

24. Shearer W. T., Schreiner R. L., Marshall R. E.: Urinary retention in a neonate secondary to maternal ingestion of nortriptyline. *J Pediatr* 81:570–572, 1972.

25. Cree J. E., Meyer J., Hailey D. M.: Diazepam in labour: Its metabolism and effect on the clinical condition and thermogenesis of the newborn. *Br Med J* 4:251–255, 1973.

26. Hill R. M., Desmond M. M., Kay J. L.: Extrapyramidal dysfunction in an infant of a schizophrenic mother. *J Pediatr* 69:589–595, 1966.

27. Tamer A., McKey R., Arias D., et al: Phenothiazine-induced extrapyramidal dysfunction in the neonate. *J Pediatr* 7:479–480, 1969.

28. Levy W., Wisniewski K.: Chlorpromazine causing extrapyramidal dysfunction in newborn infants of psychotic mothers. *NY State J Med* 74:684–685, 1974.

29. Gelenberg A. J.: Pregnancy, psychotropic drugs, and psychiatric disorders. *Psychosomatics* 27:216–217, 1986.

30. Gelenberg A. J.: Psychotropic drugs during pregnancy and the perinatal period. *Biol Ther Psychiatry* 2:41–42, 1979.

31. Nahas C., Goujard J.: Phenothiazines, benzodiazepines and the fetus, in Scarpelli E. M., Cosmi E. V. (eds): *Reviews in Perinatal Medicine*. New York, Raven Press, 1978, pp 243–280.

32. Rivera-Calimlin L.: The significance of drugs in breastmilk: Pharmacokinetic considerations. *Clin Perinatol* 14:51–70, 1987.

33. Anderson P. O., McGuire G. G.: Neonatal alprazolam withdrawal—possible effects of breast feeding. *Drug Intell Clin Pharm* 23:614, 1989.

34. Gelenberg A. J.: Antidepressants in milk. *Biol Ther Psychiatry* 10:1, 1987.
35. Charney D., Heninger G.: Abnormal regulation of noradrenergic function in panic disorder. *Arch Gen Psychiatry* 43:1042–1054, 1986.
36. Siever L., Uhde T.: New studies and perspectives on the noradrenergic receptor system in depression: Effects of the alpha two adrenergic agonist clonidine. *Biol Psychiatry* 19:131–156, 1984.
37. Cohen L. S., Rosenbaum J. F., Heller V. L.: Panic attack-associated placental abruption: A case report. *J Clin Psychiatry* 50:266–267, 1989.
38. Avant K.: Anxiety as a potential factor affecting maternal attachment. *J Obstet Gynecol Neonatal Nurs* 10:416–419, 1981.
39. Brazelton T. B.: Mother infant reciprocity, in Klaus M. H., Leger T., Trause M. A. (eds): *Maternal Attachment and Mothering Disorders: A Roundtable.* North Brunswick, NJ, Johnson & Johnson, 1975.
40. Zahn-Waxler C., Cummings E. M., Ianoff R. J., et al: Young offspring of depressed patients: A population of risk for affective problems and childhood depression, in Cichetti D., Schneider-Rosen (eds): *Childhood Depression.* San Francisco, Jossey-Bass, 1984.
41. Cogill S. R., Caplan H. L., Alexandra H., et al: Impact of maternal depression of cognitive development of young children. *Br Med J* 292:1165–1167, 1986.
42. Sapolsky R.: Stress glucocorticoids and the rate of brain aging. Presented at Annual Meeting of the American College of Neuropsychopharmacology, December, 1988, Nashville, TN.
43. Protheroe C.: Puerperal psychoses: A long term study 1927–1961. *Br J Psychiatry* 115:9–30, 1961.
44. Kopelman A. E., McCullar F. W., Heggeness, L.: Limb malformations following maternal use of haloperidol. *JAMA* 231:62–64, 1975.
45. Scokel P. W., Jones W. D.: Infant jaundice after phenothiazine drugs for labour: An enigma. *Obstet Gynecol* 20:124–127, 1962.
46. Falterman L. G., Richardson D. J.: Small left colon syndrome associated with maternal ingestion of psychotropics. *J Pediatr* 97:300–310, 1980.
47. Spear L. P., Shalaby I. A., Brick J.: Chronic administration of haloperidol during development: Behavioral and psychopharmacological effects. *Psychopharmacology* 70:47–58, 1980.
48. Desmond M. M., Rudolph A. J., Hill R. M., et al: Behavioral alterations in infants born to mothers on psychoactive medication during pregnancy, in Farrell G. (ed): *Congenital Mental Retardation.* Austin, University of Texas, 1967.
49. Platt J. E., Friedhoff A. J., Broman S. H., et al: Effects of prenatal exposure to neuroleptic drugs on childrens growth. *Neuropsychopharmacology* 1:205–212, 1988.
50. Kris E. B.: Children of mothers maintained on pharmacotherapy during pregnancy and postpartum. *Curr Ther Res* 31:690–695, 1965.
51. Zajicek E.: Psychiatric problems during pregnancy, in Wolkind S., Zajicek E. (eds): *Pregnancy: A Psychological and Social Study.* London, Academic Press, 1981, pp 57–73.
52. O'Hara M. W., Neunaber D. J., Zekoski E. M.: Prospective study of postpartum depression: Prevalence, course, and predictive factors. *J Abnorm Psychol* 93:158–171, 1984.
53. O'Hara M. W.: Social support, life events, and depression during pregnancy and the puerperium. *Arch Gen Psychiatry* 43:573, 1986.
54. O'Hara M. W.: Postpartum blues, depression, and psychosis: A review. *J Psychosom Obstet Gynecol* 7:205–227, 1987.
55. Targum S. D., Gershon E. S.: Pregnancy, genetic counselling and the major psychiatric disorders, in genetic diseases, in Schulman J., Simpson J. (eds): *Pregnancy—Maternal Effects and Fetal Outcome.* New York, Academic Press, 1981.
56. Akiskal H. S., Walker P., Puzantian V. R., et al: Bipolar outcome in the course of depressive illness: Phenomenologic, familiar, and pharmacologic predictors. *J Affect Disord* 5:115–128, 1983.
57. McBride W. G.: Limb deformities associated with iminodibenzyl hydrochloride (letter). *Med J Aust* 1:492, 1972.

58. Crombie D., Pinsent R. J., Felming D.: Imipramine and pregnancy. *Br Med J* 1:745, 1972.

59. Coyle I. R., Singer G.: The interaction of post-weaning housing conditions and prenatal drug effects of behavior. *Psychopharmacology* 41:237–244, 1975.

60. Giller E., Bialos D., Riddle M., et al: Monoamine oxidase inhibitor responsive depression. *Psychiatry Res* 6:41–48, 1982.

61. Liebowitz M. R., Quitkin F., Stewart J., et al: Phenelzine vs. imipramine in atypical depression. *Arch Gen Psychiatry* 41:669–677, 1984.

62. Poulson E., Robson J. M.: Effect of phenelzine and some related compounds in pregnancy. *J Endocrinol* 30:205–215, 1964.

63. Quitkin F. M.: The importance of dosage in prescribing antidepressants. *Br J Psychiatry* 147:593–597, 1985.

64. Schou, M. Concerning the start of the Scandinavian Register of Lithium Babies. *Acta Psychiatr Scand* 207(suppl.):96–97, 1969.

65. Ackerman D., Jefferson J., Griest J., et al: The lithium index: An innovative approach to consultation by computer. *Am J Psychiatry* 141:415–417, 1984.

66. Weinstein M. R., Goldfield M. D.: Cardiovascular malformations with lithium use during pregnancy. *Am J Psychiatry* 132:529–531, 1975.

67. Kirklin J. W., Barratt-Boyes B. G.: Ebstein's malformation, in Kirklin J. W., Barratt-Boyes B. G. (eds): *Cardiac Surgery.* New York, John Wiley & Sons, 1986, pp 889–907.

68. Elia J., Katz I. R., Simpson G. M.: Teratogenicity of psychotherapeutic medications. *Psychopharmacol Bull* 23:531–585, 1987.

69. van Gent E. M., Nabarro G.: Haloperidol as an alternative to lithium in pregnant women (letter). *Am J Psychiatry* 144:1241, 1987.

70. Post R. M., Uhde T. W., Ballenger J. C., et al: Prophylactic efficacy of carbamazapine in manic-depressive illness. *Am J Psychiatry* 140:1602–1604, 1983.

71. Prien R. F., Gelenberg A. J.: Alternatives to lithium for preventive treatment of bipolar disorder, *Am J Psychiatry* 146:840–848, 1989.

72. Nidely J. R., Blake D. A., Freeman J. M., et al: Carbamazapine levels in pregnancy and lactation. *Obstet Gynecol* 53:139–140, 1979.

73. Nakne Y., Okuma T., Takahashi R., et al: Multi-institutional study on the teratogenicity and fetal toxicity of antiepileptic drugs: A report of a collaborative study group in Japan. *Epilepsia* 21:663–680, 1980.

74. Paulson G. W., Paulson R. B.: Teratogenic effects of anticonvulsants. *Arch Neurol* 40:140–143, 1981.

75. Sullivan F. M., McElhatton P. R.: A comparison of the teratogenic activity of the antiepileptic drugs carbamazepine, clonazepam, ethosuximide, phenobarbital, phenytoin, and pyrimidone in mice. *Toxicol Appl Pharmacol* 40:365–378, 1977.

76. Jones K. L., Lacro R. V., Johnson K. A., et al: Pattern of malformations in the children of women treated with carbamazepine during pregnancy. *N Engl J Med* 320:1661–1666, 1989.

77. Dallesio D. J.: Seizure disorders in pregnancy. *N Engl J Med* 312:559–564, 1985.

78. Cohen L. S., Heller V. L.: Manic depressive illness in pregnancy and the peurperium (letter). *Psychosomatics,* 1990, in press.

79. McElroy S. L., Keck P. E., Pope H. G., et al: Valproate in the treatment of rapid cycling mood disorder. *J Clin Psychopharmacol* 8:275–279, 1988.

80. Lindhout D., Schmidt D.: In utero exposure to valproate and neural defects. *Lancet* 1:329–393, 1986.

81. Schou M.: What happened later to the lithium babies: A follow-up study of children born without malformations. *Acta Psychiatr Scand* 54:193–197, 1976.

82. Cohen L. S., Rosenbaum J. F., Heller V. L.: Prescribing lithium for pregnant women (letter). *Am J Psychiatry* 145(6):773, 1988.

83. Weinstein M. R., Goldfield M. D.: Cardiovascular malformations with lithium use during pregnancy. *Am J Psychiatry* 132:529–531, 1975.

84. Impasato D. J., Gabriel A. R., Lardara M.: Electric and insulin shock therapy during pregnancy. *Dis Nerv Syst* 25:542–546, 1964.
85. Wise M. G., Ward S. C., Townsend-Parchman W., et al: Case report of ECT during high-risk pregnancy. *Am J Psychiatry* 141:99–101, 1984.
86. Repke J. T., Berger N. G.: Electroconvulsive therapy in pregnancy. *Obstet Gynecol* 63(suppl):39–40, 1984.
87. Remick R. A., Maurice W. L.: ECT in pregnancy (letter). *Am J Psychiatry* 135:761–762, 1978.
88. Boyd J. H., Weissman M. M.: Epidemiology, in Paykel E. S. (ed): *Handbook of Affective Disorders*. New York, Guilford Press, 1982, p 6.
89. Istvon J.: Stress, anxiety, and birth outcome: A critical review of the evidence. *Psychol Bull* 100:331–348, 1986.
90. Rementaria J. L., Blatt K.: Withdrawal symptoms in neonates from intrauterine exposure to diazepam. *J Pediatr* 90:123–126, 1977.
91. George D. T., Ladenheim J. A., Nutt D. J.: Effect of pregnancy on panic attacks. *Am J Psychiatry* 144:1078–1079, 1987.
92. Cohen L. S., Rosenbaum J. F., Heller V. L., et al: Course of panic disorder in 24 pregnant women. Presented at the 142nd annual meeting of the American Psychiatric Association, San Francisco, May 11, 1989.
93. Rosenbaum J. F.: The drug treatment of anxiety. *N Engl J Med* 306:401–404, 1982.
94. Klein D. F.: Anxiety reconceptualized, in Klein D. F., Raskin J. (eds): *Anxiety: New Research and Changing Concepts*. New York: Raven Press, 1981.
95. Safra M. J., Oakley G. P.: Association between cleft lip with or without cleft palate and prenatal exposure to diazepam. *Lancet* 2:478–480, 1975.
96. Saxen I.: Association between oral clefts and drugs taken during pregnancy. *Int J Epidemiol* 4:37–44, 1975.
97. Laegreid L., Olegard R., Wahlstrom J., et al: Abnormalities in children exposed to benzodiazepines in utero. *Lancet* 1:108–109, 1987.
98. Winter R. M., Czeizel A., Lendvay A., et al: In-utero exposure to benzodiazepines. *Lancet* 1:627, 1987.
99. Ballenger J. C.: Pharmacotherapy of panic disorders. *J Clin Psychiatry* 47(suppl 6):27–32, 1986.
100. Cohen L. S., Rosenbaum J. F.: Clonazepam: New uses and potential problems. *J Clin Psychiatry* 48(suppl):50–55, 1987.
101. Barry W. S., St. Clair S. M.: Exposure to benzodiazepines in utero (letter). *Lancet* 1:1436–1437, 1987.
102. Spier A. S., Tesar G. E., Rosenbaum J. F., et al: Clonazepam in the treatment of panic disorder and agoraphobia. *J Clin Psychiatry* 47:238–242, 1986.
103. Pollack M. H., Tesar G. E., Rosenbaum J. F., et al: Clonazepam in the treatment of panic disorder and agoraphobia: A one year follow-up. *J Clin Psychopharmacol* 6:302–304, 1986.
104. Tesar G. E., Rosenbaum J. F., Pollack M. H., Sachs G. S., Herman J. B., Cohen L. S., Sidari J: Panic Disorder: Clonazepam versus alprazolam and placebo. New Research Section, American Psychiatric Association, 141st Annual Meeting, Montreal, May 12, 1988.
105. Herman J. B., Rosenbaum J. F., Brotman, A. W.: The alprazolam to clonazepam switch for treatment of panic disorder. *J Clin Psychopharmacol* 7:175–178, 1987.

11

Temporolimbic Epilepsy

JEFFREY B. WEILBURG, M.D., and
GEORGE B. MURRAY, M.D.

I. INTRODUCTION

Generalized seizures involve both cerebral hemispheres simultaneously and produce convulsions (tonic, clonic, tonic–clonic, or atonic) or periods of absence. Partial seizures involve limited areas of the brain and produce alterations in mood, cognition, perception, sensation, autonomic function, or behavior.

Because partial seizures may secondarily generalize, partial seizures and generalized convulsions may occur in the same patient. However, the neuropathological processes that produce partial seizures are thought to be different from those that produce generalized seizures. Partial and generalized seizures are thus distinct problems.

The brain site from which partial seizures originate is called the focus. Seizure foci may be located in the temporal lobes or related structures, such as the frontal lobes or the septum. The amygdala, hippocampus, septum, parts of the frontal lobes, and the cingulum, fornix, and hypothalamus are considered to make up the limbic system. The temporal lobes and the limbic system are extensively interconnected. Many writers use the term "temporolimbic epilepsy" (over the older "temporal lobe epilepsy" or "psychomotor epilepsy") to describe the syndrome of partial seizures whose focus is in the limbic system.

A particular set of mental and emotional changes and psychiatric problems may appear in some patients with temporolimbic epilepsy. This chapter discusses

JEFFREY B. WEILBURG, M.D. • Neuropsychiatry Section, Psychopharmacology Unit, Massachusetts General Hospital, Boston, Massachusetts 02114; Department of Psychiatry, Harvard Medical School, Boston, Massachusetts 02115. **GEORGE B. MURRAY, M.D.** • Department of Psychiatry, Harvard Medical School, Boston, Massachusetts 02115.

the psychiatric problems associated with temporolimbic partial seizures and provides guidelines for the use of anticonvulsants and psychotropic medications in patients with epilepsy.

II. PARTIAL SEIZURES

A. Etiologic Factors

Various congenital brain lesions, such as hamartomas or arteriovenous malformations, may produce partial seizures. Acquired lesions such as sclerosis of the temporal tips from anoxia, brain injury secondary to head trauma or infection, tumors (primary or metastatic), strokes, and hemorrhages can also produce seizure foci. Sometimes the lesion is idiopathic.

B. Characteristics of Partial Seizures

There is as yet no perfect way to classify temporolimbic seizures. Most workers currently use the descriptive terms "simple partial" and "complex partial" seizures. Simple partial seizures involve no loss of consciousness, whereas complex partial seizures do (see Table 1).[1]

1. Simple Partial Seizures

Nonmotor simple partial seizures present as paroxysmal alterations of cognition, thinking, perception, sensation, affect, or autonomic function. For example, the sudden experience of fear is a common expression of simple partial seizure activity. Fear coupled with autonomic changes such as tachycardia and diaphoresis may mimic a panic attack. Affective changes, such as depression that "washes up over," occur. Changes in the perception of space (described by patients as a feeling of "falling as in a dream" or "standing up in a canoe") may be reported. Other perceptual changes such as micropsia, macropsia, or hyperacusis can result. A particular abdominal sensation likened to going over the hump on a roller coaster may be felt. Some patients describe a sense of *déjà vu* or *jamais vu*. Others hallucinate objects, colors or light flashes, sound or voices, tastes or smells, or complete scenes. Headache or interference with talking, thinking, or moving may be the only symptom of a simple partial seizure. Although most seizure-induced experiences are dysphoric, orgasm may also occur.

Some authors consider an "aura" to be the simple partial seizure (that later can generalize).

If symptoms occur together with loss of consciousness or in association with alteration or loss of consciousness, the seizure may be called a complex partial seizure.

TABLE 1. Classification of Seizures

Simple partial seizures (consciousness not impaired)
  With motor signs
    Focal motor without march
    Versive (generally contraversive)
    Phonatory (vocalization or arrest of speech)
    Focal motor with march (Jacksonian)
    Postural
  With somatosensory or special-sensor symptoms (simple hallucinations, e.g.,
    tingling, light flashes, buzzing)
    Somatosensory
    Gustatory
    Olfactory
    Auditory
    Visual
    Vertiginous
  With autonomic symptoms or signs
    Explosive diarrhea
    Vomiting
    Diaphoresis
  With psychic symptoms (disturbances of higher cortical function)
    Dysphasic
    Cognitive (e.g., forced thinking)
    Illusions (e.g., macropsia)
    Dysmnesic (e.g., déjà vu)
    Affective (e.g., fear, anger)
    Structured hallucinations (e.g., music, scenes)
Complex partial seizures (generally with impaired consciousness; may sometimes
  begin with simple symptoms)
  Simple partial seizures followed by impaired consciousness
    With simple partial features (see above) and impaired consciousness
    With automatisms
  With impaired consciousness at onset
    With impaired consciousness only
    With automatisms
Partial seizures evolving to generalized tonic–clonic (GTC) seizures (GTC with
  partial or focal onset)
  Simple partial seizures (see above) evolving to GTC seizures
  Complex partial seizures (see above) evolving to GTC seizures
  Simple partial seizures evolving to complex partial seizures evolving to GTC

2. Complex Partial Seizures

    Complex partial seizures include automatisms that may appear when full
awareness is lost. Classically, automatisms were defined as gross motor behaviors
such as fumbling with fingers, walking, running, laughing, crying, or other vocaliz-
ation occurring in the setting of confusion, that is, without conscious intent. More
recent definitions of automatisms deemphasize the motor component and focus on
the confusion. Thus, any behavior associated with impaired consciousness during a

partial seizure may be considered an automatism. The behavior may be an inappropriate (e.g., confused) response to external events or may be driven by internal cognitive or affective factors arising during the seizure.

Seizures typically last from 30 sec to a few minutes and may be followed by a postictal period of variable duration of mental dullness or of feeling poorly. Partial status epilepticus has been reported. Seizures may occur during waking or sleeping. Some female patients report changes in seizure frequency or quality in the perimenstrual period (catamenial epilepsy).

3. Ictal Experience and Pseudoseizures

Ictal phenomena often have a bizarre, hard-to-describe quality. Olfactory hallucinations may be like, but not exactly the same as, the smell of urine or burning rubber. A sense of perplexity about these experiences and an initial reluctance to describe them are not uncommon.[2]

Seizure-induced mental changes tend to be paroxysmal and stereotypical. The running commentary of auditory hallucinations experienced by some schizophrenic patients and the sustained sad mood of the depressed patient are not likely to represent discrete seizure activity. However, almost any kind of experience or feeling may be evoked by a partial seizure, and relatively complicated behaviors may occur during periods of prolonged or frequent seizures.

Patients with or without epilepsy may manifest seizure-like behaviors (pseudoseizures) when emotionally upset. Since emotional stress may lower the seizure threshold, it may be difficult to determine if a particular behavior or mental state is truly ictal or not. Careful electroencephalographic monitoring coupled with videotaped monitoring of behavior as well as clinical evaluation of the context and the quality of the behavior may be required. Ictal activity in subcortical areas may escape detection when surface, nasopharyngeal, or sphenoidal electrodes are used. Therefore, recording with depth electrodes may be required for differential diagnosis in some cases.[3]

C. Interictal Phenomena

1. Interictal Behavior Syndrome

In addition to the experiences that occur during seizures, most authors assert that some patients with temporolimbic epilepsy develop changes in personality and behavior that are enduring and can be seen between seizures. A particular constellation of personality and behavior traits, called the interictal behavior syndrome (IBS) (of temporolimbic epilepsy), has been described.[6] The hallmark of this syndrome is the deepening of affect: previously unimportant matters become invested with great significance. Hypergraphia, preoccupation with excessive detail, tendency toward cosmic philosophical speculation, and hyperreligiosity may appear. Irritability and lability may also appear. A quality called "stickiness" or "viscosity," manifested by the inability to end a task or to leave at the end of an interview, may also arise. Hyposexuality is common, but hypersexuality or change in choice of libidinal stimulus can occur. Not all patients with temporolimbic epilepsy have the IBS. The IBS may appear years after the onset of the seizures.[3,7]

2. Kindling

Sometimes features of the IBS are seen in patients with abnormalities in the temporal or frontal lobes or limbic system who do not have actual seizures. Thus, the IBS may arise not simply from seizures but may involve some other process. This process is probably related to kindling.[4]

Kindling is observed in laboratory animals when repeated, intermittent subseizure-threshold stimulation (electrical pulses or iontophoretic application of a drug such as cocaine) is applied to susceptible brain areas. The amygdala is an area particularly amenable to kindling. The initial stimulus produces little change, but repeated stimulation produces increasingly large electrical changes (called afterdischarge) in widening nets of neurons, eventually culminating in a full seizure. Kindled foci are active for long periods of time and may generate spikes on the EEG similar to those seen interictally in epileptic patients. The firing of these foci may be influenced by experience, as it is easier to produce a seizure in the environment in which kindling took place than in a novel environment.[4,5]

Kindling-like mechanisms might produce the novel, aberrant "sensory–limbic hyperconnections" that may underlie the IBS, but this has not been firmly established. The role kindling may play in human subjects with seizures remains unclear.[8]

3. Psychosis

a. Acute

The appearance of acute psychosis in a patient with partial seizures should prompt a search for etiologic factors. Some possible causes include:

1. Medication toxicity. High levels of anticonvulsants can produce agitation, confusion, behavioral disinhibition, and disordered thinking. Nystagmus, diplopia, dysarthria, or ataxia may be present. Anticonvulsant blood levels should be checked if acute psychosis supervenes.[9]
2. Delirium. Disrupted thinking or bizarre behavior may be the earliest sign of delirium. Patients with seizures may be particularly sensitive to infection and fever, electrolyte imbalance, and medication (other than anticonvulsant) toxicity or adverse interaction; central nervous system (CNS) trauma or lesions should be ruled out.
3. Seizures. If seizures become frequent, an agitated confusional state that mimics a psychosis may appear interictally. Bizarre confusional or psychotic

states associated with partial status have also been reported. Metrizamide, lidocaine (Xylocaine® and others), procaine penicillin (Bicillin® and others), and abrupt withdrawal of alprazolam (Xanax®), triazolam (Halcion®), and other benzodiazepines or barbiturates may precipitate such states.

Conversely, some patients may become psychotic if their seizures are too tightly controlled and may need to be allowed to have an occasional seizure for mental stability to be maintained. Psychosis without worsening seizures has been reported to occur when anticonvulsants are abruptly withdrawn.[10]

b. Chronic

In some patients, features of the IBS assume psychotic proportions. Chronic psychosis in such patients has been called the "schizophreniform psychosis of epilepsy" by some investigators. Patients with this psychosis may be different from patients with idiopathic chronic schizophrenia because the seizure patients remain more "affectively related."

III. DIAGNOSIS OF TEMPOROLIMBIC EPILEPSY

A. History

Epilepsy is a clinical diagnosis. Careful history taken from the patient or those around him can establish the presence of seizures. The electroencephalogram (EEG) can be used to confirm the diagnosis, but lack of abnormality on EEG in the face of a convincing history should not rule out epilepsy.

A family history of seizures or a past history of birth trauma, head injury, brain injury (e.g., from infection), or childhood febrile convulsions is often present. Seizures may appear at any age; the incidence of seizure onset peaks in adolescence. Partial epilepsy is the most common epilepsy of adulthood.

Presence of IBS features alone cannot confirm a diagnosis of epilepsy.

There are medical, emotional, and social ramifications associated with a diagnosis of epilepsy; thus, the diagnosis should not be made unless the evidence is clear.

B. Neuropsychiatric Examination

Focal findings on neuropsychiatric examination may be present and may indicate the presence of a structural brain lesion. Asymmetry of face, hands, or of involuntary facial expression, dystonic posturing on stressed gait, or asymmetry in dichotic listening may be present. However, elementary neurological examination is often normal.

Complete neurological workup should be done when a patient first presents with seizures or if a patient with known seizures develops drop attacks, new focal signs, or new changes in mental status. The computerized tomography (CT) scan, magnetic resonance imaging (MRI) scan, and lumbar puncture may aid in identifying treatable lesions in such cases.

C. Electroencephalogram

The fortuitous observation of a seizure during EEG recording confirms the diagnosis of epilepsy. Seizures (detectable by depth electrodes in deeper structures) may, however, produce no changes on surface (scalp) recording. Nasopharyngeal or sphenoidal electrodes may help detect these abnormalities.

On the interictal EEG, spikes or epileptiform sharp waves over the temporal areas may be seen. Recording during sleep, telemetered recording over 24–48 hr, and computer-analyzed recording may also be helpful.

IV. GENERAL THERAPEUTIC MEASURES

A. Psychotherapy

It is often difficult to separate the psychodynamic components of a problem from those driven by neurobiological (ictal and interictal) forces. However, such a separation is useful. Exploratory, analytic psychotherapy is generally not helpful in treating the problems that arise directly from seizures or the IBS. Rather, supportive cognitive strategies are best suited to help patients recognize and manage problems caused by seizures and the IBS. Psychodynamically based analytic approaches can be invoked to help foster personal growth as patients cope with interpersonal, family, and developmental issues. Group therapy with other epileptic patients can provide valuable information, support, and reality testing.

Some patients can be taught to abort their partial seizures by mental concentration, relaxation, or behavior modification.

Even if formal psychotherapy is not used, patients and their families are able to make optimal use of treatment if dynamic factors are considered during interventions.

B. Surgery

Sometimes partial seizures continue despite maximal medical treatment. If intractable seizures produce significant disability and can be shown to emanate from a single, well-defined focus, surgical excision of this focus may be indicated. The implantation of depth electrodes may be required to ensure that the focus is isolated.

The surgical technique often involves block resection of the anterior temporal lobe. Mental and emotional function may improve if seizures are controlled, whereas psychiatric problems may improve, worsen, or remain unchanged.[11]

C. Pharmacotherapy

Psychotropic medication can play a central role in treating patients with temporolimbic seizures. However, these patients often have multiple problems, and dispensing medication alone is not enough to ensure proper care. The clinician must pay attention to the psychodynamic factors that influence the patient's experience of himself and his world and that may influence his interactions with family and caregivers. Ignoring or misunderstanding these factors may add to the confusion and frustration of a complicated clinical situation.

Some patients have social and vocational problems that need attention. Others have mental retardation or neurological handicaps. These additional problems may be treated with medications that have primary CNS effects or interact with the patient's psychotropic or anticonvulsant medications. Collaboration with the patient's other caregivers is an important part of treatment.

The importance of tailoring the drug regimen to the specific needs of each individual cannot be overemphasized. The principles outlined below are useful guidelines for planning a rational therapeutic strategy. However, an individual patient's response to these medications is never totally predictable, so the clinician must be attentive and flexible. Awareness of the adverse behavioral effects of medications, the potential complications with polypharmacy, attention to the overall clinical picture, and sensitivity to the human needs of each patient are critical parts of adequate drug treatment. (See Section VI for discussion of clinical applications.)

V. MEDICATIONS

A. Carbamazepine

1. Chemistry

Carbamazepine (Tegretol® and others) is a dibenzazepine derivative structurally related to the tricyclic antidepressants [e.g., imipramine (Tofranil® and others)]. Its structural formula is:

2. Adverse Reactions

a. Neuropsychiatric

Sedation, dizziness, and ataxia may appear on initiation of therapy unless starting doses are low and the dose is raised slowly. Fatigue, headache, dizziness, ataxia, and diplopia may occur during maintenance therapy in some patients. There may be poor correlation between these problems and dose or blood level.

Cognitive slowing, perhaps secondary to sedation, may arise in some patients. The cognitive impairments produced by carbamazepine are usually less than those produced by the barbiturates or phenytoin (Dilantin® and others). Cognitive improvements may occur in epileptic patients if carbamazepine is added to or substituted for other anticonvulsants.[12]

Depression has been reported to occur in patients with borderline personality disorder given carbamazepine for behavioral dyscontrol who had histories of affective disorders. Positive effects on mood in patients with epilepsy often appear when carbamazepine is used.[13,14]

b. Gastrointestinal

Nausea is sometimes a problem when the drug is started or at higher blood levels. Hepatitis is rare but can occur, so baseline laboratory studies should include SGOT, alkaline phosphatase, and bilirubin levels. These studies should be repeated monthly for the first 2 to 3 months and then every 3–4 months or if symptoms of hepatitis appear.[15]

c. Ocular

Increased intraocular pressure may occur, so patients with glaucoma should be examined ophthalmologically before and during treatment.

d. Hematological

Carbamazepine routinely produces statistically significant but clinically insubstantial decreases in erythrocyte, leukocyte, and platelet counts. The routine neutropenia is not correlated with agranulocytosis. Severe or fatal aplastic anemia, agranulocytosis, or other blood dyscrasias are rare, appearing in approximately one out of 125,000 cases. Early cases of agranulocytosis may have been caused by impurities, which have been reduced as manufacturing techniques have improved. Concomitant use of myelosuppressive agents is, however, not recommended.

There is controversy over the need for close hematological monitoring in patients on carbamazepine.[16,17] It is prudent to **obtain pretreatment complete blood count (CBC) and platelet count. Patients with significant abnormalities on these tests or history of blood dyscrasia should receive treatment only after hematological consultation. The CBC and platelet counts may be repeated weekly for the first 2–3 months of treatment and then at 1-to-3-month intervals thereafter.** Some centers monitor the hematocrit and white count every 4–8 weeks for the first year in those with initially normal indices and check a differential only if the white count is less than 4000 cells/mm³.

Serious blood dyscrasias may appear suddenly. Periodic routine laboratory assessment, however frequent, may not predict the onset of blood dyscrasias. Patients should be carefully educated about signs of impending problems. Patients should be instructed to contact their physician if fever, sore throat, petechiae, bruising, or bleeding appears. Patients should be seen, hematological assessment performed, and carbamazepine discontinued or continued with caution and expert hematological consultation if testing reveals:

Erythrocytes $< 3.0 \times 10^6/mm^3$
Platelets $< 100,000/mm^3$
Leukocytes $< 4000/mm^3$ (or neutrophils $< 1500/mm^3$)

e. Dermatologic

Rashes of various sorts appear in approximately 5–15% of patients on carbamazepine. Minor, nonspecific rashes may be observed or treated with antihistamines. Skin changes such as urticaria, exfoliative dermatitis, or toxic epidermal necrolysis are indications for immediately stopping the drug. Stevens–Johnson syndrome, lupuslike reaction, or any skin problem associated with fever, chills, arthralgia, myalgia, lymphadenopathy, and eosinophilia may herald impending bone marrow suppression. The drug should be stopped and the patient closely monitored if these problems appear. Sensitivity to sunlight may be increased.[17]

f. Cardiovascular

There are recent reports indicating that complete A–V block or atrial fibrillation may be induced by carbamazepine, but this is uncommon.

g. Renal

Carbamazepine has a vasopressinlike effect on the kidney (retention of free water). This may cause the serum sodium to fall to low or slightly subnormal values in many patients, but this rarely causes clinical difficulty.[18] Renal damage leading to oliguria and hypertension has been reported after carbamazepine use in very rare instances.

h. Cross-Sensitivity

Cross-sensitivity to the allergic effects of other tricyclics may occur.

i. Pregnancy

Teratogenic effects of carbamazepine have been noted in animals and in humans. Until further research is completed, very careful consideration must be used before the drug is given to pregnant women. Carbamazepine does appear in breast milk.

j. Drug Interactions

See Table 2 for a list of adverse interaction and recommendations for treatment.

TABLE 2. Drug Interactions

Drug	Adverse effects	Recommendations
Carbamazepine, with:		
Anticoagulants, oral	Decreased anticoagulant effect (metabolism)	Monitor prothrombin time.
Antidepressants, tricyclic	Possible toxicity of both drugs (possibly decreased metabolism)	Monitor serum concentration of drugs.
Barbiturates	Difficulty in evaluating carbamazepine concentration (increased metabolism to the epoxide)	Low or normal carbamazepine concentration may not be significant if epoxide is increased; however, relative antiepileptic potencies of the two compounds are unknown; if possible, monitor concentration of both carbamazepine and its epoxide.
Cimetidine	Increased toxicity of carbamazepine (decreased metabolism)	Experimental studies indicate that temporary increase lasting about 7 days in carbamazepine concentration can occur after starting cimetidine; ranitidine probably does not interact.
Contraceptives, oral	Possible contraceptive effect (increased metabolism)	Use alternative contraceptive.
Corticosteroids	Decreased corticosteroid effect (probably increased metabolism)	Monitor corticosteroid effect.
Cyclosporine	Decreased cyclosporine effect (increased metabolism)	Monitor cyclosporine concentration.
Danazol	Increased carbamazepine toxicity (decreased metabolism)	Avoid concurrent use.
Diltiazem	Neurotoxicity (probably decreased metabolism)	Single case report, well documented.
Erythromycins	Increased carbamazepine toxicity (decreased metabolism)	Avoid concurrent use, if possible.
Furosemide	Hypnonatremia (mechanism not established)	Single case report; monitor sodium concentration.
Haloperidol	Decreased haloperidol effect (increased metabolism)	Monitor haloperidol concentration.
Influenza vaccine	Possible carbamazepine toxicity (decreased metabolism)	Monitor carbamazepine concentration.
Isoniazid	Increased toxicity of both drugs (altered metabolism)	Avoid concurrent use, if possible.
Lithium	Increased neurotoxicity (possibly synergism)	Monitor neurological status; patients with hypothyroidism may be especially susceptible.
Metoclopramide	Neurotoxicity	Single case report.
Narcotics	Increased carbamazepine toxicity	Choose another analgesic.

(continued)

TABLE 2. (Continued)

Drug	Adverse effects	Recommendations
	with propoxyphene (decreased metabolism)	
Neuromuscular blocking agents	Accelerated recovery from neuromuscular blockade (mechanism not established)	Monitor neuromuscular status.
Phenytoin	Decreased carbamazepine effect (increased metabolism); altered phenytoin effect (mechanism not established)	Monitor carbamazepine and phenytoin concentrations; both increased concentrations have been reported.
Primidone	Decreased primidone effect and increased phenobarbital effect (increased conversion of primidone to phenobarbital)	Monitor primidone and phenobarbital concentrations.
	Decreased carbamazepine effect (increased metabolism)	Monitor carbamazepine and carbamazepine epoxide concentrations; both increased and decreased concentrations have been reported.
Propoxyphene	Increased carbamazepine toxicity with propoxyphene (decreased metabolism)	Choose another analgesic.
Stimulants	Methylphenidate may increase carbamazepine toxicity	Single case report.
Tetracyclines	Decreased doxycycline effect (increased metabolism)	Choose another antibiotic.
Theophyllines	Decreased theophylline effect (increased metabolism)	Monitor theophylline concentration.
Thiazide diuretics	Hyponatremia (mechanism not established)	Single case report; monitor sodium concentration.
Troleandomycin	Increased carbamazepine toxicity (decreased metabolism)	Choose another antibiotic.
Valproic acid	Decreased valproic acid effect (increased metabolism)	Monitor valproic acid concentration.
Valpromide	Increased carbamazepine toxicity (increased epoxide metabolite levels)	Use valproic acid rather than valpromide with carbamazepine.
Verapamil	Increased carbamazepine toxicity (decreased metabolism)	Avoid concurrent use.
Barbiturates, with:		
Acetaminophen	Increased acetaminophen hepatic toxicity (mechanism not established)	Case report in one epileptic maintained on barbiturates after taking acetaminophen for 6 months.
Alcohol	Decreased sedative effect with chronic alcohol abuse (increased barbiturate metabolism, increased CNS depression with acute intoxication	Avoid barbiturates in alcohol abusers, if possible.

TABLE 2. (*Continued*)

Drug	Adverse effects	Recommendations
	(addictive; decreased barbiturate metabolism)	
Anticoagulants, oral	Decreased anticoagulant effect (increased metabolism)	Avoid barbiturate hypnotics; stable anticoagulant dose can be established for epileptics on maintenance barbiturates.
Antidepressants, tricyclic	Decreased antidepressant effect (increased metabolism)	Avoid concurrent use unless barbiturate essential as anticonvulsant.
Benzodiazepines	Decreased clonazepam effect with phenobarbital (increased metabolism)	Monitor clonazepam concentration.
Beta-adrenergic blockers	Decreased beta-blocker effect (increased metabolism) except with sotalol and probably atenolol and nadolol (minimal hepatic metabolism)	Avoid concurrent use unless barbiturate essential as anticonvulsant.
Chloramphenicol	Increased barbiturate effect (decreased metabolism), decreased chloramphenicol effect (increased metabolism)	Monitor barbiturate and chloramphenicol concentrations in epileptic; avoid concurrent use in others; adjust barbiturate dosage when used as anesthetic in epileptics.
Cimetidine	Possible decreased cimetidine effect (increased metabolism)	Small effect; significance not established.
Contraceptives, oral	Decreased contraceptive effect (increased metabolism)	Avoid concurrent use.
Corticosteroids	Decreased corticosteroid effect (increased metabolism)	Monitor corticosteroid effect or concentration; avoid concurrent use.
Cyclosporine	Possible decreased cyclosporine effect (increased metabolism)	Single case report; monitor cyclosporine concentration.
Digitoxin	Decreased digitoxin effect (increased metabolism)	Use digoxin.
Guanfacine	Decreased antihypertensive effect; rebound hypertension on withdrawal (increased metabolism; increased guanfacine concentration on withdrawal)	Taper barbiturate before withdrawing guanfacine
Haloperidol	Decreased haloperidol effect (increased metabolism)	If barbiturate essential as an anticonvulsant, monitor haloperidol concentration.
Influenza vaccine	Possible barbiturate toxicity (decreased metabolism)	Monitor barbiturate concentration.
Methsuximide	Possible increased toxicity of both drugs (increased metabolism)	Monitor methsuximide and barbiturate concentrations.
Metronidazole	Decreased metronidazole effect	Double dose of metronidazole if

(*continue*

TABLE 2. (*Continued*)

Drug	Adverse effects	Recommendations
	with phenobarbital (probably increased metabolism)	phenobarbital essential as anticonvulsant.
Narcotics	Increased CNS depression with meperidine (increased meperidine metabolites)	Avoid concurrent use, except for anesthesia.
	Precipitate methadone withdrawal (increased metabolism)	Avoid concurrent use, if possible; monitor methadone for withdrawal.
Phenothiazines	Decreased phenothiazine effect (increased metabolism)	Avoid concurrent use; monitor phenothiazine concentration if barbiturates required.
Probenecid	Increased effect of IV thiopental (displacement from binding)	Avoid concurrent use.
Pyrodoxine	Decreased barbiturate effect (possibly increased metabolism)	Monitor barbiturate concentration; avoid concurrent use in others.
Quinidine	Decreased quinidine effect (increased concentration)	Monitor quinidine concentration; avoid concurrent use in others.
Reserpine	Hypotension during thiopental anesthesia (additive)	Discontinue reserpine two weeks before anesthesia.
Rifampin	Decreased barbiturate effect (increased metabolism)	Monitor barbiturate concentration if required as anticonvulsant.
Sulfonamides	Increased thiopental effect (decreased albumin binding)	Decrease thiopental dosage, if necessary.
Tetracyclines	Decreased doxycycline effect (increased metabolism)	Avoid concurrent use; choose another antibiotic.
Theophyllines	Decreased theophylline effect (increased metabolism)	Monitor theophylline concentration; avoid concurrent use; individual response varies.
Valproic acid	Increased phenobarbital effect (decreased metabolism)	Monitor phenobarbital concentration if essential as anticonvulsant.
Valproic acid, with:		
Antacids	Increased valproic acid concentration (increased absorption)	Clinical significance not established; give at least one hour apart.
Barbiturates	Increased phenobarbital effect (decreased metabolism)	Monitor phenobarbital concentration if essential as anticonvulsant.
Benzodiazepines	Increased IV diazepam effect (displacement from binding and decreased metabolism)	Use IV diazepam with caution.
	Clonazepam may precipitate absence status (mechanism not established)	Avoid concurrent use.
‥zepine	Decreased valproic acid effect (increased metabolism)	Monitor valproic acid concentration.

TABLE 2. (Continued)

Drug	Adverse effects	Recommendations
Cimetidine	Possible increased valproic acid effect (decreased metabolism)	Based on studies in nonepileptic ulcer patients; avoid concurrent use; ranitidine did not interact.
Ethosuximide	Possible increased ethosuximide effect (decreased metabolism)	Highly variable; clinical significance not established; monitor ethosuximide concentration.
Phenothiazines	Possible increased valproic acid effect with chlorpromazine (decreased metabolism)	Clinical significance not established; monitor valproic acid concentration.
Phenytoin	Increased phenytoin toxicity (displacement from binding)	Conflicting reports; monitor phenytoin concentration and clinical status; saliva and free serum concentration may be helpful.
Salicylates	Increased valproic acid effect (displacement from binding and possibly metabolism)	Use alternative analgesic or antipyretic drugs.

4. Kinetics

Carbamazepine is rapidly absorbed following oral administration. Peak plasma levels are reached within 2 to 6 hr and vary widely in different individuals.

Eighty percent of the drug is bound to plasma proteins. It is metabolized by the liver and induces its own metabolism. The initial half-life ranges from 25 to 65 hr, whereas the steady-state half-life ranges from 12 to 17 hr.

5. Toxicity

Toxicity is very similar to that of imipramine.

6. Preparation and Dosage/Monitoring

Carbamazepine is available as scored tablets of 200 mg. Chewable tablets of 100 mg are also available. A liquid suspension has been recently released. There is no parenteral preparation. Generic forms are available but may vary in terms of bioavailability and amounts of impurities.

Treatment may begin with 200 mg twice a day, but some patients who are pharmacologically naive, particularly anxious, or prone to complain of adverse subjective effects may do better starting with 100 mg at bedtime. Dosage may be raised in 200-mg increments per week, or less often, based on the patient's ability t

.olerate side effects. Adults often require 800–1600 mg per day, administered two or three times per day.

Blood levels can be drawn approximately 12 hr after the last dose and are useful guides for managing seizures. Values of 6–12 µg/ml are considered "therapeutic." Some patients respond well at lower blood levels and may be maintained at these levels. There is poor correlation between carbamazepine blood levels and control of psychiatric symptoms. The correlation between serum levels of the 10,11-epoxide (one of carbamazepine's major metabolites) and control of symptoms of affective disorder may be meaningful, but this remains to be clarified.[19]

Carbamazepine stimulates hepatic microsomal activity. The blood levels of other drugs (and their metabolites), as well as the level of carbamazepine itself, are therefore affected as carbamazepine is introduced (or withdrawn). Approximately 90% of autoinduction occurs within the first month of therapy. Blood levels of carbamazepine and other drugs should be monitored and dosage adjusted as needed during the several months surrounding initiation or discontinuation of treatment.

B. Phenytoin

Phenytoin is used in the management of generalized convulsive seizures and some cases of partial seizures.[20,21] This agent may worsen absence seizures. Phenytoin generally does not appear to have antikindling properties.

1. Chemistry

Phenytoin's structure is:

The two benzene (aromatic) substituents at the 5 position appear to be essential ᴐ the antiepileptic properties of the compound.

\dverse Reactions

₊ Neuropsychiatric

ᶃnitive dulling and affective blunting can occur at any dose but become ᵐinent at higher blood levels. A patient's mental retardation, dyslexia, or ᶦtive or learning problems may be especially sensitive to this.[12]

Nystagmus is common and often appears at blood levels of 10–20 μg/ml. Ataxia, sedation, and dysarthria may appear at blood levels of 30 μg/ml. Lethargy and confusion appear at higher blood levels. Hallucinations, agitation, worsening seizures, delirium, and psychotic symptoms may be produced by phenytoin toxicity.[17] The mechanism involved is unclear; phenytoin-induced folate deficiency may play a role. Peripheral neuropathy, which is usually asymptomatic, occurs in 10–30% of patients with long-term therapy. This neuropathy may be related to phenytoin-induced interference with folate metabolism. Cases of cerebellar and frontal lobe damage secondary to long-term phenytoin use have been reported.

b. Cardiovascular

Rapid intravenous administration of phenytoin may precipitate asystole and cardiovascular collapse. Phenytoin should therefore never be given at a rate greater than 50 mg/min IV push. During IV administration, vital signs should be monitored.

c. Cutaneous

Hirsutism is a frequent and annoying unwanted effect that can usually be kept at tolerable levels by careful dosage management.

Gingival hyperplasia occurs in about 20% of patients on long-term phenytoin therapy and is even more common in adolescents and children. Careful oral hygiene can control this problem.

Morbilliform (measleslike) rashes may sometimes appear, often during the first 10 days of treatment, and indicate that phenytoin should be stopped. The drug may be cautiously restarted when the rash clears and may be continued if the rash does not recur. Exfoliative, purpuric, or bullous skin rashes indicate a drug allergy; the clinician should stop the drug in these patients and not restart it.

A lupuslike syndrome accompanied by a positive antinuclear antibody and lupus erythematosus prep sometimes occurs.

Coarsening of facial features may occur with long-term use.

d. Gastrointestinal

Nausea, vomiting, epigastric pain, and anorexia sometimes accompany phenytoin treatment. Administering the drug with meals or milk and in divided doses may minimize the patient's discomfort.

e. Endocrine

Osteomalacia, accompanied by hypocalcemia and elevated alkaline phosphatase, sometimes occurs and may be a result of altered metabolism of vitamin D and calcium. It is relatively resistant to vitamin D therapy. The clinician should check the serum calcium every 6 months.

Phenytoin binds to thyroid-binding globulin, resulting in an artificially lowered protein-bound T_4. No treatment is indicated. Hyperglycemia and glycosuria, probably secondary to phenytoin-induced inhibition of insulin secretion, are rare effects of this drug.

f. Hematological

Macrocytic and megaloblastic anemia, secondary to phenytoin's interference with folate and B_{12} metabolism, may appear. The clinician should draw a complete blood count every 6 months to screen for this problem and administer folate and B_{12} if the patient develops an anemia.

g. Pregnancy

The still incomplete data on the human teratogenicity of phenytoin suggest an increased incidence of birth defects in babies born to epileptic mothers on phenytoin. However, maternal seizures themselves may pose a risk to the fetus. Thus, although the clinician should avoid using this drug during pregnancy, the patient, her family, and obstetrician should assess and weigh the risks and benefits in each individual.

3. Drug Interactions

Several drugs can influence phenytoin's metabolism (see Table 1):

1. Barbiturates may increase the rate of metabolism of phenytoin, although this effect is variable and unpredictable.
2. Carbamazepine increases the rate of metabolism of phenytoin.
3. Oral anticoagulants, isoniazid (INH® and others), and possibly salicylates may inhibit the metabolism of phenytoin, thereby leading to toxicity.

4. Toxicity

The lethal dose of phenytoin in adults is 2 to 5 g. Initial symptoms of nystagmus, ataxia, and dysarthria progress to obtundation. Death is generally a result of respiratory depression.

Supportive measures, such as maintenance of airway, assisted ventilation, and use of vasopressors after the stomach has been emptied, may be adequate. However, some patients may require hemodialysis.

5. Kinetics

Phenytoin can be given orally and intravenously.

Phenytoin is well absorbed from the gastrointestinal (GI) tract. Peak blood levels occur 2 to 6 hr following oral administration. The average half-life is about 22 hr, so the drug may be given only once a day.

Phenytoin is extensively bound to plasma proteins, mainly albumin. It is metabolized by the hepatic microsomal enzymes into inactive metabolites, which are reabsorbed by the GI tract and excreted in the urine.

Some patients show genetically determined differences in their ability to meta-bolize phenytoin. Both hyper- and hypometabolizers have been reported. The drug weakly induces the microsomal system and may induce its own metabolism.

6. Preparation and Dosage/Monitoring

The therapeutic range of phenytoin in the serum is 10 to 20 $\mu g/ml$. A dose of 5 mg/kg or 300 mg/day in the average adult usually is sufficient to keep blood levels within the therapeutic range. Because of the long half-life of the drug at steady state, the maintenance dose may often be given on a once-a-day basis, usually at bedtime.

The rate at which therapeutic blood levels are attained depends on the clinical situation. A stable patient may be given 300 mg PO per day and will attain therapeu-tic levels in about 5 days (longer if patient is a rapid metabolizer). In emergencies, when seizures need to be controlled rapidly, "dilantinization" can be rapidly achieved by giving 1000 mg IV push via syringe **at rates never exceeding 50 mg/min** (10 to 15 mg/kg or 25 mg/kg for children) while vital signs are closely monitored. These amounts can produce therapeutic blood levels in 20 min.

Phenytoin sodium is available as 30- and 100-mg capsules for oral use and as a sterile solution of 50 mg/ml for IV use. Preparations of phenytoin, USP are avail-able in 50-mg tablets and oral suspensions of 30 and 125 mg/5 ml. **Significant differences in bioavailability exist among the preparations of different manufac-turers; therefore, the clinician should select one product and continue it.**

C. Valproic Acid

1. Chemistry

$$CH_3—CH_2—CH_2$$
$$\diagdown$$
$$CH—C{=}O$$
$$\diagup \qquad \diagdown$$
$$CH_3—CH_2—CH_2 \qquad OH$$

Valproic acid (Depakene® and others) differs from more conventional anti-epileptics in being a simple molecule with no mitogens or ring moiety. It was first approved for use in the United States in 1978. Although the mechanism of action is not clearly known, one of its major actions appears to be to increase levels of brain γ-aminobutyric acid (GABA).

2. Adverse Reactions

a. Neuropsychiatric

Sedation with valproic acid alone is rare, occurring in about 2% of patients. Valproic-acid-induced tremor has been reported. Behavioral disturbances such as hyperactivity or increased aggression in children have also been reported with the drug, but one double-blind study showed no effect. Fine tremor of the hands may appear in some patients. Preliminary evidence indicates that little if any cognitive impairment is produced by valproic acid.[12]

b. Gastrointestinal

Indigestion, anorexia, nausea, and vomiting are the most common problems and are often temporary. Diarrhea appears rarely. The incidence of GI disturbance is reduced when enteric-coated divalproex (Depakote®) is used.

Asymptomatic elevation of SGOT may appear. Monitoring laboratory and clinical indices of liver function until the abnormality spontaneously resolves and lowering the dose (which may produce normalization) are appropriate strategies. Severe and idiopathic hepatic dysfunction is rare.

No hepatic fatalities from valproic acid monotherapy have been reported in patients over age 10 years. Liver function tests should be checked before drug is started and repeated immediately if problems suggestive of liver disease appear or at 3- to 4-month intervals. Interference with ammonia detoxification has been reported, and stupor caused by elevated serum ammonia levels has been noted. Case reports of pancreatitis, some fatal, have appeared.[15]

Increased appetite and subsequent weight gain may occur in some patients.

c. Hematological

Transient neutropenia has been noted but is rare. Thrombocytopenia and impaired platelet aggregation may produce bleeding problems; this tends to occur with doses at or above 2 g per day. Some centers discontinue valproic acid 7–10 days before elective surgery if possible.

d. Cutaneous

Thinning of the hair in 2–5% of patients has been reported; this effect may be transient.

e. Pregnancy

The effects of valproic acid on the fetus have not been clearly established, but neural tube defects may appear in babies when this drug was used during pregnancy. Teratogenicity does appear in animals. Attempts to avoid use of valproic acid, especially during the first trimester of pregnancy, should be made.

3. Drug Interactions

Valproic acid does not induce hepatic enzymes, so medications such as birth control pills may be used concomitantly without difficulty.

Valproic acid may inhibit the metabolism of phenobarbital (Tedral® and others), so plasma phenobarbital and primidone concentrations may increase when valproic acid is added.

Valproic acid can displace phenytoin from its binding proteins. Adding the two usually produces a transient decrease in total plasma phenytoin, but free phenytoin levels may increase. Phenytoin toxicity may be seen even when plasma levels are reported within the normal range, so the phenytoin dose must be carefully monitored.

Primidone (Mysoline®), phenobarbital, phenytoin, or carbamazepine can increase the clearance and metabolism of valproic acid, so adding these agents can lower valproic acid levels.

4. Kinetics

Valproic acid is rapidly and almost completely absorbed by the gastrointestinal tract. Peak plasma concentrations are usually reached in from 30 to 120 min. Enteric-coated divalproex sodium reaches peak plasma concentration in about 4 hr. It is about 90% protein bound in the human; concentrations of valproic acid in cerebrospinal fluid suggest that unbound drug may diffuse from the serum across the blood–brain barrier. The plasma half-life is about 8–16 hr but may be shorter when drugs that stimulate hepatic microsomal systems are used concurrently. Amitriptyline (Elavil® and others) may lengthen the plasma half-life of valproic acid and may therefore raise valproic acid blood levels.

The drug is primarily metabolized in the liver and is excreted as the glucuronide conjugate. It is eliminated principally in the urine with a small amount in fecal material. Valproic acid does not stimulate hepatic metabolism.

5. Toxicity

The toxicity of valproic acid is low. The most serious risk is hepatotoxicity. Valproic acid monotherapy was connected with fatal hepatotoxicity in one in 7000 children 0–2 years of age. Children over 2 years of age had a risk of about one in 37,000 with monotherapy. No hepatic fatalities appeared in adults with monotherapy.

In polytherapy with valproic acid the risk of fatal hepatotoxicity in adults 21–40 years of age is about one in 17,000. In those 40 years or older, it is about one in 38,000.

One suicidal fatality had a valproic acid serum level of 1970 mg/liter. Naloxone (Narcan® and others) has been reported to reverse valproic acid-associated coma in some cases.

Valproic acid is excreted in the breast milk.

6. Preparation and Dosage

The drug is available as sodium valproate (liquid), valproic acid (250-mg capsules), and the enteric-coated divalproex sodium, a stable compound of valproic acid and sodium valproate (250-mg and 500-mg capsules).

The usual adult dose is started at 15 mg/kg per day and gradually increased at 1-week intervals by 5 to 10 mg/kg per day. Maximum recommended dose is 60 mg/kg per day, but some feel that no additional benefit will be obtained from >2.5 g/day.

The therapeutic range of valproic acid as an anticonvulsant is approximately 50–100 μg/ml, although some patients respond well at levels above or below these limits. Plasma albumin binding sites for valproic acid become saturated at levels around 50 μg/ml.

D. Phenobarbital

1. Chemistry

Phenobarbital was the first effective anticonvulsant drug and continues to be a potent, widely used antiepileptic. Many anticonvulsants were developed based on the structure of phenobarbital:

2. Adverse Reactions

a. Neuropsychiatric

Sedation occurs commonly at therapeutic levels of phenobarbital; this limits the dose level. Tolerance to the sedative effect may develop, but tolerance to the anticonvulsant effect may appear concomitantly, necessitating dosage increase. Patients who remain drowsy after the dose is increased are unable to use this drug. Compromised cognition (slowed thinking) is not uncommon even when sedation is minimal and may be a serious, often overlooked, side effect.

Nystagmus commonly develops at the low end of the toxic range (above 20 μg/dl), and ataxia appears at higher levels.

Behavioral disinhibition and paradoxical excitement may be exhibited by some patients, especially the elderly or those with compromised brains. Depression, psychosis, and delirium may also appear in some patients.

b. Hematological

Megaloblastic anemia similar to that seen with phenytoin occurs only rarely. The anemia usually responds well to therapy with folate or B$_{12}$.

c. Endocrine

Osteomalacia and vitamin D deficiency are seen rarely. Hypoprothrombinemia and hemorrhage have been reported in the newborns of mothers who are using phenobarbital.

3. Drug Interactions

Barbiturates interact with a wide variety of drugs. Most commonly, phenobarbital decreases the effects of selected medications by inducing microsomal enzymes (see Table 1). Thioridazine (Mellaril® and others) may lower phenobarbital levels.

4. Toxicity and Withdrawal

Overdosage of phenobarbital can produce profound respiratory depression and death (see Chapter 7).

Withdrawal syndromes may occur, and the seizures that accompany these syndromes may be very refractory to treatment. Large doses of barbiturates may be required to control such withdrawal seizures. Other antiepileptics are generally of little value.

Dependence, addiction, and abuse are all potential dangers with phenobarbital (see Chapter 7).

5. Kinetics

Oral absorption is slow but complete. Peak plasma concentrations appear several hours after oral administration. The plasma half-life is about 90 min in adults.

Phenobarbital is 40% to 60% bound to plasma proteins. Approximately 25% of the drug is excreted unchanged in the urine. The remainder is converted by the hepatic microsomal enzymes into inactive metabolites and excreted in the urine.

6. Preparation and Dosage/Monitoring

Blood levels of 10 to 25 μg/ml are usually required for control of seizures. The drug is available in 15- and 30-mg tablets. Doses of 60 to 120 mg/day are generally sufficient to produce adequate blood levels in most patients. Sedation can be minimized by using a twice- or three-times-a-day dosage schedule. Hospitalized patients or those with refractory seizures may need higher blood levels; sedation or other unwanted effects remain the limiting factors.

E. Primidone

1. Chemistry

Primidone is a congener of phenobarbital. Its structure is:

2. Adverse Reactions

a. Neurological

Sedation is the major unwanted effect of primidone. It may be severe and is less likely to disappear than the sedation caused by phenobarbital. Sedation may be minimized by beginning with low doses and increasing the dosage only gradually.

Vertigo, ataxia, and an experience of feeling intoxicated may occur during the first few days of therapy but often resolve spontaneously.

Although rare, some patients with temporolimbic epilepsy may become acutely psychotic soon after primidone is started. If this occurs, the drug should be immediately stopped, and the psychosis treated with antipsychotic agents.

b. Hematological

Megaloblastic anemia secondary to folate and B_{12} deficiency may appear. This usually responds to folate and B_{12} replacement. Patients on primidone should have a complete blood count checked every 6 months.

c. Endocrine

Osteomalacia similar to that seen with phenobarbital may occur.

d. Cutaneous

Allergic skin rash is rare, but if it occurs, the drug should be immediately stopped and not reused.

3. Drug Interactions

Primidone has properties similar to the barbiturates. Therefore, the clinician generally should assume that drug interactions resemble those of phenobarbital (see Table 1).

4. Toxicity

Primidone is very similar to phenobarbital in its ability to produce life-threatening respiratory depression following overdosage. The development of tolerance and withdrawal syndromes and their treatment are also similar to those for phenobarbital.

5. Kinetics

Primidone can only be given orally. It is rapidly and completely absorbed from the GI tract. Peak plasma levels are observed about 3 hr after ingestion. The plasma half-life of this drug is about 8 hr.

Primidone is converted into two active metabolites: phenobarbital and phenylethylmalonamide (PEMA). Only a small portion of primidone and PEMA is bound to plasma proteins.

6. Preparation and Dosage/Monitoring

Primidone is available as 50- and 250-mg tablets and as an oral suspension of 250 mg/5 ml. The usual daily adult dosage is 300 to 750 mg given in two or three divided doses. The usual starting dose of primidone is 50 to 125 mg per day. Primidone blood levels are available; the therapeutic range is between 10 and 25 µg/ml.

F. Benzodiazepines

Clonazepam is a sedating nitrobenzodiazepine which is marketed as an anticonvulsant but appears to be useful for managing anxiety. (See Chapter 5 for a complete discussion of clonazepam.) Clonazepam may be useful for patients with partial seizures who also have anxiety. In patients whose seizures present as fear spells or anxiety attacks, clonazepam may be used as a primary drug; in other patients, it may be used adjunctively. If it is used alone, tolerance to the anticonvulsant effect may develop.

Rare cases of coma possibly related to an adverse interaction between clonazepam and valproate have been reported. Care must be taken if these agents are used in combination.

Clorazepate (Tranxene® and others) has also been reported to have been particularly useful in managing anxiety in patients with partial seizures and may be used adjunctively with carbamazepine or valproate.[22]

Intravenous injection of diazepam (Valium® and others) or lorazepam (Ativan® and others) has been used to abort partial status epilepticus.

G. Other

Ethosuximide (Zarontin®) and methsuximide (Celontin®) have been used as adjuncts in the management of partial seizures, and preliminary reports indicate that they may be used in some cases of affective disorder and disorders of impulse control.

VI. CLINICAL APPLICATIONS

A. Use of Anticonvulsants

1. Control of Seizures

Anticonvulsants are used to reduce the frequency and severity of auras, ictal symptoms, and secondary major convulsions.

Carbamazepine is now considered to be the first drug of choice for the control of partial seizures. Phenytoin, valproic acid,[23] the barbiturates and their congeners, e.g., primidone, valproate, and clonazepam, may be used as second-choice drugs if carbamazepine is not effective or is not tolerated. Methsuximide may be used if seizures are refractory to all other agents. Clorazepate may be useful as an adjunct, especially if anxiety is present.

It is preferable to use only one drug, but since partial seizures are not infrequently difficult to control, combination therapy may be required.[21] It is important to document, by the careful use of anticonvulsant blood level measurements and a seizure log, that a particular drug has been given a fully adequate trial before a second drug is added. It is then important to follow blood levels, seizure frequency, and the impact of adverse effects on overall function as a decision regarding the usefulness of combination therapy is made.

Starting with low doses and increasing slowly will help minimize side effects and improve compliance. Seizure control is generally maximum when therapeutic levels are attained (for carbamazepine, phenytoin, and barbiturates), but levels should serve as guides rather than goals. Some patients may achieve relief at low plasma levels or may have good control with acceptable side effects at high levels.

Complete suppression of partial seizures is thought in some patients to produce worsening in mental function, and in some cases to precipitate psychosis. Some patients feel and function better if they have an occasional partial seizure. The goal of treatment is therefore to seek a balance between seizure control and overall daily functioning.

All anticonvulsants may interfere with cognitive function. Barbiturates and phenytoin may produce more problems in this area than carbamazepine or valproic acid.[12] Cognitive compromise may be subtle but should not be overlooked.

2. Control of IBS Features

Carbamazepine may promote affective brightening and lessening of irritability in patients with the IBS. This effect may occur independently of seizure control.[3] Patients with temporolimbic abnormalities and features of the IBS but without frank seizures may benefit from a careful clinical trial of carbamazepine if IBS features present problems.[14] Phenytoin and barbiturates do not seem to have this effect.

3. Psychosis

If psychosis is related to frequent, uncontrolled seizures or status epilepticus partialis, anticonvulsants (especially carbamazepine) may be helpful. Carbamazepine may modulate irritability but often will not improve the chronic psychosis associated with the IBS. Antipsychotic agents, typically in small doses, sometimes moderate psychotic symptoms. (See Section VI.B for discussion of antipsychotic agents and seizures).

4. Use of Anticonvulsants in Nonepileptic Conditions

There is a growing literature on the use of carbamazepine, clonazepam, and more recently valproic acid alone and in combination with standard psychotropic medications for managing various psychiatric conditions including mania, atypical or treatment-resistant mania, depression, treatment-resistant or atypical depression, treatment-resistant or atypical panic disorder, episodic dyscontrol, atypical and treatment-resistant psychosis, bulimia, chronic pain, and others.[24] Recent data suggest that carbamazepine may be less efficacious over long periods than originally hoped.[25] See Chapters 2, 3, 5, and 12 for further discussion of this complex and unsettled topic.

In general, it is often advisable to follow both anticonvulsant and psychotropic blood levels until steady state is reached, as the anticonvulsants tend to increase the metabolism of many psychotropics. In particular, significant reductions in haloperidol (Haldol® and others) level, resulting in clinical worsening, may appear when carbamazepine is added. Thioridazine levels may also be lowered. Carbamazepine may decrease the blood levels of clonazepam, coumarin (Coumadin®), doxycycline (Vibramycin® and others), heparin, phenytoin, theophylline (Primatene® and others), and valproate; it may interfere with pregnancy tests. Symptoms of carbamazepine toxicity, related to the inhibition of carbamazepine metabolism, may appear when erythromycin, propoxyphene (Darvon® and others), isoniazid (INH® and others), nicotinamide, verapamil (Calan® and others), diltiazem (Cardizem®), or danazol (Danocrine® and others) is added to the regimen. Cases have been reported in which seizures occurred in previously nonepileptic patients with affective disorder when carbamazepine and clonazepam were withdrawn,[26,27] and psychosis worsened when carbamazepine was abruptly discontinued.[28] The clinician should remain alert when anticonvulsants are stopped; no specific guidelines exist for discontinuing anticonvulsant treatment of psychiatric disorders.

The literature on the use of phenytoin and barbiturates for treating psychiatric problems is incomplete, but it does not appear that these agents are frequently useful as primary drugs or adjuncts in the way carbamazepine and valproic acid may be. It is possible that the special propensity of carbamazepine and valproic acid to inhibit kindling in limbic structures (phenytoin and barbiturates apparently do not have this property) may play a role. Patients on phenytoin may require increased doses of antidepressants. A case of phenytoin-induced lithium toxicity has been reported.

The astute reader may wonder at this point if antikindling agents (such as carbamazepine and valproic acid) may be called for in a patient who has behavior problems akin to those of the IBS and relevant abnormal brain electrical activity (such as temporal spiking or abnormal sharp rhythms) but does not appear to have actual seizures. There is as yet no definitive answer to this question, but it may be reasonable to proceed with treatment after a careful analysis of the risk–benefit ratio.[29]

The combination of valproic acid with neuroleptics has been reported to induce stupor and slowing on the EEG. It may be advisable to use reduced doses of neuroleptics for patients also receiving valproic acid.

Patients treated with benzodiazepines may develop seizures after rapid drug withdrawal, even if only low doses of benzodiazepines have been used. Psychosis may appear in some patients during or after anticonvulsant withdrawal.

B. Use of Psychotropic Medication in Patients with Seizures

Antidepressants and antipsychotic drugs produce EEG changes and may precipitate convulsions when used in overdosage. Some patients with partial and generalized epilepsy may have worsening of seizure control while using these drugs. It has therefore been suggested that antidepressants and antipsychotic drugs lower the seizure threshold.

However, recent work with animal seizure models has suggested that haloperidol may diminish hippocampal but exacerbate cortical kindled seizures. Amitriptyline, imipramine, and desipramine may also be weak anticonvulsants at hippocampal kindled sites; maprotiline (Ludiomil® and others) shows a proconvulsant effect in this model.[30] These observations correlate with clinical experience, which shows that patients with depression and partial seizures often experience relief of depression with no change or even improvement in seizures when using certain antidepressants. A similar response may be seen in some patients with psychosis and partial seizures who use antipsychotics. Therefore, statements about the effects of psychotropics on seizures must take into account the particular drug in question, the type of seizures present, and the location of the focus (for partial seizures).

Concern about epileptogenesis need not influence drug decisions in patients with no history of seizures or EEG abnormality. Neurological investigation is indicated if seizures appear while patients are on psychotropics.[31]

1. General Clinical Guidelines

General guidelines for the safe usage of psychotropics when specifically indicated in patients with seizures or abnormal EEGs include:

1. Use the lowest possible doses to avoid sudden dose changes (up or down).
2. Pick agents with low potential for epileptogenesis (see below).
3. Avoid conditions that lower seizure threshold, such as exhaustion, fever, hypoglycemia, or electrolyte imbalance.
4. Instruct patients to avoid hazardous situations until stabilized (driving, swimming, etc.).
5. Use concomitant anticonvulsants as needed.

2. Antidepressants

The overall frequency of antidepressant-related seizures in the general population has been estimated to be approximately 0.1%. Limited available clinical data suggest that patients with abnormal EEGs, epilepsy, or coarse brain disease are probably at greater risk.

Maprotiline and amoxapine (Asendin®) may have clinically significant proconvulsant properties. Clomipramine (Anafranil®) appears to be associated with increased risk of seizures at high doses (>250 mg/day). The monoamine oxidase inhibitors may have a slight proconvulsant effect. Trazodone (Desyrel® and others)-related seizures have been reported but are probably infrequent. Doxepin (Sinequan® and others) has been suggested to have some anticonvulsant properties, although the clinical relevance of this is unclear. Methylphenidate (Ritalin® and others) and D-amphetamine (Dexedrine® and others) do not appear to be proconvulsant.

The overall incidence of seizures associated with bupropion (Wellbutrin®) has been reported to be approximately 0.4%. The incidence of seizures may be higher in bulimics and anorectics and at doses greater than 450 mg/day. The incidence of fluoxetine (Prozac®)-related seizures is approximately 0.2%, comparable to the overall rate for tricyclics.

Data from animal suggest that amitriptyline, imipramine, and desipramine (Norpramin® and others) actually weakly inhibit kindled seizures. Other models indicate that imipramine, amitriptyline, nortriptyline (Pamelor® and others), and desipramine (in that order) increase spike activity in isolated cortex, whereas doxepin, protriptyline (Vivactil®), and trimipramine (Surmontil®) decrease spiking.

In practice, amitriptyline, nortriptyline, desipramine, doxepin, fluoxetine, or stimulants are reasonable choices for patients in whom seizures are a concern.

Depressive symptoms are common in patients with seizures, and antidepressants generally should not be withheld because of concerns about precipitating seizures. When prescribing antidepressants, however, clinicians should note that the risk of precipitating a seizure may be greater in patients with generalized seizures than in those with partial seizures. The risk of seizure also increases with increasing antidepressant dose. Practitioners should always choose the lowest effective dose, exercise caution when changing drugs or drug dosages, and continually assess the potential for suicide or adverse drug interactions (discussed elsewhere in this chapter).[32]

3. Lithium

Conflicting reports regarding the effects of lithium (Eskalith® and others) on seizures appear in the literature. Some reports indicate that lithium can exacerbate seizures and may produce delirium (even when used in low dose) in some patients with abnormal EEGs. However, others find that lithium may have anticonvulsant properties in some patients. Lithium may be used successfully in some patients with seizures and mood cycling or concomitant affective illness.

4. Antipsychotics

The lower-potency phenothiazines appear to be somewhat more epileptogenic than the higher-potency phenothiazines. Chlorpromazine (Thorazine® and others) may worsen or induce seizures in predisposed patients, so it may be best to use alternative agents. Thioridazine and thiothixene (Navane® and others) appear to have moderate to low epileptogenic potential. Fluphenazine (Prolixin® and others) and pimozide (Orap®) probably have a low propensity to worsen seizures.[31,33]

Haloperidol has been reported to suppress limbic seizures but to facilitate motor cortex seizures in rats. Clinical reports regarding haloperidol conflict; it probably has low to moderate epileptogenic properties.

Molindone (Moban®) has been noted to have very low epileptogenicity.

Neuroleptic combinations may have synergistic epileptogenic potential and so are best avoided.

Low doses of antipsychotic agents are often useful for patients with psychosis who also have epilepsy, and, as with antidepressants, these agents should not be withheld because of fears of worsening seizures. In acutely psychotic patients with seizures, a physiological etiology of the problem (e.g., drug toxicity, delirium) should be actively sought. Drug interactions are discussed elsewhere in this chapter.

VII. CONCLUSION

Abnormalities in temporolimbic electrophysiology can be associated with emotional and behavioral changes. Anticonvulsants may modulate these emotional and behavioral changes by their influence on temporolimbic electrical function. Further study in this area may advance the understanding of the neurobiological basis of psychiatric problems.

REFERENCES

1. Gasteau H.: Proposal for revised clinical and electroencephalographic classification of epileptic seizures. *Epilepsia* 22:489–301, 1980.
2. Gloor P., Olivier A., Quesney L. F., et al: The role of the limbic system in experiential phenomenon of temporal lobe epilepsy. *Ann Neurol* 12:129–144, 1982.
3. Mesulam M. M. (ed): *Behavioral Neurology,* New York, Plenum Press, 1986.
4. Engel J., Cahn L.: Potential relevance of kindling to human partial epilepsy, in Wada J. A. (ed): *Kindling III.* New York, Raven Press, 1986, pp 37–51.
5. Albright P. S., Burnham W. M.: Development of a new pharmacologic seizure model: Effects of anticonvulsants on cortical-amygdala-kindled seizures in the rat. *Epilepsia* 21:681–689, 1980.
6. Bear D. M.: Behavioral changes in temporal lobe epilepsy: Conflict, confusion, challenge, in Trumble M., Bolwig T. G. (eds) *Psychiatry and Epilepsy.* New York, John Wiley & Sons, 1986.
7. Bear D. M.: Temporal lobe epilepsy—A syndrome of sensory-limbic hyperconnection. *Cortex* 15:357–384, 1979.
8. Adamec R. E., Stark-Adamec C.: Limbic kindling in animal behavior—Implications for human psychopathology associated with complex partial seizures. *Biol Psychiatry* 18:269–293, 1983.
9. Collaborative Group for Epidemiology of Epilepsy: Adverse reactions to antiepileptic drugs: A multicenter survey of clinical practice. *Epilepsia* 27:323–330, 1986.
10. Sironi V. A., Franzini A., Ravagnati L., et al: Interictal acute psychoses in temporal lobe epilepsy during withdrawal of anticonvulsant therapy. *J Neurol Neurosurg Psychiatry* 42:730–742, 1979.
11. Rasmussen T. B.: Surgical treatment of complex partial seizures: Results, lessons, and problems. *Epilepsia* 24(suppl 1):S65–S76, 1983.
12. Thompson P. J., Tremble M. R.: Anticonvulsant serum levels: Relationship to impairments of cognitive functioning. *J Neurol Neurosurg Psychiatry* 46:227–233, 1983.
13. Emrich H. M., Oquma T., Mull A. A.: *Anticonvulsants and Affective Disorders.* Amsterdam, Excerpta Medica, 1984.
14. Penry J. K., Daly, D. D.: *Complex Partial Seizures and Their Treatment: Advances in Neurology 11,* New York, Raven Press, 1975.
15. Rizack M. A., Hillman C. D. M., Dreifuss F. E., et al: Hepatic considerations in the use of antiepileptic drugs. *Epilepsia* 28(suppl 2):S23–S29, 1987.
16. Hart R. G., Easton J. D.: Carbamazepine and hematological monitoring. *Ann Neurol* 11:309–312, 1982.
17. *The Medical Letter:* Drugs for epilepsy. *Med Lett* 28:91–94, 1986.
18. Yassa R., Iscandar H., Nastase C., et al: Carbamazepine and hyponatremia in patients with affective disorder. *Am J Psychiatry* 145:339–342, 1988.
19. Post R. M., Uhde T. W., Bellenger J. C., et al: Carbamazepine and H-10,11-epoxide metabolite in plasma and CSF: Relationship to antidepressant response. *Arch Gen Psychiatry* 40:673–676, 1983.
20. Turnbull D. M., Howell D., Rawlins M. D., et al: Which drug for the adult epileptic patient: Phenytoin or valproate? *Br Med J* 290:815–819, 1985.
21. Callaghan N., Kenny R. A., O'Neill B., et al: One perspective study between carbamazepine, phenytoin and sodium valproate as monotherapy in previously untreated and recently diagnosed patients with epilepsy. *J Neurol Neurosurg Psychiatry* 48:639–644, 1985.
22. Griffith J. L., Murray G. B.: Clorazepate in the treatment of complex partial seizures with psychic symptomatology. *J Nerv Ment Dis* 173:185–186, 1985.
23. Bruni J., Albright P.: Valproic acid therapy for complex partial seizures. *Arch Neurol* 40:135–137, 1983.

24. Post R. M.: Mechanisms of action of carbamazepine and related anticonvulsants in affective illness, in Meltzer H. Y. (ed): *Psychopharmacology: The Third Generation of Progress*. New York, Raven Press, 1987, pp 567–594.
25. Frankenburg F. R., Tohen M., Cohen B. M., et al: Long-term response to carbamazepine: A retrospective study. *J Clin Psychopharmacol* 8:130–132, 1988.
26. Garbutt J. C., Gillette G. M.: Apparent complex partial seizures in a bipolar patient after withdrawal of carbamazepine. *J Clin Psychiatry* 49:10, 1988.
27. Ghadirian A. M., Gauthier S., Wong T.: Convulsions in patients abruptly withdrawn from clonazepam while receiving neuroleptic medication. *Am J Psychiatry* 144:686, 1987.
28. Heh C. W. C., Sramek J., Herrera J., et al: Exacerbation of psychosis after discontinuation of carbamazepine treatment. *Am J Psychiatry* 145:878–879, 1988.
29. Post R. M., Uhde T. W., Putnam F. W., et al: Kindling and carbamazepine in affective illness. *J Nerv Ment Dis* 170:717–731, 1982.
30. Clifford D. B., Rutheford J. L., Hicks F. G., et al: Acute affects of antidepressants on hippocampal seizures. *Ann Neurol* 18:692–697, 1985.
31. Itil T. M., Soldatos C.: Epileptogenic side effects of psychotropic drugs. *JAMA* 244:1460–1463, 1980.
32. Davidson J.: Seizures and bupropion: A review. *J Clin Psychiatry* 50:256–261, 1989.
33. Oliver A. P., Luchins D. J., Wyett R. J.: Neuroleptic induced seizures. *Arch Gen Psychiatry* 39:206–209, 1982.

12

Eating Disorders

ANDREW W. BROTMAN, M.D., and
DAVID B. HERZOG, M.D.

I. INTRODUCTION

A. What Are the Eating Disorders?

The eating disorders discussed in this chapter include two syndromes: anorexia nervosa (AN) and bulimia nervosa (BN). Both have increased in prevalence over the last 20 years and are characterized by disturbed eating behavior and body image. See Tables 1 and 2 for the third edition of the *Diagnostic and Statistical Manual of Mental Disorders* (DSM-III) and DSM-IIIR criteria.

Eating disorders affect 5–10% of young women, with bulimia nervosa the more common of the two.[1] The prevalence of AN in young women is estimated to be 1%[2]; its incidence has doubled since the 1960s.[3,4] The lifetime prevalence rate of BN is in the range of 10%, making the eating disorders a serious public health problem.

Approximately 90% of the eating-disordered population is female, but as the disorders have become more common, men are also presenting with these syndromes. The affected population crosses race, class, and educational barriers.

B. Clinical Presentation

1. Anorexia Nervosa

The anorectic usually seeks help reluctantly, often being asked by family, friends, or employer to see a physician. She frequently denies any problems. Typically she describes herself as a slightly overweight teen-ager (or perceives herself

ANDREW W. BROTMAN, M.D. • Social and Community Psychiatry, Massachusetts General Hospital, Boston, Massachusetts 02114; Department of Psychiatry, Harvard Medical School, Boston, Massachusetts 02115; Outpatient Services, Erich Lindemann Mental Health Center, Boston, Massachusetts 02114. **DAVID B. HERZOG, M.D.** • Eating Disorders Clinic, Massachusetts General Hospital, Boston, Massachusetts 02114; Department of Psychiatry, Harvard Medical School, Boston, Massachusetts 02115.

TABLE 1. Defining Criteria for Anorexia Nervosa

DSM-III criteria
1. Refusal to maintain normal body weight.
2. Loss of more than 25% of original body weight.
3. Disturbance of body image.
4. Intense fear of becoming fat.
5. No known medical illness leading to weight loss.

DSM-IIIR criteria
1. Refusal to maintain body weight over a minimal normal weight for age and height, e.g., weight loss leading to maintenance of body weight 15% below expected.
2. Intense fear of becoming obese, even when underweight.
3. Disturbance in the way in which one's body weight, size, or shape is experienced, e.g., the individual claims to "feel fat" even when emaciated, believes that one area of the body is "too fat" even when obviously underweight.
4. In females, absence of at least three consecutive menstrual cycles when otherwise expected to occur (primary or secondary amenorrhea). A woman is considered to have amenorrhea if her periods occur only following hormone (e.g., estrogen) administration.

TABLE 2. Defining Criteria for Bulimia Nervosa[a]

DSM-III criteria
1. Recurrent episodes of binge eating.
2. At least three of the following:
    a. Consumption of high-calorie, easily ingested foods during a binge.
    b. Termination of binge by abdominal pain, sleep, or vomiting.
    c. Inconspicuous eating during a binge.
    d. Repeated attempts to lose weight.
    e. Frequent weight fluctuations of more than 4.5 kg.
3. Awareness of abnormal eating patterns and fear of not being able to stop voluntarily.
4. Depressed mood after binge.
5. Not caused by anorexia nervosa or any physical disorder.

DSM-IIIR criteria
1. Recurrent episodes of binge eating (rapid consumption of a large amount of food in a discrete period of time).
2. During the eating binges there is a feeling of lack of control over eating behavior.
3. The individual regularly engages in either self-induced vomiting, use of laxatives, strict dieting, fasting, or rigorous exercise in order to prevent weight gain.
4. A minimum average of two binge-eating episodes per week for at least 3 months.
5. Persistent overconcern with body shape and weight.

[a]Referred to in DSM-III as bulimia.

as such) who in the early midteens began a diet that escalated into an obsession.[5] She continued to lose weight by severely restricting food intake, by exercising excessively, or by purging. She has a morbid fear of obesity and perceives herself as fat despite persistent weight loss.

Anorectics can appear cheerful and energetic or depressed and isolated; they can be controlling and rigid in their thinking and are frequently achievement oriented. Peculiar food rituals are common and tend to persist in over half of patients even if weight gain is achieved.

Anorexia nervosa may begin quickly or insidiously and can be self-limited, cyclical, or chronic. Morbidity and mortality rates for AN are among the highest of all psychiatric disorders. At some point in the course of the illness, more than half these patients meet the criteria for clinical depression.[6,7] One-third have recurrent affective illness, and one-fourth never regain menses or attain 75% of ideal body weight.[8]

2. Bulimia Nervosa

Bulimia usually begins in late adolescence or young adulthood, when the patient becomes aware that self-induced vomiting, laxative use, ipecac, or diuretics help to cause weight loss or to maintain weight in the face of binging. Bulimics may binge on huge quantities of food, usually in secret, followed by purging or restriction. Unlike the anorectic, the bulimic who comes for help is usually distressed and ashamed of her symptoms. Bulimics usually look much better than they feel. They may appear strong, sociable, and assertive but are frequently tormented by low self-esteem, fear of intimacy, and aggression.

Bulimia can be self-limited but is usually chronic and episodic. Recovery rates vary, and relapse rates can be up to 40%.[9] More than half of bulimics meet criteria for affective disorder at some point in their lives, but most studies have shown that depression at intake does not affect the probability of recovery from bulimia.

C. Medical Complications

The medical sequelae of AN are similar to those of chronic starvation, whereas those of BN cause complications primarily related to purging and binging. Anorexia nervosa signs and symptoms include hypotension, bradycardia, amenorrhea, hypothermia, anemia, growth failure, lanugo, and osteoporosis.[10] Delayed gastric motility can produce abdominal pain, constipation, or stomach fullness.

Bulimics may have electrolyte abnormalities from purging, dental enamel erosion, parotidomegaly, and abrasion of the knuckles from self-induced vomiting.[11] About 40% of patients have menstrual abnormalities, even with normal weight. Cardiac arrhythmias, particularly when the patient is hypokalemic, can occur. Chronic ipecac use can cause toxic myocardial dysfunction as a result of emetine poisoning.

More complete information on medical complications of eating disorders can be found elsewhere.[10,11]

II. ETIOLOGICAL THEORIES

A. Biologic Framework

The search for a neurochemical abnormality that "causes" AN has been frustrating and is clouded by the fact that starvation can lead to decreases in 3-methoxy-4-hydroxyphenylglycol (MHPG), homovanillic acid (HVA), tyrosine, and 5-hydroxyindoleacetic acid.[11] Similarly, increased rates of cortisol production and nonsuppression of the dexamethasone suppression test cannot be solely attributed to AN.

However, some changes are not fully explained by starvation. One-fifth to one-third of anorectics lose their menses before weight loss occurs, and in up to one-third menses are not restored even after weight gain. Patients with AN also have decreased levels of cerebrospinal fluid (CSF) arginine vasopressin even after weight restoration.[12] These changes may be specific to AN and point to possible neurochemical changes that could contribute to the onset of the disorder.

Many studies of bulimic patients suggest serotonergic and noradrenergic abnormalities. Increases in serotonin produce satiety and reduce carbohydrate craving.[13] Bulimics may have a low level of serotonin, causing binging and perhaps explaining why antidepressant medication may be useful in treatment.[13,14]

B. Sociocultural Framework

The physical ideal for women in Western society as portrayed in the media and the fashion world is a thin body. The cultural value of slimness has contributed to the development of eating disorders. Over the last 25 years, the average American has become heavier, but the measurements of *Playboy* centerfolds and of Miss America winners have significantly decreased.[15] Women appear to be at greater risk for eating disorders, in part because our culture overemphasizes a woman's appearance. Most designers create fashions that the majority of women couldn't possibly wear, thereby contributing to high rates of eating disorders among models. Some investigators suggest that the focus on the thin, "athletic" body in Western culture may be a major factor contributing to the increase of eating disorders over the last quarter century.

C. Psychological Framework

A full discussion of the psychological theories of eating disorders is beyond the scope of this chapter. However, impaired psychological development is frequently viewed as the core problem in eating disorders, though there is little agreement on a specific developmental problem.

Hilde Bruch[16] posited that if appropriate responses from the mother are chronically lacking, the child cannot learn how to differentiate her needs from those imposed by others. Thus, if a mother frequently feeds a child to keep her quiet, puts her to sleep to have some free time, or otherwise persistently responds to her own needs rather than those of the child, distorted self-image can develop, leading to a vulnerability to eating disorders. The child achieves a feeling of competence and control only by rejecting her own hunger and becoming thin. Selvini Palazzoli suggested that the anorectic perceives her body as a threat that must be controlled.[17] She describes an overprotective narcissistic mother who cannot tolerate a child's attempts to separate. The child experiences her body as the maternal object from which the ego must separate.

Geist further suggested that the daughter turns to the father for mirroring and empathy after feeling unable to separate from mother. The father–daughter bond is threatened by the daughter's sexual maturation, explaining the onset of eating disorders in adolescence.[18]

These theories implicitly define autonomy (e.g., self-sufficiency) as a major developmental goal in our society, a value increasingly being questioned by researchers studying normal female development.[19] Steiner-Adair proposed that eating disorders may be prevalent in this culture because of an unrealistic overemphasis on autonomy in women as opposed to validation via interpersonal relationships.[20] If the rounded female body is associated with mothering and dependency, the anorectic may be rejecting this symbol, thereby colluding with the cultural norm of control and autonomy.

Many clinicians report that a history of sexual abuse is common in patients with eating disorders and may contribute to their development. Therefore, appropriate exploration of this subject during the initial psychiatric evaluation and during follow-up is important.

III. TREATMENT MEASURES

A. Introduction

Unfortunately, the treatment of AN and BN is frequently frustrating and sometimes unsatisfactory. Most patients have been suffering from these disorders for years before seeking treatment, and many are unwilling participants in treatment. Treatment modalities include individual, group, and family psychotherapy, pharmacotherapy, and inpatient treatment. Theoretical orientations include psychodynamic approaches, cognitive–behavioral models (individual and in groups), self-help forums, strategic family work, and biological interventions. Frequently a combination of approaches is used, and no single modality has emerged as the treatment of choice. Matching a patient with a particular treatment is in its infancy.

The type of treatment one receives may be more a function of the provider's philosophy than the patient's clinical presentation.

B. Outpatient Treatment

1. General Approach

Outpatient treatment should be designed to meet the needs of the individual patient. Ideally, the patient's initial assessment should include physical examination, psychiatric evaluation, nutritional history, and in many cases a family evaluation. Although some patients can be managed by a single clinician, multiple treatment programs are often indicated. The severely ill patient may require a team of clinicians, including a psychiatrist and internist or pediatrician. For all patients, initial therapy should include medical monitoring and behavioral treatment for weight recovery or control of binging and purging.

2. Psychodynamic Treatment

Psychodynamic or cognitive–behavioral therapies are commonly used. Psychodynamic approaches are usually long-term and consider eating disorders as a compromise solution to unsolvable psychological conflicts. These can include a concurrent desire for autonomy and dependence and the need to be in complete control of a body that refuses to cease having appetites. During therapy, erroneous attitudes and assumptions are recognized, defined, and questioned within the context of a nonjudgmental, firm, yet empathic therapeutic relationship.

3. Cognitive–Behavioral Treatment

Cognitive–behavioral treatment is usually time-limited and focuses on directly eradicating symptomatology.[21,22] Making short-term goal contracts, journal keeping, constructing lists of alternative activities, time-delaying tactics, relaxation, meal planning, and weight monitoring may all be employed. Cognitive approaches modified from Beck's technique are used to treat low self-esteem, body image distortion, hopelessness, and other self-destructive styles.

Treatment modalities are not mutually exclusive, so that a bulimic patient might be in individual, long-term therapy or time-limited, cognitive–behavioral group therapy and be receiving antidepressants. A comprehensive review of outpatient treatments can be found elsewhere.[23]

C. Inpatient Treatment

Although there are no specific guidelines for hospitalization, it can provide a necessary bridge or catalyst to outpatient treatment. Inpatient treatment should be viewed as an emergency measure and not a routine part of treatment. Therefore, the decision to admit an anorectic patient should be based on various criteria.

Suggested Criteria for Hospitalization of an Anorectic Patient

1. Weight loss greater than 30% over 3 months.
2. Severe metabolic disturbances as manifested by pulse less than 40 per minute, temperature less than 36°C, systolic blood pressure less than 70 mm Hg, or serum potassium less than 2.5 mmol per liter despite oral potassium replacement.
3. Severe depression or suicide risk.
4. Severe binging and purging.
5. Failure to maintain outpatient weight contract.
6. Complex differential diagnosis.
7. Psychosis.
8. Family crisis.
9. Need for confrontation of individual and family denial and initiation of individual therapy, family therapy, and pharmacotherapy.

Patients with eating disorders may be hospitalized on a pediatric or psychiatric ward, depending on several factors. Patients with significant suicide risk or family crises are usually better managed on a psychiatric ward. Those hospitalized primarily for nutritional rehabilitation can be adequately treated on a medical ward if the staff feels comfortable treating these disorders and a behavior protocol exists.

IV. PHARMACOTHERAPY

A. Relationship of Eating Disorders to Depression: Use of Antidepressants

There is substantial evidence in the literature suggesting a link among anorexia nervosa, bulimia nervosa, and affective illness. Eating-disordered patients have:

1. Frequent current and previous symptoms of affective illness.
2. Higher than expected current and previous episodes of major depressive disorder.
3. Familial aggregation and genetic predisposition to affective disorder.
4. Neuroendocrine abnormalities similar to those found in affective probands.
5. Some response to antidepressant medications.

Based on this evidence, antidepressants have been commonly used to treat eating disorders. These agents are intended to increase weight gain, reduce binging and purging behaviors, diminish anxiety, improve eating attitudes, and decrease depression.

Others have suggested that eating disorders resemble anxiety disorders, obsessive compulsive disorders, seizure disorders, or the addictive disorders and have tried medications used to treat those problems.

B. Drug Therapies of Anorexia Nervosa

As shown in Table 3, various medications have been tried in the treatment of anorexia nervosa, most prominently antipsychotic drugs, antidepressant medication, cyproheptadine (Periactin® and others), and antihistamine. Although there are no controlled studies investigating the effectiveness of chlorpromazine (Thorazine® and others) in anorexia, a number of reports suggest that it may reduce resistance to weight gain by directly stimulating appetite.[24] Side effects include weight gain,

TABLE 3. Controlled Pharmacological Trials for Anorexia Nervosa

Drug	Number of completers	Duration (weeks)	Dose	Results[a]	Comments
Amitriptyline[27]	25	5	3 mg/kg	− for AN − for depression	Mostly inpatients
Clomipramine[28]	16	10	50 mg	− for weight gain, increased hunger	Inpatients
Lithium[30]	16	4	0.9 mg	− for depression + + for weight gain at week 3–4	All patients received behavior modification
Cyproheptadine[40]	81	5	12–32 mg	− except in a subgroup of very severe AN	Multicenter study
Amitriptyline[29]	36	8	40–160 mg	+	3-arm study of cyproheptadine, amitriptyline, and placebo; all inpatients; cyproheptadine was better than placebo and amitriptyline for depression and weight gain
Cyproheptadine[29]	36	8	32 mg	+	
Cyproheptadine[41]	24	8	12 mg	− for weight gain	All inpatients

[a]Symbols used: −, no better than placebo; +, statistical trend better than placebo; + +, statistically significant, better than placebo.

hypotension, akathisia, and tardive dyskinesia. Chlorpromazine is rarely used today except transiently in low doses or for anorectic patients who have concomitant psychotic illness.

A number of higher-potency antipsychotics have been used in placebo-controlled crossover studies for anorexia nervosa, including pimozide (Orap®), sulpiride (not available in the United States), and thiothixine (Navane ® and others).[25] The short duration of treatment makes it difficult to interpret the results of these studies, except to say that the use of antipsychotics generally produces clinically insignificant weight gain and that they are not usually indicated. These drugs might be used in patients with concomitant psychotic illness or in patients with overwhelming anxiety when minor tranquilizers fail. Moreover, the anorectic's distorted body image does not respond to antipsychotic medications.

The literature describing the treatment of anorexia nervosa with tricyclic and heterocyclic antidepressants is sparse. This is true for both open and controlled studies. In the late 1970s case reports on successful treatment of anorectic patients with amitriptyline (Elavil® and others) and imipramine (Tofranil® and others) appeared. These anecdotal reports noted that patients seemed to respond to modest doses by gaining weight and by experiencing a decrease in depressive symptoms and abnormal attitudes toward food.

Hudson and colleagues recently described a series of nine patients with anorexia nervosa, several of whom also had concomitant bulimia; all were treated with various antidepressant medications including tricyclics, trazodone (Desyrel® and others), and monoamine oxidase (MAO) inhibitors.[26] The authors found that four of these patients became significantly less depressed and gained weight. All four responders had concomitant affective disorder.

In open studies, most patients were engaged in other psychotherapeutic treatment modalities, and most had been inpatients during some portion of the medication trial. There is a suggestion that anorectics with concurrent affective illness may do better on antidepressants than anorectics without such disorders; however, the numbers are small, and no generalizable conclusions can be reached from these reports.

Controlled studies have used amitriptyline,[27] clomipramine (Anafranil®),[28] cyproheptadine,[29] and lithium carbonate (Eskalith® and others).[30] In one study, amitriptyline was found to be somewhat better than placebo, but in a second study there was no difference between amitriptyline and placebo. There have been several studies of cyproheptadine, with conflicting results. The latest report shows that it may be moderately helpful for inducing weight gain and decreasing depressive symptoms in anorectics. However, this was an inpatient study; the authors recommend the medication only as an adjunctive treatment and do not believe it has a substantial impact outside of a comprehensive treatment program.

Lithium carbonate was found superior to placebo in an 8-week study, but the results were not clinically meaningful, and investigators do not generally advocate the use of lithium for anorectics unless they also suffer from manic–depressive illness.

Lacey and Crisp treated 16 anorectic patients for 10 weeks using 50 mg of clomipramine, and found the medication had no substantial effect on weight gain when compared with placebo[28] (see Table 3).

When taken together, the results of these studies suggest that thymoleptic medications do not generally show superiority over placebo for the treatment of anorexia nervosa. Statistical significance has not generally translated into clinical significance, and most of these studies were short-term, inpatient-based trials in which other forms of nonpharmacological treatment were the primary modalities being used. Long-term effects of psychotropic medications with anorectics are not known, and at this point in time pharmacological treatment of the disorder is rather discouraging. Some interesting preliminary work has been done with opiate antagonists, but they remain experimental agents for the disorder.

C. Drug Therapies of Bulimia Nervosa

In the last 10 years there have been more than 30 reports on the pharmacological treatment of bulimia, of which approximately one-third have been controlled, double-blind studies. Most of the medications studied have been antidepressants, and the results to date are much more encouraging than studies conducted for anorexia nervosa.

When pharmacological research on bulimia and compulsive eating began in the 1970s, most of the literature concerned phenytoin (Dilantin® and others). Although early results were encouraging, further research proved equivocal, and this medication is now rarely used in clinical practice.

As shown in Table 4, six placebo-controlled, double-blind studies with antidepressants have been completed, of which four were positive [desipramine (Norpramin® and others),[31] imipramine,[32,33] phenelzine (Nardil®)[34]], one was marginally positive (amitriptyline[35]), and one was negative (mianserin, not available in the United States[36]). Fewer than 100 patients have completed controlled trials with any medication for bulimia, but preliminary results tend to show that in therapeutic doses antidepressants can decrease binging and purging over the short term, even without the presence of concomitant depression. Several other studies are currently in progress including a major multi-center study of more than 300 subjects using fluoxetine (Prozac®). The results suggest that 60 mg/day is superior to 20 mg/day, which is in turn superior to placebo. Fluoxetine seems to be as effective as tricyclics, with a more benign side-effect profile.

Other double-blind studies with very few subjects have used amphetamine and carbamazepine (Tegretol® and others),[37] but the methodological weaknesses of these studies prevent any generalization about their clinical utility. One group has reported on an open trial of naltrexone (Trexan®), an opiate antagonist, which must still be viewed as an experimental intervention.[38] Fenfluramine (Pondimin®), a serotonin-enhancing drug that promotes satiety via central serotonergic mechanisms, may be effective in bulimia and supports the notion that serotonin metabolism may be an etiological factor in bulimia.[39]

TABLE 4. Controlled Pharmacological Trial for Bulimia

Drug	Number of completers	Duration (weeks)	Dose	Results[a]	Comments
Phenelzine[34]	50	10	60–90 mg	+ +	Side effects, especially orthostatic hypotension common; high drop-out rate, 7 patients sustained a response at follow-up
Imipramine[32]	19	6	200 mg	+ + +	At 2-year follow-up most patients required multiple drug trials in order to sustain or improve response
Desipramine[31]	19	6	200 mg+	+ + +	Limited follow-up data; blood levels improved response
Amitriptyline[35]	32	8	150 mg	+	AMI group had a 70% reduction in binging; placebo group had a 52% reduction
Mianserin[36]	36	8	60 mg	–	Both groups substantially improved
Imipramine[33]	20	16	180 mg (mean)	+ +	IMI group had a 72% reduction; placebo a 43% reduction in binging

[a]Symbols used: –, no better than placebo; +, statistical trend better than placebo; + +, statistically significant, better than placebo; + + +, greater than 80% difference between active drug and placebo.

D. Critique of Studies

Fewer than 250 bulimic patients taking tricyclic antidepressants have been reported on; fewer than 150 have taken part in a placebo-controlled double-blind study. The low n is compounded by other problems that may be even more significant. The placebo response rate has been widely variable, spanning 0% to more than 50%. It is very difficult to explain this disparity except to suggest that more patients need to be treated in more centers so that the spread between placebo and active drug can be further documented. The disparity could be caused by the use of different criteria to define bulimia, but it is unlikely that this could explain such major differences. The use of the third revised edition of the *Diagnostic and Statistical Manual of Mental Disorders* (DSM-IIIR) diagnostic criteria will probably help to reduce discrepancies related to inclusion criteria.

Most studies do not report how many patients need to be screened in order to reach their n. The studies reporting better results may have stricter inclusion criteria that will eliminate people who are medically unstable, actively suicidal, or previously treatment resistant. Future studies should specifically report on the number of patients who need to be screened in order to reach n and the number of dropouts occurring during the study. This would be helpful so clinicians could predict what their success rate might be in the office as compared with the efficacy data of the tricyclics in a strict protocol with rigid inclusion and exclusion criteria.

One of the most difficult problems in assessing the outcome of bulimia is the lack of a verifiable outcome measure; essentially, every outcome study using outpatients ultimately resorts to self-report. The degree to which patients are distorting their clinical situation in various treatment settings is unknown. The patients who wish to please the researcher may initially report a more positive response.

Perhaps the most crying need in these studies is for long-term outcome data. Over the short term, a significant percentage of patients have decreased bulimic behavior with antidepressants. However, over time these improvements may not persist, leading to multiple sequential trials of medication. We hope that longitudinal studies currently in progress will document the outcome of patients with the bulimia syndrome and the outcome of patients who have been treated with psychotropic medication.

Long-term outcome studies of anorexia nervosa have consistently shown that behavioral therapy and psychotropic medication trials, although helpful in the short term, have little effect on the long-term outcome of the illness. Some investigators suggest that rates of relapse and the need for sequential medication trials are not much different from the treatment of patients with major depression. It could turn out that patients with bulimia nervosa and depressive symptoms or syndromes could represent a population requiring aggressive, ongoing psychopharmacological intervention. Further studies stratifying depressed versus nondepressed eating-disordered patients should be done to clarify the relationship of major depression to bulimia nervosa and anorexia nervosa.

V. CLINICAL CONSIDERATIONS

A. Anorexia Nervosa

Anorectic patients often refuse pharmacotherapy because they fear loss of control, think that taking medication means they are crazy, or have misinformation about the effects of medication. Families may also wish to deny that their child has a drug-responsive psychiatric disorder and collude with the patient in her refusal to take medication. In addition, these patients may be exquisitely sensitive to side effects and frequently report excessive sedation, insomnia, or dizziness even at very low doses. It is important to inform the patient about the medication and its uses. Most anorectics want to be well informed, and the therapist should ally with their cognitive style and respect their wish for information.

Psychotropics should not be employed as the sole treatment modality for anorexia nervosa but only as an adjunct in the context of a psychotherapeutic relationship (Table 5). Antidepressants should be the first choice for the anorectic with a concomitant major depressive disorder. Additionally, the anorectic with neurovegetative signs or recurrent preoccupations or rituals may benefit from a tricyclic antidepressant or MAO inhibitor. The choice of a specific agent involves several factors. Although amitriptyline is the best-studied tricyclic in anorexia nervosa, it frequently produces uncomfortable side effects, particularly excessive sedation and anticholinergic effects. Imipramine and desipramine have not been well studied in anorexia nervosa but may be clinically useful. Generally treatment can be initiated with either imipramine or desipramine. Desipramine produces fewer side effects than other tricyclics; however, it is more expensive.

As discussed previously, anorectic patients may present with medical complications.[40] Therefore, before initiation of pharmacotherapy, routine laboratory tests should be performed, including complete blood count (CBC), serum electrolytes, liver function tests, blood urea nitrogen, creatinine, thyroid functions, and electrocardiogram. Metabolic and physiological abnormalities should be at least partially corrected prior to beginning antidepressant treatment. In general, start medication at a low dose (not more than 25 mg) and slowly titrate upwards. Plasma

TABLE 5. Pharmacotherapy of Bulimia and Anorexia Nervosa: Antidepressant Treatment Steps

Steps	Clinical considerations
1. Evaluate the patient (and family).	Identify depressive symptoms. Depression is not a requirement for the initiation of medication.
2. Inform the patient about medication, potential benefits, and adverse reactions.	Enlist patient as an ally in pharmacotherapy.
3. Perform appropriate laboratory tests.	Correct laboratory abnormalities to within medically acceptable limits.
4. Treat signficant medical problems prior to a drug trial.	Administer imipramine or desipramine, 25 mg at bedtime, increasing by 25 mg q 2 days as
5. Start medication at low doses and increase dosage slowly.	tolerated up to 200 mg for normal-weight bulimics and 75 mg for emaciated anorec-
6. Determine drug plasma levels to ensure adequate dosage.	tics. Plasma levels q 2–3 weeks until compliance is established, condition improved,
7. Follow symptoms such as weight gain, increased interest in eating, decreased obsessions and compulsions, decreased depression and anxiety, increased cooperation in treatment.	or level is greater than 125 ng/ml. Alternatively, fluoxetine, 20–60 mg, can be used as a first line agent.
8. Ensure concomitant psychotherapy.	
9. Administer other medications only where a full trial of antidepressant has been completed, severe symptoms persist, or a trial of psychotherapy has failed.	For AN consider cyproheptadine, fluoxetine, MAOI, or opiate antagonist. For BN consider fluoxetine, lithium augmentation, MAOI, fenfluramine, opiate antagonist, or pemoline.

blood levels should be drawn to determine whether the patient has taken medication and whether the dose is adequate. Successful outcome is indicated by weight gain, increased interest in eating, fewer obsessional thoughts and compulsive rituals, reduced depressive symptomatology, decreased anxiety, and willingness to participate in a treatment program.

Other medications including cyproheptadine, MAO inhibitors, and adjunctive lithium could be tried with the anorectic patient who is not responding to psychotherapy or traditional antidepressant treatment. Antianxiety agents have not been well studied but can be useful, especially around meal time, to reduce food-related anxiety. Fluoxetine has not specifically been studied for anorexia, but it may be useful in low doses, provided that its appetite suppressant side effect is not an issue.

B. Bulimia Nervosa

Bulimic patients are also often resistant to accepting medication. They express concerns about getting "hooked" on the drug, that drugs are an artificial way to treat the problem, that they may get fat, or that taking medication means they are extremely disturbed. Before beginning therapy bulimic patients should be thoroughly psychiatrically evaluated, particularly because of the common association with affective illness, anxiety disorders, suicidal behaviors, substance abuse disorders, and severe character pathology. Furthermore, the patient deserves a complete medical examination because of the potential for complications.

We recommend pharmacotherapy for the bulimic who has concomitant major depressive disorder, substantial depressive or anxiety symptomatology, significant obsessive compulsive symptomatology, or who has been resistant to the usual psychotherapeutic interventions. We would recommend using extreme caution in the outpatient setting with bulimics who are unreliable, suicidal, drug abusers, or who have severe character pathology when there is a poor therapeutic alliance.

Antidepressants plus psychotherapy are more effective than either modality alone (Table 5). Psychotropic medication is usually not recommended as the sole intervention in bulimia nervosa except in selected patients who are not appropriate for therapy, and only under close supervision. Drug treatment for bulimia nervosa can be started with desipramine or imipramine taken entirely at bedtime. Excessive weight gain and carbohydrate cravings are not common side effects among bulimics treated with antidepressants. The clinicians should prescribe a low dose of 25 mg and gradually increase to about 3.5 mg/kg of body weight per day by the third week. To avoid the likelihood of losing the drug through purging, it should be taken at bedtime. To ensure that the medication is being taken and to determine if the dose is adequate, tricyclic plasma levels can be monitored. Responses to medication are evaluated by a lessening of binging and purging behaviors and a decrease in depressive symptomatology and obsessive or anxiety symptoms. Fluoxetine, in doses up to 60 mg/day, can also be recommended as a first-line treatment.

In patients who do not respond to an adequate trial of a tricyclic, another class of antidepressant, such as an MAO inhibitor, should be administered. The MAO inhibitors should be used with caution in these patients. For anorectics, prescribing a drug that requires a diet can potentially exacerbate food preoccupation. For bulimics, impulsiveness around food can potentially lead to a binge on tyramine-containing food. These risks must be fully explored and the risk/benefit ratio evaluated. Most bulimics can accurately assess their ability, or inability, to stay away from tyramine-containing foods.

Some depressed bulimics respond with a lessening of their depressive symptoms, whereas others experience a decrease in their bulimic symptoms. Most show some improvement even though fewer than a third will be "cured," and the long-term outcome is currently not known. Bulimics who have not responded to nondrug therapy should be given a trial of medication regardless of the presence of depression. Antidepressant "resistant" cases might be tried on lithium as an adjunct to the antidepressant, fenfluramine, or naltrexone if other agents have failed.

VI. CONCLUSION

Anorexia nervosa and bulimia are poorly understood disorders that have multi-determined etiologies. Extensive research describing these disorders has been completed; in the next decade there should be substantial advances in their epidemiology and treatment. As the medical profession and lay public have become increasingly informed about these disorders, the mortality rates for anorexia nervosa have declined. Evaluation of the eating-disordered patient should be comprehensive and include a nutritional history, medical examination, and psychiatric interview. Treatment should be multimodal and may include a combination of individual psychotherapy, pharmacotherapy, group therapy, nutritional counseling, ongoing medical management, and family therapy. The treating clinician should be well informed about the disorder, flexible in his approach, and modest in setting treatment goals. Furthermore, idiosyncratic eating behaviors and attitudes are prevalent in young women and merit assessment in all patients in this category.

Finally, there has been much controversy in the literature about the use of psychotropic medication for eating disorders, especially bulimia, and its relationship to depression. The controversy is not whether medication should be used but rather a philosophic one of what place medication has in the overall treatment of these disorders. Some investigators have advocated the use of medication as the primary treatment, with other treatments to be used only for secondary indications. We suggest that pharmacotherapy should not be prescribed as the only treatment, but rather that medication has an important place, particularly in bulimia, for managing specific symptoms in these often treatment-resistant disorders.

REFERENCES

1. Pope H. G., Hudson J., Yurgium-Todd D.: Anorexia nervosa and bulimia among 300 women shoppers. *Am J Psychiatry* 141:292–294, 1984.

2. Crisp A. H., Palmer R. L., Kalucy R. S., et al: How common is anorexia nervosa? A prevalence study. *Br J Psychiatry* 128:549–554, 1976.

3. Jones D. J., Fox M. M., Babigiar H. M., et al: Epidemiology of anorexia nervosa in Monroe County, New York: 1960–1979. *Psychosom Med* 42:561–558, 1980.

4. Willi J., Grossman S.: Epidemiology of anorexia nervosa in a defined region of Switzerland. *Am J Psychiatry* 140:564–567, 1983.

5. Halmi K. A., Casper R. C., Eckert E. D., et al: Unique features associated with the age of onset of anorexia nervosa. *Psychiatr Res* 1:209–215, 1979.

6. Kirstein L.: Diagnostic issues in primary anorexia nervosa. *Int J Psychiatry Med* 11:235–244, 1982.

7. Rollins N., Piazza E.: Diagnosis of anorexia nervosa: A critical reappraisal. *Am J Child Psychol* 17:126–137, 1978.

8. Hsu L. K. G., Crisp A. H., Harding B.: Outcome of anorexia nervosa. *Lancet* 1:61–65, 1979.

9. Mitchell J. E., Davis L., Goff G., et al: A follow-up study of patients with bulimia. *Int J Eat Disord* 5:441–450, 1986.

10. Brotman A. W., Rigotti N. A., Herzog D. B.: Medical complications of eating disorders. *Comp Psychiatry* 26(3):258–272, 1985.

11. Herzog D. B., Copeland P. M.: Eating disorders. *N Engl J Med* 313:481–487, 1985.

12. Kaye W. H., Ebert M. H., Raleigh M., et al: Abnormalities in CNS monoamine metabolism in anorexia nervosa. *N Engl J Med* 308:1117–1123, 1983.

13. Wurtman R. J.: Behavioral effects of nutrients. *Lancet* 1:1145–1147, 1983.

14. Kaye W. H., Ebert M. H., Gwirtsman H. E., et al: Differences in brain serotonergic metabolism between nonbulimic and bulimic patients with anorexia nervosa. *Am J Psychiatry* 141:1598–1601, 1984.

15. Garner D. M., Garfinkel P. E., Schwartz D., et al: Cultural expectations of thinness in women. *Psychol Rep* 47:483–491, 1980.

16. Bruch H.: *Eating disorders: Obesity, Anorexia Nervosa, and the Person Within.* New York: Basic Books, 1973.

17. Selvini Palazolli M.: *Self-Starvation.* London: Chaucer Publishing, 1974.

18. Geist R. A.: Psychotherapeutic dilemmas in the treatment of anorexia nervosa: A self-psychological perspective. *Contemp Psychother Rev* 2(fall), 1984.

19. Gilligan C.: *In a different voice.* Cambridge: Harvard University Press, 1982.

20. Steiner-Adair C.: The body politic: Normal female adolescent development and the development of eating disorders. *J Am Acad Psychoanal* 14(1), 1986.

21. Fairburn C. G.: Cognitive behavioral treatment for bulimia, in Garner D. M., Garfinkel P. E. (eds): *Handbook of Psychotherapy for Anorexia Nervosa and Bulimia.* New York: Guilford Press, 1985, pp 160–193.

22. Garner D. M., Bemis K. M.: Cognitive therapy for anorexia nervosa, in Garner D. M., Garfinkel P. E. (eds): *Handbook of Psychotherapy for Anorexia Nervosa and Bulimia.* New York: Guilford Press, 1985, pp 107–147.

23. Garner D. M., Garfinkel P. E. (eds): *Handbook of Psychotherapy for Anorexia Nervosa and Bulimia.* New York: Guilford Press, 1985.

24. Dally P. J., Sargant W.: A new treatment for anorexia nervosa. *Br J Med* 1:1770–1774, 1960.

25. Hsu L. K. G.: The treatment of anorexia nervosa. *Am J Psychiatry* 143:573–581, 1986.

26. Hudson J. I., Pope H. G., Jonas J. M., et al: Treatment of anorexia nervosa with antidepressants. *J Clin Psychopharmacol* 5(1):17–22, 1985.

27. Biederman J., Herzog D. B., Rivinus T.: Amitriptyline in anorexia nervosa: A double blind study. *J Clin Psychopharmacol* 5(1):10–16, 1985.

28. Lacey J. H., Crisp A. H.: Hunger, food intake and weight, the impact of clomipramine on a refeeding anorexia nervosa population. *Postgrad Med J* 56:79–85, 1981.

29. Halmi K. A., Eckert E., Falk J. R.: Cyproheptadine for anorexia nervosa. *Lancet* 1:1358–1376, 1982.

30. Gross H. A., Ebert M. H., Faden V. B. et al: A double-blind controlled trial of lithium carbonate in primary anorexia nervosa. *J Clin Psychopharmacol* 1:376–381, 1981.

31. Hughes P. L., Wells L. A., Cunningham C. J., et al: Treating bulimia with desipramine: A double-blind, placebo-controlled study. *Arch Gen Psychiatry* 43:182–186, 1986.

32. Pope H. G., Hudson J. I., Jones J. M., et al: Bulimia treated with imipramine: A placebo-controlled double-blind study. *Am J Psychiatry* 140:554–558, 1983.

33. Agras W. S., Dorian B., Kirkley B. G., et al: Imipramine in the treatment of bulimia: A double-blind controlled study. *Int J Eat Disord* 6(1)29–38, 1987.

34. Walsh B. T., Stewart J. W., Roose S. P., et al: Treatment of bulimia with phenelzine: A double-blind placebo controlled study. *Arch Gen Psychiatry* 41:1105–1109, 1984.

35. Mitchell J. E., Groat R.: A placebo controlled double blind trial of amitriptyline in bulimia. *J Clin Psychopharmacol* 41:186–193, 1984.

36. Sabine E. T., Yonaie A., Forringten A. T., et al: Bulimia nervosa: A placebo controlled double-blind trial of mianserin. *Br J Clin Pharmacol* 16:1955–2025, 1983.

37. Pope H. G., Hudson J. I.: Antidepressant drug therapy for bulimia: Current status. *J Clin Psychiatry* 47:339–345, 1986.

38. Jonas J. M., Gold M. S.: Treatment of antidepressant resistant bulimia with naltrexone. *Int J Psychiatry Med* 16(4):305–310, 1986–1987.

39. Blouin A. G., Blouin J. E., Perez E. L., et al.: Treatment of bulimia with fenfluramine and desipramine. Presented at the APA meeting, Chicago, 1987.

40. Goldberg S. C., Halmi K. A., Eckert E. D., et al: Cyproheptadine in anorexia nervosa. *Br J Psychiatry* 134:67–70, 1979.

41. Vigersky R. A., Loriaux D. L.: The effect of cyproheptadine in anorexia nervosa: A double blind trial, in Vigersky R. A. (ed): *Anorexia Nervosa*. New York: Raven Press, 1977, pp 349–356.

SELECTED READING

1. Garfinkel P. E., Garner D. M. (ed): *The Role of Drug Treatments for Eating Disorders*. New York: Brunner/Mazel, 1987.

13

Borderline Disorders

PAUL H. SOLOFF, M.D.

I. INTRODUCTION

Definition of borderline disorder and its place in psychiatric nomenclature remain controversial topics. Historically, psychoanalytic psychiatrists used the term to define a "borderland" between neurotic and psychotic functioning where patients manifested primitive character styles, impulsive behavior, and disturbed interpersonal relations.[1] The disorder was considered either a severe neurosis, a transitional prepsychotic disorder, or an infantile, impulse-ridden personality disorder. In the course of psychoanalytic treatment for various neurotic complaints, these patients would at times have difficulty reality testing, regress, and develop "transference psychoses." In everyday life, they manifested "micropsychotic" episodes in response to specific psychodynamic stressors. These episodes were generally characterized by brief periods of referential thinking and paranoid ideation but responded to structure and direction from the therapist.

Descriptive psychiatrists used the term "borderline" to define nonpsychotic, odd, eccentric individuals who demonstrated peculiarities of thinking, especially under stress. Such individuals commonly demonstrated magical and referential thinking, odd speech, suspiciousness, and paranoid ideation, symptoms that were felt to represent a *forme fruste* or latent form of schizophrenia. The many diagnostic formulations of the borderline concept encompass a heterogeneous spectrum of symptom presentations that are bounded by syndromes of affective instability at one extreme and mild thought disorders at the other. The wealth of diagnostic terms for these patients arises from the many possible combinations of state and trait manifestations. Terms such as emotionally unstable character disorder in the first edition of the *Diagnostic and Statistical Manual of Mental Disorders* (DSM-I) and hysteroid dysphoria capture the affective instability of these patients and their dramatic, flamboyant personal style, while latent, ambulatory, and pseudoneurotic

PAUL H. SOLOFF, M.D. • Department of Psychiatry, University of Pittsburgh, Western Psychiatric Institute and Clinic, Pittsburgh, Pennsylvania 15213.

schizophrenia reflect distorted thinking in a context of odd, eccentric, or perverse functioning. More comprehensive terms such as psychotic character or borderline personality organization emphasize the stable but vulnerable character styles from which reactive symptom states arise.

Despite many references to borderline symptomatology in the early psychiatric literature, empirical studies of the borderline concept were not attempted until 1950, when Grinker et al. developed diagnostic criteria for borderline syndrome.[2] Four subtypes captured the heterogeneity of the syndrome: group I, the psychotic border; group II, the "core" borderline syndrome; group III, the adaptive, affectless, defended, "as if" persons, and group IV, the border with the neurosis. Modern empirical efforts have further clarified descriptive criteria for borderline disorders, recognizing the importance of the underlying character style and reducing the defining characteristics to two related sets of criteria: an unstable borderline personality disorder (DSM-III, BPD) and a schizotypal personality disorder (DSM-III, SPD). Extensive field trials among patients commonly diagnosed as borderline in clinical practice indicate that over half meet both unstable and schizotypal criteria.[3]

II. BORDERLINE DISORDERS

For making decisions about pharmacotherapy, borderline disorders are best viewed as **chronic, maladaptive personality disorders** characterized by a vulnerability to loss of affective, cognitive, and behavioral controls in the context of acute and chronic life stresses. Since borderline disorders primarily represent severe character pathology, they may have clinical presentations of varying severity. For some individuals, the severity of the personality disorder can predispose them to severe symptoms requiring active pharmacological intervention.

A. Clinical Presentation

Table 1 lists the diagnostic criteria for borderline disorder arranged by content area. The most discriminating diagnostic criteria for borderline disorders are:

1. Impulse-action patterns (i.e., self-destructive acts such as wrist cutting and overdosing or dynamically related behaviors such as episodic promiscuity, and binge eating).
2. Chaotic, intense, unstable interpersonal relationships characterized by intolerance of being alone, manipulation, idealization or devaluation, and pathological dependency.

Less discriminating but equally important diagnostic features include:

3. Affect dominated by dramatic lability, reactive depressed moods (especially in the context of perceived rejection), hostility, demandingness and entitle-

TABLE 1. Diagnostic Criteria for Borderline Disorder by Symptom Class[a]

Affective
  Inappropriate, intense anger, lack of control
  Affective instability
  Chronic depression, anhedonia, emptiness, loneliness
  Demanding/entitled
  Rejection sensitivity, hypersensitivity to real or imagined criticism

Impulsive-behavioral
  Suicidal threat/gesture
  Self-mutilation
  Overdose
  Promiscuity, sexual deviance
  Drug, alcohol, food binges
  Assaults/threats/antisocial acts
  Property destruction, temper tantrums

Cognitive
  Episodic distortions of reality
  Derealization, depersonalization, illusions
  Transient psychotic episodes (especially in psychotherapy)
  Brief paranoid experience, referential thinking, magical thinking, suspiciousness

Interpersonal/intrapsychic
  Intense unstable relationships
  Devaluation, manipulation
  Pathological dependency
  Intolerance of being alone
  Masochistic/sadistic behavior
  Unstable perception of others
  Disturbance in self-perception and identity

[a]From Gunderson, Kolb, and Austin[4]; Perry[5]; and Sheehy, Goldsmith, and Charles.[6]

ment, complaints of chronic dysphoria, anhedonia, emptiness, and loneliness.

4. Micropsychotic experiences, principally referential thinking, and paranoid ideation (under duress), with derealization/depersonalization and illusions in the realm of perception.
5. Marginal social–vocational functioning with repeated episodes of decompensation or hospitalization. Borderline patients can usually work at the level of their intellectual capability in highly structured settings.
6. Regression and even transient psychosis in the context of treatment.
7. Identity diffusion (difficulty maintaining a stable sense of self) and unstable perception of others.

Borderline disorders account for about 7–10% of admissions to acute general psychiatric units and up to 25% in longer-term psychoanalytic facilities.[7,8] Estimates of prevalence range from 10% to 30% among persons seeking psychiatric care and from 1.8% to 4% of the general population.[9] Generally the diagnosis is first considered in young adults, ranging in age from approximately 16 to 36 years. It occurs more in women than men (4 : 1 ratio). This sex bias appears to be more an artifact of sampling than a biological determinant. Because of cultural differences in symptom expression, female borderline patients more often present to mental hospitals for self-destructive behavior, whereas male borderlines may be criminally charged for aggression toward others.

Borderline patients generally present for help during an acute episode precipitated by a significant emotional loss or perceived rejection. The full range of affective, cognitive, and impulsive-behavioral symptoms may be present acutely. Patients are often seen in an emergency room setting, having recently attempted suicide or committed other extreme behaviors.

B. Differential Diagnosis

The differential diagnosis at the time of initial intake may run the full gamut of axis I and II disorders. Because the symptoms of borderline disorder overlap with its diagnostic "next neighbors," and comorbidity with other disorders is common, differential diagnosis is an unusually difficult exercise. Among the axis I affective disorders commonly masquerading as—or coexisting with—borderline disorder are major depression, minor depression, chronic and intermittent depression (Research Diagnostic Criteria), and dysthymic disorder. Lability of mood suggests a need to rule out cyclothymia and bipolar II disorders. At the psychotic end of the spectrum, schizophreniform and paranoid disorders must be considered. The "polysymptomatic neuroses" found in borderline patients may include phobias, obsessive ruminations, conversion symptoms, somatization, and hypochondriasis. Anxiety and panic disorders as well as specialized symptom presentations, such as bulimia, drug and alcohol abuse, and sexual deviance, may be prominent. The differential diagnosis for impulsive-aggressive behaviors, especially in the presence of dissociative symptoms, must include temporal lobe epilepsy, attention-deficit disorder/minimal brain dysfunction syndromes, and drug- and alcohol-related behavior. Borderline disorder generally overlaps with other axis II disorders such as schizotypal, histrionic, narcissistic, and antisocial personality disorders.

C. Course of the Illness

Borderline disorders are chronic, relatively unremitting personality disorders. The clinical manifestations of severe pathology begin early in life as low self-esteem, intolerance of being alone, superficial relationships, loneliness (even in settings with adequate social supports), depression and emptiness, and problems tolerating frustration and anxiety. In adults, borderline disorders have myriad symptom presentations that wax and wane, generally correlating with the state of their relationships. Although borderline individuals exhibit persistent problems with relationships (especially rejection sensitivity), self-image, depression, anger, and self-control, they may develop more profound symptoms such as transient lapses in reality testing, referential and paranoid ideation, illusions, and various sensory distortions. Perhaps, the most difficult problem for the clinician is diagnosing possible medication-responsive syndromes in individuals who demonstrate severe and fluctuating symptom presentations and who can provoke anger, vulnerability, and various countertransference responses.

III. GENERAL ISSUES RELATED TO PHARMACOTHERAPY

A. General Nonbiological Therapeutic Approaches: Psychotherapy

Just as controversy has surrounded the pharmacotherapy of borderline disorders, the psychotherapy of this heterogeneous syndrome continues to spark heated debates. The longest-standing and most focused debate is between advocates of supportive psychotherapy and therapists favoring intensive exploratory psychoanalysis. The goals and methods of the two therapies are dramatically different, though each recognizes the potential for the borderline patient to regress, use primitive defenses, and develop transient psychotic symptoms under stress, even in the context of treatment.

Supportive therapy of borderline patients emphasizes "psychoeducation," that is, clarifying the patient's dynamic conflicts in clear, simple language and appreciating the patient's trait vulnerabilities to loss of control of affect, cognition, and impulses under stress. The therapist participates **actively** in treatment, reviewing conflicts in the "here and now" and in terms of the patient's characteristic defenses and styles of relating. In essence, the patient is taught to identify and label pathological dynamics and helped to find more appropriate solutions. The therapist uses clarification as a principal tool and may use family or friends to enhance effectiveness or even to manipulate the patient's social world. The therapist is active in setting limits and in setting firm expectations with clear behavioral contingencies. Hospitalization is viewed as an emergency measure only, to be utilized during crises and for brief periods—days to a few weeks. Therapy hours are limited to once weekly with exceptions clearly defined and limited to emergencies. Although weekly visits are well defined, the duration of supportive therapy is open-ended. Borderline patients tend to drop in and out of treatment as the need arises. The therapist must think in terms of completing a "piece of work" with each treatment series.

In contrast, the intensive approach utilizes exploration of the past and present to make comprehensive interpretations about the origins of the patient's character style. The therapist is generally neutral, providing the needed ambiguity to allow transference to develop. The method is nondirective and time intensive, with the patient being seen three or more times weekly. The goal is reconstruction of character "from the inside out." Because some degree of regression is anticipated with this approach, hospitalization may be part of the overall treatment plan and may be of many months' duration. Limit setting is utilized primarily around the mechanics of treatment, e.g., keeping appointments, paying bills, no violence in office, while a greater degree of neutrality is maintained toward behavior in the community.

Although therapists argue the merits of both approaches, the decision of which therapy to use generally depends on factors such as severity of illness, prior treatment experience, physician's training, and socioeconomic realities. Neither method has been studied sufficiently to be considered a "treatment of choice."

B. Hospitalization

Leaving aside the question of **duration** of hospitalization, one may clearly define **indications** for admission to a hospital. These include behaviors injurious to self or others, especially when social controls are inadequate to prevent **significant** physical or social consequences. Such impulsive-destructive behaviors can include overdosing, self-mutilation, or assaultiveness or socially destructive behavior such as promiscuity and shoplifting. When such behavior is either seriously dangerous to self or others or clearly out of control, admission to a hospital is warranted. Hospitalization and inpatient care are important treatment modalities in their own right and have clear therapeutic effects.

Group therapy, family therapy, and the supportive milieu have a profound effect on the acute symptoms of borderline patients. Although the effects of these psychosocial interventions may be less dramatic than those of medication, they produce improvement in suspiciousness, paranoia, schizotypal symptoms, anger–hostility, observed depression, and interpersonal sensitivity.

Pharmacotherapy of the borderline patient is directed against state symptoms while psychotherapy properly addresses the patient's trait vulnerability through exploring and modifying maladaptive character. Coupling medication with effective crisis-oriented psychotherapy is the optimal treatment for the acutely symptomatic borderline patient.

C. Special Risks

Overdose, abuse, and noncompliance with medication are special risks of pharmacotherapy in borderline outpatients. Although these problems are not restricted to this patient population, the borderline's penchant for "testing" or "acting out" conflicts with medication makes these risks greater. The borderline inpatient may "act out" conflicts through abusive or destructive behavior on the ward, "splitting" staff into "good" and "bad" and testing the limits of the milieu. Because "splitting" of staff is an external consequence of an internal psychodynamic process, it may be managed by effective communication among staff (e.g., making the split apparent), setting a unified behavioral standard for the patient, and also teaching the patient about the splitting process. Behavioral regression should be dealt with quickly by setting firm but fair limits, even including administrative discharge from the hospital for repeated intentional violations of rules. Transfer from acute to longer-term care may be required if regression persists and involves serious physical risks. In a long-term setting, a strict behavioral program should be implemented in order to set limits and reverse regression (e.g., staff attention should be strictly contingent on responsible behavior). "Fair" medication trials 4–6 weeks long may be systematically evaluated in this setting.

Borderline patients generally respond to firm limits and a collaborative approach to pharmacotherapy. A supportive doctor–patient relationship is the sine qua non of any pharmacotherapy, more so when the method of the treatment is empirical and the goals are narrowly defined and modest.

Evaluation is often complicated by the patient's lack of cooperation. Clinicians are often surprised by the entitled, demanding, and hostile presentation of a patient referred acutely after a suicide attempt. Indeed, this "characterologic" presentation should strongly suggest borderline disorder in a patient complaining of depression. The most pressing immediate question is to determine the need for hospitalization. The patient's impulsiveness, suicidal ideation, and motivation must be thoroughly explored. If social support is available in the form of family and friends and the acute crisis appears resolved, outpatient treatment is feasible. Borderline patients are often "well known to the system." For such patients, suicidal or self-abusive behaviors may be experienced often enough and be understood well enough to make it possible to trust the patient to family and friends. If the patient is not well known, social support is lacking, and crisis intervention does not resolve self-destructive impulses, hospitalization is clearly indicated.

IV. PHARMACOTHERAPY

A. Approach to Treatment

Research studies among inpatients suggest that some acute stress symptoms resolve in the hospital without having to resort to medication. A medication-free period of 7–10 days is strongly urged for evaluating borderline patients. With observations available to staff and further history from patient and family, the diagnosis may also be determined.

The most difficult part of the differential diagnosis is ruling out a DSM-III axis I affective disorder. Depressed mood is ubiquitous in patients with borderline disorder. Empirical study indicates that 50–70% of hospitalized borderline patients meet diagnostic criteria for major depression.[10,11] However, careful study indicates that the same patients may meet criteria for atypical depressive disorder or hysteroid dysphoria when these syndromes are systematically investigated.[12] One perspective (whether DSM-III, Schedule for Affective Disorders and Schizophrenia-Research Diagnostic Criteria, (SADS–RDC), or Hamilton Rating Scale for Depression) is often insufficient to characterize accurately the borderline patient's depression.

Some investigators have shown that the diagnosis of major depression in patients with concomitant borderline disorder does not "predict" suppression on the Dexamethasone Suppression Test, blunting of thyroid-stimulating hormone to thyrotropin-releasing hormone stimulation, or favorable clinical response to tricyclic antidepressants.[13-15] Many studies use cross-sectional evaluation (by SADS-RDC, DSM-III, etc.) based on brief interview and patient report to establish the diagnosis of major depression. Observing the patient on the ward will quickly reveal if the patient's mood is reactive or autonomous, the depression labile or pervasive. All too often the research finding is grossly discrepant from the ward clinician's experience. Diagnoses should be based on a combination of clinical and research indicators.

In general, when a clearly diagnosed and treatable axis I disorder is defined, it should be given priority and treated separately. For example, when major depression is diagnosed, with clear endogenous, neurovegetative features, pharmacotherapy must be directed toward the axis I disorder. Because other depressive diagnoses—e.g., dysthymic disorder, adjustment disorder with depressed mood, chronic and intermittent depression—are insufficiently characterized and can not be easily discriminated from borderline pathology, pharmacotherapy of the axis II disorder should be considered. Indeed, some authorities argue that the borderline disorder is, itself, an atypical affective disorder.[16] Patients with a clear diagnosis of BPD and episodic alcohol abuse, food binges (bulimia), or anxiety attacks may also be considered candidates for pharmacotherapy of the axis II disorder, especially when these symptom presentations are secondary to the borderline pathology and do not reflect independent axis I disorders.

During the medication-free period of study, some improvement may be expected in overall symptom severity including suspiciousness, anger–hostility, schizotypal symptoms, and depressed mood (i.e., the "placebo" effects found in systematic studies). The remaining sustained symptoms are appropriate targets for pharmacotherapy.

B. General Clinical Considerations in Pharmacotherapy

There is no proven "treatment of choice" for borderline disorders. Pharmacotherapy is directed against all reactive state symptoms of the disorder, including affective, cognitive, and impulsive behavioral states. Empirical study has shown that many of the presumed stable character traits of the borderline patient also improve with relief of acute state symptoms. For example, the classic "emptiness" of the borderline patient and "intolerance of being alone" have been shown to improve with successful antidepressant treatment.[17]

Methodological problems in determining how to choose appropriate medications include (1) the overlap of symptoms of borderline disorder with "next neighbors," including dysthymic, cyclothymic, and bipolar II disorders, hysteroid dysphoria, atypical depression, and phobic anxious states; (2) true comorbidity with axis I disorders, especially major depression; and (3) lack of a clear etiology. Subtypes of borderline disorder have been proposed, such as subaffective disorders, subschizophrenic disorders, organic or epileptic disorders, attention-deficit and learning disorders, and psychosocial deficits in early childhood. As more is learned about this syndrome, it appears that the borderline disorder may well represent "a final common pathway" of several psychopathological states. Faced with a reliable diagnosis whose validity remains controversial, the psychopharmacologist's approach must be purely empirical. This discussion focuses on pharmacological approaches to medicating borderline patients; it is suggested by a slowly developing body of empirical knowledge. We first discuss how specific target symptoms found in borderline patients are medication-responsive (the "main effects" of systematic studies) and then move toward a cohesive pharmacological strategy that is directed toward empirically defined subtypes of the syndrome.

C. Main Effects of Medications on Target Symptoms

1. Low-Dose Neuroleptic Strategy

In borderline patients, low-dose neuroleptics are indicated principally to treat acute symptoms including (1) anger and hostility; (2) suspiciousness, referential thinking, and paranoid ideation; and (3) anxiety, derealization, and depersonalization (Table 2). These medications also have significant effects against depressed mood and behavioral impulsivity. The spectrum of response is broad, suggesting a lack of specificity. Overall symptom severity and global functioning are improved. These effects are independent of actual DSM-III subtype (BPD versus mixed BPD–SPD). Although effects are statistically significant, clinical improvement is modest in magnitude. Patients change from severely to moderately impaired. Treatment effects appear within 2 weeks. Duration of treatment must be determined by the patient's needs but generally may be between 6 and 12 weeks. Prophylactic efficacy against symptom recurrence has not been established by research.

The clinician must weigh the risks of neuroleptic treatment (including the very small risk of tardive dyskinesia) against the potential benefits of any prolonged maintenance strategy. Such prolonged treatment should be reserved for the especially

TABLE 2. Indications for Low-Dose Neuroleptics

1. Anger, hostility, belligerence
2. Referential thinking, paranoid ideation, suspiciousness
3. Derealization/depersonalization/illusions
4. Anxiety, agitation, somatization
5. Global symptom severity (incl. depressed mood)

vulnerable schizotypal patient. The danger of abuse or overdose of neuroleptic medication poses less medical risk than that of any other medication considered here. This may be a deciding factor in choosing medication for the seriously impulsive outpatient.

The uncontrolled use of street drugs and alcohol by the borderline patient also poses less medical danger of drug interaction with low-dose neuroleptic treatment.

Typical treatment protocols include haloperidol (Haldol® and others) 1–6 mg daily, thiothixene (Navane® and others) 2–20 mg daily, or perphenazine (Trilafon® and others) 4–16 mg daily.

2. Antidepressants

As noted above, most depressed borderline patients demonstrate a mixed profile of symptoms including both neurovegetative and atypical reactive patterns. If a melancholic endogenous pattern can be clearly demonstrated, the treatment is directed toward axis I major depression and may include tricyclic antidepressant medications. In the more common borderline presentations dominated by reactive labile mood and rejection sensitivity, monoamine oxidase inhibitor (MAOI) antidepressants are the medications of choice (Table 3). Phenelzine (Nardil®) or tranylcypromine (Parnate®) is indicated for (1) atypical pattern depressed mood with reactivity and rejection sensitivity and (2) behavioral impulsivity, self-destructiveness, and suicidality.

Phenelzine is generally used in a range of 45–90 mg daily, although efficacy depends on achieving greater than 80% inhibition of (platelet) MAO. This is usually achieved when the daily dose is approximately 1 mg/kg. Tranylcypromine in doses of 20–60 mg daily has been used successfully against depressed mood, suicidality, impulsivity, anger, and rejection sensitivity in borderline outpatients.

Factors limiting the use of MAOIs in borderline patients include insomnia, excessive stimulation, agitation, racing thoughts, and potentiation of mild thought

TABLE 3. Indications for Monoamine Oxidase Inhibitors

1. Reactive, dramatic mood crashes in response to perceived rejection
2. Atypical pattern depression: mood reactivity plus
   a. Hypersomnia
   b. Hyperphagia, weight gain
   c. Anergia, amotivational state, "leaden paralysis"
   d. Rejection sensitivity
3. Behavioral impulsivity, self-destructiveness
4. Suicidality

disorders. Tranylcypromine is the more arousing of the two, with less (initial) sedative effect and orthostatic hypotension. Phenelzine is associated with (initial) sedation, orthostatic hypotension, edema, and anorgasmia, all disquieting symptoms in borderline patients.

Dietary adherence is a concern in patients who have a pattern of "testing limits" or "acting out" aggression with their medication (Table 4). Minor dietary indiscretions in the service of testing the clinician appear to be the rule, but hypertensive crises are exceedingly rare in our own large borderline population. If the consequences of dietary indiscretion are explained in clear medical detail, including the rare but defined risks of stroke, heart attack, and death, patients will generally use less painful means for testing the relationship, such as overdosing on medications other than the MAOI!

Typical treatment protocols include (1) phenelzine 45–90 mg daily (where available, platelet MAO inhibition should exceed 80% for optimal efficacy) or (2) tranylcypromine 20–60 mg daily.

3. Carbamazepine

Neuroleptics and antidepressants comprise the first line of treatment because of safety and proven efficacy. Carbamazepine (Tegretol® and others) and lithium (Eskalith® and others) represent useful second-line treatments but are limited in use because of difficulties in management as well as a paucity of clear empirical support for their effectiveness.

TABLE 4. Special Risks of Pharmacotherapy of Borderline Disorders

I. Low-Dose Neuroleptics
　　Extrapyramidal symptoms, especially dystonia/akathesia, lead to noncompliance
II. MAOI Antidepressants
　　MAOI diet may be problematic, especially with concurrent street drug usage, over-the-counter drugs
　　Side effects of sedation, orthostatic hypotension, edema, anorgasmia, and arousal/insomnia particularly limiting
III. Carbamazepine (Tegretol® and others)
　　Sedation, ataxia, nausea, blurred vision may be limiting factors
　　Risk of bone marrow depression requires weekly CBC, differential, and platelet count for first 3 months
　　Melancholic depression may be precipitated
IV. Lithium (Eskalith® and others)
　　Weekly blood sampling required until stable, then monthly
　　Tremor, polyuria/polydipsia, gastric irritation especially annoying
V. Minor Tranquilizers
　　Abuse, addiction, overdose potential are high
　　May disinhibit suicidal or self-destructive behavior, assaultiveness

Carbamazepine is indicated for impulsivity, behavioral dyscontrol, anger, suicidality, and anxiety. Management is complicated by significant risk of bone marrow depression and the need to follow hematological parameters as well as blood levels of the drug. Carbamazepine is given as 400–1200 mg daily with dose guided by an optimal blood level range of 4–10 μg/ml. Limiting factors include sedation, ataxia, nausea, and blurred vision. Carbamazepine has precipitated melancholic depression in borderline patients with a history of this disorder.[18]

4. Lithium Carbonate

Lithium carbonate is indicated for mood lability and episodic explosive aggressiveness. Daily doses are given in a range of 900–1200 mg to achieve therapeutic blood levels, e.g., 0.8–1.2 mEq/liter. Blood must be sampled weekly until stability is achieved. Limiting side effects may be tremor, polyuria–polydipsia, and gastric irritation. Maintenance lithium therapy poses a risk of hypothyroidism, which must be periodically assessed. As with carbamazepine, frequent blood sampling may lead to patient resistance and noncompliance.

V. CLINICAL APPLICATIONS

Since individuals diagnosed as borderline disorders can exhibit such a wide array of symptoms, what approach can the clinician adopt for developing the most reliable medication strategies? Factor-analytic techniques have been used to reduce the diversity of symptom presentations into a few recognizable clinical patterns. The base-line symptoms for various borderlines can be organized into four clinical patterns:

1. Global/depression.
2. Hostile/depression.
3. Schizotypal symptom pattern.
4. Impulsive symptom pattern.

A. The Globally Symptomatic–Depressed Borderline Patient

This symptom pattern is defined by a preponderance of chronic affective complaints, generally more subjective than objective, that are expressed in a demanding and often dramatic manner. Patients often report multiple somatic or hypochondriacal concerns, anxiety symptoms, and a mix of "classical" and atypical depressive symptoms. Careful diagnostic study may reveal criteria suggesting either major depression or atypical depression or both depending on the diagnostic method. These patients complain of a pervasive low mood, guilt, suicidal ideation, sleep and appetite disturbance, and loss of interest in social and sexual activity. Among their atypical features are mood reactivity, hypersomnia, hyperphagia, and anergia with "leaden paralysis" and rejection sensitivity. They are depressed when alone but brighten with attention and psychological support.

The monoamine oxidase inhibitors are the drugs of choice for the global depressed patient. Phenelzine (45–90 mg daily) has been shown to be effective against the depression and mood lability of these borderline patients, especially those termed "hysteroid dysphoric" for their dramatic quality and their atypical depressive symptoms. Among such patients, complaints of chronic emptiness, problems being alone, and behavioral impulsivity—presumed character traits— improve with phenelzine, indicating a mood-dependent quality to these diagnostic criteria.

B. The Hostile/Depressed Borderline Patient

In reference to this presentation, Grinker wrote that "anger is the main or only affect of the borderline patient."[2] The angry, demanding, entitled borderline patient shares with the global/depressed patient many affective features, primarily the subjective psychological complaint of depressed mood. There are few observed signs of clinical depression. These patients endorse a multiplicity of neurotic symptoms on self-rated clinical questionnaires (e.g., HSCL-90), which are more impressive for their subjective severity than for specific findings. The defining characteristics, however, are their subjective reports of anger, resentment, and suspiciousness of others and their defiant, hostile, provocative behavior.

Low-dose neuroleptics (e.g., haloperidol 1–6 mg, perphenazine 4–16 mg, thiothixene 2–10 mg) are effective against anger, hostility, suspiciousness, and many of the associated features of this pattern. Overlapping with the global/depression symptom pattern, the anxiety, somatization, obsessive, and phobic symptoms of the hostile/depressed borderline respond to low-dose neuroleptic treatment. Surprisingly, low-dose neuroleptics are also clearly beneficial for the depressed mood, suicidality, and rejection sensitivity of the depressed borderline patient. For the hostile/depressed pattern, low-dose neuroleptics have been shown superior to traditional tricyclic antidepressants [e.g., amitriptyline (Elavil® and others)] among borderline inpatients.[19]

C. The Schizotypal Borderline Pattern

This symptom pattern is defined by the presence of multiple schizotypal **state** symptoms during periods of severe interpersonal stress. Most prominent among these are paranoid ideation, referential thinking, derealization, and depersonalization. Such symptoms generally represent acute responses to emotional stressors and often seem dynamically meaningful and "mood congruent." They are not expressions of chronic or residual thought disorders or mild expressions of schizophrenic thinking. Research has shown that even patients with the "pure" unstable BPD diagnosis have an average of six acute schizotypal symptoms when examined carefully using a structured format.[20] In addition to the schizotypal state symptoms noted above, mixed BPD/SPD patients may demonstrate sensory distortions, illusions, magical thinking, muddled thinking, thought blocking and intrusive thoughts, unusually violent or perverse sexual fantasies, preference to be alone, hypersensitivity to criticism, ambivalence, anhedonia, and pananxiety. These symptoms define the so called "micropsychotic episodes" of the borderline patient and are the principal reason the diagnosis was historically considered a subschizophrenic disorder.

Much of the symptom severity associated with acute schizotypal symptoms will resolve spontaneously with supportive care alone. However, the efficacy of low-dose neuroleptic treatment is clearly superior to that of hospitalization or psychotherapy alone in the treatment of severe schizotypal symptoms. Specifically, neuroleptics are strongly indicated for sustained ideas of reference, paranoid ideation, derealization/depersonalization, and illusions.

Since this pattern often coexists with depression (especially in the "mixed" BPD/SPD patient), the clinician may be tempted to utilize tricyclic antidepressants, perhaps viewing the mood-congruent micropsychotic symptoms as evidence of psychotic depression. Research indicates that the presence of schizotypal features in the borderline patient may be a predictor of poor outcome with tricyclic antidepressants. (In this case, amitriptyline was the study drug.[19]) Amitriptyline caused an increase in suicidality, assaultiveness, and referential thinking in more schizotypal borderline patients.

D. The Impulsive Borderline Symptom Pattern

This presentation, although often associated with affective or schizotypal features, may be based in etiological factors that are independent of these associated symptoms. Borderline patients are characterized by trait impulsiveness and generally express impulse behavior in the specific context of interpersonal stress. These behaviors include overdose, self-mutilation, suicidal threat, episodic food or alcohol binges, dystonic promiscuity, and assaultive or antisocial acts. The self-destructive behaviors are more typical of female borderline patients, whereas the other-directed violence is more common among male borderline patients.

Impulsive behavior in the borderline patient may have diverse etiologies that reflect (1) functional disinhibition analogous to loss of control of affect and cognition under stress, (2) impulsive temperament, expressing a biologically heritable trait, (3) residual deficit of early life insult to the central nervous system (as in attention-deficit hyperactivity disorder syndromes), or (4) ongoing epileptiform disorders. Careful electroencephalogram (EEG) and neuropsychological study may help resolve this differential and confirm the association of "cerebral dysrhythmia" (abnormal EEG findings) and impulsive behavior in some patients. A neurobehavioral etiology for borderline disorder has been proposed for such patients but represents, at best, a small and atypical sample (e.g., male borderlines with attention-deficit disorder). In consecutively admitted adult patients (predominantly women), EEG abnormalities are not significantly more frequently associated with borderline disorder than with nonborderline personality disorder controls.[21,22] In most cases of impulsive behavior, the etiology remains unknown and is assumed to be "characterologic."

Because the etiology of behavioral impulsiveness generally remains unclear in the majority of borderline disorders, the choice of drug is often suggested by other features of the syndrome. Recent empirical work indicates some efficacy against impulsive behaviors for low-dose neuroleptics, MAOI antidepressants, carbamazepine, and lithium carbonate.

Low-dose neuroleptics result in a general improvement in overall symptom severity, including actual impulsive ward behaviors. This effect appears nonspecific and is part of a broad spectrum of efficacy that includes affective and schizotypal symptoms as well. Side effects of the neuroleptics, as well as their nonspecific action, suggest that they may be of limited usefulness in treating the impulsiveness of these nonpsychotic patients.

Behavioral impulsivity associated with cerebral dysrhythmia may be responsive to carbamazepine. However, carbamazepine is also effective against behavioral impulsivity in patients with normal EEG, suggesting a role for the mood-stabilizing actions of this drug.[23] Carbamazepine has also been reported to precipitate melancholic depression in borderline patients.[18] An impulsive–aggressive pattern (as found in impulsive sociopathic patients) may also respond to treatment with lithium carbonate, another drug with proven mood-stabilizing efficacy. Similarly, in patients with concomitant atypical depressive patterns, tranylcypromine has been useful in treating behavioral impulsiveness. Minor tranquilizers must be used with caution in severely impulsive patients. Alprazolam (Xanax®) has been shown to "disinhibit" serious behavioral dyscontrol in such patients.[24]

VI. CONCLUSION

The goal of pharmacotherapy in treating patients with borderline disorders is to relieve their acute and chronic affective, cognitive, and impulsive behavioral symptoms. Expectations should be modest. Medication does not cure borderline disorder. Psychosocial interventions are most certainly needed to deal with the longstanding maladaptive character traits of these patients. These modalities augment each other and offer the best chance of progress in a difficult clinical syndrome.

ACKNOWLEDGMENT. This work is supported by NIMH grants MH35392, MH00658, and CRC MH30915.

REFERENCES

1. Mack J. E.: Borderline states: An historical perspective, in Mack J. E. (ed): *Borderline States in Psychiatry.* New York, Grune & Stratton, 1975, pp 1–28.
2. Grinker R. R. Sr., Werble B., Drye R. C.: *The Borderline Syndrome.* New York, Basic Books, 1968.
3. Spitzer R. L., Endicott J., Gibbon M.: Crossing the border into borderline personality and borderline schizophrenia: The development of criteria. *Arch Gen Psychiatry* 36:17–24, 1979.
4. Gunderson J. G., Kolb J. E., Austin V.: The diagnostic interview for borderline patients. *Am J Psychiatry* 138:895–903, 1981.
5. Perry J.: *The Borderline Personality Disorder Scale.* Cambridge, MA, Cambridge Hospital, 1982.
6. Sheehy M., Goldsmith L., Charles E.: A comparative study of borderline patients in a psychiatric outpatient clinic. *Am J Psychiatry* 137:1374–1379, 1980.
7. Soloff P. H.: Pharmacotherapy of borderline disorders. *Compr Psychiatry* 22:535–543, 1981.
8. Andrulonis P. A., Glueck B. C., Stroebel C. F., et al: Organic brain dysfunction and the borderline syndrome. *Psychiatr Clin North Am* 4:61–66, 1981.
9. Swartz M., Blazer D., George L., et al.: Estimating the Prevalence of Borderline Personality Disorder in the Community. *J. Personality Disorders* 4(3): 257–272, 1990.
10. Pope H. G., Jonas J. M., Hudson J. I., et al: The validity of DSM-III borderline personality disorder. *Arch Gen Psychiatry* 40:23–30, 1983.
11. Baxter L., Edell W., Gerner R., et al: Dexamethasone suppression test and axis I diagnoses of inpatients with DSM-III borderline disorder. *J Clin Psychiatry* 45:150–153, 1984.
12. Soloff P. H., George A., Nathan R. S., et al: Characterizing depression in borderline patients. *J Clin Psychiatry* 48:155–157, 1987.
13. Soloff P. H., George A., Nathan R. S.: The dexamethasone suppression test in patients with borderline personality disorder. *Am J Psychiatry* 139:1621–1623, 1982.
14. Nathan R. S., Soloff P. H., George A., et al: DST and TRH in borderline personality disorder, in Shagass C., Josiassen R., Bridger W. H., et al (eds): *Biological Psychiatry, 1985.* New York: Elsevier, 1986, pp 563–565.
15. Soloff P. H., George A., Nathan R. S., et al: Progress in pharmacotherapy of borderline disorders. *Arch Gen Psychiatry* 43:691–697, 1986.
16. Akiskal H. S.: Subaffective disorders: Dysthymic, cyclothymic and bipolar II disorders in the "borderline" realm. *Psychiatr Clin North Am* 4:25–60, 1981.
17. Liebowitz M. R., Klein D. F.: Hysteroid dysphoria. *Psychiatr Clin North Am* 2:555–575, 1979.
18. Gardner D. L., Cowdry R. W.: Development of melancholia during carbamazepine treatment in borderline personality disorder. *J Clin Psychopharmacol* 6:236–239, 1986.
19. Soloff P. H., George A., Nathan R. S., et al: Patterns of response among borderline patients with amitriptyline and haloperidol. Presented at the NCDEU meeting, Key Biscayne, FL, May 1987.
20. George A., Soloff P. H.: Schizotypal symptoms in patients with borderline personality disorders. *Am J Psychiatry* 143:212–215, 1986.
21. Cornelius J. R., Brenner R. P., Soloff P. H., et al: EEG abnormalities in borderline personality disorder: Specific or non-specific. *Biol Psychiatry* 21:974–977, 1986.
22. Cornelius J. R., Schulz S. C., Brenner R. P., et al: Changes in EEG mean frequency associated with anxiety and amphetamine challenge in BPD. Presented at the 41st Annual Meeting of the Society of Biological Psychiatry, Chicago, May, 1986.
23. Gardner D. L., Cowdry R. W.: Positive effects of carbamazepine on behavioral dyscontrol in borderline personality disorder. *Am J Psychiatry* 143:519–522, 1986.
24. Gardner D. L., Cowdry R. W.: Alprazolam-induced dyscontrol in borderline personality disorder. *Am J Psychiatry* 142:98–100, 1985.

14

Medicolegal Psychopharmacology

THOMAS G. GUTHEIL, M.D.

I. INTRODUCTION

Clinicians in all fields of medicine, including psychiatry, have become acutely conscious of the problems posed by the threat of liability. In psychopharmacology, courts and attorneys are able to understand and see some of the concrete actions of the pharmacological agents, in contrast to the relatively cloudy effects of psychotherapy, thus increasing the likelihood of litigation. In this context, clinicians must understand the medicolegal issues that surround, limit, and influence their practice of psychopharmacology.

This chapter discusses three critical legal issues: the informed consent process, the problem of liability, and the right to refuse treatment. Although this list is not exhaustive, it encompasses the major medicolegal questions and difficulties confronting clinicians.

II. INFORMED CONSENT

Treating anyone without his or her consent is a battery, which is both a crime and a civil wrong. Treating someone without **informed** consent may be a form of malpractice. Because the informed consent process is commonly misunderstood by clinicians, this section attempts to clarify some of the confusion.

Informed consent has three components: information, voluntary, and competence. The information in question refers to what a patient would wish to know to

THOMAS G. GUTHEIL M.D. • Department of Psychiatry, Harvard Medical School, Boston, Massachusetts 02115; Program in Psychiatry and the Law, Massachusetts Mental Health Center, Boston, Massachusetts 02115.

make a reasonable (i.e., informed) decision about the particular procedure. In this case, it refers to the use of medication in psychiatric treatment.

The legal definition of "information" varies among jurisdictions and is based on rulings in case law that establish how much information should be given or how explicit the information should be. Despite this diversity, the following guidelines concerning the information to be conveyed should be usable for almost all treatments:

- Risks and benefits of the proposed treatment.
- Risks and benefits of alternative treatments.
- Risks and benefits of no treatment (i.e., the consequences of not taking medication at all).

"Voluntarity" is the "consent" in informed consent. This means that compliance based on threat or coercion is not an example of informed consent, nor are involuntary emergency interventions, though they may be justified on other grounds. Perhaps most importantly true, voluntary consent, by definition, can be revoked at any time. For example, "living will" arrangements (in which a competent patient attempts to provide for being treated over his/her own objection during an expected future state of incompetence) are notoriously unsuccessful in medication cases, since the patient's present refusal revokes even a previously arranged consent.

Finally, competence represents the patient's capacity or ability to weigh and process information necessary to make a reasonable decision about treatment. Competence may be impaired by organic or functional illness. Also, minors are considered incompetent by law in most areas.

A. Consent Forms and Alternatives

Many clinicians assume that the hurdle of informed consent has been overcome once the patient has signed a consent form prepared by the hospital's lawyer. Unfortunately this approach sometimes fails to meet the goals envisioned by law in the informed consent process. Practitioners, however, who work in institutions or settings requiring the use of consent forms should employ them, since it is unwise to practice at variance with institutional policy. But consent forms have several problems of their own.

First, consent forms may be either too long and detailed or so terse as to be incomplete. Second, since those forms are often designed by attorneys, their goal is usually to protect, not to inform; thus, they may fail in their instructive purpose. Such failure is augmented when the patient suffers from a cognitive disability, speaks English as a second language, or speaks no English at all.

What is the alternative to an informed consent form? One model for informed consent that has withstood courtroom testing is to combine a standardized discussion with the patient and a progress note that the physician records. The standardized discussion includes a summary of major or common risks and benefits for each major category of medications; (e.g., one for phenothiazines, for tricyclics, and for lithium). A progress note, written at the time of the discussion with the patient, should record that the discussion occurred, list any cautions (alcohol, driving, heavy machinery), and note any questions asked by the patient as well as the answers given. This documents the patient's active participation in the consent process. It is not necessary to write down the standard discussion verbatim, since it represents the usual or customary one given by the clinician. A typical consent form, read to a jury, can often sound stilted and unintelligible. By contrast, a physician describing his usual patient information to the jury, and reading his notes of conversation with the plaintiff, can be very persuasive.

B. Informed Consent as Process

Informed consent is not merely record keeping. Rather, it is a collaborative process that occurs as a result of the interaction between the physician and patient. The goal of the informed consent process is for the physician to create an atmosphere of cooperation and negotiation that is predicated on two fundamental principles. First, both the physician and patient should jointly recognize and accept the uncertainty underlying clinical practice. Second, consent should be viewed as an open and ongoing dialogue, in which the physician makes every effort to leave the "door" open for the patient to discuss, at any time during the treatment, any questions, concerns, fears, or possible side effects. An open atmosphere of collaboration militates in favor of good clinical outcome. It is a strong liability preventative as well.

III. LIABILITY

Liability is one of the major concerns facing the clinician who practices or specializes in psychopharmacology. Although a theoretical understanding of this issue is not essential to the practitioner, the following brief review provides a background.

A. Definition

For malpractice to be present the plaintiff must prove, by preponderance of the evidence (i.e., that it is more likely than not), that four elements are present. These may be mnemonically summarized as the "four Ds": duty, dereliction, damages, and direct causation.

First, there must exist a **duty**, usually in the form of a doctor–patient relationship. In simplest terms this means that, based on the existence of the professional relationship with that person, only your patient (or your patient's estate) can sue you. This duty may exist under the circumstances of treatment, of consultation, and—at times—of an offer of help.

Second, this duty must be breached by the clinician's **dereliction:** a deviation from the standard of care established by a professional peer group. The clinician is usually measured against "the average prudent professional in that specialty," though the wording may vary by jurisdiction. Such deviation represents negligence, the imputation of a "sin of omission" in standard medical practice. The court is informed about the standard of care by the testimony of expert witnesses for both sides.

Third, **damages** or harms must have resulted, in the form of actual death, injury, pain, suffering, emotional distress, and the like.

Last, **direct causation** must be present. This indicates that the harms must be directly related to the negligence. The law requires that the negligence be the "proximate cause" (immediate, direct precipitant) of the damages.

In sum, malpractice does not exist until the patient/plaintiff can prove, by preponderance of the evidence, the four Ds: **dereliction of a duty directly causing damages.** However, this technical, legalistic description, though constituting an accurate definition, does not accurately reflect the realities of malpractice litigation. Our experience teaches that the actual occurrence of lawsuits for malpractice has little to do with the standard of care or actual bad practice; instead, lawsuits are generally caused by **bad outcomes in the context of bad feelings.**

Bad outcomes include death, injury, and disability. The bad feelings in question may include guilt (in the survivors of a suicide, hoping to alleviate their guilt by transferring some of the blame to the physician's negligence), rage (at the arrogant conduct of a physician), grief (over the loss of a person or even the loss of an ideal of a perfect outcome), and surprise (about an unexpected side effect).

One bad feeling that may be present in most malpractice suits is the feeling of abandonment, of being left alone with one's calamity. Here the clinician's availability—throughout the patient's illness and despite a bad outcome—is an essential element in preventing litigation.

B. Some Common Forms of Liability

This section of the chapter briefly describes some of the common grounds for liability claims in psychopharmacology. An orientation, rather than an exhaustive listing, is provided.

1. Lack of Informed Consent

As suggested above, medical intervention without informed consent is broadly recognized as deviating from the standard of care. Thus, failure to obtain informed consent might represent malpractice, assuming all other elements are present and proven. In psychiatry especially, plaintiff's attorneys routinely claim at the outset of litigation that informed consent was not obtained, since a cause of action in this area is always plausible. The attorneys may then later decide to drop this as a claim if warranted by the evolving case.

2. "Informed Dissent"

"Informed dissent," a term coined by A. J. Gelenberg (personal communication), refers to instances in which a patient refuses a recommended treatment regimen. This situation commonly stirs liability fears in the clinician, and it raises several important considerations.

First, a competent patient can refuse even life-saving treatment; that is the essence of informed consent. The physician bears the burden, however, of documenting that the risks of "no treatment" were fully explored and detailed. For the competent patient with schizophrenia who refuses phenothiazines, for example, the risks to be outlined might include the following: decompensation, rehospitalization, deterioration, job loss, and death.

Second, if the patient is incompetent, the physician bears the affirmative duty to arrange for guardianship or a similarly appropriate process, depending on jurisdiction.[1] This action clearly affirms that the patient was competent when he refused treatment, either because the patient has been found competent by the probate judge or because the court-appointed guardian (defined as competent by this appointment) becomes the recipient of the informed consent discussion. Arranging guardianship usually involves urging the patient's family to consult an attorney or for the physician to mobilize legal aid resources.

Third, if a patient persistently fails to follow a physician's prescribed regimen, the physician must question how long he can collaborate in this noncompliance. If the doctor reaches a point where he cannot ethically continue treatment without the patient following his preferred treatment (e.g., antipsychotic drugs for schizophrenia), an endpoint for treatment should be set and the matter discussed further. If the patient remains intransigent, the clinician can begin to plan for a therapeutic termination. Referral to another treatment resource must be made, to preclude the charge of abandonment. A letter to the patient (with copy to the chart) should review the complete history and circumstances in detail. (See Section IV, "The Right to Refuse Treatment.")

3. Misdiagnosis

Clinicians first encountering this term sometimes fear that they will be cross-examined on the details of the third revised edition of the *Diagnostic and Statistical Manual of Mental Disorders* (DSM-IIIR), where much could be made of uncertainties differentiating between various diagnoses (e.g., bipolar affective from schizoaffective disorders); misdiagnosis does not refer to this type of diagnostic dilemma. Instead, as a medicolegal claim, it alludes to the allegation that the defendant clinician failed to detect something that the average clinician would have detected (i.e., that a diagnosis was "missed"). Common examples include missing the patient's suicidal state; missing an underlying medical condition that causes or worsens the presenting symptoms; and missing a medication side effect that goes undetected and causes harm.

4. Inadequate Treatment

Inadequate treatment refers to cases in which the plaintiff alleges that the treatment itself deviated from the standards of care. Typical examples include overtreatment, undertreatment, and inappropriate treatment for the specific problem.

Although clinical judgment can and should prevail in these cases, clinicians attempting more innovative and pioneering (more intrusive or, in some situations, less aggressive) courses of treatment than usual, or using novel and experimental therapies, are urged to obtain consultation and to document their own risk–benefit reasoning with care. Competent patients can consent to necessary, novel, or even experimental treatment as long as the full implications of such treatment are explained and understood. Exceeding the doses for various medications given in the *Physicians' Desk Reference* (*PDR*) occurs almost universally. In response, plaintiff's attorneys sometimes suggest that the *PDR* is equivalent to the standard of care, a ploy that commonly evokes clinician's anxiety. In actuality, the professional literature and standards of practice are determinative. Thus, novel indications, nonmainstream prescribing of a drug, or unusually high dosages (contrasted with *PDR* limits) should be justified and supported by consultation as needed. Citing specific literature in the record is worth its weight in "liability-preventive gold."

A useful rule of thumb is that deviation from mainstream practice carries with it the possibility of deviation from the standard of care; this possibility need not be realized, but the plaintiff's side may suggest that deviation from common practice equals deviation from standard of care. The further a clinician strays from mainstream practice, the stronger should be the support for the treatment. Use of consultants, team review, and, in some circumstances, an institutional review board might be involved to approve a test protocol.

5. Side Effects and Bad Reactions

Although all drugs have side effects, this truism may clash with our wish for a perfect world. Sometimes, when a patient encounters a particular side effect (bad outcome) in a context triggering surprise, anxiety, and betrayal, these feelings may spark litigation. In contrast, if the same patient has been suitably prepared, the same side effect may spark only the response, "Oh, yes, that is the muscle stiffness the doctor warned me about." Common claims in this area are in response to extrapyramidal or anticholinergic side effects, hypertension and other cardiovascular effects, and allergies to the medication.

6. Tardive Dyskinesia

Although this side effect could logically be included in the preceding section, for empirical reasons it merits a section of its own. In the vast number of cases, tardive dyskinesia (TD) lies well within the risk–benefit ratio favoring treatment. Furthermore, it is usually not particularly painful or harmful.

Tardive dyskinesia, however, suffers from three contextual problems that make it a potential major liability for the late 1980s and 1990s. First, it produces a harmful effect that is visible: a judge and jury can see it. Psychologically, it thus borrows from the public stigma of the mentally ill: here is a visible sign, an even worse fate, perhaps, than being ill in the first place, in the jury's eyes.

Second, advocates seeking judicial establishment of a mental patient's right to refuse treatment (see Section IV) have pointed to this side effect. Although TD is generally not as harmful as some side effects or, for that matter, as being floridly psychotic, the law has become most familiar with TD.

Third, the courts are concrete as well as inherently risk-aversive. (No one litigates a benefit.) Tardive dyskinesia is a visible harm. (Courts have a poor concept of the effects of untreated mental illness.) Moreover, the benefits of phenothiazines are comparatively invisible: one cannot bring in and display before a judge and jury the abstract fact of "ten years without symptoms and without rehospitalization."

Despite these considerations, TD requires the same informed consent process as any other risk. A competent patient with TD may decide to continue medications and assume this risk, usually because discontinuing them is more destructive and dangerous. Similarly, a patient may relapse if medications are discontinued at the onset of detected TD; a competent patient may sometimes assume that risk also.

In such challenging judgment calls, the best response is for the clinician to obtain a second opinion, validating the indications for continuing medication and the patient's competence to participate in this decision. If the patient is believed incompetent, consent of a guardian or judge should be sought, depending on relevant local legislation or court holdings in this area. Before or after TD develops, whether a patient agrees or disagrees with the physician's recommendation of antipsychotic medication, the key is the informed consent process we have been discussing.

7. Special Problems with Antidepressants

The problems addressed in this section do not differ fundamentally from those discussed above, but their particular form and common occurrence merit separate consideration.

Work with antidepressants takes place in the shadow of an inescapable paradox. Tricyclics are among the most lethal medications used in modern psychiatry. Yet suicidal overdoses are most likely to occur when a patient is seriously depressed and when tricyclics are most useful. This paradox cannot be resolved by actual clinical practice. The possibility of plaintiffs' attorneys second-guessing such treatment, however, has prompted clinicians to attempt certain strategies for dealing with this problem.

A popular approach involves prescribing only a limited amount of medication for the patient at any given time. For example, a clinician may prescribe a week's supply in an attempt to prevent the patient from accumulating a lethal total amount. This approach may indeed be useful, especially when the patient tends to be impulsive, but it is far from a panacea. It has several limitations.

First, a suicidal patient is under no "contractual obligation" to use only these medications for suicide; they may be mixed with other substances, or the patient may use completely different methods (e.g., hanging, jumping). Second, some patients view short-term prescribing as coercive, as a sign of physician distrust of the patient or as an approach calculated to protect the doctor's interests but not necessarily the patient's. Third, both physician and patient may be lulled into a false sense of security about the matter of suicide. Any patient can save the weekly supply of medication or obtain more from other sources.

The author recommends that the physician acknowledge that these medications are not only potentially harmful but also tempting for depressed individuals. Agreements not to overdose on these medications and to call the physician if the patient is feeling impulsive may establish or restore a collaborative atmosphere of open discussion about suicidal thoughts.

The decision to prescribe limited amounts of medication should be presented and discussed with the patient with the intent of forming a therapeutic alliance. For example, "You have told me that in the past you have been troubled by your sometimes impulsive decisions; it would be sad if that led you to take too much medication before it could finish helping you. Would you feel that you were more in control if I prescribed only a small amount at a time, each week, that would be safe for you?" In this example the physician is allying with the part of the patient that wishes to retain control and to resist impulsive action.

C. Fundamentals of Liability Prevention

Anyone can sue anyone else for anything. Thus, it is impossible to offer absolute prevention against liability. However, certain tested principles improve the likelihood of not being sued and of prevailing if sued. This section discusses some basic principles.

1. Sharing Uncertainty

To the surprise of some clinicians, defensive practice is not only ineffective but may also be provocative and lead to litigation. Defensive practice converts the patient into an enemy against whom one "defends," and many patients can readily detect this quality in the relationship. The patient feels, "This doctor is doing this for his own interests, not for mine. This is self-protection, not patient care."

An alternative model recognizes the experience of the sick patient in relation to the physician. Illness fosters regression, often expressed as the return to childhood magical wishes for certainty and perfection in caregivers, wishes certain to be disappointed. The physician must implicitly recognize this universal desire for a guaranteed outcome, empathize with it, yet gently counter it. This can be done by sharing uncertainty.

The core problem facing both doctor and patient (and thus, potentially, uniting them) is the uncertainty inherent in clinical practice. After all, when a patient takes a medication for the first time, the physician is, in effect, performing an experiment; this particular combination of person and drug is unique. A realistic appreciation of this uncertainty reflects the doctor's awareness of unknowable dimensions of clinical care. When this uncertainty is shared by doctor and patient, the uncertainty is "halved," as many feelings are when shared.[2] This sharing also alerts us to another powerful factor in decreasing liability risk—the therapeutic alliance.

2. The Therapeutic Alliance

The therapeutic alliance has been described as "participant prescribing," in which a patient is recruited into this collaboration with the physician.[3] Even with psychotic patients, trust may be decisive in ensuring a positive outcome. Most important, the alliance reassures the patient against feared abandonment in the face of a bad outcome.

3. Selection

Another important liability prevention technique is the physician's careful selection of patients whose illnesses lie within his or her field of experience and competence. Readiness to obtain supervision and consultation and to refer patients to specialists when indicated marks the capable clinician.

4. Documentation and Consultation

The "twin pillars" of liability prevention in psychopharmacology (and all medical practice) are documentation and consultation.[1] Documentation reflects the law's attitude toward written materials: "if you didn't write it, it didn't happen." The legal system is socialized to value and accept written material over other forms of communication.

Why is consultation an important element of liability prevention? The answer relates to the standard of care. Each time a clinician consults a peer in the same discipline, a brief sampling of the standard of care occurs. When a consultant agrees with your plan, then **two** reasonable practitioners have felt that the treatment plan lived up to the standard of care. The burden is on the plaintiff to prove that the care still deviated from the standard.

Even more importantly, the very act of obtaining consultation—submitting one's clinical judgment to the scrutiny of a peer—conveys to the average juror that the clinician is not the sort of person who should be sued for negligence.

IV. THE RIGHT TO REFUSE TREATMENT

This complex topic cannot be comprehensively addressed here, but some brief comments about the central issues may aid clinicians to make their way through the conceptual brambles. A more detailed assessment may be found elsewhere.[4] Readers should notice that this particular medicolegal issue is treated differently in different jurisdictions, so it is necessary to know the local rulings or regulations.

A. Historical Considerations

Historically, patients committed to an institution were considered to be under the care of that institution and had to be treated, most often with medication, willing or not. Commitment without treatment would have represented "warehousing" or a form of preventive detention, conditions that are both illegal and unethical.[1]

In the evolution of the law's conceptualization of this process, however, there occurred what has been termed the "commitment–treatment schism"—the separation of the question of the patient's involuntary residence in the hospital from the issue of involuntary treatment (usually in the form of medication). The arbitrariness of this separation still puzzles clinicians, since the patient's rejection of treatment—not uncommonly on delusional grounds, grandiosity, denial of illness, or other symptoms—accounts in many cases for that patient's refusal of hospitalization and of medication as treatment. Nevertheless, this essential distinction underlies the medicolegal issue.

Another important development was a shift in interest from the distinction between voluntary and involuntary patients to the distinction between competent and incompetent patients. In general, a voluntary patient could refuse treatment and/or leave the hospital; an involuntary patient could do neither. Later, experts proposed that a competent patient could decide to give or withhold informed consent (q.v.) to medication, but an incompetent patient, whether voluntary or involuntary, required an alternative decision maker. By default and through a sense of medical responsibility, the physician usually served as the decision maker, treating the patient according to his/her best medical judgment. Of course, this represents the ideal; in many state institutions, patients were fortunate merely to see a physician, much less have their competence or consent assessed.

The right to refuse treatment developed in close temporal proximity to the right to treatment itself. The latter appeared to be a response to neglect of the patient's needs, the former a response to overly paternalistic treatment of the patient. These ostensibly conflicting issues have a common foundation.

B. Basic Issues

The core question posed under the rubric of the right to refuse treatment might be summarized as: "Under what circumstances may patients refuse treatment (usually antipsychotic medication) prescribed by the physician responsible for their care?" Two aspects of the response to this question are emergencies and vicarious consent.

1. Emergencies

Emergencies are the exception to many rules, this use among them. In emergency situations, no court has taken the position that intervention is proscribed; in fact, all known judicial holdings have made exceptions for emergencies. The emergency itself is defined in various ways in different jurisdictions—imminent harm, urgent need, "requires immediate intervention." One basic element of the emergency does not differ: in true emergencies, the clinician **must act.** To fail to do so out of fear of being second-guessed by the law is empirically unrealistic and highly detrimental to the welfare of patients. **In any case of acute danger to self or others posed by a patient's condition, the need for intervention reliably overcomes the right to refuse treatment.**

2. Vicarious Consent

If competent patients can refuse treatment, and if no patient can refuse treatment in a true emergency, the remaining "subject population" of the right to refuse treatment issue should be incompetent, or questionably competent, nonemergency patients; indeed, this is the case. Thus, we are dealing with decision making for the incompetent patient.

a. Proxy Decision-Makers

Historically, a guardian appointed by the court has been the most frequently chosen individual to decide about treatment on behalf of the incompetent patient.[1] In right-to-refuse litigation, however, the courts have chosen diverse individuals or groups as the appropriate vicar of the decision: judge, guardian, clinical review panel, independent psychiatrist, or treating psychiatrist.[4]

b. Two Current Models

Two models of vicarious decision making have guided efforts to give voice to the incompetent patient's wish: the best-interests model and the substituted-judgment model.[1] As its name suggests, the first model holds that the decisions made on behalf of the incompetent ward should be made in that person's best interests. Predictably, problems with this model arise from contention as to what is in the best interests of the ward and whether those interests can be accurately distinguished from the best interests of the decision maker. In short, the problems of this model flow from its contrary-to-fact premise: "If I were you. . . ."

The substituted-judgment model attempts to improve on this by divining: "If you were yourself. . ." This model requires the decision maker to determine what the ward, now incompetent, would want if he were competent. Thus, while being no less contrary to fact, this model aims at individualizing the decision: best interests aside, what would **this** individual choose?

The concept of substitute judgment for a mentally ill patient poses a paradox. Since the incompetence is caused by illness, deciding what medication the patient would want if competent is tantamount to asking: "If you were well, what medicine would you take for your illness?" This paradox is not easily resolved in the courtroom.

C. Some Modern Developments

In many states the basic issues of informed consent, competence, and guardianship (q.v.) apply in standard fashion; in a few states, however, some distinction has been made between treatment in general and treatment of psychoses in particular. Several lower courts have found no explicit right to refuse treatment for psychoses.[5] The U.S. Supreme Court has expressed significant deference to medical judgment.[6] Appellate courts have come down more narrowly on the need for judicial decisionmaking in the context of maximum procedural expression of due process.[7]

Recent empirical study of preliminary findings in a very restrictive holding (*Rogers v Commissioner*, Massachusetts, which requires a full judicial hearing about the patient's treatment) has revealed a surprising fact: in comparison with other, more lenient models involving informal review by multidisciplinary boards in the hospital, the more restrictive method actually reviews fewer cases and overturns more refusals than less stringent models![8] The future of this issue is unclear. Equally unclear is whether the rights of patients gain any meaningful protection from such rulings. The issue is not as simple as "patients' rights"; in reality it is "some rights at the cost of other rights," such as the right to good treatment, the right to rapid release from the hospital, the right to prompt treatment, and the right to an end to suffering.

This last point touches on the core issue underlying both the right to refuse treatment and the right to treatment: the quality of care.[9] Much of the debate around the principles of the right to refuse treatment masks issues related to the poor care often encountered in state hospital systems. When care is fundamentally inadequate, the questions of whether to offer or withhold treatment, to leave it accepted or refused, become purely academic. The issue of quality of care for the mentally ill lies beyond the reach of most doctors and patients; it is a matter for the legislature, for funding by departments of mental health, and for patient advocates.

V. GUIDELINES FOR CLINICIANS

Some general guidelines can be extracted for the clinician's benefit. First, attempt to deliver the highest quality of care. This apparently self-evident goal may involve, in addition to individual practice, political actions such as boycotting of substandard institutions, working with parent groups such as the Alliance for the Mentally Ill, and acting through the American Psychiatric Association's district branches.

Second, note that treatment refusal is a clinical issue, whatever its legal implications.[10] First, intervene clinically, using usual methods of exploration and discussion, education, information, and permission. Legalizing the intervention may cloud its clinical core.

Third, know the rulings in your jurisdiction. If there are no rulings, consider designing regulations that are protective of patients' needs through legislative action by the district branch of the American Psychiatric Association.

Fourth, carefully document the appropriate clinical evidence (e.g., presence of an emergency, patients' incompetence) to offer the best likelihood that patient needs will pevail in the legal setting.

VI. CONCLUSION

No discussion of the medicolegal aspects of psychopharmacology can guarantee success in coping with the vagaries of the law. Aiming at a more realistic goal, this chapter has attempted to equip the clinician with practical and tested advice that may serve to improve the odds in the clinician's favor. In conjunction with other chapters in this book, the material may also help to diminish the clinician's anxiety, thus improving clinical practice and patient care.

REFERENCES

1. Gutheil T. G., Appelbaum P. S.: *Clinical Handbook of Psychiatry and the Law.* New York, McGraw-Hill, 1982.
2. Gutheil T. G., Bursztajn H., Brodsky A.: Malpractice prevention through the sharing of uncertainty: Informed consent and the therapeutic alliance. *N Engl J Med* 311:49–51, 1984.
3. Gutheil T. G.: The psychology of psychopharmacology. *Bull Menninger Clin* 46:321–330, 1982.
4. Gutheil T. G.: The right to refuse treatment: Paradox, pendulum and quality of care. *Behav Sci Law* 4:265–277, 1986.
5. For example, see *Stensvad v Reivitz* (D.C. Wisc. 84-C-383-5, Jan. 10, 1985).
6. *Youngberg v Romeo* 50 U.S.L.W. 4681 (June 15, 1982).
7. For example, see *Rivers v Katz* 495 N.E.2d 337 (N.Y., 1986).
8. Hoge S. K., Gutheil T. G., Kaplan E.: The right to refuse treatment under *Rogers v. Commissioner:* Preliminary empirical findings and comparisons. *Bull Am Acad Psychiatry Law* 15:163–169, 1987.
9. Appelbaum P. S., Gutheil T. G.: The right to refuse treatment: The real issue is quality of care. *Bull Am Acad Psychiatry Law* 9:199–202, 1981.
10. Appelbaum P. S., Gutheil T. G.: Clinical aspects of treatment refusal. *Compr Psychiatry* 23:560–566, 1982.

Index